Submicroscopic Cytochemistry

VOLUME I

Proteins and Nucleic Acids

Contributors

PETER S. AMENTA

EILEEN S. GERSH

ISIDORE GERSH

PETER J. HAND

G. L. ROSSI

VOUCHER ISSUE

Submicroscopic Cytochemistry

Volume I

PROTEINS AND NUCLEIC ACIDS

Edited by

Isidore Gersh

Laboratories of Anatomy
Department of Animal Biology
School of Veterinary Medicine
University of Pennsylvania
Philadelphia, Pennsylvania

ACADEMIC PRESS New York and London 1973

A Subsidiary of Harcourt Brace Jovanovich, Publishers

ACADEMIC PRESS, INC.
111 Fifth Avenue, New York, New York 10003

United Kingdom Edition published by
ACADEMIC PRESS, INC. (LONDON) LTD.
24/28 Oval Road, London NW1

Library of Congress Cataloging in Publication Data

Gersh, Isidore, DATE
Submicroscopic cytochemistry.

Includes bibliographies.
CONTENTS: v. 1. Proteins and nucleic acids.–v. 2.
1. Cytochemistry. 2. Ultrastructure. 3. Connective tissues. I. Title. [DNLM: 1. Histocytochemistry. 2. Molecular biology. QS504 G381s]
QH613.G47 574.8'76 73-904
ISBN 0-12-281401-0

PRINTED IN THE UNITED STATES OF AMERICA

To Dr. Louis B. Flexner
for his inspiration and support over a period of forty years

Contents

List of Contributors

Numbers in parentheses indicate the pages on which the authors' contributions begin.

PETER S. AMENTA (365), Department of Anatomy, Hahnemann Medical College, Philadelphia, Pennsylvania

EILEEN S. GERSH (98,145,163,338,365), Laboratories of Anatomy, Department of Animal Biology, School of Veterinary Medicine, University of Pennsylvania, Philadelphia, Pennsylvania

ISIDORE GERSH (1,16,53,98,145,163,195,221,251,269,283,303,338,365), Laboratories of Anatomy, Department of Animal Biology, School of Veterinary Medicine, University of Pennsylvania, Philadelphia, Pennsylvania

PETER J. HAND (303), Laboratories of Anatomy, Department of Animal Biology, School of Veterinary Medicine, University of Pennsylvania, Philadelphia, Pennsylvania

G. L. ROSSI (53), Institut für Tierpathologie, Universität Bern, Bern, Switzerland

Preface

In this age of specialization and fragmentation the opportunity for an individual to carry out a comprehensive laboratory study is rare indeed. Through a series of accidental happenings, I found myself in a position which impelled me to continue a long-term study of a general type. It is difficult to express briefly the fascination and absorbing interest in developing subtle interrelationships of methods and ideas and in molding complex materials into a coherent view of some aspects of cellular activity. The first eight years of this study were devoted largely to developing and testing cytochemical methods to the point of satisfaction, and the next five years chiefly to applying them to biological studies. The analytic processes continued up to and even after the time of writing of the last chapter.

Three major topics are considered in this two-volume work: (1) the pattern of distribution of nucleic acids at the molecular level in various cells and states of activity and in relation to protein synthesis; (2) the molecular and macromolecular organization of cellular membranes; and (3) the origin and distribution of the major macromolecular aggregates of connective tissue. The first major topic comprises Volume I; the remaining ones, Volume II. All three are integrated in the last chapter of Volume II.

There is a high premium in this work on morphology, that is, on the distribution of cell components and their organization in cells at the macromolecular level. Such studies required that new methods be developed to preserve these molecules and molecular aggregates very nearly in the position they occupied in the living state and to identify them more or less selectively. The crucial tests were chiefly for proteins, nucleic acids, lipids, and acid mucopolysaccharides. A glance at the Table of Contents will indicate the wide range of biological problems which could be approached with the new methods. The last chapter is the most general one, and serves to bind together all the others. Here morphology, through cytochemistry, is integrated with parts of genetics, biochemistry, cytophysiology, developmental biology, and pathology.

The main emphasis in this work is on the presentation of laboratory findings and a theoretical aspect involved or derived from them. These are related, whenever possible, to the main body of thought on the various topics discussed by specific or general references. The latter are so varied, competent, and numerous that it was felt they would suffice in most instances to relate the original work to contributions from the literature available to most workers. Since there was no need for encyclopedic reviews of the various fields, the references cited are minimal, except in certain specific instances where the reviews were incomplete or otherwise unsatisfactory. An effort was made to update the literature citations up to the time the manuscript went to the publisher.

A few words should be written about prints. Nearly all negatives were printed directly at a magnification of two times. Where the contrast was marked, as with apatite crystals (Chapter 9, Volume II), some prints were magnified five times. Only a few negatives were printed with some dodging or burning. A small number of negatives were double printed to give enhanced, selective contrast to some specific component. These were printed from the original negative, usually on a soft paper. This was then photographed on a second EKTAPAN (Kodak) film, which was printed again usually on a soft paper. The final magnification was two to four times the original electron micrograph. Such prints are labeled "G2 print" in the figure legend. In taking some of the electron micrographs for such prints, an appreciable astigmatism was introduced in order to improve resolution and contrast. This method was suggested by F. S. Sjöstrand ("Electron Microscopy of Cells and Tissues," Volume I, p. 127, Academic Press, New York, 1967).

Isidore Gersh

Acknowledgments

The original research reported in this book was supported in part by grants from the Commonwealth Fund of New York, the U.S. Public Health Service, and the American Cancer Society. Two small intramural grants helped to tide me over a difficult period. The financial and moral support of Drs. Lester Evans, Roderick Heffron, and Charles O. Warren, executive officers of the Commonwealth Fund, encouraged the initiation of the long search for cytochemical methods adequate for solution of some biological problems. A separate grant by the Fund made possible the preparation of the manuscript.

I am indebted to the following for their essential and loyal technical work: Faustina Manelis and Elizabeth Vilkas, of Chicago; Vivien Catlin, Eugénie Ford, Patsy Jo Terrell, and Eliana Muñoz, formerly of Philadelphia; and George J. Grigonis, Jr., also of Philadelphia. It was a great comfort when Diane Weinstock took over the typing. Diagrams were developed by Mary Jo Larsen, who was of great aid, in addition, in connection with other illustrations. I wish to thank Janice Heald for facilitating work in the library and for checking the references.

I am pleased to recall the perceptive advice of Dr. Herbert S. Anker, Department of Biochemistry, University of Chicago, and numerous other chemists. For advice on matters pertaining to Chapter 2, Volume II, I am indebted to Dr. Donald A. Abt, of the Department of Medicine, and to Dr. James A. Dvorak, when he was in the Department of Pathobiology in the School of Veterinary Medicine, University of Pennsylvania.

In addition to working jointly on the material described in five chapters, my wife has helped me significantly to interpret observations in other chapters from a genetic point of view. She also helped me clarify the analyses of the various problems studied and to express the thoughts involved. I enjoyed the collaboration of several colleagues, past and present, who are coauthors of several chapters: Dr. Peter S. Amenta, Department of Anatomy, Hahnemann Medical College, Philadelphia; Dr. Peter J. Hand, Department of Animal Biology, University of Pennsylvania, Philadelphia; Dr. Zelma Molnar, Department of Pathology, University of Chicago, Chicago; and Dr. Giovanni L. Rossi, Institut für Tierpathologie, Universität Bern, Bern.

This work was started when I was Professor in the Department of Anatomy at the University of Chicago, and was continued with very little interruption at the University of Pennsylvania.

Contents of Volume II

Membranes, Mitochondria, and Connective Tissues

Introduction

The technical ideas described in this book were developed since 1958 after publication of a sheaf of papers from my laboratory on the preparation of specimens for electron microscopy by freezing and drying. There seemed to be relatively little connection between the descriptions of cell structure of freeze-dried specimens with the great body of description based on the use of a limited number of fluid fixatives. So it was no surprise when the reports from my laboratory were ignored by most workers in the field. One of the difficulties with freeze-drying as a method of preparation, as was true also of other methods of preparation, was the rather limited range of staining techniques, for cytochemistry especially. I searched for other possibilities and developed the idea of the use of vapor reagents *in vacuo* for postfixation and staining of freeze-dried material.

The logic of the technical system seemed compelling. Specimens were to be frozen ultrarapidly to avoid the formation of ice crystals, and water in the specimen was to be sublimed from the frozen tissue in a vacuum system at a sufficiently low temperature to prevent ice crystals from forming. (There were several reasons which convinced me that ice crystals were not in fact formed—the chief one being that the "spaces" contain structured material, some of which can be identified cytochemically.) Then, without breaking the vacuum, reagents were to be introduced which would react in the vapor state with appropriate groups in the protoplasm in the dry, solid state. These compounds would act as cross-linking reagents, and such action would tend to retain solid constituents of cells during subsequent manipulations very nearly in the position occupied in life. In this way, it should be possible to preserve and at the same time stain proteins, lipids, some polysaccharides, and nucleic acids. With few guidelines for the selection of vapor reagents, or in determining the optimal vacuum, reaction time, and temperature, the work progressed slowly. But after nine years, during which all the methods were being tested simultaneously, I had accumulated a large body of observations which convinced me that the methods had merit. It has taken another five years to complete the collection of materials, to work them up, to analyze the electron micrographs, and to plan and write the greater part of this work.

The studies covering a wide range of materials centered on two main interrelated topics treated in this book: (1) morphological aspects of the genetic control of cellular activities, and (2) the cellular role in the formation of the extracellular components of the body. The probes used in these studies were largely cytochemical in nature, with a very strong morphological bias.

It might be useful to indicate some of the major problems and results included in this book. The key method was designed to preserve and stain nucleic acids in cells. The DNA double helix of Watson and Crick appears in electron micrographs as a line about 45 Å in thickness when treated in this way. In most unstimulated cells, the DNA is in the form of second-order helices which are aggregated to form chromosomes or chromatin. The densely packed second-order helices, for steric and chemical reasons, would be impermeable to nucleotide precursors and to particulate RNA for which they may have served as template. Such compacted DNA, aggregated as chromatin, must thus be regarded as very largely inactive genetic material. Moreover, as the chromatin masses are situated primarily near the nuclear membrane, they could be expected to act as barriers to the exchange of nucleotides and polynucleotides between the nucleus and the cytoplasm. Such nucleocytoplasmic exchange could however be expected to take place in the gaps between chromatin clumps. Thus these two morphological factors control the expression of genic action. Together they constitute the coarse adjustments of genetic control of cellular activities.

The degree of compactness of the second-order helices and the size of the compacted aggregates, as well as the number and size of the nuclear gaps, vary in a variety of conditions in cells, as during marked stimulation and depressed activity, during differentiation, metaplasia, maturation, pyknosis, karyolysis, and mitosis. These coarse adjustments are presumably supplemented by finer adjustments of the type represented by repressor and derepressor substances, which are not amenable to study by the methods used. A mechanism is postulated for achieving the coarse adjustments. It consists of proteins having three actions: protein which imposes a second order of coiling on the DNA double helix of Watson and Crick, protein which causes and maintains aggregation of the second-order helices, and protein which attaches the DNA helices to the nuclear membrane. All these actions are potentially reversible.

The more dispersed second-order DNA coils are considered to be potentially genetically active, while the compact DNA aggregates are considered inactive. The dispersed DNA coils are restricted to a small fraction of the nucleus in most unstimulated somatic cells. The pattern of distribution of the dispersed and compact DNA helices as well as of the distribution and size of nuclear gaps is probably constant for each cell type, and differs from the corresponding patterns of all other cell types. Stated in another way, the active DNA helices must form a three-dimensional mosaic in the nucleus of each cell type which is specific for that cell type and different from the active DNA mosaic of other cells. Such a mosaic could be the basis for the organization of gene hierarchies and systems in the nucleus which would serve to integrate cytoplasmic activities. Combined with the concomitant pattern of nuclear gaps, the nuclear mosaic could be the basis of a functionally mosaic organization of cytoplasmic activity. This would imply that in a cell like the exocrine pancreatic cell (as in all others) an apparent uniformity in the RER obscures an underlying functional mosaicism in the cytoplasm.

The products of transcription from DNA appear in the nucleolus and nucleoplasm as particulate RNA. When these particles pass through the nuclear gaps into the cytoplasm, they are recognized in electron micrographs as ribosomes and messenger RNA; both the nucleic acid and protein moieties are stained separately or together. It is possible that aminoacyl tRNA may also be observed in electron micrographs. In addition, the nascent polypeptide may be seen to be wound around the ribosome, or free in the periribosomal protein (plus associated lipid). The intracisternal space of the RER contains some intracisternal lipid, probably interspersed in relation to the intracisternal (globular) molecules. Much more lipid is present in the immediately adjacent wall of the RER, and it forms a narrow, discontinuous layer of irregular thickness. Extending through this and a little beyond it is a layer of readily stained protein (hydrophilic or lipophobic).

When the nascent protein is shed from the messenger–ribosomal complex, it passes from the wall of the RER and enters the intercisternal space, where its conformation is progressively altered. On leaving the hydrophilic layer of protein of the wall of the RER, it is immediately enclosed by hydrophobic (lipophilic) protein molecules presumably synthesized by other nearby polysomes. Despite the uniform appearance of the RER in different parts of a single cell, or in different cells, there is reason to suggest that the RER is functionally, if temporarily, compartmented. The demonstration of such compartmentation requires the identification of specific nascent polypeptide molecules on and adjacent to polysomes. According to the hypothesis proposed, these specific proteins should be in different parts of the RER, not permanently, but for the moment. At a later time, some other protein may be synthesized in the same site, the messenger RNA, and perhaps some ribosomes, being different from those formerly in the same site. Such molecules may among other possibilities be secreted directly or may be stored as secretory granules. In either case, the nascent protein, if enzymatic in nature, is enveloped in lipophobic molecules which restrict access to the enzyme molecule. Hydrophobic (lipophilic) globular protein molecules of this sort have been observed in addition in the walls of mitochondrial compartments, in the outer mitochondrial membrane, and in the intracisternal space of the RER.

By contrast with the above is the clear evidence of compartmentation in mitochondria, where the compartment walls are composed primarily of hydrophobic (lipophilic) protein molecules, with some lipid. The walls of each compartment enclose readily stainable (hydrophilic) protein with some lipid, which together constitute the contents of each compartment. These protein molecules are aggregated to some extent. The compartments are stacked one on another, and together each stack constitutes a mitochondrion. It seems highly probable that a basic requirement for mitochondrial structure is that most of the mitochondrial proteins have the property of self-assembly. There seems to be a sharing of structural and enzymatic functions by the proteins which has not yet been clearly defined. Self-assembly is an important property probably characteristic of the protein and lipid molecules of the RER and of the protein–polysaccharide complexes of cartilage cells.

In connective tissue cells, two types of presumptive secretory structures are described: the first for tropocollagen, the second for the protein–polysaccharide complex. In cartilage cells, the process of secretion is related to the DNA molecular pattern and to RNA granules in the nucleolus and nucleoplasm. The possible precursor for tropocollagen is found to be in the intercisternal space, perhaps on its way to the Golgi apparatus or the cell surface. The structure of the microscopically visible granule reflects the possible presence of lipophobic molecules between groups of presumed tropocollagen molecules as well as between them and their intercisternal environment. The possible precursor for the protein–polysaccharide complex is also in the intercisternal space in close relation to the RER. It seems to be already preassembled in the intercisternal space in the form of small and large compartments, which are probably equivalent to the small and large compartments of the extracellular matrix. The protein–polysaccharide complex, preassembled and stored in direct contact with the ribosomal protein synthetic system in the intercisternal space, seems to be secreted centrifugally directly through the surface of the cell through small temporary apertures to join the large matrix compartments of extracellular cartilage matrix. The walls of the latter compartments consist chiefly of the polysaccharide moiety of the protein–polysaccharide complex as well as collagen fibrils. The large compartments in the cell, as in the cartilage matrix, comprise smaller compartments, each of which consists of a protein center enclosed by polysaccharide. The walls of the large matrix components of extracellular compartments seem to be the chief site of transport of diffusible substances in cartilage. A similar transport site seems to be present also in tendon, which like cartilage, may be avascular, or minimally vascularized. The walls of the large matrix compartments of calcifying cartilage matrix are the site of earliest deposition of calcium and of the formation of the earliest crystals of apatite.

A legitimate question may arise in the mind of the prospective reader: What is new and different about the material in this book to justify publication of still another book? One of the new things is the cytochemical methodology used as an analytical tool in the investigation of cell structure. There is no doubt that basically the structure of mitochondria, Golgi apparatus, secretory granules, RER, etc., as described in this book, does resemble closely some accepted descriptions in the literature. Such differences as there may be between the two descriptions are referable to differences between the strictly morphological approach and the cytochemical viewpoint. Despite the similarities, there is little doubt that many new features are described, such as the identification and arrangement of the DNA double helix under a variety of conditions, as first-, second-, third-, and fourth-order helices; the description of RNA granules in the nucleoplasm and the variation in their numbers in different functional states; the relations of helical DNA compactions and nuclear gaps in different states of activity of cells; the description of mRNA and (perhaps) of aminoacyl tRNA in relation to ribosomes in the cytoplasm; the internal structure of ribosomes in terms of their RNA and protein components; the appearance and maturation of nascent protein molecules; the apparently stereotyped nature of compartmentation in mitochondria; the possible precursors of tropocollagen and of the protein–polysaccharide complex in cartilage cells and their method of release from the cell; the unit structure of extracellular matrix of hyaline cartilage and the relation of their various components to the transport of substances in the extracellular matrix and to calcification; the two- or multiphase structure of the ground substance of protoplasm.

Isidore Gersh

1

Submicroscopic Cytochemistry*

Isidore Gersh

The subject matter of this work includes the chemical anatomy of cell structures at the molecular and supramolecular level. This broad definition includes a large variety of topics: the chemical composition and arrangement of the components in cellular membranes and their relation to cell functions; the chemical composition and spatial relations of RNA granules of the nucleus and cytoplasm and their relation to messenger and transfer RNA and the synthesis of protein; the identification and arrangement of mitochrondrial and nuclear DNA and the relation of DNA molecular patterns to the action of genes during differentiation, maturation, activity cycles, and cell death; the nature and origin of secretory and storage precursors; the organization of the major chemical components of cell organelles and the ground substance of the cytoplasm, as well as of the nucleus; and numerous other topics. Some products of cellular activity are extruded extracellularly, and many of these are incorporated in the connective tissues of the body. Therefore, consideration of their origin and organization also must be regarded as a part of submicroscopic cytochemistry. Some aspects of all of these topics are considered in this work.

For many purposes, the separation of cell (or intercellular) components is the method of choice, as in studies of the chemical, biophysical, and kinetic properties of, for example, mitochondria or mitochondrial membranes, or of ribosomes or nuclear DNA. Eventually, the question arises in such studies, as in studies of bacteria, as to how far the analytical findings can be extended to the activities of the intact cell. The integration of the activities in eukaryotic cells and organisms is so complex that it requires the action of a very large proportion of nuclear genes to

* Throughout this work, the term "microscopic" is used in describing objects which are visible with the light microscope. The term "submicroscopic" is used to characterize objects which cannot be resolved with the light microscope, either because they are too small or because the spaces between them are too small.

achieve smooth operation. It is necessary to make an effort to relate the ideas and data of disembodied cytochemistry back to the living cell. One of the ways of achieving this (the morphological way) is to locate these chemical substances in cells in order to determine their interrelations, and to learn how they vary during development, maturation, and activity at the molecular or supramolecular level. This involves an analysis of the distribution of chemical substances of normal and abnormal cells in the living state. This aspect of cytochemistry is rather poorly developed, being most advanced when dealing with glycogen and certain enzymes, chiefly those with hydrolytic functions. There are few sound cytochemical methods for use with the electron microscope in the identification of proteins, lipids, and nucleic acids.

Cytochemisty is thus of interest for shedding light on the behavior of cells. Cytochemistry also serves as a kind of bridge between cytology, on the one hand, and biochemistry and genetics, on the other. Cytochemistry gives substance to the theoretical and chemical formulations of the two latter subjects by referring them to cellular structures which are, in turn, related to cellular activities. Finally cytochemistry may serve as a rational basis for the fixation and staining of cells and intercellular substances. The study of the submicroscopic structure of cells with the electron microscope has approached the supramolecular and molecular level, as in the study of skeletal muscle. It seems only a small step to consider the possibility that knowledge of the fixation and staining of chemical substances in other cells may aid in resolving them also for study at the molecular level. Some of the studies reported in this work are at the molecular level, and most of the remainder are at the supramolecular (but still submicroscopic) level. They supplement empirically derived cytological studies, where compromises to an unknown extent have had to be made because of uncertainty or ignorance of the effects of fixation and because of the nonspecificity of the staining reaction. The cytochemical methods developed and reported in this work require special, though not very complicated, equipment, and are unfortunately not suited for general use in routine laboratories.

Cytochemical methods must be more or less specific, sensitive, and relatively free of significant displacements or diffusion, and the end product must be visible. These aims can sometimes be achieved by physical methods, such as, phase contrast or interference microscopy, fluorescence microscopy, X-ray absorption, X-ray diffraction, or polarization microscopy. For most problems, chemical methods are required. The chemical requirements enumerated above are difficult enough to meet in light microscopy, where displacements of 2000 A or more may not even be detectable, and where intense coloration and a wide range of possible chemical treatments are very advantageous. They are even more difficult to meet in electron microscopy, where the resolution is much finer, and the chemical identification must be made with gentle procedures whose effectiveness depends chiefly on increase (or decrease) in mass. The criteria for cytochemical methods at the level of the electron microscope will be discussed under two general headings: fixation and staining.

Fixation

Fixation is the preservation of cell or intercellular constituents or structures in a manner approximating their state and relationships in the living cell. For example, chloride can be preserved by immersing a tissue specimen in an aqueous acidified solution of silver nitrate. The distribution of chloride observed in such specimens gives little useful information, since all the chloride in the specimen has diffused to a narrow zone at the periphery of the tissue block. In fact, fixation of chloride by this method is poor, because information on chloride distribution in such preparations is misleading. A long list can be compiled of substances and structures which are poorly preserved because of the extraction and consequent disorganization of structures by the solvent in the fixative solution. These include chromatin and DNA, ergastoplasm and ribosomal RNA, soluble RNAs, presecretory granules, proteins, membranes and lipids, cartilage matrix and protein–polysaccharide complexes, and inorganic substances, including even the most insoluble of all inorganic substances, amorphous calcium phosphate and crystalline apatite. This deleterious solvent effect of fluid fixatives can be attributed to two major factors: (1) the relatively large volume of solvent in the fixative as compared with that of the specimen, and (2) the dilution effect of water in the tissue specimens. To the first factor should be added the additional volumes of solvent used during washing, dehydration, embedding, and section floating. The second factor refers to the process whereby after the first few micra, on the outer surface of the specimen, the fixative is diluted by the water in the tissues. The end result is that the composition of the advancing front of the fixative always differs from that of the original solution, and at any single point in a given cell, the concentration of the fixative increases progressively. Two deleterious consequences may be deduced from this analysis: (1) the advancing front of the fixative consists of a layer of solvent free of other fixative components capable of drastically affecting local (submicroscopic) pH, osmotic changes, and extraction of cell constituents, and (2) the dilute, advancing front of the fixative exerts, in addition, a toxic effect before reaching a fully effective concentration as fixative components.

The mechanisms of action of fixatives vary, depending on the class of substances to be fixed. With proteins, the chief action seems to be through cross-linking of a variety of reactive groups of adjacent molecules and parts of molecules. With lipids, the chief actions seem to be through combination with the phosphate groups or with the hydrophilic part of phospholipids as a whole, through oxidation of the unsaturated bonds of fatty acids, or simply by maintaining the phase separation of the hydrophobic structure. With nucleic acids, the chief actions seem to be insolubilization of the associated proteins and inactivation of depolymerases. With carbohydrates, the chief actions are (1) maintenance of insolubility of the polysaccharide (glycogen) and inactivation of depolymerase, and (2) precipitation of the polysaccharide, as with certain metal and organic ions, for chondroitin sulfate. With inorganic substances the chief action of the fixative is to form a precipitate with a low solubility product, or an insoluble chelate. For light microscopy, the precipitate or chelate is nearly always colored; for electron microscopy, the increased mass caused by the precipitating agent is registered as increased contrast, which may be sufficient to affirm the presence of the substance tested for. Obviously the effectiveness of all of these mechanisms may be limited by the important consequences of solvent action mentioned briefly in the preceding paragraph. In all instances, fixatives which coagulate protein components of protoplasm cause the separation of the solids from the aqueous phase, which is removed in subsequent steps of specimen preparation *(14)*.

Some of the disadvantages of fixation by immersion in fluid fixatives could be avoided if it were possible to remove tissue water instantaneously, first temporarily by freezing it as ice and then more permanently by sublimation or replacement with another more favorable solvent. These are the ideal aims of freeze-drying, freeze-substitution and, to a certain limited extent, of freeze-etching and freeze-fracturing. The latter two methods are designed to analyze and describe relations of relatively undisturbed submicroscopic structure of cells. Specimens are frozen ultrarapidly, with or without pretreatment with a protective agent, and the frozen specimen is fractured. The cleaved surface may be shadowed directly to form a replica, or it may be etched minimally by sublimation of surface ice and then shadowed, all processes taking place at very low temperatures. Finally, the last traces of the original specimen are removed from the replica before it is examined in the electron microscope, and there remains no possibility of deriving chemical information from the replica *(5)*.

The methods of freeze-drying and freeze-substitution are frequently juxtaposed as alternate methods, although their aims are largely identical *(5, 23)*. In both methods, the first step is the same, i.e., ultrarapid freezing to remove the water in the specimen as ice. This may be preceded by partial replacement of water with (or dehydration with) glycerol, dimethyl sulfoxide, or some other protective agent. The exact conditions for achieving the phase transformation without ice crystal formation in the specimen, and the exact conditions within the specimen during the freezing, are so complicated and largely unknown that it can no longer be stated that no cell constituents are displaced during the process, especially at the molecular level. In the second step, water is permanently removed from the specimen by sublimation at a low temperature by the method of freeze-drying, or by solution at a low temperature in the method of freeze-substitution. Again, the exact conditions of sublimation and solution of water are unknown, and it is not possible to state with certainty that no displacement of other cell constituents takes place during these processes. Regardless of how the water has been removed, even under the best conceivable conditions, there are certainly changes in charge distribution, in intra- and intermolecular rearrangements and displacements, and, in addition, in the case of freeze-substitution, in solution of some substances and denaturation of others. How significant these changes are at the level of the electron microscope is not known.

Two other methods for using frozen sections have been proposed for the preparation of specimens for study with the electron microscope. The first method involves the freezing of rather large specimens, cutting frozen sections, and then allowing them to thaw while they are immersed in a fixative solution *(1)*. This method combines the disadvantages of both methods of preparation; i.e., fixation by freezing and by immersion in a fluid fixative. It is claimed that the structure of the protoplasm is the same as after immersion of large specimens in the fluid fixative without freezing, and this is regarded as a measure of the success of the method. On the other hand, this could be regarded equally as a measure of the inadequacy of fixation by immersion as usually practiced, since it is quite certain that structure has been disrupted by the freezing and thawing of the preliminary step, and subjected to extraction and coagulation displacements, albeit more rapidly than usual.

The second method *(7–9)* is the preparation of sections of ultrarapidly frozen blocks directly, without embedding the preparations. It is not yet clear whether the frozen specimens were free of ice crystals or that displacements of small molecules did not take place in subsequent steps.

Bullivant *(5)* has revived the question of the importance of eutectic mixtures in tissues as a determinant of the drying temperature. The lowest known binary eutectic for tissue constituents is that for calcium chloride and water, with a melting point of $-54.9°C$. Above this temperature, some water could be fluid, and hence displacements of cell constituents could be possible; below this temperature, all water supposedly would be solid. This theoretical difficulty was first raised by Bayliss *(2)* and revived by Scott *(24)*, only to be abandoned later by Scott and Hoerr *(25)*, on the grounds that the eutectic temperatures for protoplasm would be difficult (or impossible) to estimate. Moreover, the very concept is inappropriate for rapidly frozen specimens "since it is not likely that eutectics will be present in quenched tissues as they represent an equilibrium condition, and the constituents of the tissue do not have time to reach equilibrium during the very rapid quenching process" *(3)*. As with so many aspects of the freeze-drying process, empirical findings take precedence over theoretical concepts which can define simpler systems than those found in protoplasm. The reader is urged to consult recent review articles on this subject *(15, 20, 23, 32)*.

Staining

After water has been removed by fixation and subsequent transfers, specimens may be stained for morphological studies or for cytochemical purposes. Here the term stain is considered to denote any substance, which when added to living cells or fixed structures or structural components makes them more clearly visible or detectable with the electron microscope. The exact mechanism of staining is not always known. One consistent factor is change in mass, which results in a corresponding change in electron scattering, which is registered as alterations in contrast on the photographic image. To the extent that change in mass is selective, contrast in electron micrographs is changed selectively. For light microscopy, selectivity (morphological or cytochemical) is based on color, solubility product, charge, differential solubility, cross-linking reagents, steric relations, oxidation–reduction changes, and other factors. Of all the numerous possible reactions in cytochemistry at the level of the light miscroscope, only a handful do not call for a variety of controls. For example, few methods are as straightforward as the Prussian blue test for ferrocyanide or ferric ions, and the blue color is a sufficient identification of either ion in the tissue specimen. At the level of the electron microscope, selectivity depends on differences in contrast caused by the differential addition of mass which lifts the mass of the reaction product above the background control level, or by differential loss of a substance normally stained but now removed by chemical action or, more specifically, enzymic action. No cytochemical reaction studied with the electron microscope is as simple, direct, and specific as the Prussian blue reaction in light microscopy, and controls are required to assure a certain degree of specificity. In addition, when specimens are stained by immersion, or during freeze-substitution, the solvent action of the water limits the sensitivity of the cytochemical reaction in ways which have been described above. Finally, the need to "stabilize" tissue components by use of a second fixative commonly adds sufficient mass to the specimen so as to decrease the differential between background or control mass and that of the component being studied.

For material prepared by freezing and drying, the difficulties are reduced by the device of treating specimens with regents in vapor form *in vacuo*. The reaction products are subject only to the solvent action of the monomer mixture and of the water in the sectioning trough. This is the basis of all but two of the methods to be described in this work. The reagents used in vapor form, and the substances tested for are listed below:

1. *Difluoro organic compounds*—cross-linking reagents for polypeptides and proteins (Vol. I, Chapters 2, 3, 5; Vol. II, Chapters 4–7). The stabilization of nuclear and cytoplasmic proteins by use of these compounds makes possible the use of gallocyanin–chromalum for staining of nuclear DNA and RNAs, mitochondrial DNA and RNA, and ribosomal and soluble RNA (Chapters 2, 3, 6–13 in this volume, and Chapters 3 and 5 in Volume II).
2. *Acetylacetonates*—cross-linking reagents for proteins; make possible subsequent staining of nucleic acids as in (1) above and, in addition, staining of the polysaccharide moiety of protein–polysaccharide complexes (Vol. II, Chapters 7 and 10).
3. *Diiodophenylhydrazine*—test for phospholipids (Vol. II, Chapters 1–3).
4. *Benzethonium acetate*—test for protein–polysaccharide complexes (Vol. II, Chapter 7).
5. *Heptafluoroheptanedione*—test for di- and polyvalent cation, chiefly calcium, as well as for proteins (Vol. II, Chapter 9). In addition, alcohol or alcoholic solutions were used as described in earlier publications *(6, 11, 12, 22)*.
6. *Simple embedding, with no stain (to reduce background density to the minimum)*—to determine amorphous and crystalline calcium (Vol. II, Chapter 9), as used by Durning *(12)* and Molnar *(22)*.
7. *Alcoholic platinum tetrabromide*—test for proteins (Vol. I, Chapters 2 and 5), as used by Gersh *et al. (16, 17)*.

8. *Alcoholic ferric chloride*—test for ferrocyanide (Vol. II, Chapter 8) as used by Chase *(6)*.

Finally, sections were stained with uranyl acetate to enhance contrast in studies on proteins and lipids.

It has been assumed that, when vapors react with the solids of freeze-dried specimens, the reactions of the vapor–solid system (when they are detectable) are of the same sort as take place in solution. Double bonds of unsaturated aliphatic compounds of freeze-dried specimens, for example, will combine with diiodophenylhydrazine in the gas phase as they do when both are in solution, while at the same time amine groups of proteins will not react with the reagent. On the other hand, amine groups of proteins in freeze-dried specimens will react with difluorodinitrobenzene in the gas phase, as they do when both are in solution, while nucleic acids and lipids will not react with the reagent to any appreciable extent. In other words, when freeze-dried specimens are treated with a vapor reagent and localized increases in mass are observed in sections with the electron microscope, it is assumed that a reaction has taken place and that the reagent has reacted chemically in accordance with its properties (in solution) with an appropriate substrate in the freeze-dried specimen. Any other interpretation requires substantiation. I wish to point out that, in the unlikely event that further research proves that none of the vapor reagents act as cytochemical stains, there is some merit in adding to the meager list of reagents in suitable solvents used by electron microscopists as general stains.

Freeze-dried specimens are regarded as solids. With the rare exceptions of crystalline substances like apatite, the solid components of freeze-dried specimens could be considered as very finely divided amorphous powders and/or surfaces, with the various types of reactive groups by and large available for reaction with reagents in the gas phase. The gas phase generally penetrates to all surfaces and is partly depleted locally where the reaction has taken place. The penetration of the reagent is probably progressively limited by the increase in mass which takes place, and by the progressive cross-linking, when this is possible.

In very general terms, the solid phase–vapor phase reactions within freeze-dried specimens resemble those of heterogeneous catalytic systems which are readily and irreversibly poisoned. In both types of reactions, the solid component enters into the reaction with the gaseous reagent, with a certain degree of selectivity. The course of the reaction in freeze-dried specimens cannot be predicted or described fully because of the extreme chemical and physical heterogeneity of the reacting surface. In addition, there is an absence of any quantitative information on the diffusion of the reagent, the adsorption of the reagent, the reactions with the adsorbed layer, the desorption and diffusion of the products of the reaction *(31)*, especially during infiltration with the plastic components of the embedding medium, and the floating of the ultrathin sections in the sectioning trough. Moreover, the reagents must cause only minimal distortions of the structure of protoplasm. The choice of the most favorable reagent for any test, and of the optimal conditions for the test is, and is likely to remain, empirical.

Other Preparatory Steps

In order to retain the emphasis on a rational approach to submicroscopic cytochemistry, efforts were made to maintain all other technical aspects as constant and unvarying as possible. These include the freezing and drying processes, the use of a single plastic monomer mixture in nearly all instances, and the conditions for infiltration and polymerization of the plastic. The only exceptions are the bone preparations which were embedded in methacrylate (Vol. II, Chapter 6). The nearly total restriction of embedding plastic to Durcupan carries with it certain penalties, such as limits to the use of certain enzymes. However, it was judged that the correlation of the various images of cell constituents and their reconstitution as a whole cell (as presented in the last chapter of Vol. II) would be promoted under these conditions.

Consideration of Artifacts of Freeze-Drying and Freeze-Substitution

Artifacts are, at present, inevitable in morphological or cytochemical studies with the electron microscope. The important question is whether the artifacts promote the understanding of the nature of protoplasm or whether they are misleading. There are three main categories or degrees of certainty (or utility) of artifacts.

Category I includes those features studied with the electron microscope which can be confirmed by some other different, reliable technique. These features are probably real. Included are structures in three subgroups: (A) Structures that can be seen in living cells or tissues with the light microscope (the shape, size, and relations of nucleus, mitochondria, etc., and cell processes of various types). (B) Structures that show periodicity (lamellae of retinal rods, myelin sheath, and chloroplasts, and banding of collagen fibers) or are crystalline (apatite and DNA). (C) Structures which can be separated by relatively mild procedures (glycogen granules and ribosomes).

Category II includes all structures observed with the electron microscope, not confirmed by other methodologies. These features may be real, but there is less certainty of their reality. They include single or double membranes of the cell, various vacuoles and fibrils, nonfibrillar components of connective tissues, etc.

Category III includes structures which clearly differ in appearance using other more reliable observations. Such election microscopic observations are probably artifacts and are conceptually misleading. These include the pericellular spaces around cartilage cells which can be shown by direct observation with the light microscope to arise from differences in tonicity, and DNA granules in chromatin, whose DNA must be linearly arranged.

In category II, one may distinguish several subgroups in relation to the techniques employed. Subgroup A includes structures which are reasonably similar when specimens are fixed by immersion or freeze-substitution and by freeze-drying. This group includes rough endoplasmic reticulum, internal and external membranes of mitochondria, nuclear membranes, glycogen, Golgi apparatus, apatite crystals, etc. Subgroup B includes structures observed after preservation by freeze-drying but not by other methods, chiefly because this method prevents or reduces solution during the preparation of the specimen. Like the structures in category I, they may be real, but there is less certainty about their reality than the structures in the first category. These include nonfibrillar components of connective tissues, euchromatic and heterochromatic DNA supercoils of somatic and mitotic cells, nuclear proteins, possible protropocollagen granules in fibroblasts, chondrocytes and osteocytes, and a variety of other structures.

The submicroscopic vacuoles of protoplasm are also in this uncertain category. These constitute, in fact, the chief morphological differences between protoplasm prepared by freeze-drying and by freeze-substitution. Several observations support the thesis that submicroscopic vacuoles are real structures in the protoplasm of living cells. The first is that submicroscopic vacuoles are of approximately the same size in all parts of all specimens. Where ice crystals have formed, they may or may not be detectable at the periphery, but develop and become progressively larger toward the center of the specimen where they are approximately 1 μ in diameter *(18, 21, 32)*. It is also significant that the dimensions of the submicroscopic vacuoles are the same when the specimens have been dried at $-30°$ or $-78°C$, or when they were stored before drying for 15 days at $-30°C$ or in liquid nitrogen *(13,16)*. The significance of these findings is that although some of these conditions are thought to encourage ice crystal formation, in actual experiment they failed to do so. The second set of observations which bear on the reality of submicroscopic vacuoles of protoplasm is exemplified by the distribution of proteins in the nuclei of salivary glands of *Drosophila.* (See Vol. I, Chapter 5.) The chromosomal and nucleoplasmic proteins may appear uniformly distributed, or in a variety of highly structured patterns, depending on the postfixation agent used. After five methods of postfixation, the nucleoplasm was observed to be homogeneously distributed (with no "spaces" attributable to ice crystals, which could be resolved with the electron microscope). On the other hand, submicroscopic vacuoles are visible after certain other methods of postfixation. Moreover, when present, the pattern of submicroscopic vacuoles varies according to the postfixative employed; i.e., different postfixatives preserve and stain different components of nuclear proteins and when all proteins are stained the nucleus appears homogeneous and free of ice crystal spaces. Yet the process of freezing and drying was the same in all instances. This suggests that submicroscopic vacuoles of protoplasm, far from being empty, are actually filled. Another example is glycogen, whose second-order granules or β granules (*4, 10, 11*) fill submicroscopic vacuoles. Still another example is the preservation and staining of pseudovacuolar proteins in liver cells by vapors of monofunctional reagents which do not act as cross-linking agents. These include *p*-iodophenylsulfonyl chloride, 3,4,5-triiodobenzoyl chloride, and 1-fluoro-2,4-dinitrobenzene *(19)*. The first two reagents are solid at room temperature and atmospheric pressure, the third is fluid. In specimens treated with each of these three reagents, the submicroscopic vacuoles are filled with a protein-reacting material which resists extraction by reagents used in the process of embedding. The same is true also when vapors of certain cross-linking reagents are used in certain conditions. There are at least eight different cross-linking reagents which were used as vapors *in vacuo*, after which the contents of the submicroscopic vacuoles were stained (see this volume, Chapter 2 for details). Although it is not capable of being proved, evidence suggests that submicroscopic vacuoles comprise a constant feature of protoplasm after freezing and drying, and that the contents of the vacuoles may be chiefly glycogen or protein in nature. Thus the submicroscopic vacuoles of constant appearance under most conditions may be regarded as pseudovacuoles. The underlying significance of the two preceding statements is that the contents of the vacuoles are different (less readily stained) from those found in the walls. This view is contrary to the one commonly held by many researchers, who believe that the submicroscopic vacuoles are caused by the expansion of ice crystals formed during the process of freezing or drying *(5, 23)*. Submicroscopic vacuoles can be formed or prevented at will by varying only the degree of denaturation and staining of the contents of the vacuoles, and not by varying the water content of the specimen, the degree of the vacuum, the temperature of drying, etc. The only preconditions for the formation of submicroscopic vacuoles are related to postfixation treatments, and not to the process of freezing and drying per se as carried out in actual practice.

Before coming to a conclusion as to whether submicroscopic vacuoles reflect an underlying organization, the possibility should be considered that freeze-substitution obliterates preexisting submicroscopic vacuoles. Most specimens prepared by freeze-substitution were so large that it is certain that many large ice crystals had formed in the protoplasm. Yet no traces of the deformations and displacements accompanying the process could be detected when sections were viewed with the electron microscope. One must conclude that signs of ice crystals had been obliterated during the substitution process, possibly because of a swelling and solution process at some point in the gradient between ice and acetone as the water is dissolved. This is somewhat similar to the process whereby ice crystals are obliterated in frozen (and thawed) sections of large specimens during the process of staining them *(1)*. It is possible, then, to consider that submicroscopic vacuoles of protoplasm are really present in protoplasm, but are not visible after freeze-substitution because they are obliterated during the fixation process. They are obliterated in the same way in which growing ice crystals known to be formed during the freezing process are also eliminated *(15, 23)*.

It cannot be proved that submicroscopic vacuoles (as they appear in freeze-dried specimens) reflect a real, underlying structure of protoplasm, since no other techniques are available to test the proposition. This morphological uncertainty has little bearing on the value of cytochemical studies of specimens prepared by freeze-drying. Part of this work is devoted to descriptions of new cytochemical methods. Such methods, whether old or new, have merit only insofar as they are useful in the study or solution of significant problems. A general statement of the problems and fields investigated with the new methods is briefly summarized in the first paragraph of this chapter. More specific statements can be gleaned by perusal of the table of contents, the chapter summaries, or the last chapter of Volume II. The extent to which light is shed on these problems is a measure of the value of the new methods used. Most of the problems could not have been attacked without the methods, and, to some degree, the definition of the problems was determined by the nature of the methods.

Perhaps this can be restated in a different way, which is essentially in agreement with the views expressed by Bullivant *(5)* and Christensen *(8)*. At the present time, it is not possible to prove that ice crystals of a significant size for cytochemistry do not occur in specimens even under optimal conditions of freezing, as it is impossible to prove that ice crystals of a significant size do form under the most favorable conditions. But this is, in a way, an academic question; the more important questions are whether cytochemical information gathered from such specimens throws significant light on important questions in cell biology, and whether artifacts of preparation are minimized.

Although the topics have been separated in this analysis, actually fixation and staining constitute a single process. It is of little consequence that a substance can be stained for electron microscopy, if it has been displaced significantly during fixation. Similarly, good preservation is of little consequence if the substance in question is not more or less selectively stained.

Identification of Submicroscopic Structure in Cytochemical Electron Micrographs

The structure of protoplasm as developed in this work differs in some respects from that derived by use of fluid fixatives, and this raises questions as to the identification of the various structures and organelles in freeze-dried specimens and their equivalence with structures as commonly described in the literature. There is no real difficulty in transferring morphological identifications to freeze-dried specimens at the level of the light microscope, i.e., nucleus, chromatin, nucleolus, nucleoplasm, mitochondria, secretory granules, ergastoplasm, etc. Within chromatin, nucleolus, and nucleoplasm, distinctions between DNA and all other nuclear components can be made, in most cases, by the use of nucleases. Within mitochondria, the compartmentalized structure so well described by Sjöstrand and his colleagues *(26–30)* and by Malhotra *(20)* is, in the main, very close to that described in Volume II, Chapter 3. Secretory granules (in pancreas) are always spherical, lack the characteristic compartmentalized structure of mitochondria (which may appear spherical in sections), and are usually restricted to certain parts of cells. The rough endoplasmic reticulum (RER) does not differ significantly from the descriptions of Sjöstrand and his colleagues *(26–30)*, and these are not very far off from conventional descriptions of the RER.

Perhaps questions of identification become more difficult when the structures are considered cytochemically. For example, when mitochondria are stained selectively for lipids or for proteins, each component separately may appear more or less homogeneously distributed. When the same organelles are stained simultaneously for both components, the compartmentalized structure of mitochondria is clarified (see Vol. II, Chapter 3), and there is no doubt about the identification of these structures as mitochondia. Another example is the RER, which can be stained selectively for ribosomal and soluble RNA, for protein including ribosomal and membrane protein, and for unsaturated lipid. (See this volume, Chapters 2; Vol. II, Chapters 1, 2, and 5.)

This difficulty of morphological identification of cytochemically reacted structures is no greater than cytologists have always faced when observing with the light microscope sections stained differentially for RNA components of nucleus and cytoplasm, DNA components, proteins, lipids, polysaccharides, enzymes, calcium, etc., as well as ergastoplasm, mitochondria, secretory granules, chromosomes, and chromatin, nucleoli, etc. By a feat of synthesis, each component and constitutent is related to every other one. In the same way, the protein, nucleic acid, and lipid constituents of the RER are studied separately in this work by methods which are more or less selective, and these are then integrated and reconstituted into the RER. The same general outlook is applied to all the topics studied in these volumes. The findings and interpretations presented in the key chapters require that the reader apply the same process of synthesis and integration which has characterized the work of cytologists and cytochemists for generations.

With few exceptions, cytochemical electron micrographs in this work lack high contrast and seem fuzzy. This gray quality is, I believe, intrinsic to cytochemical work with the electron microscope, and can be caused by two main factors: (1) the relative specificity of cytochemical tests and (2) the retention of more protoplasmic solid components after freezing and drying. Certainly, a stain for RNA will enhance contrast less than one which also stains proteins and lipids. The former, when stained, has less mass than the latter. The same also applies to stains of proteins, lipid, etc. This lesser contrast (as compared with situations when less specific morphological stains are used) is characteristic of cytochemical electron micrographs despite the advantage that the ambient background stain is reduced because of the very selectivity or specificity of the stain. This is less of an advantage in electron microscopy than in light microscopy, where image contrast is usually caused by the colored reaction product. In electron microscopy, on the other hand, the mass of the reaction product competes in contrast with the mass of the rest of the section, including the plastic and contained proteins, whether stained or unstained. The contrast of the cytochemical reaction is thus reduced to a certain degree, though not as much as when osmium tetroxide is used as a counterstain. No matter what the cytochemical reaction, it must take place under conditions of minimal distortion or extraction. This is in contrast with morphological studies where some protoplasmic constituents may be diluted, extracted, or displaced, and where the contrast of the remaining structure may be relatively increased by the reduction of ambient mass. This combination of circumstances is such that, except in rare instances [such as studies on amorphous and crystalline inorganic calcium and phosphate (Vol. II, Chapter 9)] cytochemical electron micrographs are of lower contrast than those electron microscopists are accustomed to view. This phenomenon is unrelated to resolution; since, in favorable sections studied at high magnifications, the resolution with apatite crystals in cartilage was about 7 A, while in tests of largely organic parts of protoplasm the resolution was about 10 A. It is obvious, therefore, that the lower contrast of the electron micrographs is not caused by poor resolution.

Summary

Submicroscopic cytochemistry includes studies on the chemical anatomy of cell structures and of some of their products of secretion such as extracellular connecive tissues at molecular and supramolecular levels. It is thus of interest (1) for the light it sheds on cell behavior, (2) as a bridge between cytology, on the one hand, and biochemistry and genetics, on the other, and (3) as a rational basis for the fixation and staining of intercellular substances. Fixation and staining are important, separately and jointly, in the development of cytochemical methods. Each topic is treated separately, and what emerges is that water is the greatest obstacle in the development of sensitive, delicate, complete, selective or specific, and anatomically correct cytochemical methods, whether this water originates in the cell or is added during the fixation and staining process. Accordingly, a general method was developed which essentially removes water first temporarily by ultrarapid freezing and then permanently by sublimation at low temperature. The cytochemical reagents are applied in vapor form in the absence of fluid water. By this means, methods were developed for nucleic acids in cells (nuclear and mitochondrial DNA, and ribosomal, soluble, nuclear, and mitochondrial RNA), for polypeptides, for acidic polysaccharides, for phospholipids, and for bi- and polyvalent cations including calcium. These vapor methods supplement methods developed earlier for use of alcoholic solutions for protein and for ferrocyanide. Evidence is cited for the conclusion that ice crystals of a significant size (for example, 10–20 A or more) are not formed during the process of freezing or drying. There is a general discussion of artifacts, especially as they pertain to freezing and drying, and identification of cytochemical components in relation to submicroscopic structure, with a consideration of specific examples.

The methods were applied to a wide range of problems concerned with: nucleocytoplasmic relations; the relation of DNA molecular patterns to the action of genes during differentiation, maturation, activity cycles, and cell death; the nature and origin of secretory and storage precursors; the chemical structure of membranes; compartmentalization in mitochondria, the possible cellular origin of protropocollagen and the protein–polysaccharides of cartilage, and the physical basis of transport in cartilage and tendon. These and other topics are treated more fully in the remainder of this work.

References

1. Baker, R. F. (1962). Freeze-thawing as a preparatory technique for electron microscopy. *J. Ultrastruct. Res.* **7**, 173–184.

2. Bayliss, W. M. (1915). "Principles of General Physiology." Longmans, Green, London.
3. Bell, L. G. E. (1952). The application of freezing and drying techniques in cytology. *Int. Rev. Cytol.* **1**, 35–63.
4. Bondareff, W. (1957). Morphology of particulate glycogen in guinea pig liver revealed by electron microscopy after freezing and drying and selective staining *en bloc. Anat. Rec.* **129**, 97–114.
5. Bullivant, S. (1970). Present status of freezing techniques, *In* "Some Biological Techniques in Electron Microscopy" (D. F. Parsons, ed.), pp. 101–146. Academic Press, New York.
6. Chase, W. H. (1959). Extracellular distribution of ferrocyanide in muscle. *AMA Arch. Pathol.* **67**, 525–532.
7. Christensen, A. K. (1967). A simple way to cut frozen thin sections of tissue at liquid nitrogen temperature. *Anat. Rec.* **157**, 227 (abstr.).
8. Christensen, A. K. (1969). A way to prepare frozen thin sections of fresh tissue for electron microscopy. *In* "Autoradiography of Diffusible Substances." (L. J. Roth and W. E. Stumpf, eds.), pp. 349–362. Academic Press, New York.
9. Christensen, A. K. (1970). Frozen thin sections of fresh tissue for autoradiography of diffusible substances. *7th. Congr. Int. Microsc. Electron.*, Grenoble.
10. Drochmans, P. (1960). Mise en évidence du glycogène dans la cellule hépatique par microscopie électronique. *J. Biophys. Biochem. Cytol.* **8**, 553–558.
11. Drochmans, P. (1962). Morphologie du glycogène. *J. Ultrastruct. Res.* **6**, 141–163.
12. Durning, W. C. (1958). Submicroscopic structure of frozen-dried epiphyseal plate and adjacent spongiosa of the rat. *J. Ultrastruct. Res.* **2**, 245–260.
13. Gersh, I. (1956). The preparation of frozen-dried tissue for electron microscopy. *J. Biophys. Biochem. Cytol. Suppl.* **2**, 37–43.
14. Gersh, I. (1959). Fixation and staining. *In* "The Cell" (J. Brachet and A. E. Mirsky, eds.), Vol. I, pp. 21–66. Academic Press, New York.
15. Gersh, I. (1965). Freeze-substitution. *Fed. Proc. Suppl.* **15**, S233–S234.
16. Gersh, I., Isenberg, I., Stephenson, J. L., and Bondareff, W. (1957). Submicroscopic structure of frozen-dried liver specifically stained for electron microscopy. I. Technical. *Anat. Rec.* **128**, 91–111.
17. Gersh, I., Isenberg, I., Bondareff, W., and Stephenson, J. L. (1957). II. Biological. *Anat. Rec.* **128**, 149–169.
18. Gersh, I., and Stephenson, J. L. (1954). Freezing and drying of tissues for morphological and histochemical studies. *In* "Biological Applications of Freezing and Drying" (R. J. C. Harris, ed.), pp. 329–384. Academic Press, New York.
19. Gersh, I., Vergara, J., and Rossi, G. L. (1960). Use of anhydrous vapors in post-fixation and in staining of reactive groups of proteins in frozen-dried specimens for electron microscopic studies. *Anat. Rec.* **138**, 445–460.
20. Malhotra, S. K. (1968). Freeze-substitution and freeze-drying in electron microscopy. *In* "Cell structure and Its Interpretation" (S. M. McGee-Russell and K. F. A. Ross, eds.), pp. 11–21. St. Martin's Press, New York.
21. Menz, L. J., and Luyet, B. J. (1961). An electron microscope study of the distribution of ice in single muscle fibers frozen rapidly. *Biodynamica* **8**, 261–294.
22. Molnar, Z. (1959). Development of the parietal bone of young mice. I. Crystals of bone mineral in frozen-dried preparations. *J. Ultrastruct. Res.* **3**, 39–45.
23. Rebhun, L. I. (1965). Freeze-substitution: fine structure as a function of water concentration in cells. *Fed. Proc. Suppl.* **15**, S217–S232.
24. Scott, G. H. (1933). A critical study and review of the method of microincineration. I. Introduction. *Protoplasma* **20**, 133–151.
25. Scott, G. H., and Hoerr, N. L. (1950). Drying: Frozen-dehydration method for histologic Fixation. *In* "Medical Physics" (O. Glasser, ed.) Vol. 25, pp. 292–296. Year Book, Chicago, Illinois.
26. Sjöstrand, F. S. (1968). Ultrastructure and function of cellular membranes. *In* "The Membranes" (A. J. Dalton and F. Haguenau, eds.), pp. 151–210. Academic Press, New York.

27. Sjöstrand, F. S. (1968). Molecular structure and function of cellular membranes. *In* "Regulatory Functions of Biological Membranes" (J. Järnefelt, ed.), pp. 1–20. Elsevier, Amsterdam.
28. Sjöstrand, F. S. (1969). Morphological aspects of lipoprotein structures. *In* "Structural and Functional Aspects of Lipoproteins in Living Systems." (E. Tria and A. M. Scanu, eds.), pp. 73–128. Academic Press, New York.
29. Sjöstrand, F. S., and Baker, R. F. (1958). Fixation by freezing-drying for electron microscopy of tissue cells. *J. Ultrastruct. Res.* **1**, 239–246.
30. Sjöstrand, F. S., and Barajas, L. (1968). Effect of modifications in conformation of protein molecules on structure of mitochondrial membranes. *J. Ultrastruct. Res.* **25**, 121–155.
31. Trapnell, B. M. W. (1955). "Chemisorption." Butterworth, London.
32. Van Harreveld, A., Crowell, J., and Malhotra, S. K. (1965). A study of extracellular space in central nervous tissue by freeze-substitution. *J. Cell Biol.* **25**, 117–137.

2

Distribution of Protein in Hepatic and Exocrine Pancreatic Cells of the Mouse

Isidore Gersh

While there was some reluctance to accept the possibility that the freezing and drying process could preserve structure at the submicroscopic level, this has diminished somewhat with the general acceptance of freeze-etching and freeze-fracturing as effective techniques *(3, 4, 6, 12, 13, 19, 20, 27, 29, 30).* Subsequent manipulations for cytochemical studies could, of course, distort structure. It was in an effort to reduce extraction and dislocation to a minimal or inappreciable extent that the method of vapor postfixation was developed.

The strongest objections to freeze-drying as a method suitable for the adequate preservation of the submicroscopic structure of protoplasm center chiefly around the possibility that ice crystals are formed during the process of freezing and drying, and that their growth causes physical displacements of protoplasm large enough to vitiate meaningful studies with the electron microscope. "Spaces" which I have called "submicroscopic vacuoles" have been alleged to be occupied before the process of drying by ice crystals. Evidence is summarized in Chapter 1 (this volume) that suggests strongly that ice crystals resolvable with the electron microscope do not form during freezing or drying, and, moreover, that "submicroscopic vacuoles" appear only because the contents of the vacuoles are unstained in the course of most postfixation treatments. The most cogent evidence is that when specimens are frozen and dried under specified conditions, and then treated *in vacuo* in the absence of fluid water with vapors of certain cross-linking bifunctional agents, there are no empty spaces free of protein which are large enough to be visible with the electron microscope. Examples of such vapors are heptafluorodimethyloctanedione, a reagent which reacts with proteins and with bi- and polyvalent cations (Volume II, Chapter 9), zirconium hexafluoroacetylacetonate (Table 2.1), and osmium tetroxide. Other methods of postfixation

TABLE 2.1

VAPOR REAGENTS AND ABBREVIATED CONDITIONS OF USE

Reagent; abbreviation	Reaction temperature (°C)/exposure time (hours)[a]	Evacuation temperature (°C)/time (hours)[a,b]	Source of reagent[c]
1, 5-Difluoro-2, 4-dinitrobenzene; FFDNB	50°/5	60°/ON	1
4, 4′-Difluoro-3, 3′-dinitrophenyl sulfone; FF sulfone	45°/5	55°/ON	1
Heptafluorodimethyloctanedione; HEFDOD	0°/5	0°/2, RT/ON	2
Barium acetylacetonate • $2H_2O$; Ba acac	45°/5	RT/ON	3
Calcium acetylacetonate; Ca acac	RT/5	RT/ON	3
Chromium (III) hexafluoroacetylacetonate; Cr hfac	−40°/7	−40°/3, RT/ON	4
Cobalt (II) acetylacetonate; Co acac	70°/6	80°/ON	3
Cobalt (III) trifluoroacetylacetonate; Co tfac	RT/5	RT/ON	4
Copper (II) acetylacetonate; Cu acac	80°/5	90°/ON	3
Hafnium (IV) hexafluoroacetylacetonate; Hf hfac	0°/3	0°/2, RT/ON	2
Iron (III) acetylacetonate; Fe acac	−40°/7	−40°/2, RT/ON	3
Lanthanum (III) acetylacetonate; La acac	RT/5	RT/ON	3
Mercury (II) hexafluoroacetylacetonate; Hg hfac	RT/5	RT/ON	2
Nickel (II) hexafluoroacetylacetonate; Ni hfac	−40°/7	−40°/9, RT/ON	2
Rhodium (III) hexafluoroacetylacetonate; Rh hfac	−40°/5	−40°/9, RT/ON	2, 3
Thorium (IV) acetylacetonate; Th acac	0°/5	0°/9, RT/ON	3
Zinc (II) hexafluoroacetylacetonate; Zn hfac	−40°/5	−40°/2, RT/ON	4
Zirconium hexafluoroacetylacetonate; Zr hfac	−40°/5	−40°/2, RT/ON	2

[a] RT = room temperature (22°–28°C).

[b] ON = overnight (12–18 hours); a comma separates two successive stages of evacuation.

[c] Sources: (1) Prof. H. Zahn, Deutsches Wollforschungsinstitut, Aachen, West Germany. (2) Mr. Roy Oliver, Pierce Chemical Co., Rockford, Illinois 61105. The mercury reagent was not pure; one sample had too much mercury, the other too little. The results with both samples were the same. (3) Purchased from MacKenzie Chemical Works, Inc., Central Islip, Long Island, New York 11722. (4) Dr. R. E. Sievers, Aerospace Research Laboratories, Wright-Patterson Air Force Base, Ohio 45433. Some of the properties of fluoro derivatives of acetylacetonates are described in reference *(21)*. I am indebted to the gentlemen named for the samples and advice furnished.

and staining which are effective in the same way are given on p. 36–42. Apparently the proteins in the submicroscopic vacuoles are stained when the denaturation is sufficiently extensive or of such a nature as to prevent significant solution during subsequent manipulations as well as to make the protein amenable to section staining with uranyl acetate. Only when the denaturation is less complete or the protein remains averse to staining with uranyl acetate in sections are submicroscopic vacuoles observed.

Almost as cogent evidence is the staining of protein in the submicroscopic vacuoles by another series of monofunctional vapor reagents which react with proteins *(10)*. These are 1-fluoro-2,4-dinitrobenzene, iodophenylsulfonyl chloride, and 4,5-triiodobenzoyl chloride. Each of the three reagents when used *in vacuo* in vapor form cause the retention and staining of proteins in the spaces occupied by submicroscopic vacuoles.

Still another example of apparent homogenization is the nucleus of the salivary gland of third instar larvae of *Drosophila melanogaster* (Vol. I, Chapter 5). Here postfixation with 95% alcohol was sufficient to retain protein in the submicroscopic vacuoles. The protein was also preserved and stained by four different cross-linking reagents *in vacuo* in vapor form, and remained so after sections were stained with uranyl acetate. In addition, the vacuolar pattern was found to vary with every vapor used as a postfixative, probably because different types of proteins were made insoluble and stainable by the different vapor reagents used.

The fact is, then, that when specimens are postfixed in certain ways, the regions that otherwise appear as empty submicroscopic spaces are shown to contain stainable protein. There appear to be spaces (or submicroscopic vacuoles) after most methods of postfixation and staining, because the proteins in the walls are stained very much more than those in the vacuoles. These findings can be interpreted to mean that submicroscopic vacuoles are artifacts or pseudovacuoles, whose contents are preserved but are less readily stained than the walls which enclose them. In any case, the result of this use of a wide range of postfixatives is that a difference in composition or state of aggregation or both are demonstrated between the contents of the pseudovacuole and the wall surrounding it. The submicroscopic pseudovacuoles observed consistently throughout our studies from 1954 to the present *(8–10)* turn out to be probably real, in the sense that they indicate the presence of two major types of proteins in protoplasm: (1) materials which are less readily stained and which characterize the contents of submicroscopic vacuoles, and (2) materials which are more readily stained and which characterize the denser, spongelike walls. This fortunate artifact gives some insight into a complexity of protoplasmic organization which would have been entirely missed were all proteins preserved and stained equally everywhere in the cell. (See discussion of artifacts in this volume, Chapter 1.)

Methods

Specimens were cut in a moist box to a thickness of 0.1 to 0.2 mm, and placed on thin aluminum foil. They were frozen ultrarapidly by immersion in Freon 13, cooled to $-180°C$ with liquid nitrogen, and vigorously swirled to prevent significant ice crystal formation. They are stored in liquid nitrogen in vials which are marked by colored glass beads melted on their edge. As needed, specimens were dried at $-40°C$ or lower *(8)*. When the specimens were to be postfixed in an alcoholic solution, they were dried in tube C (Fig. 2.1); when they were to be postfixed with cross-linking vapors, they were dried in tube S (Fig. 2.1).

After drying, some of the specimens were infiltrated directly *in vacuo* with 95% alcohol and then treated with a 5% alcoholic solution of platinum tetrabromide. Others were infiltrated directly with the 5% alcoholic solution of platinum tetrabromide. This salt was used since it seemed the simplest of a long list of salts tested previously *(8, 9)*. If the specimens are to be heat and alcohol denatured, the lower part of tube C is immersed in an oil or steam bath for several hours. The tube is then removed, allowed to cool to room temperature, and then treated with 95% alcohol as above.

Still other freeze-dried specimens were treated *in vacuo* with organic and organometallic vapors and then infiltrated with alcohol. However, some vapor-treated specimens bypassed treatment with alcohol, and were infiltrated directly with the components of water-insoluble or water-soluble Durcupan. All specimens were eventually infiltrated with Durcupan which was slowly polymerized. Sections were made with a Porter-Blum or a Huxley microtome, and mounted on Formvar-coated grids, some of which were carbon-coated. Some sections were photo-

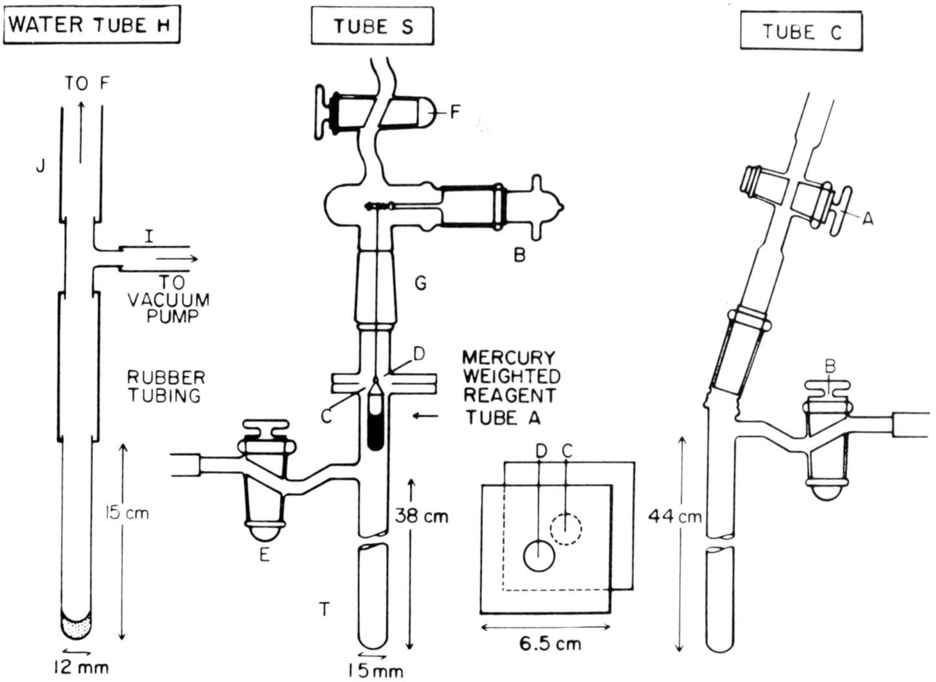

FIG. 2.1 Diagram of freeze-drying apparatus. See text for explanation. The glass apparatus was blown by James D. Graham, Jr., Scientific Glassblower, Department of Physiology, University of Pennsylvania.

graphed directly with the electron microscope, while others were stained with uranyl acetate before being photographed. The rationale for section staining is discussed on pp. 7 and 18. The chief points of interest for cytochemical work (as well as for morphological study) are that the reactions of the vapors *in vacuo* take place in the absence of fluid water, and that the reaction products tend to become insoluble through cross-linking of reactive groups of polypeptides. Water is added to the system in vapor form as a catalyst and to reduce shrinkage. (See Vol. II, Chapter 3.)

All cross-linking reagents that were tested combine in one way or another with proteins, and many combine, in addition, with nucleic acids [*(11, 17, 18, 38, 39, 40)* see this volume, Chapter 3]. Some of the reagents are probably capable of also reacting with phospholipids. The evidence presented in this chapter indicates that the reactive sites of those parts of protoplasm that are stained in common by all reagents and are amino acid, polypeptide, or protein in nature are terminal or ε-amino groups, carboxyl, sulfhydryl, imidazol, or phenolic groups *(14, 15, 22, 25, 36, 39, 40)*. Since the reagents are bifunctional, they can react with two reactive groups when the distance between them is correct, and thus cross-link them. A general treatment of cross-linking in relation to fixation is given by Wolman *(37)*. Since all the vapor reagents yielded a rather uniform picture, most of the work was done with two in particular—difluorodinitrobenzene (FFDNB) and mercury hexafluoroacetylacetonate (Hg hfac).

Pertinent information on the vapor reagents is given in Table 2.1. The second reagent in the list (FF sulfone) was useful in the study of possible precursors of tropocollagen (Vol. II, Chapters 4 and 6). The cupric compound has been used to study catecholamines in the cells of the adrenal medulla *(10)*. The experimen-

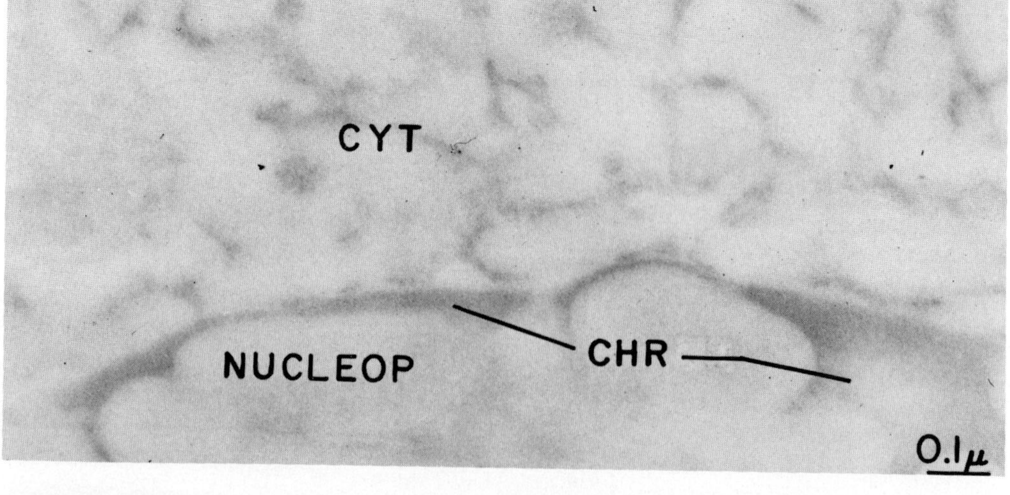

FIG. 2.2. Liver, postfixation in alcohol, no stain. Nuclear chromatin (CHR) is dense, and the walls and spaces of the nucleoplasm (NUCLEOP) are barely perceptible. Cytoplasmic (CYT) walls and spaces are indicated. × 81,000.

tal conditions were empirical—under other conditions of testing, some of the reagents might be as useful as FFDNB and Hg hfac.

Details of Procedures

1. Postfixation with 95% Alcohol. After the specimens were dried in tube C (Fig. 2.1), the thistle tube was filled with 95% alcohol, which was allowed to pass slowly into the vacuum tube through (A) while the vacuum tube continued to be connected with the drying tube. After the vials were covered with alcohol, stopcock (B) was closed and (A) opened. The vacuum was reinstituted and broken twice in the specimen chamber during about 5 minutes. When the vacuum was finally broken, the specimens were kept in alcohol at −40°C at atmospheric pressure for 1 day. The vials were then transferred to a Petri dish at room tem-

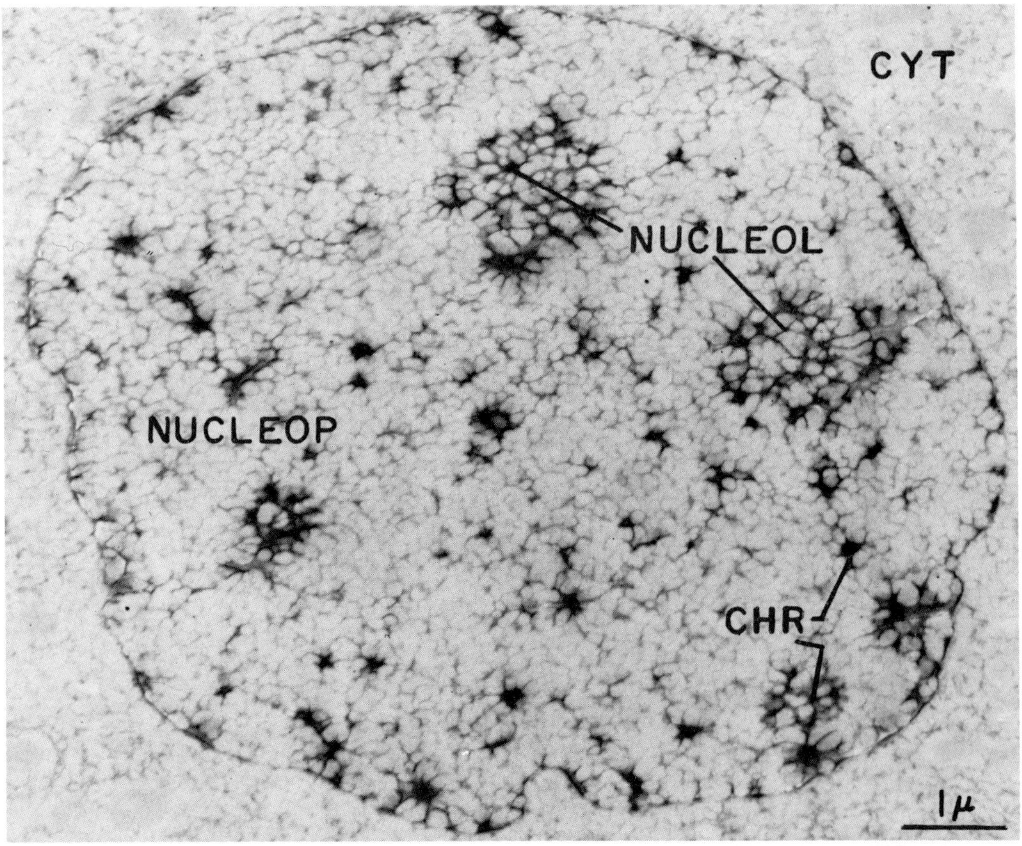

Fig. 2.3. Liver, postfixation in alcohol, section statined with uranyl acetate. The two-phase system in the nucleus is quite apparent. The denser walls are enlarged to accommodate chromatin (CHR), to the point where the intervening spaces between them are reduced or obliterated. The dense phase is continuous in chromatin, nucleolus (NUCLEOL), and nucleoplasm (NUCLEOP). × 14,000.

perature, and the specimens transferred from the foil, on which they were frozen, to a glass vial containing alcohol. They remained for 1 day, after which they were cut into smaller pieces with a razor blade. Specimens were never allowed to dry. Some of the specimens were then stained with a 5% solution of platinum tetrabromide in 95% alcohol for 5 hours and washed overnight in several changes of 95% alcohol. When the specimens were stained *in vacuo* in the alcoholic platinum tetrabromide solution, the infiltration was as with 95% alcohol alone. When the infiltration was ended, the specimens were removed from the drying tube and quickly separated from the foil, which otherwise would be dissolved by the reagent. The specimens were transferred to fresh staining solution, and after 5 hours, they were washed in several changes of 95% alcohol, and allowed to remain overnight. Specimens which were heat denatured were infiltrated *in vacuo* with 95% alcohol and stained with alcoholic platinum tetrabromide as above.

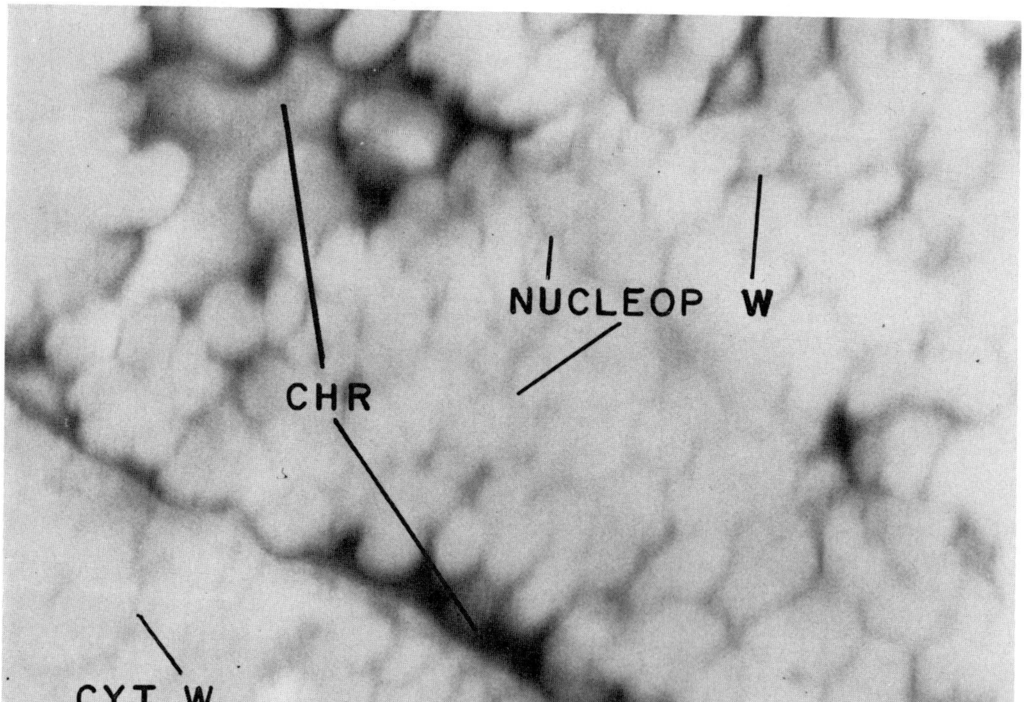

FIG. 2.4. Prepared as in Fig. 2.3 above, to be compared with Fig. 2.2 above. Staining with uranyl acetate resulted in increased contrast in chromatin (CHR), nucleoplasmic walls (NUCLEOP W), and cytoplasmic (CYT W) walls. The chromatin is homogeneous. In the region marked by an arrow, the wall between nucleus and cytoplasm cannot be distinguished from nearby nucleo- and cytoplasmic walls. × 81,000.

At this point, all the specimens, stained and unstained, were embedded in water-insoluble or water-soluble Durcupan (w-i or w-s Durcupan). For embedding in w-i Durcupan, specimens were passed through absolute alcohol, successive dilutions with freshly prepared Durcupan ABD (2:1, 1:1, 1:2), and into Durcupan ABC in a preparation dish (37 × 19 × 25 mm). The lid was sealed with parafilm, and the dish placed on an oscillator at 50°C for 1 day to facilitate diffusion of the monomer mixture. The specimens were then transferred to Durcupan ABCD on the oscillator for 30 hours and then to BEEM capsules. They were maintained at room temperature for 1 day, 37°C for 2 days, and 50°C for 2 days, after which they could be sectioned. (The oscillator is a platform which tilts 4 cm at a distance of 15 cm from its hinge, at a rate of four times a minute.) For embedding in w-s Durcupan, the following schedule was used: absolute alcohol, w-s Durcupan mixtures (2:1, 1:1, 1:2), oscillation in Durcupan ABCD at +16°C for 2 days, transfer to BEEM capsules at +16°C for 1 day, room temperature for 1 day, 37°C for 2 days, and 50°C for 2 days. When the specimens were not clearly distinguishable in the plastic, they were rolled in carbon black, prepared from burning xylene, before being transferred to fresh Durcupan at +16°C and embedded as usual.

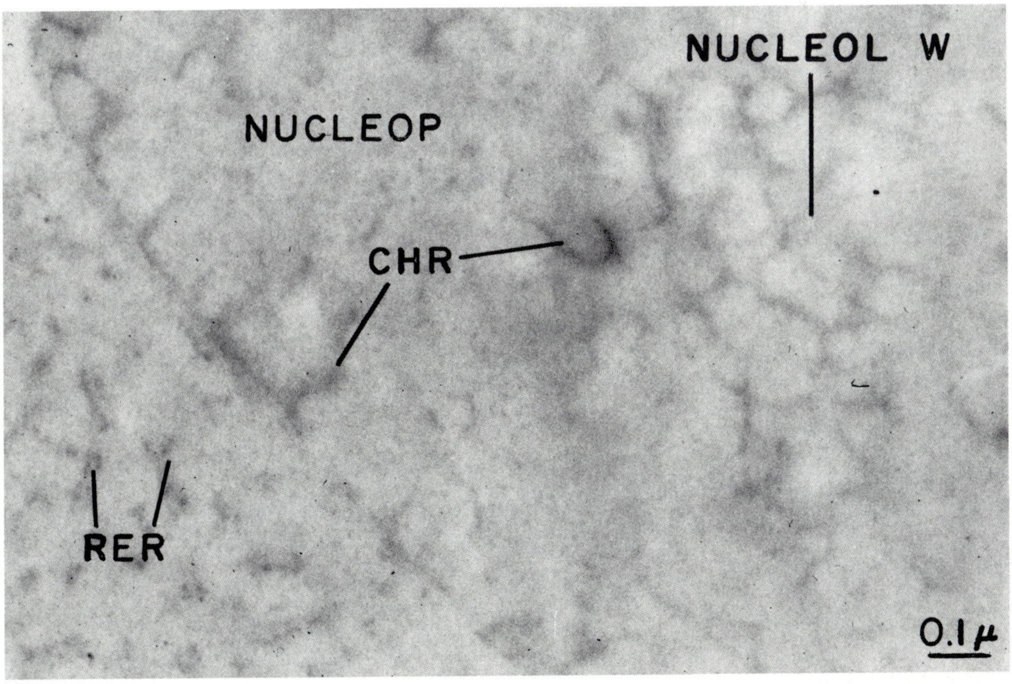

FIG. 2.5. Pancreas, postfixation with vapors of difluorodinitrobenzene (FFDNB). The walls of the chromatin (CHR) are dense, and the nucleolus (NUCLEOL W) are less dense. The walls of the nucleoplasm (NUCLEOP) are of low contrast. The rough endoplasmic reticulum (RER) is weakly stained also. × 81,000.

2. *Postfixation with FFDNB Vapor.* When specimens were to be treated with cross-linking vapors, they were placed in drying tube S (Fig. 2.1), somewhat modified from that described in an earlier paper *(10)*. The lower part of the tube was maintained at $-40°C$ or less, and the specimen vial was removed from liquid nitrogen, where it was stored, and dropped into the bottom of the tube. The plates were sealed with vacuum grease and placed as in the smaller drawing in such a way that the tube was separated from the superstructure, C and D being discontinuous. The tube was then evacuated through E. The next day, the superstructure F–G was removed at G, and the reagent was placed in the weighted reagent tube A. The superstructure was reassembled, and evacuated through F. At this point, the plates were slid so that C and D were superimposed, and F was

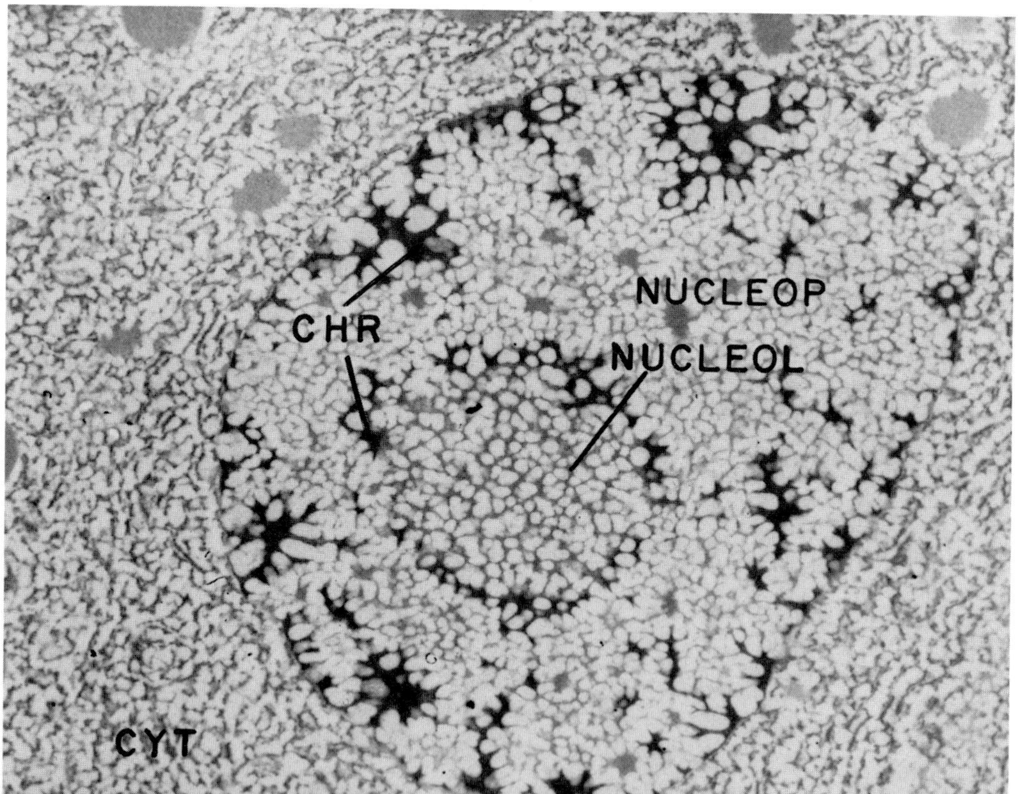

FIG. 2.6. Pancreas, postfixation in alcohol, extracted with chloroform: methanol (1:1) and stained *en bloc* with alcoholic platinum tetrabromide; sections not stained with uranyl acetate. The walls in the chromatin (CHR), nucleolus (NUCLEOL), and nucleoplasm (NUCLEOP) are continuous, and are separated by the less dense space. The cytoplasmic (CYT) two-phase system appears irregular in this plane of section. Secretory granules are homogeneous. × 14,000.

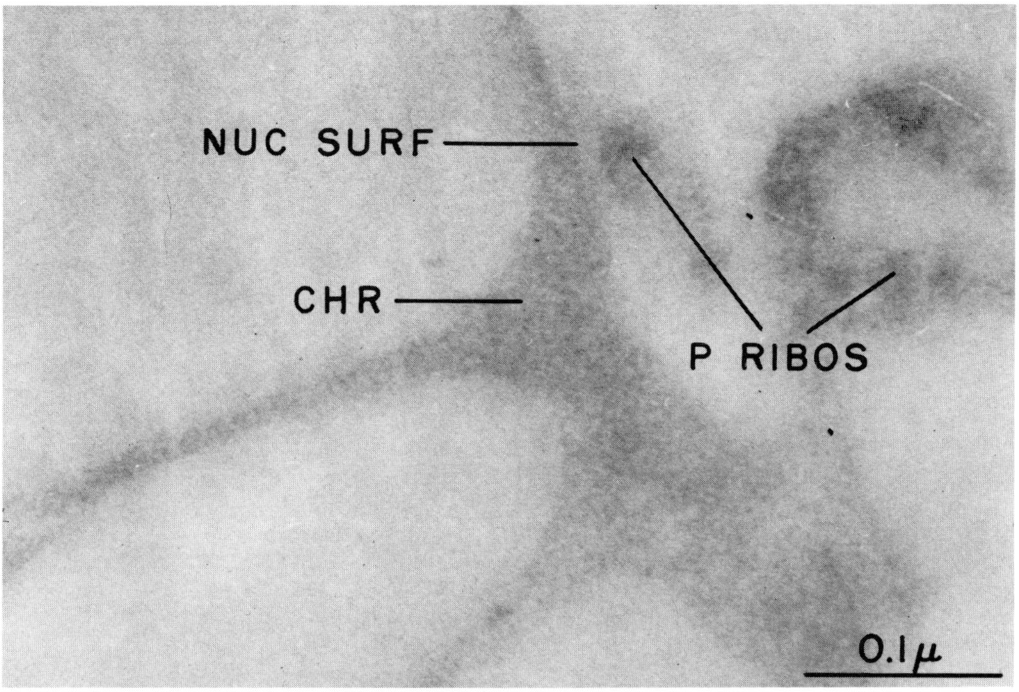

FIG. 2.7. Pancreas, postfixation in alcohol; section stained with uranyl acetate. Thin layer of homogeneous chromatin (CHR) at periphery of nucleus (NUC SURF) to show absence of negative staining of DNA coils. Ribosomes (P RIBOS) are present in cytoplasm. × 260,000.

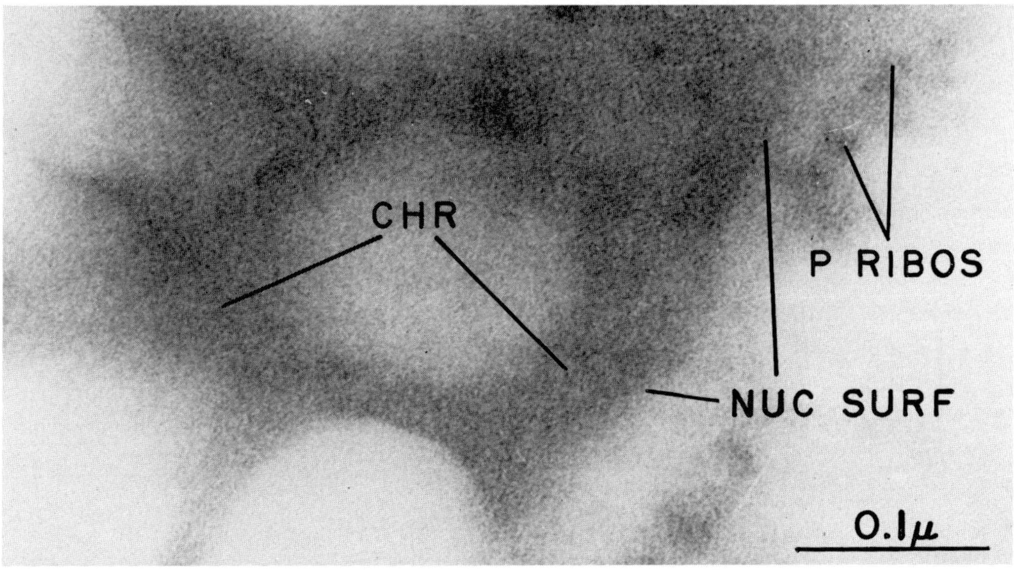

FIG. 2.8. Pancreas, postfixation with vapors of FFDNB; section stained with uranyl acetate. As in Fig. 2.7, the protein component of the chromatin is homogeneous, with no indication of negative staining of DNA coils. Abbreviations as in Fig. 2.7. × 260,000.

turned off; B was rotated so as to lower the reagent tube A to a level just above the specimen vials. Stopcock E was turned off. Water vapor was added to the column as a catalyst, in the following way. Tube H containing some water was attached with rubber tubing at F. The bottom of tube H was immersed in liquid nitrogen, and both the tube and the rubber tubing were evacuated up to F through a tee tube. When a good vacuum was established in tube H, as judged by the discharge of a Tessla coil, the rubber tube I was pinched, and tube H was warmed to room temperature. The discharge of the Tessla coil was then directed at the main tube, or the superstructure, and F was opened momentarily. When the color change of the discharge was pink, F was closed, and the rubber tube J was removed. This marked the beginning of the exposure time to the vapor. After some hours, the excess vapors were removed by raising the reagent tube A by turning stopcock B and sliding the plates so as to cut off the superstructure. The specimen tube was evacuated by connecting it to the main vacuum line at E. This was continued for some hours when the specimens were infiltrated *in vacuo* with 95% alcohol by removing the superstructure F–G, placing the alcohol in G, and slowly sliding plate D on C until a very slow stream of alcohol entered tube T. The vacu-

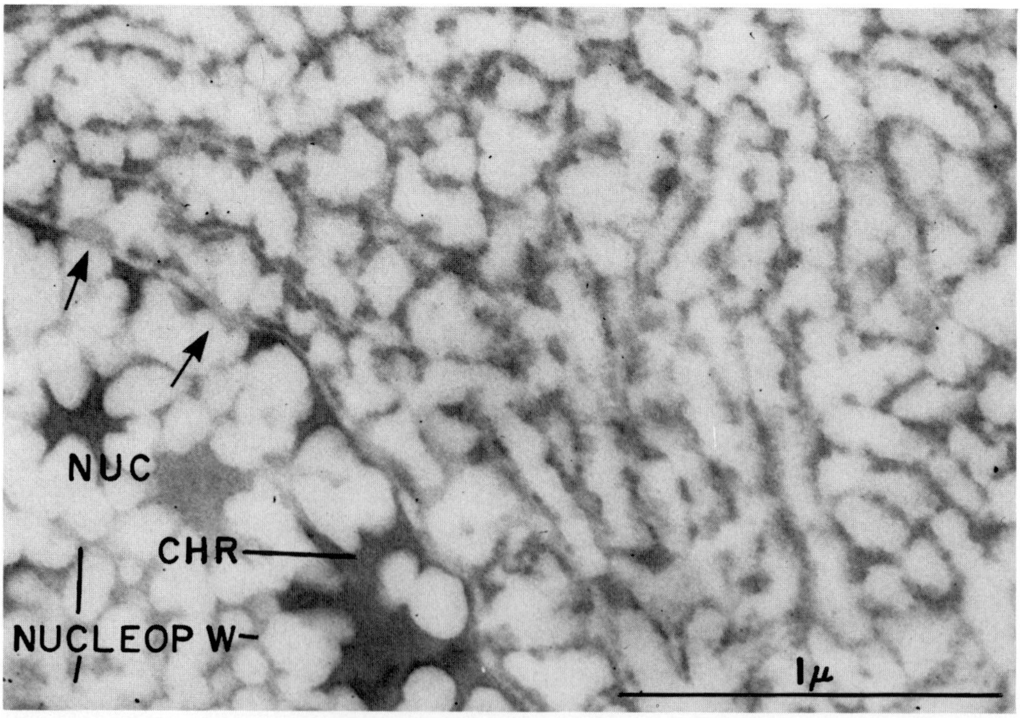

FIG. 2.9. Pancreas, prepared as in Fig. 2.6. Chromatin (CHR) at periphery of the nucleus (NUC) is dense. In minute gaps between protein (chromatin) masses lying along the nuclear surface (marked by arrows), the protein content of the nuclear wall cannot be distinguished from nucleoplasmic walls (NUCLEOP W). Ribosomes not clearly delineated. × 50,000.

um was maintained until alcohol covered the vials at the bottom of tube T. At this point, C and D were made fully continuous, breaking the vacuum in tube T. The vacuum in this tube was reinstituted and broken twice during about 5 minutes. The vials were transferred to a Petri dish, and the specimens in each vial were removed from the foil in alcohol where they remained for 1 day. Then they were cut into smaller pieces and embedded in w-s Durcupan as in procedure (1) above.

3. *Postfixation in Mercury Chelate Vapor.* Specimens were dried in tube S (Fig. 2.1) and treated with vapors of Hg hfac (with added water vapor) at the prescribed temperature and time, after which excess vapors were removed. Specimens were infiltrated with 95% alcohol at room temperature, kept in 95% alcohol for 1 day, and then cut into smaller pieces prior to infiltration with w-s Durcupan. The manipulations were the same as those described with postfixation of FFDNB vapor above.

Methods for postfixation and staining of the contents of submicroscopic vacuoles are given on pp. 36–42 and Table 2.2.

Polymerized blocks were trimmed and sections were cut with glass or diamond knives, floated on water, mounted on grids, and examined with a Hitachi HU-

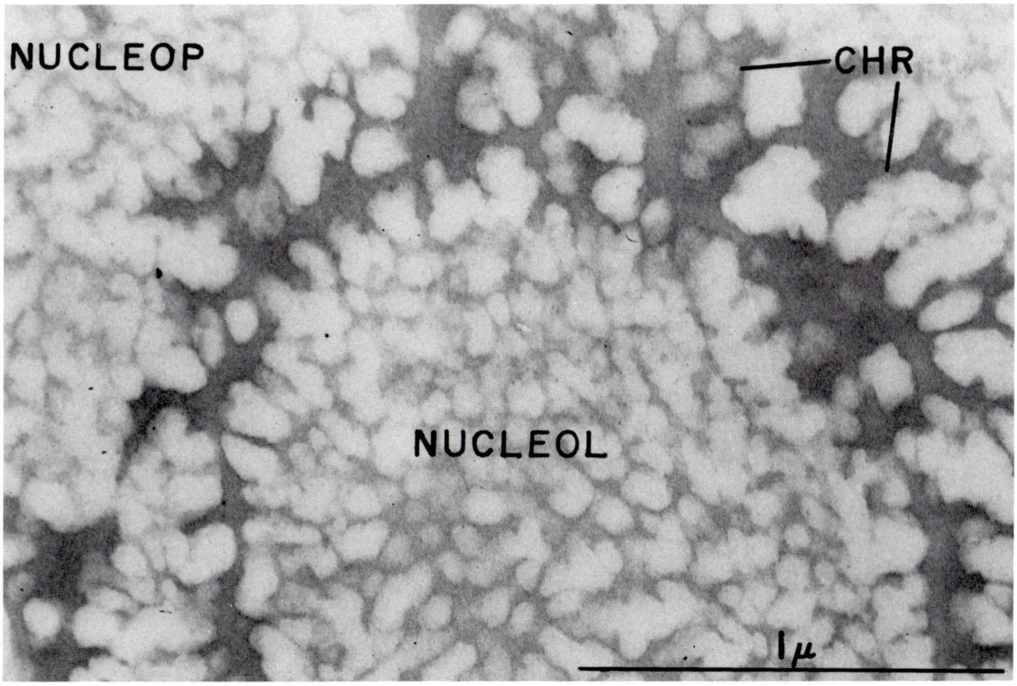

FIG. 2.10. Pancreas, prepared as in 2.6. The homogeneous protein of the peripheral and perinucleolar chromatin (CHR) is very dense. The walls of the nucleolus (NUCLEOL) are less dense, and those of the nucleoplasm (NUCLEOP) are palest. The dense phase of all three is continuous, as are the less dense spaces. × 56,000.

TABLE 2.2

DENSITOMETRIC DATA SHOWING THAT ALL REAGENTS RESULTED IN PRESERVATION AND STAINING OF PSEUDOVACUOLE CONTENTS

	Vapor reagents				
	Co tfac	Zr hfac	Hg hfac + formaldehyde	Ba acac	OsO_4
% T clear space[a]	65.2	80.6	79.9	63.8	45.2
% T pseudovacuoles[a]	51.0	68.6	70.9	49.1	27.3
n[b]	50	60	40	30	30

[a] Densitometer readings, in percentage transmission (% T), of pseudovacuolar contents (as compared with % T of clear space) of nuclei of liver cells which had been frozen-dried, treated with one or two vapor reagents and embedded in Durcupan. Ultrathin sections were stained with uranyl acetate. After readings were made of regions of electron micrographs with 100% T and 0% T, transmission of clear spaces adjacent to the nucleus (% T clear space) and of psuedo-vacuolar contents of random pseudovacuoles (% T pseudovacuoles) were made.

[b] n, number of pseudovacuoles measured.

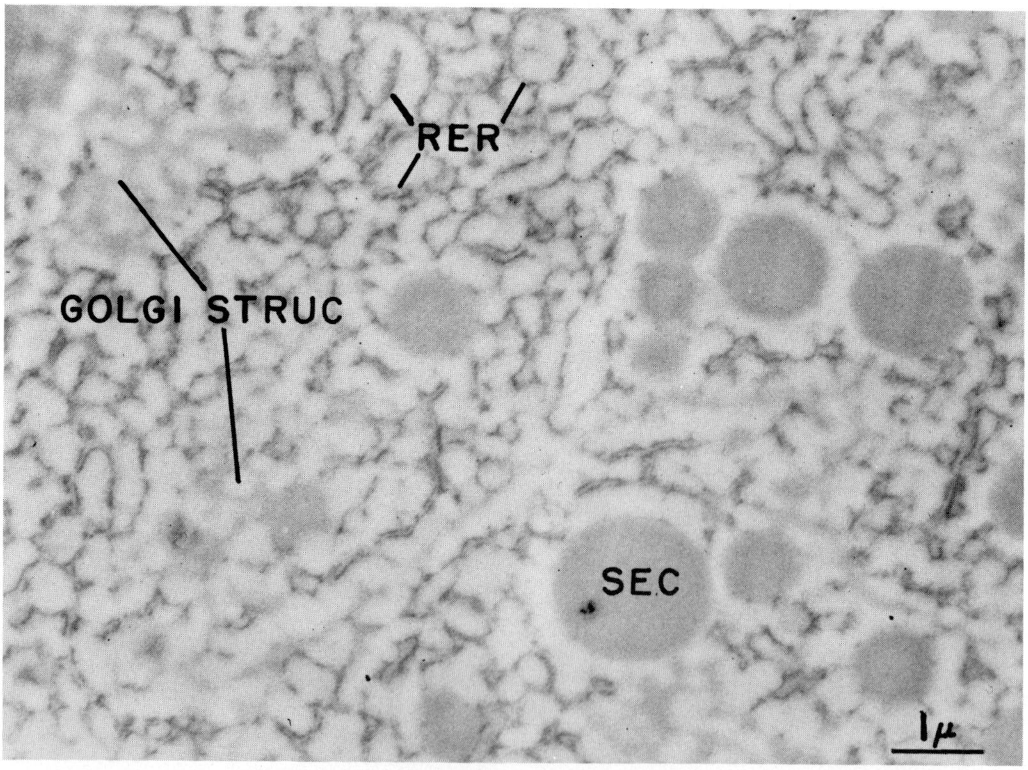

FIG. 2.11. Pancreas, postfixed with vapors of Hg hfac; not stained with uranyl acetate. In the region of secretory granules (SEC), the rough endoplasmic reticulum (RER) interdigitates with the paler, solid, irregularly branching Golgi meshwork (GOLGI STRUC). Ribosomes not clearly delineated. × 12,000.

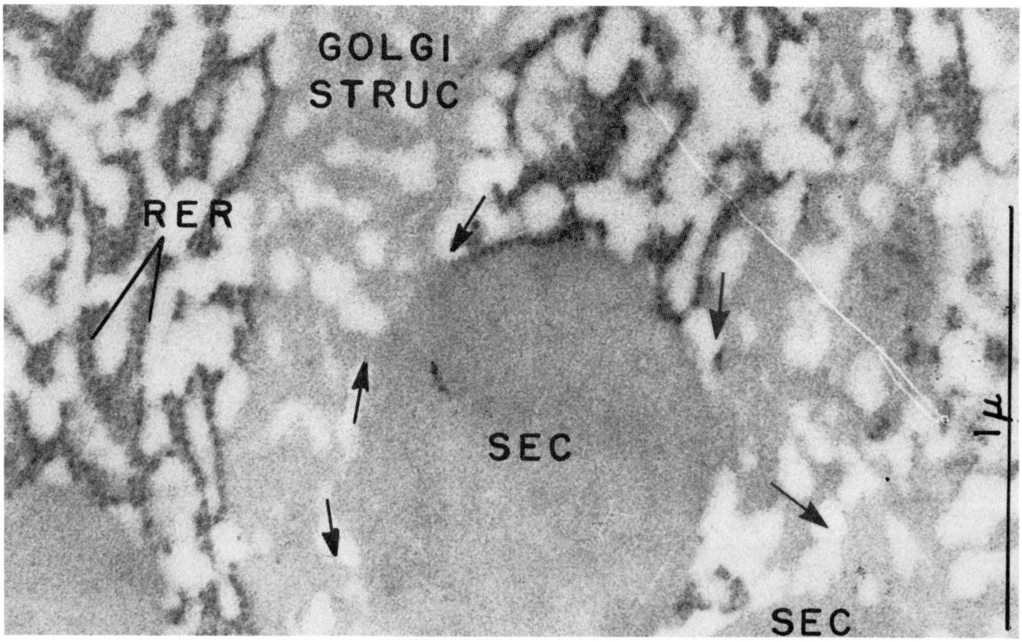

FIG. 2.12. Pancreas, prepared as in 2.6. The denser walls do not contain clearly demarcated ribosomes. Part of the paler Golgi structure (GOLGI STRUC) impinges on and seems to become continuous with the secretory granules at points marked by arrows. Abbreviations as in Fig. 2.11. × 56,000.

11A electron microscope operating at an accelerating voltage of 50 kv. Photographs were made at various magnifications and, in most cases, were enlarged two times in printing.

Observations

In order to shorten this section, some detailed descriptions are given in the figure legends only. The term "microscopic" is used in describing structures which can be seen with the light microscope; when structures cannot be resolved with the light microscope because they are too small or the spaces between them are too small, they are referred to as submicroscopic.

Most stained preparations have a consistent pattern characterized by the presence of submicroscopic or pseudovacuoles. In a small proportion of specimens, these structures can be recognized only with difficulty because their contents are stained like the walls.

The majority of the stained preparations, regardless of which stain was used, are similar in that the same structures are stained, though with differing degrees of contrast. Moreover, the structures visible in all stained sections can be recognized in unstained specimens, where the contrast may be drastically lower.

FFDNB (see Fig. 2.5 and Figs. 3.12, 3.13, and 3.34, this volume, Chapter 3) and Hg hfac (Fig. 2.11 and Figs. 3.18, 3.36, and 3.37, this volume, Chapter 3) have more contrast than unstained specimens (see Fig. 2.2 and Figs. 3.5, and 3.28, this volume, Chapter 3). Platinum staining (Figs. 2.6, 2.9, 2.10, 2.12) is similar to vapor staining but shows more contrast. Additional staining of sections with uranyl acetate enhances contrast in all specimens, regardless of which stain was used before section staining, without delineating any new features (with very few exceptions) (Figs. 2.3, 2.4, and 2.7, alcohol; 2.8 and 2.17, FFDNB; 2.13–2.16, 2.18, and 2.19 $PtBr_4$). Such specimens resemble those stained with alcoholic platinum tetrabromide. The outstanding exceptions are ribosomes, which seem to be delineated more clearly after uranyl acetate staining (Fig. 2.17).

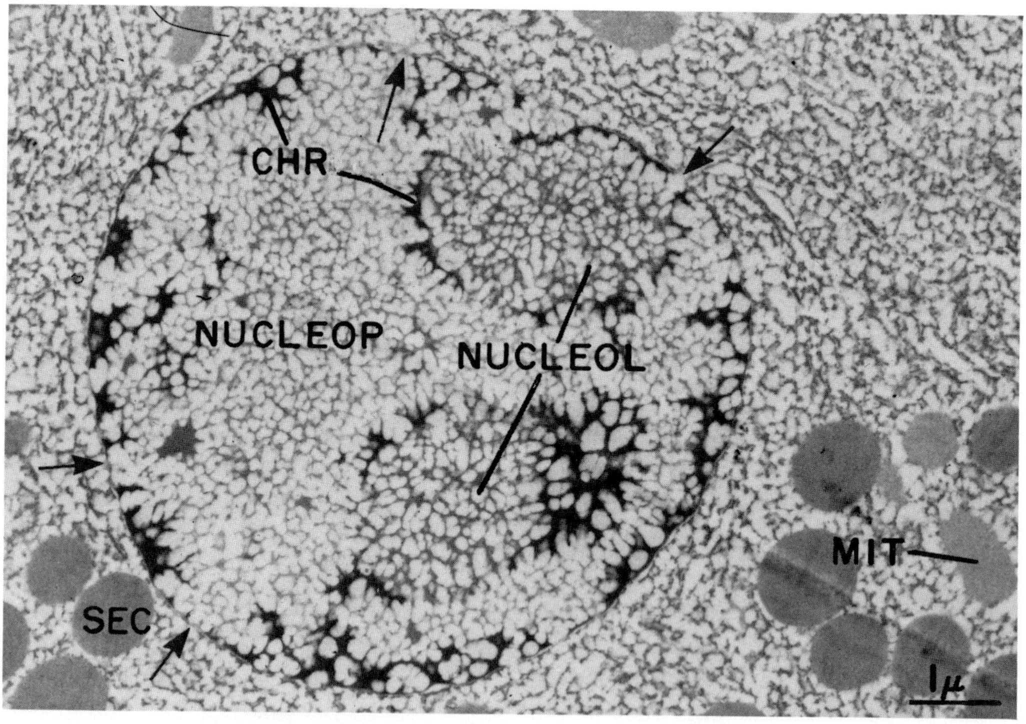

FIGS. 2.13–2.16. Pancreas, same preparation as in Fig. 2.6, sections stained with uranyl acetate. The two-phase system of protoplasm is essentially the same as in sections not stained with uranyl acetate (Figs. 2.6, 2.9, 2.10, and 2.12). The peripheral walls of the nucleus between denser protein masses of protein are not distinguishable from nearby walls in the nucleoplasm and cytoplasm. These regions are marked by arrows. Secretory granules (SEC) and mitochondria (MIT) are more or less homogeneous (Figs. 2.13 and 2.16). In Figs. 2.13 and 2.16, the walls of peripheral and perinucleolar chromatin (CHR) are very dense and homogeneous. The walls of the nucleolus (NUCLEOL) are less dense and homogeneous, and the walls of the nucleoplasm (NUCLEOP) are palest. All dense walls are continuous, as are all less dense spaces. Ribosomes are delineated somewhat in Figs. 2.9 and 2.12; they are clearly discrete in Fig. 2.17. For other abbreviations, see Fig. 2.17. Fig. 2.13, × 12,000; Figs. 2.14–2.16, × 50,000.

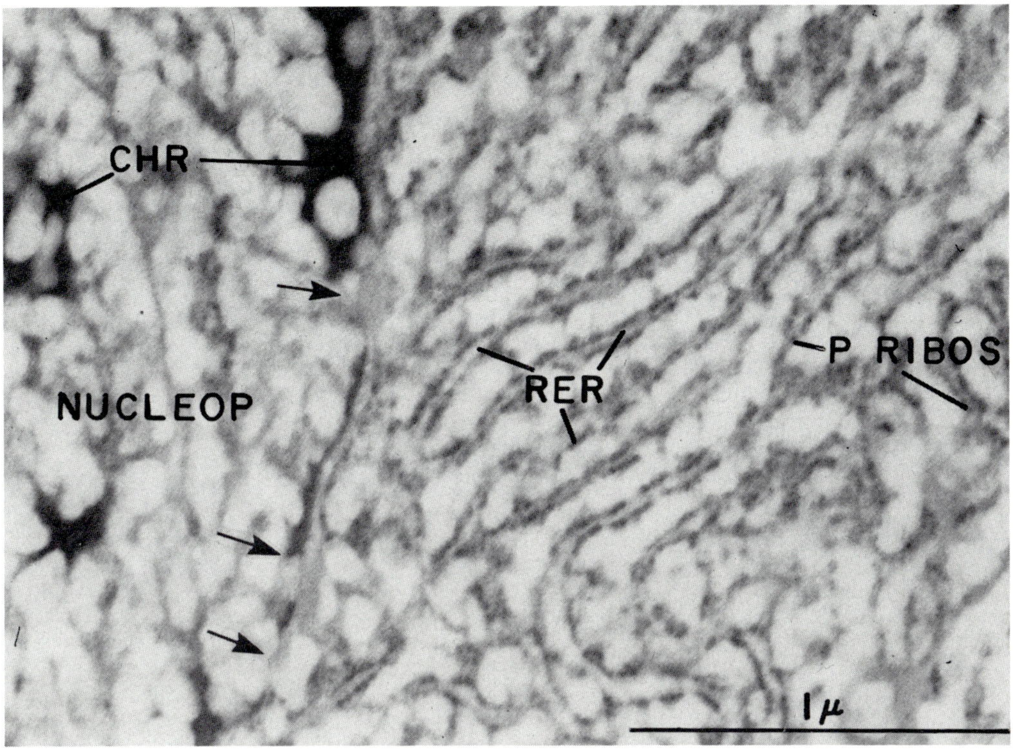

FIG. 2.14. ▲

▼ FIG. 2.15.

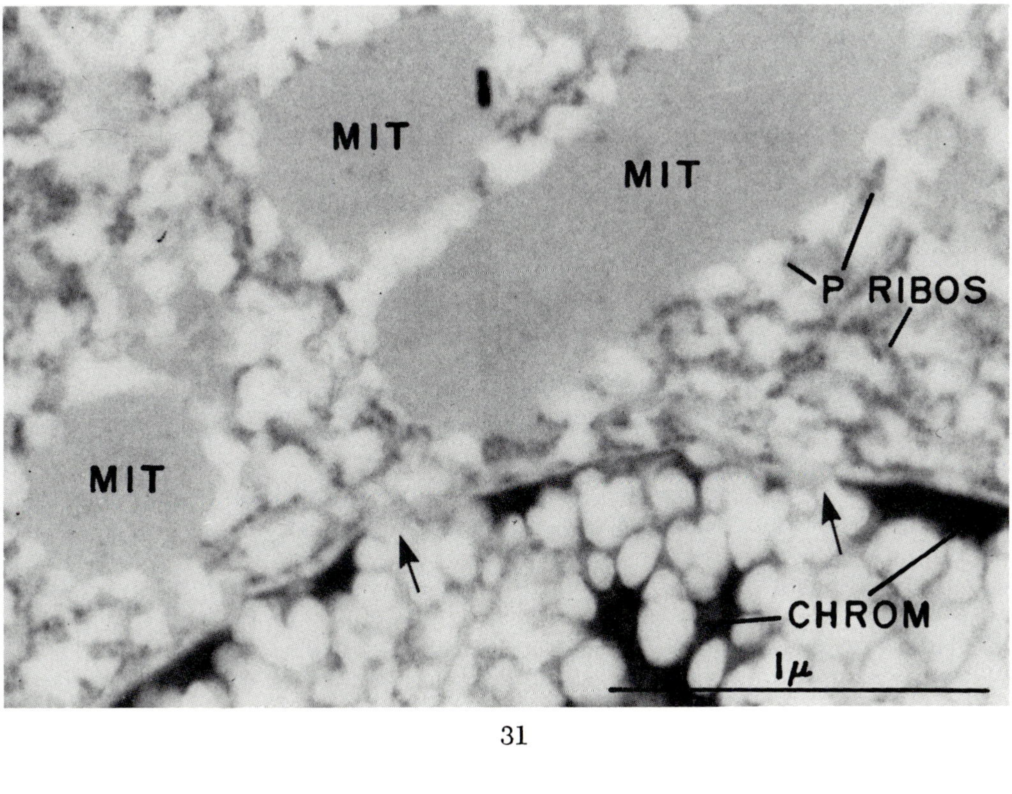

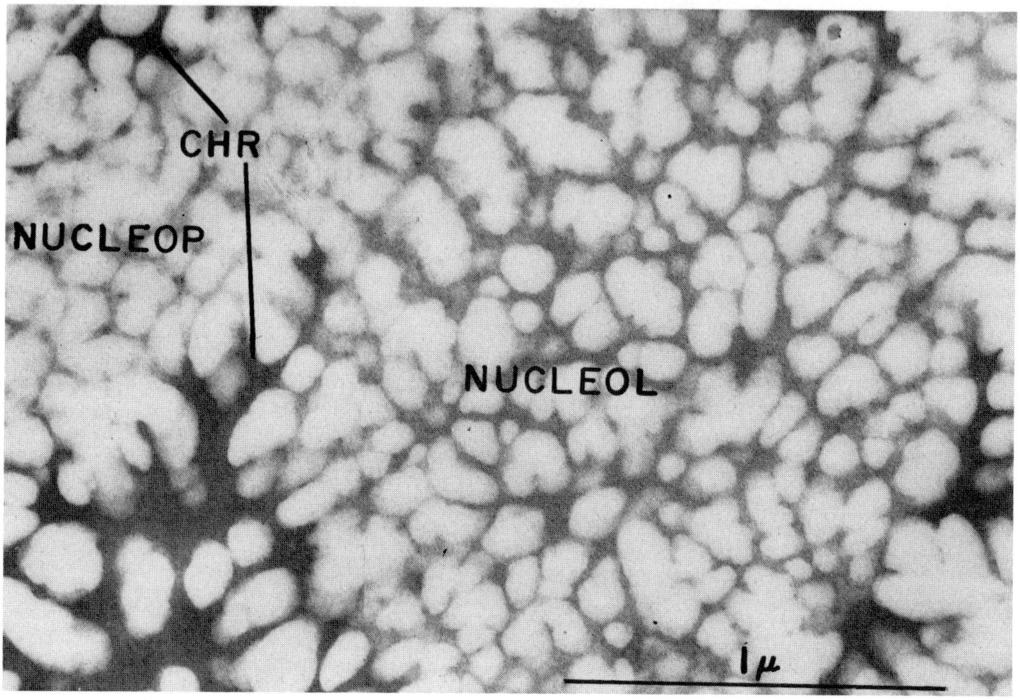

FIG. 2.16. See p. 30 for complete legend.

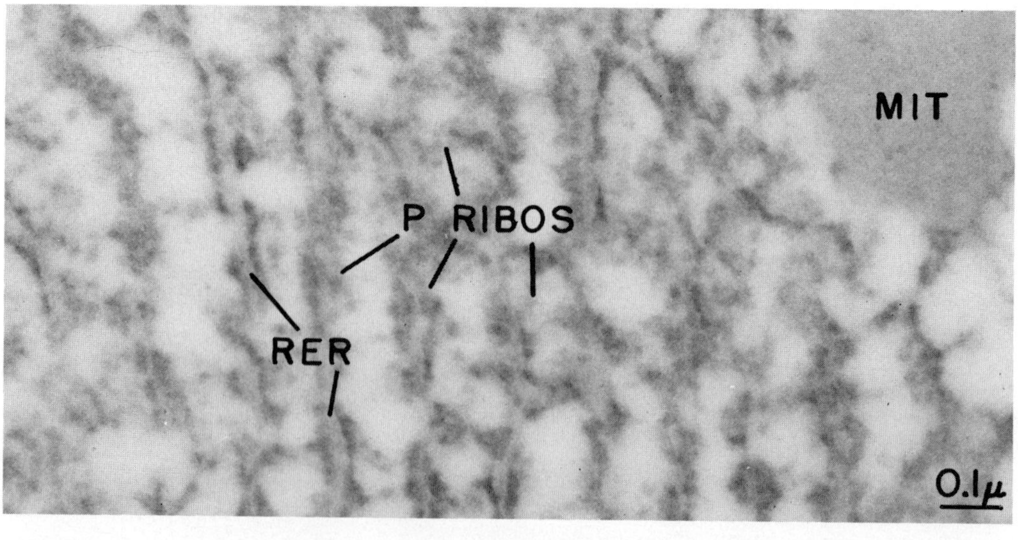

FIG. 2.17. Pancreas, postfixed in vapors of FFDNB; sections stained with uranyl acetate. Ribosomes (P RIBOS) of the rough endoplasmic reticulum (RER) are clearly delineated. MIT, mitochondria. × 76,000.

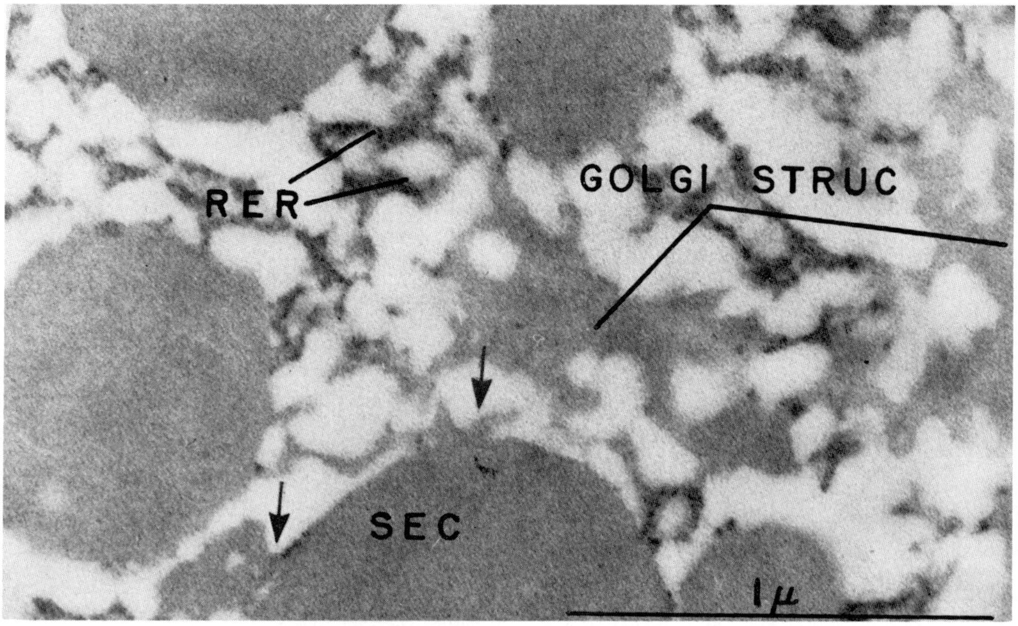

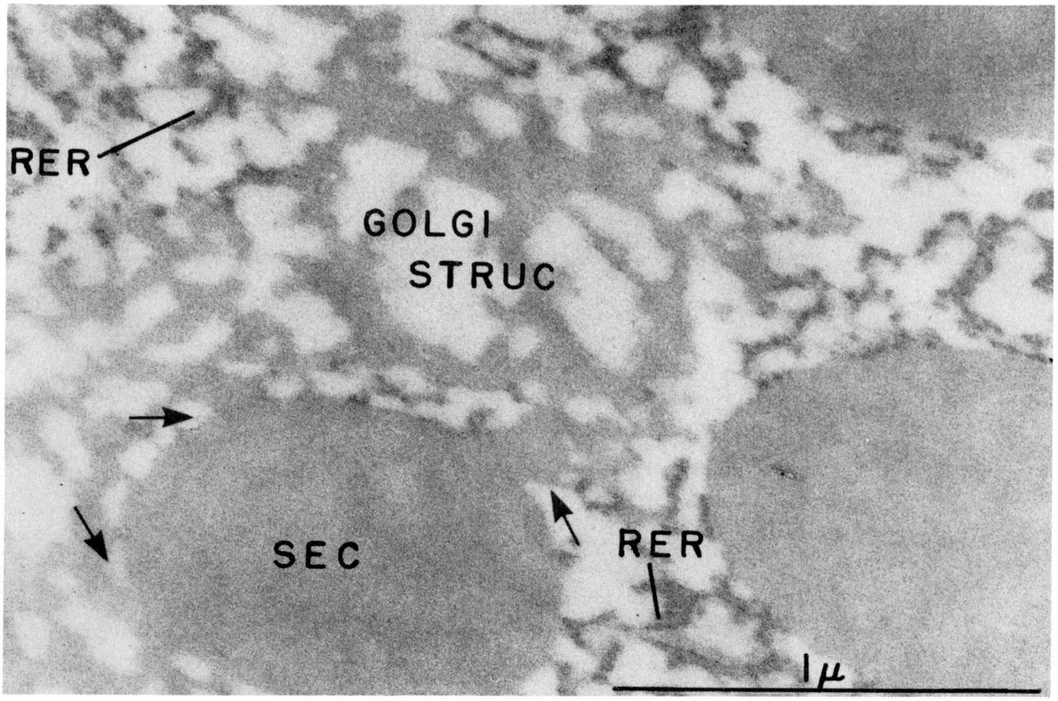

FIGS. 2.18 (TOP) AND 2.19 (BOTTOM). Pancreas, prepared as in Fig. 2.13. The pale, irregularly branching meshwork, or Golgi structure (GOLGI STRUC) interdigitates with the rough endoplasmic reticulum (RER). Part of the mesh is continuous with secretory granules (SEC) at points marked by arrows. × 56,000.

In the nucleus, in general, the chromatin is most dense and appears relatively uniform. The nucleolus is less dense, while the nucleoplasm is the least dense. In no instance have first- or second-order DNA coils [which are stained with gallocyanin-chromalum (see Chapter 3, this volume)] been detected, or have RNA granules been observed in the nucleolus or in the nucleoplasm. Even at the highest magnifications (Figs. 2.7 and 2.8) and with good resolution (about 10 A) no spaces are detectable whose measurements or morphology might signify the presence of unstained DNA or RNA (see this volume, Chapter 3). Small spaces, which appear in these high resolution electron micrographs, may be real, in part, or may represent molecular distortion and shrinkage, in part, but clearly do not correspond with unstained areas which could be occupied by nucleic acids; nor is

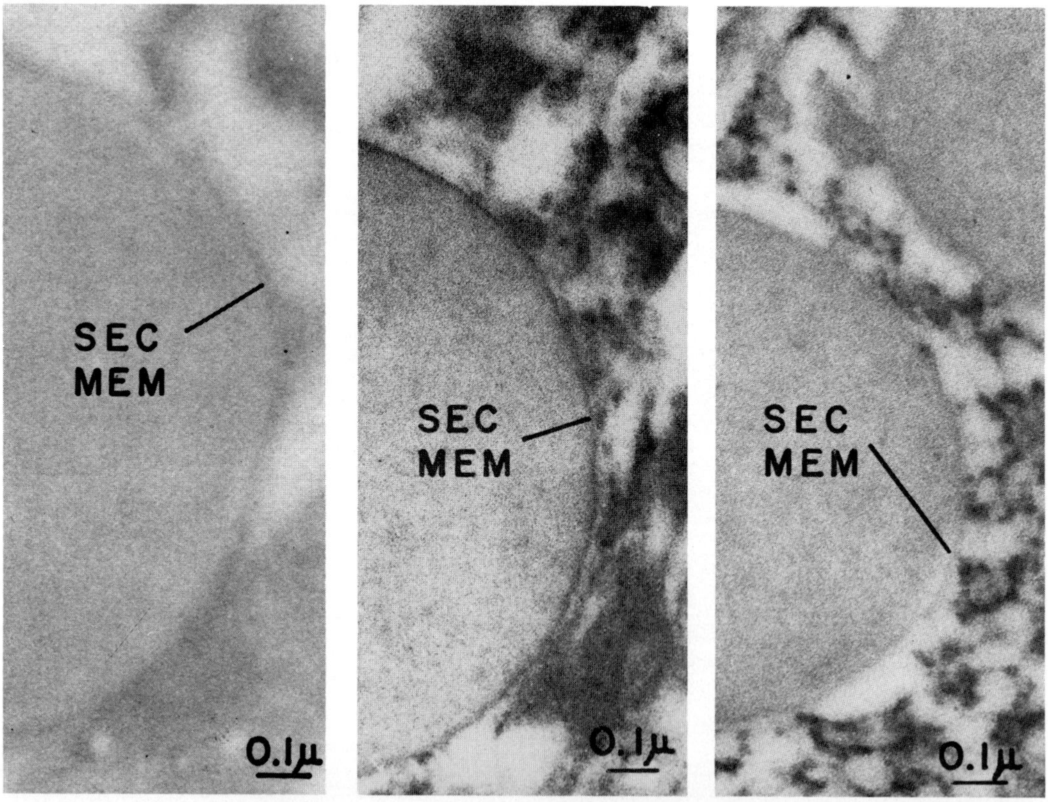

FIGS. 2.20–2.22. Portions of secretory granule of mouse pancreas. Most vapor postfixatives do not differentially stain membranes of zymogen granules, even when the ultrathin sections are stained with uranyl acetate. Membranes of secretory granules can, however, be stained differentially by a few vapor reagents, examples of which are given.

FIG. 2.20 (LEFT). A specimen postfixed with FFDNB; section stained with uranyl acetate.

FIG. 2.21 (MIDDLE). A specimen postfixed with Hg hfac; section stained with uranyl acetate.

FIG. 2.22 (RIGHT). A specimen postfixed with vapors of diiodophenylhydrazine (Vol. II, Chapter 1); section stained with uranyl acetate. SEC MEM, membrane of secretory granule. × 70,000.

there any staining which corresponds with the distribution of lipids in the nuclear membrane as illustrated in Chapter 1 (Vol. II). In addition, the structure of the nuclear surface is the same whether the specimens are stained and then infiltrated directly with the Durcupan components, whether they are stained and then treated with alcohol before infiltration, or whether they are extracted with alcohol, or with chloroform: methanol (1:1) prior to staining and subsequent infiltration with Durcupan.

In the cytoplasm, as a rule, the stains do not delineate ribosomes in the rough endoplasmic reticulum (RER). Instead, ribosomes seem to be continuous with the material in which they lie. This conforms with the earliest description, based on specimens postfixed and stained with alcoholic solutions of metal salts, in which ribosomes were described as concentrates rather then as discrete particulates *(8, 9)*. After staining for RNA with gallocyanin–chromalum, ribosomes appeared as separate granules *(7)*. They may also appear discrete after section staining with uranyl acetate (Fig. 2.17). The protein-stained wall of the RER is con-

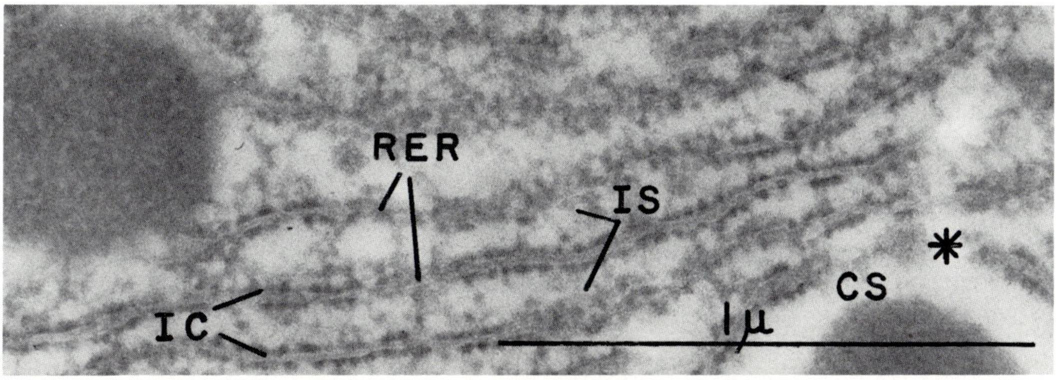

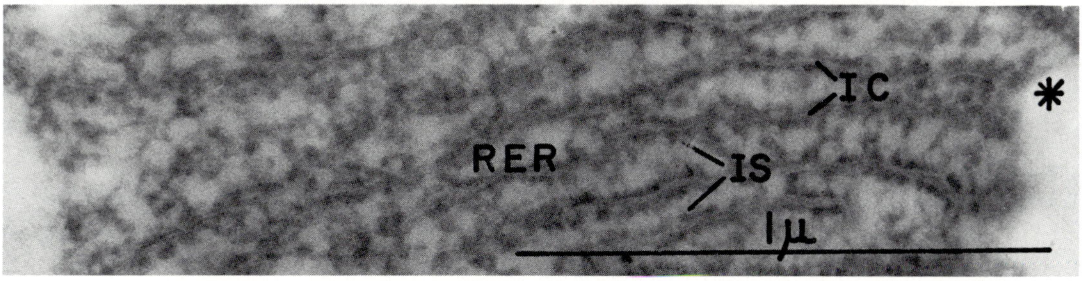

FIGS. 2.23 (TOP) AND 2.24 (BOTTOM). Liver, postfixation and staining with vapors of HEFDOD (see Vol. II, Chapter 9); sections stained with uranyl acetate. Regions of cytoplasm containing rough endoplasmic reticulum (RER). Contents of the thin intracisternal space (IC), as of the intercisternal "space" (IS) are stained, when compared with a clear space (CS) caused by mitochondrial shrinkage. In the clear space, the Formvar and embedding plastic are about as thick as in the rest of the section, except in the region marked by an asterisk, where the embedding material was lacking because of a gas bubble which developed during the infiltration with the Durcupan components. × 70,000.

tinuous and thus corresponds only in part with the discontinuous, thinner lipid-stained wall of the endoplasmic reticulum as described in Vol. II, Chapter 1.

The protein component of the intracisternal part of the RER has been postfixed and stained with a variety of vapors, with subsequent staining of ultrathin sections with uranyl acetate. These are: diiodophenylhydrazine (Vol. II, Chapter 1), Zn hfac, Ba acac, Ca acac, and Th acac (see Table 2.1). This protein component will be described in Chapter 1, Vol. II.

Mitochondria appear more or less uniformly dense after postfixation in alcohol alone, after the use of solutions of metal salts in 95% alcohol, and after postfixation with most vapor reagents, with no trace of lipid membranes or cristae. However, after postfixation with certain other (vapor) reagents followed by section staining with uranyl acetate, the protein components of mitochondrial membranes and compartments are clearly delineated. The effective vapors are: (1) diiodophenylhydrazine (Vol. II, Chapter 1), (2) Ca acac, (3) Co acac, (4) Zr hfac, (5) two-vapor reagents (Hg hfac and FFDNB, described on p. 37), and *(6)* heptafluorodimethyloctanedione (HEFDOD), a reagent which combines with bi- and polyvalent cations and protein (Vol. II, Chapter 9). The findings will be described in Chapter 3, Vol. II.

The Golgi apparatus is represented in sections mainly by pale-staining, solid cords which are irregularly branched (see Figs. 2.12, 2.18, and 2.19). Some of them bulge as if they enclose secretory granules. In the region of zymogen granules, branches of the Golgi meshwork appear to become continuous with the secretory granule. After certain postfixations, the membrane of the secretory granule is somewhat differentially stained (Figs. 2.20–2.22).

Finally, in general, the structure of both nucleus and cytoplasm appears to be spongelike, in that the protoplasm is constituted of dense, communicating walls, which enclose less dense, continuous "spaces," the submicroscopic pseudovacuoles. This underlying aspect of protoplasmic structure was described originally as a two- (or multi-) phase system *(9)*.

The contents of the submicroscopic pseudovacuoles in both nucleus and cytoplasm of freeze-dried cells remain unstained by most methods of postfixation, and accordingly these appear empty or pale in electron micrographs. That is, after most methods of postfixation, the contents of the pseudovacuoles have about the same electron-scattering capacity as the embedding material, even after the ultrathin sections are stained with uranyl acetate prior to viewing. The contents of the submicroscopic vacuoles can, however, be stained by several methods of postfixation followed by section staining with uranyl acetate. When stained in this manner, they approach the walls in density, and the contrast between both structures is consequently reduced. Because the "spaces" were filled with stainable material, the term "submicroscopic vacuole" is no longer appropriate, and has been replaced with the term "pseudovacuole," to indicate that the contents appear frequently as vacuolar, but that this condition is caused by a failure to stain the protein of the contents rather than by their absence.

The following are the conditions under which the contents of the cytoplasmic pseudovacuoles were observed to be stained. Freeze-dried liver was infiltrated *in vacuo* with 95% alcohol, or with the Durcupan components, or with both. Ultrathin sections of polymerized blocks were cut. The sections were transparent when viewed on the surface of the sectioning boat, which means they were 300–400 A thick or less. They were mounted on grids and stained with uranyl acetate.

1. Specimens were postfixed with vapors of Hg hfac for 3 hours, followed by evacuation of excess vapors at 0°C for 2 hours, and then at room temperature for 16 hours. This was followed by further postfixation with vapors of FFDNB for 5 hours at room temperature, followed by evacuation of excess vapors at room temperature for 18 hours.

2. Postfixation with vapors of HEFDOD (Figs. 2.23 and 2.24).

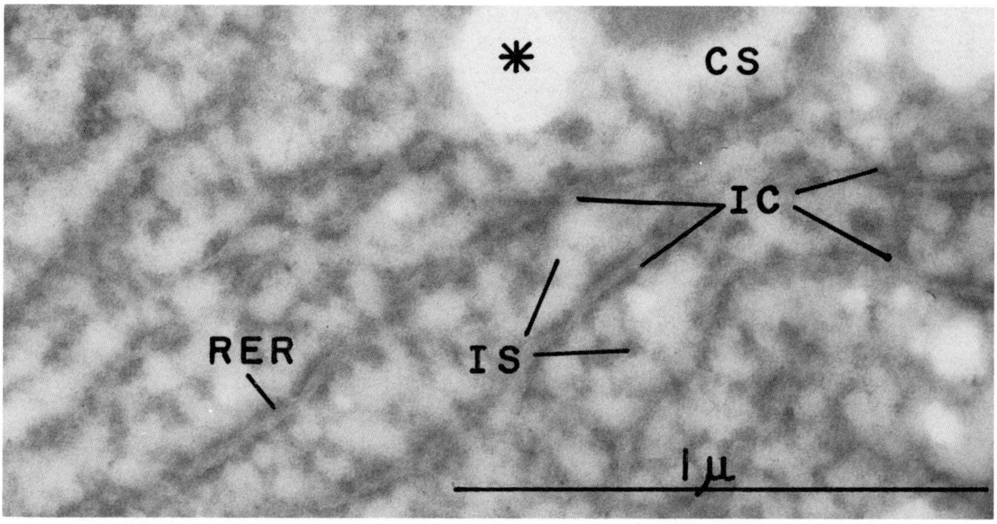

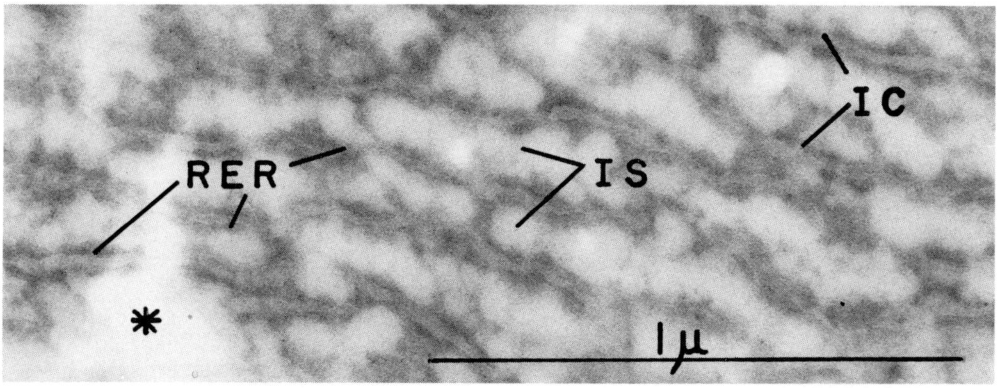

Figs. 2.25 (top) and 2.26 (bottom). Liver, postfixed and stained with vapors of FFDNB; sections stained with uranyl acetate. Asterisks and abbreviations as in Figs. 2.23 and 2.24. × 70,000.

3. Postfixation with vapors of FFDNB (Figs. 2.25 and 2.26).

4. Postfixation with vapors of Ba acac and Zr hfac under conditions described in Table 2.1 (Fig. 2.27).

5. Postfixation with vapors of osmium tetroxide at 0°C for 10 minutes at atmospheric pressure, followed by evacuation of excess vapors at 0°C for 2 hours, and then at room temperature for 16 hours.

Other vapor reagents, as noted above, did not result in consistent or complete staining of the contents of the pseudovacuoles.

No spaces in electron micrographs are present which could have represented the presence of ice crystals at some earlier stage of preparation, even at magnifications of 300,000.

Proof that the pseudovacuoles of the nucleus contain protein was obtained from freeze-dried specimens treated with a variety of vapor reagents and then embedded in Durcupan. Ultrathin sections were stained on the grid with uranyl

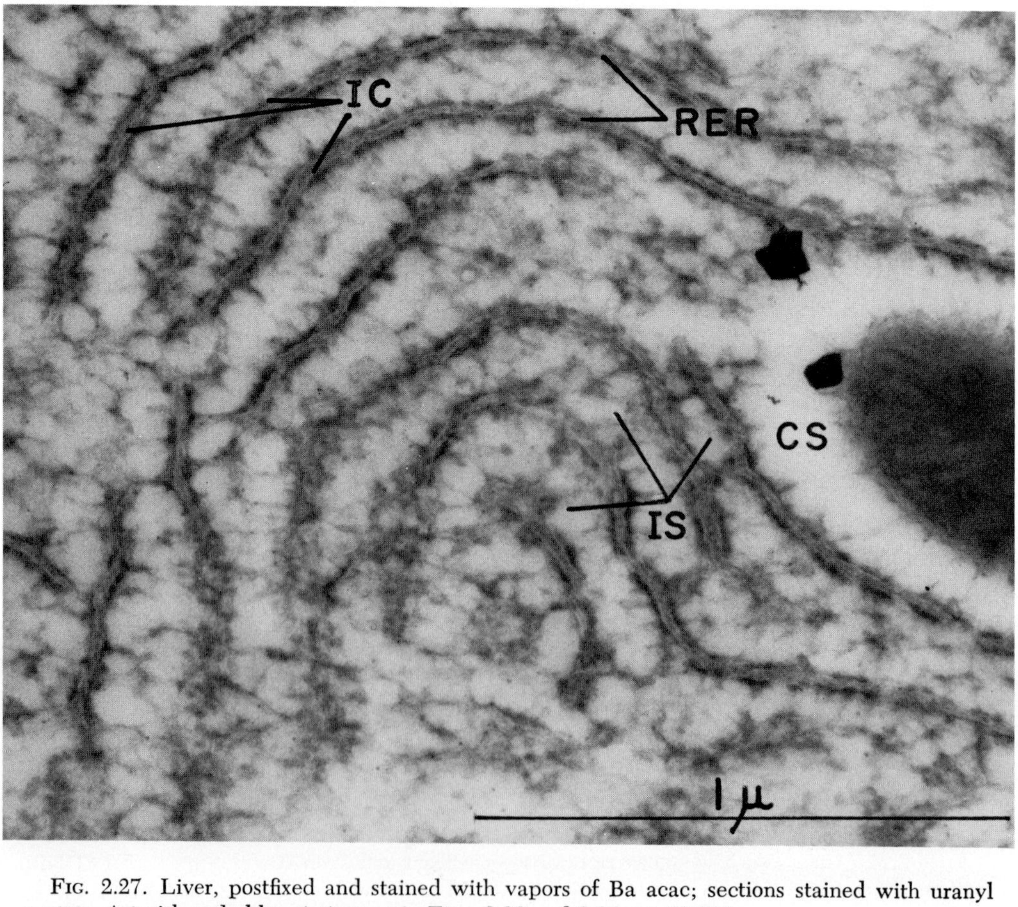

FIG. 2.27. Liver, postfixed and stained with vapors of Ba acac; sections stained with uranyl acetate. Asterisk and abbreviations as in Figs. 2.23 and 2.24. × 70,000.

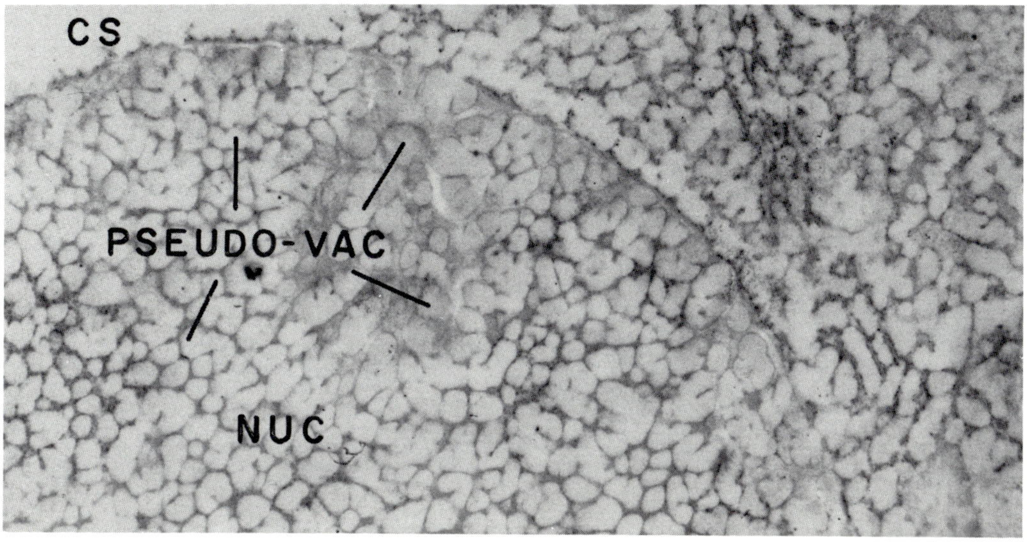

FIG. 2.28 Liver, postfixed and stained with vapors of Hg hfac and FFDNB; sections stained with uranyl acetate. The contents of the nuclear submicroscopic pseudovacuoles (PSEUDOVAC) are stained, as compared with the clear space (CS). NUC, nucleus. × 20,000.

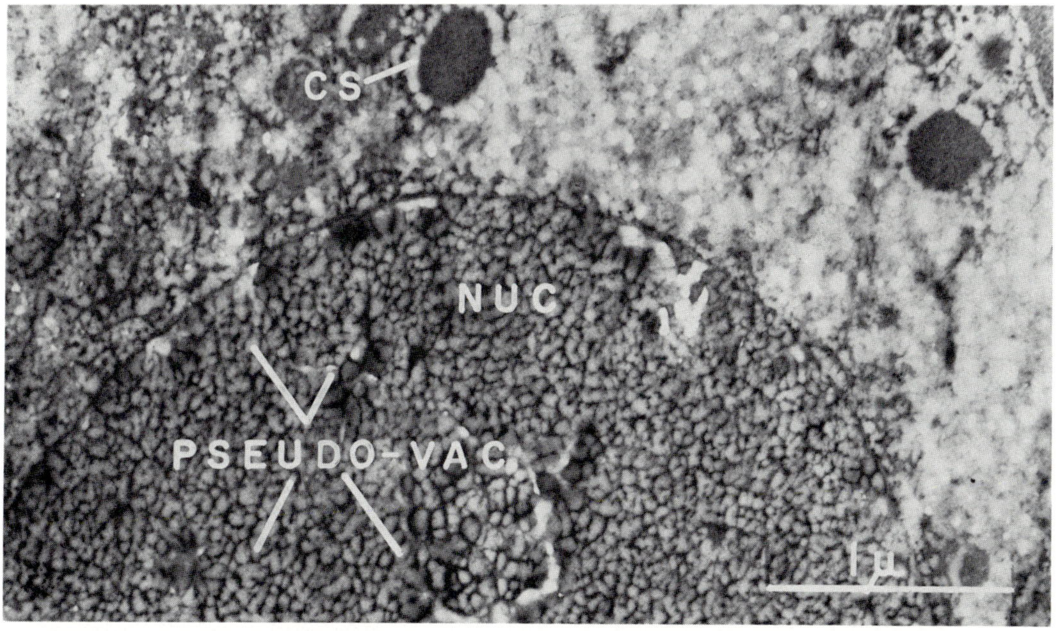

FIG. 2.29. Liver, postfixed and stained with vapors of Zr hfac; sections stained with uranyl acetate. Contents of nuclear pseudovacuoles (PSEUDOVAC) are stained as compared with clear space (CS). × 20,000.

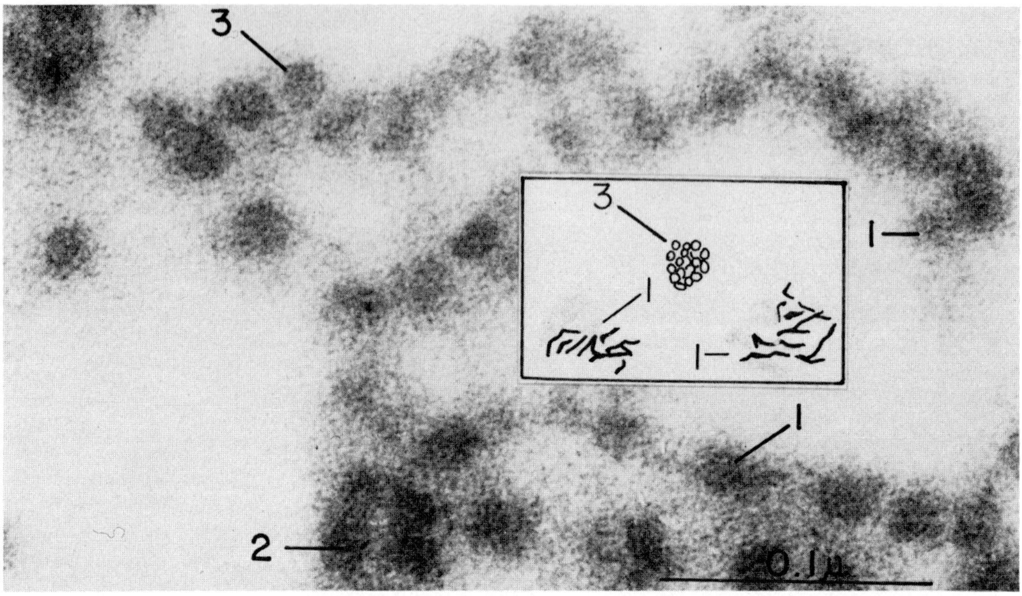

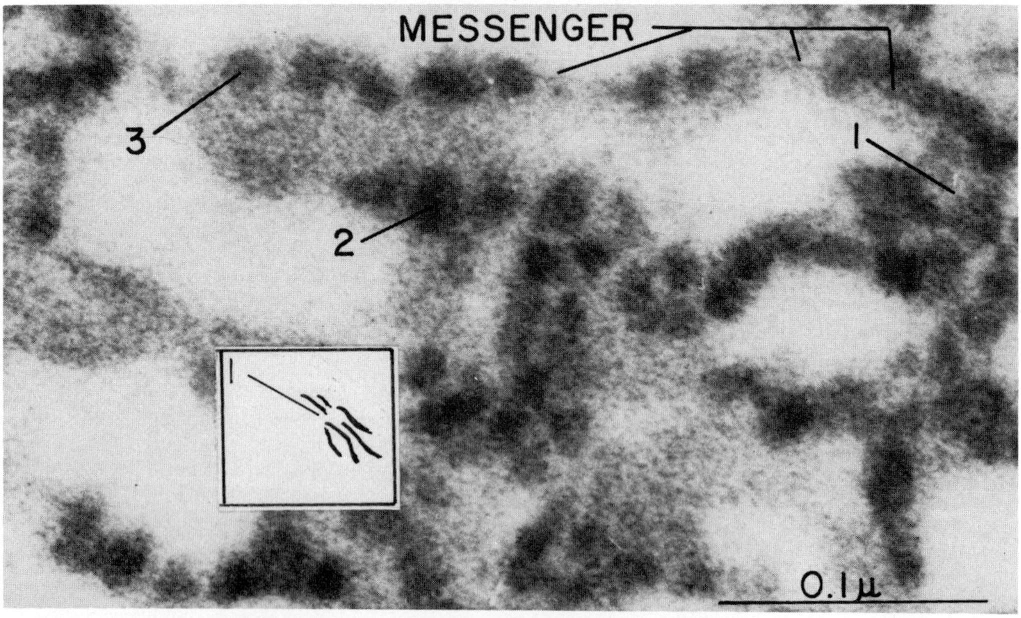

FIGS. 2.30 (TOP) AND 2.31 (BOTTOM). Pancreas, freeze-dried, postfixed, and stained with vapors of Hg hfac, and subsequently stained *en bloc* with gallocyanin–chromalum, as described in Chapter 3 (this volume). Ultrathin sections were stained on the grid with uranyl acetate. The protein stain supersedes that for nucleic acids. Some ribosomes are enclosed in protein filaments which are readily resolved (1). Around other ribosomes, the protein filaments are so dense as to be separately unresolvable (2). This could be caused also, in part, by overlay. Some ribosomes are free or nearly free of nascent protein filaments (3). These are indicated in the small inserts. Ribosomes of the third category appear as if they were aggregates of small spheroidal units, each about 30 A in diameter, clearly outlined by a dense surface enclosing a less dense core. In Fig. 2.31 there are some long unbranched dense filaments which join adjacent ribosomes. These are probably the protein components of messengers. × 350,000. G2 print. (See p. xiv for explanation of "G2 print.")

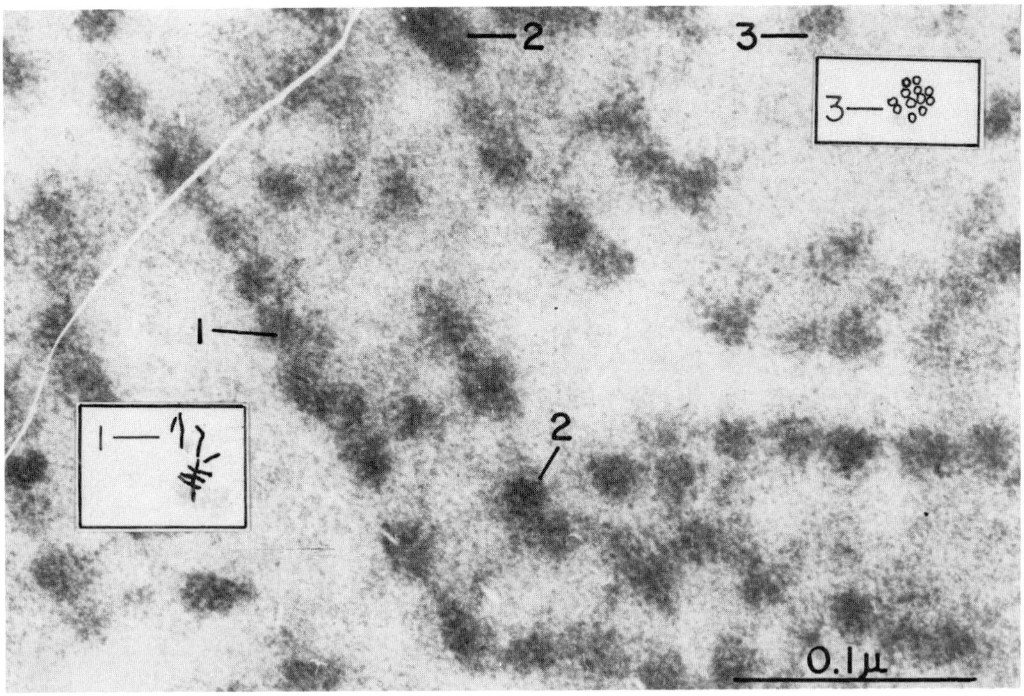

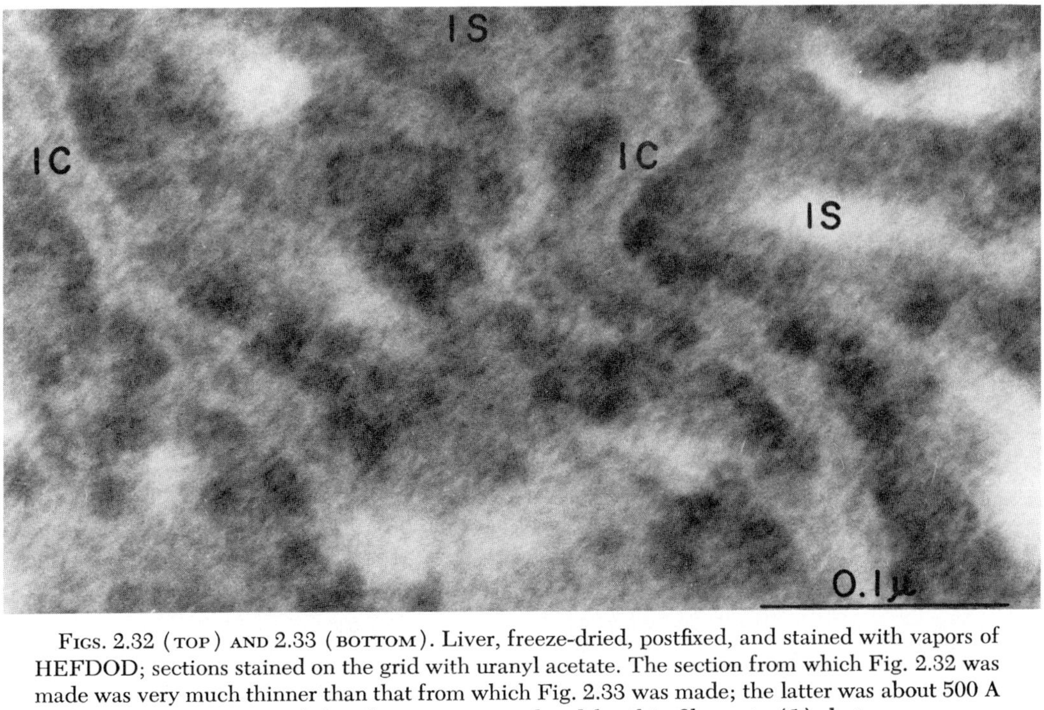

FIGS. 2.32 (TOP) AND 2.33 (BOTTOM). Liver, freeze-dried, postfixed, and stained with vapors of HEFDOD; sections stained on the grid with uranyl acetate. The section from which Fig. 2.32 was made was very much thinner than that from which Fig. 2.33 was made; the latter was about 500 A thick. In Fig. 2.32, most of the ribosomes are enclosed by thin filaments (1), but some are entirely obscured by them because they are so numerous (2). A few are clearly visible, and have the same "vacuolated" appearance as the ribosomes described in Figs. 2.30 and 2.31. These are indicated in the inserts. In Fig. 2.33, the intracisternal spaces (IC) are clearly identifiable. Between them, in the intercisternal spaces (IS) are numerous dense bodies. These are probably protein molecules in various states of structural organization after being released from ribosomes. × 321,000. G2 print.

acetate and electron micrographs were made of both nucleus and adjacent cytoplasm at a magnification of 9000. The contents of the pseudovacuoles all contained stainable material. This is illustrated in Figs. 2.28 and 2.29.

The following were the conditions for postfixation and staining of the contents of nuclear pseudovacuoles:

1. Postfixation with Co tfac for 6 hours at 70°C.
2. Postfixation with Hg hfac for 4 hours at room temperature, followed by formaldehyde vapor (Fig. 2.28).
3. Postfixation with Zr hfac for 5 hours at room temperature (Fig. 2.29).
4. Postfixation with Ba acac.
5. Postfixation with vapors of osmium tetroxide as described on p. 38.

Measurements of the percentage transmission (% T) of the pseudovacuolar contents dramatically demonstrate that the pseudovacuolar contents are stainable with protein reagents. Measurements were made in a straight line from a clear space (representing Formvar and Durcupan alone) and were not selected in any other way. Only readings from the first ten pseudovacuolar spaces were used. The data are summarized in Table 2.2. In all instances, the % T of nuclear pseudovacuoles was less than that of the clear space, which means that there was some stainable material present in all pseudovacuoles, without exception. It should be pointed out that these measurements could be recorded despite the fact that the electron micrographs as exposed and developed were too dark for quantitative work.

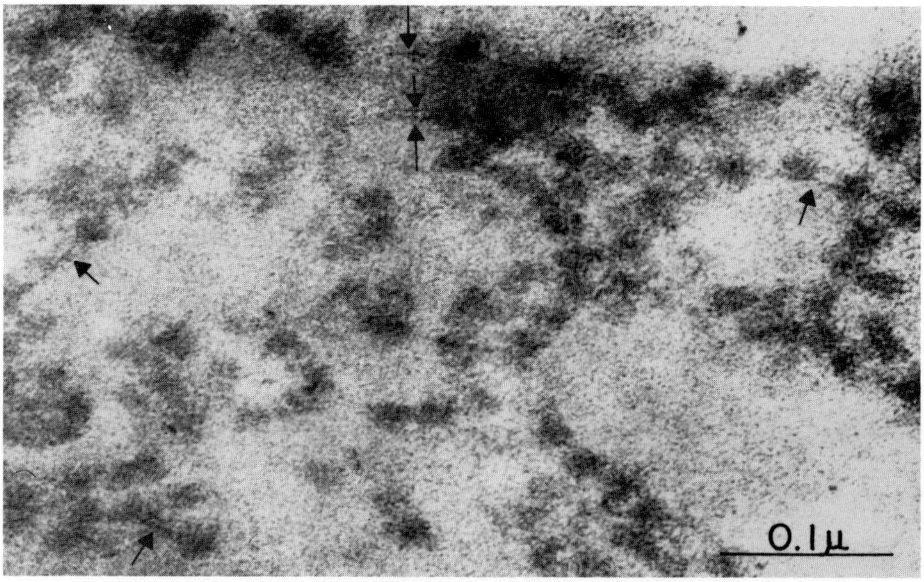

FIG. 2.34. Another extremely thin section, prepared as for the two preceding figures. The protein component of the messengers is indicated by arrows. Protein filaments are attached to most ribosomes. × 216,000. G2 print.

Many observations on the structure and relations of ribosomes were made on freeze-dried specimens treated with a variety of postfixatives and stains. All the critical observations were made on exceptionally thin sections (transparent on the water trough) stained with uranyl acetate on the grid. Most ribosomes are enclosed in a fine-meshed cage of stained linear filaments about 15 A in diameter. These may be clearly visible as discrete filaments (Figs. 2.30 and 2.31, marked "1"). The filaments may be so very dense that the ribosome is obscured by the coat enclosing it (same figures, marked "2"). However, some ribosomes are free or almost free of such filaments, and, therefore, the internal structure appears very clearly. The structure is like that of a cluster of aggregated balls, each with an outer diameter of about 30 A, and a less dense core (same figures, marked "3"). The size relations are such that each ribosome could accommodate 100 or more such "globules." The relations of these protein "globules" or molecules to the RNA of the ribosome will be analyzed in the Appendix of Chapter 3, this volume.

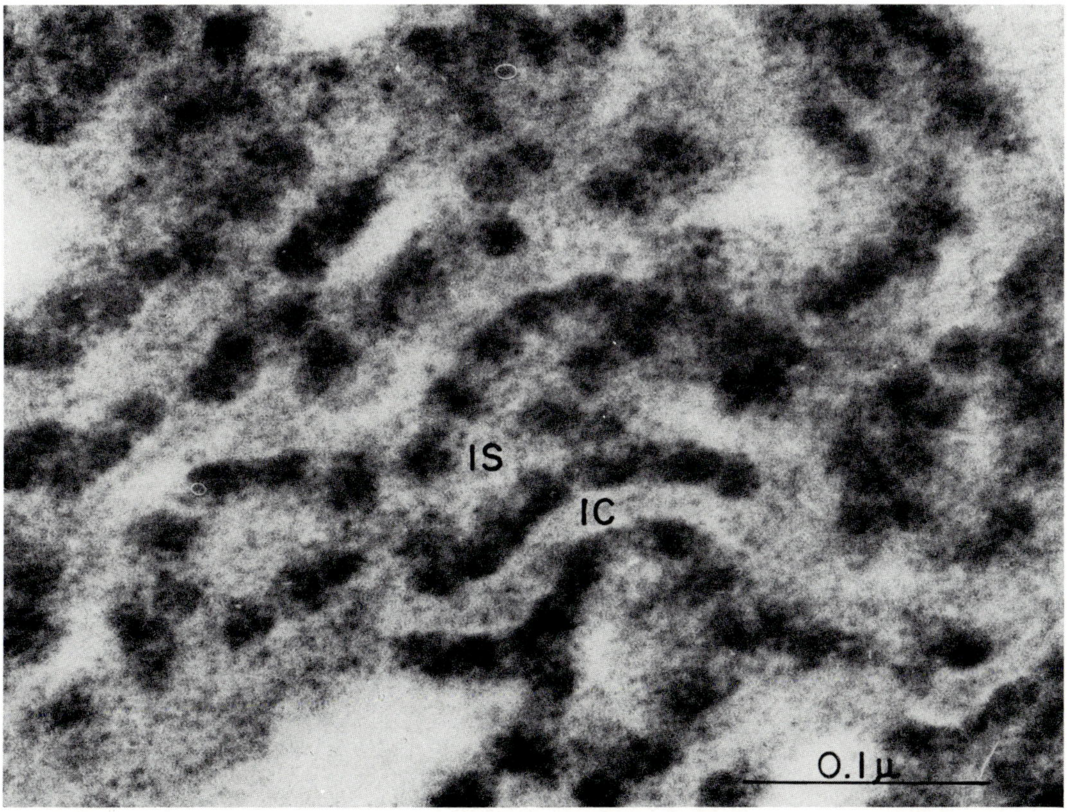

FIG. 2.35. Liver, freeze-dried, postfixed, and stained with vapors of IIPH; sections stained on the grid with uranyl acetate. The intracisternal spaces are enclosed by ribosomes which are mostly obscured to varying degrees by the filaments which enclose them. Filamentous and granular protein molecules are present in the intercisternal spaces (IS). × 310,000. G2 print.

The linear filaments are probably polypeptide chains which are in process of being synthesized. When shed from the ribosomes, the filaments appear free in the adjacent lipid–protein wall of the RER (Figs. 2.32–2.38). Some of them may be observed in the intercisternal space, weakly stained after postfixation with $PtBr_4$ (Fig. 2.37, enclosed in circle). They may also be observed in less filamentous form in the intercisternal space, after postfixation and staining with HEFDOD (Figs. 2.32–2.34), IIPH (Fig. 2.35), and Hg hfac + FFDNB (Fig. 2.36). These findings are of interest in connection with other evidence suggesting that ribosomal structure is very complex *(1, 31)*.

Filaments which extend directly between ribosomes with only gentle curvatures are observed in some preparations stained for protein (Figs. 2.31, 2.34, 2.36, and 2.37). These are probably parts of messenger units stained by virtue of their protein component. The RNA component has been observed in other preparations stained selectively for nucleic acids (this volume, Chapter 3, Appendix). Evidence for the probable association of mRNA with specific proteins, has been presented by Perry and his co-workers *(23, 24, 26)*.

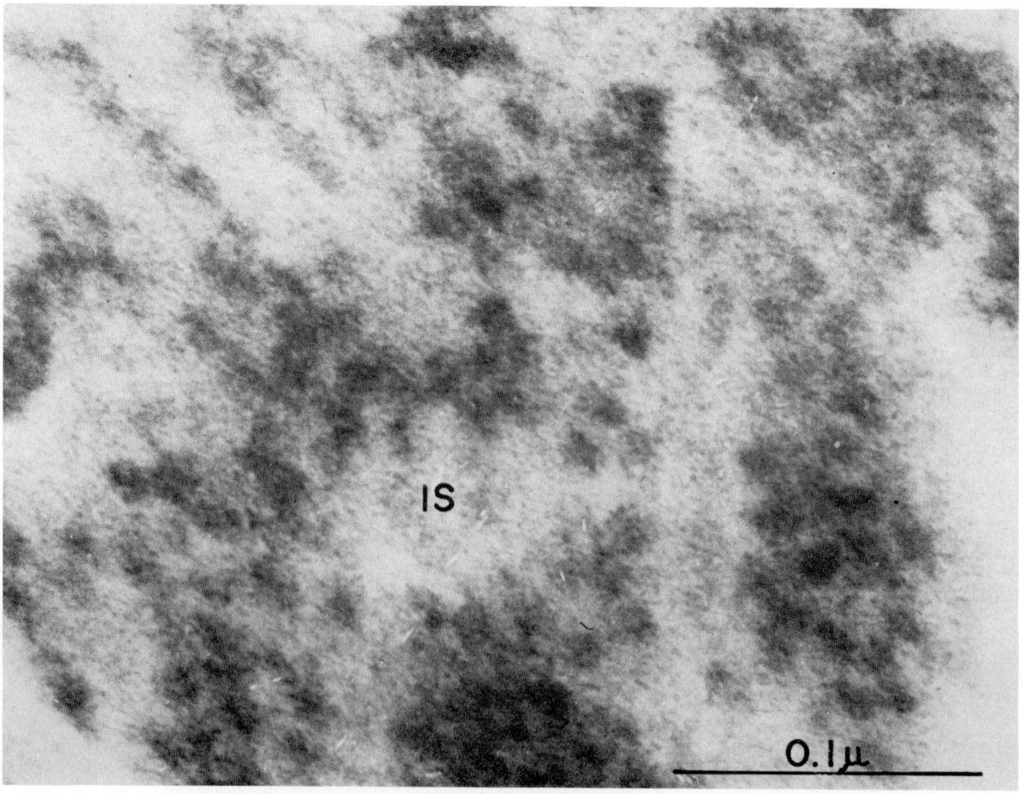

FIG. 2.36. Liver, freeze-dried, postfixed, and stained by successive exposure to vapors of Hg hfac and FFDNB. The same types of periribosomal filaments are present as those described in Figs. 2.30–2.35. Filamentous and granular protein molecules are present in the intercisternal spaces (IS), as in Fig. 2.35. × 400,000. G2 print.

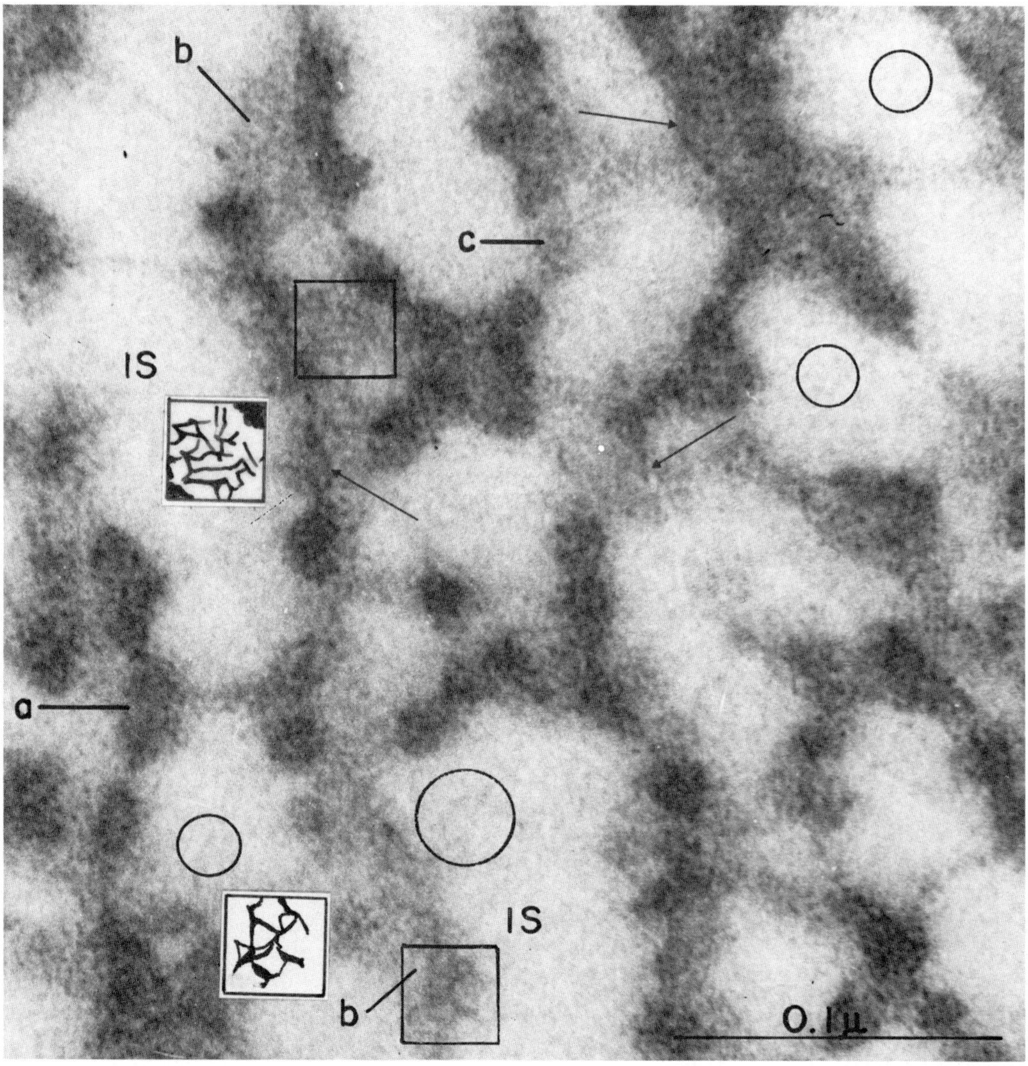

FIG. 2.37. Pancreas, freeze-dried, postfixed, and stained *en bloc* with $PtBr_4$; sections stained on the grid with uranyl acetate. Electron micrograph was printed especially to show protein filaments in the wall of the rough endoplasmic reticulum (RER). Most of them enclose ribosomes (a), but some lie free in the lipid + protein of the wall of the RER (b). Some ribosomes are free of these filaments (c), and in these ribosomes the typical internal structure described in Fig. 2.30 can be observed. Some filaments are faintly visible in the intercisternal space (IS) (enclosed in circles), in various degrees of folding. Protein components of messengers are marked by arrows. × 428,000. G2 print.

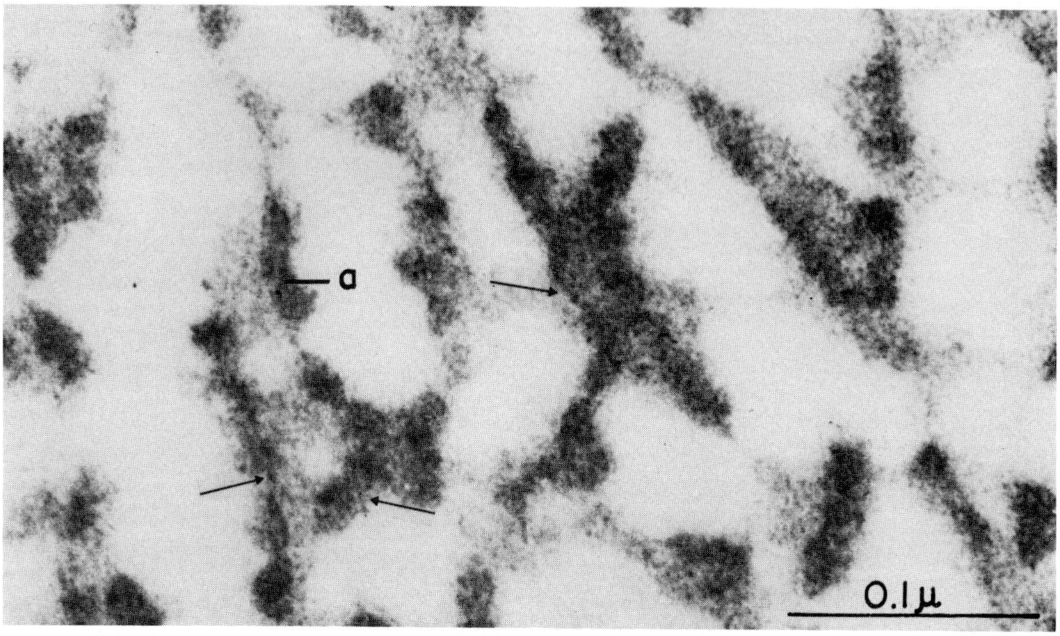

FIG. 2.38. Same electron micrograph as in preceding figure printed to emphasize the periribosomal filaments (a) and the protein components of messengers (arrows). × 321,000. G2 print.

Discussion

The evidence can now be summarized in support of the argument that most of the protoplasm stained by the procedures used here is protein in nature.

1. The metals react with amine, carboxyl, sulfhydryl or other reactive groups of proteins. Some of the vapor reagents also combine with nucleic acids, but the reaction is minimal.

2. The distribution of stainable material after metal staining does not correspond strictly with that of nucleic acids after the latter are stained by gallocyanin-chromalum. This conclusion runs counter to proposals in the literature which purport to show that certain metals stain nucleic acids *(16, 32–34).*

3. The distribution of stainable material after metal staining differs from that of lipids as described later in Vol. II, Chapter 1. Moreover, the distribution of metal stained material is the same if the specimen has been extracted with alcohol or chloroform methanol (1:1) before and/or after metal staining, or if it is simply infiltrated with the plastic components directly after metal staining. There is no indication that the metal stains discriminate between reactive groups of basic or acidic proteins. I regard the metal stains as signifying the presence of protein

(with some nucleic acid), without discriminating in any way between specific proteins or classes of protein.

As stated earlier, the submicroscopic vacuoles may be more accurately named pseudovacuoles, since their contents may be denatured and stained by suitable postfixation treatment. They may then have the same density as the walls and there may be little or no trace of vacuolar structure. The less dense and more dense parts of protoplasm have been regarded as components of a minimally two-phase system in equilibrium *(9)*. The properties of such a two-phase system were regarded as significant in maintaining homeostasis. The operation of this important property need not be considered abrogated by the fact that the vacuoles are pseudovacuoles. If it is assumed that some (even nearly all) of the stainable protein of the pseudovacuoles differs from that of the remainder of the protoplasm, and that some of the protein is the same as that in the walls, but less concentrated, then the elements of the two-phase system are present in protoplasm. The reinterpretation of the submicroscopic vacuolar structure of protoplasm does not modify significantly the theoretical importance of the pseudovacuoles in the maintenance of cellular homeostasis with respect to small molecules or atoms. It should be stated that the expression "two-phase system" is the simplest one, and that if the concept it at all applicable, it represents a series of multiphase systems.

While some displacements might take place during postfixation in aqueous or even nonaqueous fluids, and these shifts in solid phase could conceivably give rise to the appearance of a two-phase spongelike system, the distortions are less likely to take place when the specimens are exposed to cross-linking vapor-phase stains in the dry state. Evidence is summarized in Chapter 1 (this volume) and in the introduction of this chapter that the submicroscopic "spaces" are not caused by expanding ice crystals, since the "spaces" are not empty, but filled with stainable protein. It is when the protein is not stainable that the appearance of submicroscopic vacuoles is simulated.

The distribution of reactive proteins in hepatic and pancreatic cells is apparently stereotyped, at least with the methods employed in this study. In the chromatin, where the first-order (double helix) DNA is arranged as second-order coils (as illustrated in Chapter 3, this volume and Chapter 5, Vol. II), the protein is more or less uniformly distributed between and around the gyres of the latter. This is apparent regardless of the protein stain used. Protein is more or less homogeneously distributed also in the denser walls of the nucleoli, with less dense intervals between them. Both walls and the spaces between them are continuous with a similar two-phase system in the nucleoplasm. Here, again, RNA granules are not recognized, since the protein staining is more or less uniform. As pointed out above, it has not been possible to identify the stained proteins with either basic or acidic proteins. Protein components of the nuclear membrane in unstained and stained preparations appear as dense walls separated by the less dense spaces of the nucleoplasm and cytoplasm from the denser walls on each side. Certainly there is no indication of the presence of the very delicate lipid component of the nuclear membrane (demonstrated with other stains and discussed in Chapter 1, Vol. II). The clear intervals between chromosome masses attached to the nuclear membrane (the nuclear gaps) are not as prominent after protein staining as after staining for nucleic acids (see this volume, Chapter 3 and Vol. II, Chapter 5).

In the literature, it is claimed that all DNA threads and coils are enclosed in protein sheaths, regardless of the kinds of nuclei studied and of the methods of preparation. It is possible that the first- and/or second-order DNA coils are enclosed in a protein sheath, and that the interstices between them are filled with protein of the same density and stainability as that of the sheath. In such a situation, the sheath would not be detectable. This interpretation is extremely unlikely for salivary gland nuclei of *Drosophila* larvae, where the protein stains act more differentially, and where the morphological distribution of protein differs from that of DNA coils (see this volume, Chapters 4 and 5). To my knowledge, there is no theoretical or experimental requirement for such a protein sheath in relation to gene function. What is called for is that protein should be closely associated with DNA.

At first thought, one might expect that if protein were stained and DNA coils were unstained and if the section were thin enough, the DNA coils would be apparent as a negative image. Actually, there is no trace of the DNA coils in such preparations. This observation is more consistent with the fact that DNA, though unstained (or, if stained, not selectively stained), has a finite density and that its diameter is only about 20 A. Such an object is only a small fraction of à totally enveloping, moderately dense medium about 500 to 600 A thick, and would not be expected to be apparent.

The most surprising feature remains the apparent homogeneity of protein in chromatin. This is especially so in view of the known assortment of types of protein in nuclei, each class comprising closely related protein molecules which are believed to differ only subtly from each other *(2)*. If one assumes that the protein molecules of each subclass are few in number, and there is a homogeneous protein matrix, then, if this number of molecules varies or even if the whole subclass disappears, it seems extremely unlikely that such a change would be detected against the homogeneous protein background. One would predict that in random sections, there would be no detectable deviation from the homogeneous distribution of protein in chromatin.

Protein in the dense phase of the nucleolus is also more or less homogeneous, with no indication that it contains numerous RNA granules which appear discrete when stained with gallocyainin–chromalum (Vol. I, Chapter 3, and Vol. II, Chapter 5). This must be taken to mean that the average content of reactive groups in protein of the dense phase, including that of the RNA-rich granules, must be about the same even though several different proteins are known to occur in at least certain nucleoli *(5, 28, 35)*. The same also applies to nucleoplasmic walls.

In Chapter 3 (this volume), a proposal is put forward that the denser walls of the nucleoplasm could be regarded as the physical basis of the transport system of the nucleus. The argument was based on the finding that RNA granules and soluble RNA are confined to the denser nucleoplasmic walls. Unless the above mentioned RNA is synthesized *in situ* in the nucleoplasm or is stored there, it (or a substantial part of it) must be in transit, and its presence in the nucleoplasmic walls identifies them as the transport medium. Also, *prima facie,* as ribosomal, messenger, and transfer RNA are all coded by and originate from the chromosomal DNA, there must be some passage of these RNAs from chromatin to nuclear membrane. Critical information on the intimate charge distribution possibly involved in intranuclear transport deals with objects one or two orders of magnitude smaller than the resolution achieved here.

In the cytoplasm the random two-phase system is more ordered and oriented as the RER, where the intracisternal space separates the walls of the RER. The protein and lipid components of the intracisternal space will be described in Chapter 1 (Vol. II).

Ribosomes appear discrete when stained with gallocyanin-chromalum for nucleic acids (Chapter 3, Vol. I), and after postfixation with certain vapors followed by section staining with uranyl acetate (FFDNB and HEFDOD). Otherwise they are not stained preferentially by protein reagents in vapor form or in solution in 95% alcohol. Prior postfixation probably denatures the protein components of ribosomes sufficiently so as to expose new groups for reaction with the uranyl ion in sections. Some of the ribosomes are aligned along a fine thread which is suggestive of mRNA (Figs. 2.34 and 2.37).

The interpretation of the fine filaments closely associated with the ribosomes is speculative. It is suggested that the filaments are protein molecules newly synthesized or in the process of synthesis. They are wound about the ribosomes, resume a gently curved form when released, and become folded again as they pass into and along the intercisternal space. There they are enclosed by densely packed protein molecules which are stainable with great difficulty. It is suggested that the latter are synthesized in relation to, and released from ribosomes in close proximity to, the filamentous molecules. It is further speculated that these packing molecules are inert and lipophilic, and responsible for maintaining the folded proteins in an inactive form during their transport in the cytoplasm. This topic will be discussed at greater length in the last chapter of Vol. II.

In mitochondria, postfixed and stained with most reagents, the protein component appears more or less uniform and dense, and there seems to be no evidence for the compartmentation which is required for physiological activity. This apparent lack of compartmentation is belied in certain other preparations where vapors were also used. These are diiodophenylhydrazine [which is used to identify lipids in cells (see Vol. II, Chapter 1)], heptafluorodimethyloctanedione [which is used to identify bi- and polyvalent cations (Vol. II, Chapter 9)], as well as certain organometallic compounds. When freeze-dried specimens are postfixed with these vapors and the sections are stained with uranyl acetate, the mitochondria appear compartmented to a remarkable degree, with membranes whose protein molecules differ in staining properties from those of intervening matrix protein molecules. One of the features common to most of the protein reagents listed in Table 2.1 is that they cause the membrane proteins to be so denatured that they all stain similarly with uranium. It is this primary denaturation by the cross-linking vapors which is the basic cause of the homogeneous appearance of mitochondria. The same rationale lies behind the homogeneous appearance of mitochondria when the freeze-dried specimens are stained with alcoholic solutions of metal salts including platinum tetrabromide and uranyl nitrate and acetate. This subject is treated at greater length in Volume II, Chapter 3.

The spongelike nature of the Golgi apparatus is clear in electron micrographs because the walls are thicker and for the most part solid, and characteristically branched. They conform with the morphology of the lipid component described in Chapter 1 (Vol. II). The weaker staining of the Golgi structures by protein stains may support the proposal that proteins of the Golgi structures are rich in lipophilic amino acids.

The close relation of branches of the Golgi apparatus with zymogen granules in both protein and lipid preparations may be of special significance, as it suggests that the compacted proteins destined for exocrine secretion may retain their relation with the Golgi apparatus longer than suspected.

Summary

Procedures are described for the cytochemical identification of polypeptide components of protoplasm. They involve the ultrarapid freezing and drying of specimens to reduce or prevent ice crystal formation, and to expose the dried specimens *in vacuo* to vapors of cross-linking reagents which react with several active sites of polypeptides. The reagents may be organic (such as difluordinitrobenzene), or organometallic [such as mercury (II) hexafluoroacetylacetonate]. These reagents supplement a series of inorganic salts in alcoholic solution (such as platinum tetrabromide) which also stain proteins. Ultrathin sections may be further stained with uranyl acetate for additional contrast through protein staining. The protein staining is selective, in that most of the stain is referable to protein. Pseudovacuoles are a constant feature of protoplasm, but in a few instances which are enumerated, the contents of the vacuoles are preserved and stained, so that the distinction between the contents of the vacuoles and their walls is greatly reduced. Reasons are given for considering the vacuoles as real; i.e., as reflecting that a part of the protoplasm (the contents of the submicroscopic vacuoles) is different from the walls which enclose them.

The mitochondria appear homogeneous after the use of most protein stains; but postfixation and staining by several other methods and section staining with uranyl acetate results in a striking appearance of compartmentation. The Golgi apparatus appears as mostly solid, anastamotic cords of irregular dimensions, which remain attached to mature secretory granules of resting gland cells. Ribosomal protein is stained, but the ribosomes appear discrete only after section staining with uranyl acetate or after staining for nucleic acids.

In some instances, ribosomes are surrounded by protein filaments which may be nascent protein molecules in an extended state. The walls of the RER are narrowly separated by protein which forms the intracisternal space. The ribosomes lie in the walls of the RER, as do the nascent protein molecules.

References

1. Beer, M., Highton, P. J., and McCarthy, B. J. (1960). Structure of ribosomes from *Escherichia coli* as revealed by their disintegration. *J. Mol. Biol.* **2**, 447–449.
2. Bonner, J. (1967). The role of histones in the regulation of RNA synthesis. *In* "Regulation of Nucleic Acid and Protein Biosynthesis" (V. V. Koningsberger and L. Bosch, eds.), pp. 211–224. Elsevier, Amsterdam.
3. Bullivant, S. (1962). Consideration of membranes and associated structures after cryofixation. *5th Int. Congr. Electron Microsc.*, Vol. 2, p. R–2. Academic Press, New York.
4. Bullivant, S., and Ames, A., III (1966). A simple freeze-fracture replication method for electron microscopy. *J. Cell Biol.* **29**, 435–447.
5. Busch, H., and Smetana, K. (1970). "The Nucleolus." Academic Press, New York.
6. Elfvin, L.-G. (1963). The ultrastructure of the plasma membrane and myelin sheath of peripheral nerve fibers after fixation by freeze-drying. *J. Ultrastruct. Res.* **8**, 283–304.
7. Finck, H. (1958). An electron microscope study of basophile substances of frozen-dried rat liver. *J. Biophys. Biochem. Cytol.* **4**, 291–300.
8. Gersh, I., Isenberg, I., Stephenson, J. L., and Bondareff, W. (1957). Submicroscopic structure of frozen-dried liver specifically stained for electron microscopy. I. Technical. *Anat. Rec.* **128**, 91–111.
9. Gersh, I., Isenberg, I., Bondareff, W., and Stephenson, J. L. (1957). II. Biological. *Anat. Rec.* **128**, 149–169.
10. Gersh, I., Vergara, J., and Rossi, G. L. (1960). Use of anhydrous vapors in post-fixation and in staining of reactive groups of proteins in frozen-dried specimens for electron microscopic studies. *Anat. Rec.*, **138**, 445–460.
11. Gruenwedel, D. W., and Davidson, N. (1967). Complexing and denaturation of DNA by methylmercuric hydroxide. II. Ultracentrifugation studies. *Biopolymers* **5**, 847–861.
12. Hanzon, V., and Hermodsson, L. H. (1960). Freeze-drying of tissues for light and electron microscopy. *J. Ultrastruct. Res.* **4**, 332–348.
13. Hanzon, V., Hermodsson, L. H., and Toschi, G. (1959). Ultrastructural organization of cytoplasmic nucleoprotein in the exocrine pancreas cells. *J. Ultrastruct. Res.* **3**, 216–227.
14. Hu Ping-cheng, Cheng-pin, Liu, and Chih-chuan, Liang. (1962). Cited by Tien-Hsi Cheng and R. H. Doi, *Fed. Proc.* **27**, 1430–1454 (1968).
15. Hu Ping-cheng. Cheng-pin, Lui, Wu Hui, and Chih-chuan, Liang. (1964). Cited by Tien-Hsi Cheng and R. H. Doi, *Fed. Proc.* **27**: 1430–1454 (1968), and by M. Errera, *Nature (London)* **205**, 739–741 (1965).
16. Huxley, H. E., and Zubay, G. (1960). Electron microscope observations on the structure of microsomal particles from *Escherichia coli*, *J. Mol. Biol.* **2**, 10–18.
17. Katz, S. (1962). Reversible reaction of double stranded polynucleotides and Hg II: separation of the strands. *Nature (London)* **195**, 997–998.
18. Katz, S. (1963). The reversible reaction of Hg (II) and double-stranded polynucleotides. *Biochim. Biophys. Acta* **68**, 240–253.
19. Koehler, J. K. (1968). The technique and application of freeze-etching in ultrastructure research. *Adv. Biol. Med. Phys.* **12**, 1–84.
20. Moor, H. (1966). Use of freeze-etching in the study of biological structure. *Int. Rev. Exp. Pathol.*, **5**, 179–216.
21. Moshier, R. W., and Sievers, R. E. (1965). "Gas Chromatography of Metal Chelates." Pergamon Press, New York.
22. Olcott, H. S., and Fraenkel-Conrat, H. (1947). Specific group reagents for proteins. *Chem. Rev.* **41**, 151–197.
23. Perry, R. P., and Kelley, D. E. (1966). Evidence for specific association of protein with newly formed ribosomal subunits. *Biochem. Biophys. Res. Commun.* **24**, 459–465.
24. Perry, R. P., and Kelley, D. E. (1968). Messenger RNA-protein complexes and newly synthesized ribosomal subunits: analysis of free particles and components of polyribosomes. *J. Mol. Biol.* **35**, 37–59.

25. Sanger, F. (1955). The chemistry of simple proteins. *Symp. Soc. Exp. Biol.* **9,** 10–31.
26. Schochetman, G., and Perry, R. P. (1972). Characterization of the messenge RNA released from L cell polyribosomes as a result of temperature shock. *J. Mol. Biol.* **63**, 577–590.
27. Seno, S. and Yoshizawa, K. (1960). Electron microscope observations on frozen-dried cells. *J. Biophys. Biochem. Cytol.* **8,** 617–638.
28. Siebert, G. (1966). Nucleolar enzymes of isolated rat liver nucleoli. *Nat. Cancer Inst. Monogr.* **23**, 285–293.
29. Sjöstrand, F. S., and Baker, R. F. (1958). Fixation by freezing-drying for electron microscopy of tissue cells. *J. Ultrastruct. Res.* **1,** 239–246.
30. Sjöstrand, F. S., and Elfvin, L.-G. (1964). The granular structure of mitochondrial membranes and of cytomembranes as demonstrated in frozen-dried tissue. *J. Ultrastruct. Res.* **10,** 263–292.
31. Spirin, A. S. (1964). "Macromolecular Structure of Ribonucleic Acids." Reinhold, New York.
32. Swift, H. (1962). Nucleoprotein localization in electron micrographs: metal binding and radioautography. *In* "The Interpretation of Ultrastructure" (R. J. C. Harris, ed.), pp. 213–232. Academic Press, New York.
33. Swift, H. (1962). Nucleic acids and cell morphology in dipteran salivary glands. *In* "The Molecular Control of Cellular Activity" (J. M. Allen, ed.), pp. 73–125. McGraw-Hill, New York.
34. Swift, H., and Adams, B. J. (1962). Simple methods for nucleic acid and acid mucopolysaccharide localization in electron micrographs. *5th Int. Congr. Electron Microsc.,* Vol. 2, pp. 0–4. Academic Press, New York.
35. Vincent, W. S., Baltus, E., Løvlie, A., and Mundell, R. E. (1966). Proteins and nucleic acids of starfish oocyte nucleoli and ribosomes. *Nat Cancer Inst. Monogr.* **23**, 235–253.
36. Wold, F. (1960). Possible synthetic tertiary bonds in proteins. *Fed. Proc.* **19,** 339 (Abstr.).
37. Wolman, M. (1955). Problems of fixation in cytology, histology, and histochemistry. *Int. Rev. Cytol.* **4,** 79–102.
38. Yamane, T., and Davidson, N. (1961). On the complexing of desoxyribonucleic acid (DNA) by mercuric ion. *J. Amer. Chem. Soc.* **83,** 2599–2607.
39. Zahn, H. (1955). Bruckenreaktionen an Aminosäuren und Faserproteinen. *Angew. Chem.* **67,** 561–572.
40. Zahn, H. (1955). Cross-linking reactions with amino acids and fibrous proteins. *Proc. Int. Wool Textile Res. Conf.* pp. 426–451. Commonwealth Scientific Industrial Research Organization, Melbourne, Australia.

3

Methods for the Fixation and Staining of DNA and RNAs Applied to Exocrine Pancreatic and Hepatic Cells of the Mouse

Isidore Gersh and G. L. Rossi

In contrast with the tremendous advances in knowledge of the submicroscopic structure of cytoplasm, developments in nuclear morphology have been relatively unrewarding. This has been recognized by numerous researchers, among them Porter *(42)*, Swift *(52)*, Moses and Coleman *(37)*, Barnicot and Huxley *(2)*, Ris *(43)*, and Ris and Kubai *(44)*. The obstacles are probably technical, i.e., the apparent inadequacy for nuclear studies of methods of fixation and staining of nucleic acids and nucleoproteins. The present state of knowledge of the nuclear morphology is illustrated by several recently published review articles *(4, 6, 14, 23, 30)*.

Most methods for the identification of nucleic acids in cells with the aid of the electron microscope rely on metal ions in solution *(1, 5, 26, 34, 36, 50, 51, 53, 56, 57, 61, 63)*. Only Fink *(17)* has used for this purpose a dye (gallocyanin–chromalum) whose major contribution to increased contrast is organic rather than inorganic *(27)*. The identification of the original sites containing nucleic acids requires their suitable fixation as well as selective or specific and sensitive marking *(18)*. Following publication of a general statement on the use of anhydrous vapors in postfixation and staining of certain reactive groups of proteins *(20)*, Mundkur published several papers from our laboratory and later from his own, in which he showed that cross-linking agents in vapor form helped to preserve the original sites of nucleic acids *(38–40)*. These contributions, however, did not develop as fully as possible the potentialities of the method for the demonstration of nucleic acids.

This report is a fuller interpretation of the use of cross-linking vapors to preserve nucleic acids in freeze-dried specimens, and of the use of gallocyanin–chromalum as a selective and virtually specific and amplifying marker for nucleic acids in cells fixed in this way, for study with both light and electron microscopes. This report presents and analyzes methods developed for the study of nucleic acids in cells, with particular attention to pancreatic acinar and hepatic cells of the mouse. Subsequent chapters will deal with molecular DNA patterns and with soluble and insoluble RNAs in a variety of other cells of the mouse, in the giant nuclei of salivary gland cells of *Drosophila*, and in mitotic chromosomes. Studies on meiotic chromosomes will be published later.

Method

Freeze-dried specimens were treated *in vacuo* with one of two vapor reagents: 1, 5-difluoro-2,4-dinitrobenzene (FFDNB) and mercuric (II) 1,1,1,5,5,5-hexafluoro-2,4-pentanedione or mercuric (II) hexafluoracetylacetonate (Hg hfac, mercury chelate, mercury complex). These were used to preserve amino acids and polypeptides by cross-linking them and to inactivate nucleases. The mechanism of action is discussed in Chapter 2 (this volume). The aims of the cross-linking action are to preserve soluble RNAs with aminoacyl groups and to retain DNA in an extended state *(22)* for subsequent staining instead of being distorted to granular forms. Specimens thus prepared were stained for nucleic acids with gallocyanin–chromalum (GCA).

Staining of Nucleic Acids with Gallocyanin–Chromalum (GCA)

The stain, developed by Einarson *(16, 47)*, gives a blue color virtually specific for nucleic acids, at a pH of 1.9 to 2.1. Possible formulas for the lake are given by Horobin and Murgatroyd *(25)*. The simplest formula when constructed as a space-filling model shows the dye molecule to be platelike, measuring approximately 13 A long × 7–10 A wide × 3 A thick. The stain combines nearly stoichiometrically with both DNA and RNA in solution *(29, 47)*, probably through combination with phosphate groups of the backbone. The attribution of staining to a combination of the dye with amino groups of polypeptides *(54)* has been shown to be in error *(29)*. When properly prepared, the lake appears to combine selectively with nucleic acids, and, at the correct pH, this combination is specific. GCA has the advantage over other basic dyes, which also stain nucleic acids by the same mechanism, in having the property of "exquisite progressivity" *(16)*, with no possibility of overstaining as is seen with other basic dyes. The dye combines nearly stoichiometrically also with nucleic acids in sections after specified fixation, with few exceptions (sperm and nucleated red blood cells of lower vertebrates).

A basic question which had to be examined was the validity of the method of fixation and staining of nucleic acids. To resolve this question, an extensive study was made of the distribution of DNA in sperm heads.

DNA of the sperm heads of some species is known to be highly oriented from studies on their intrinsic birefringence *(41, 45, 48)* and on their dichroism in the ultraviolet *(11)* as well as on their polarized fluorescent emission after staining with acridine orange *(33)*. More intimate proof that these observations were attributable to the alignment of the DNA double helices comes from X-ray diffraction patterns of sperm heads of certain species, the patterns being identical with those of purified DNA *(58–60)*. The older observations on birefringence and X-ray diffraction are particularly noteworthy in sperm heads of *Loligo*. The extremely tight packing of the DNA, which these findings suggest, may account for the less than stoichiometric staining of the DNA with the Feulgen and gallocyanin–chromalum methods, and the reduced uptake of actinomycin D and acridine orange, as well as the progressive stabilization toward heat *(13, 21, 41)*.

The sperm of *Loligo*, rat, and mouse were prepared by the procedure used in Chapters 3 to 13 (this volume), as well as by others, including the use of acridine orange and quinocrine intravitally and *in vitro*, and the use of spermidine vapors *in vacuo*. The specimens are being studied in detail, but it is clear, even at this early stage, that gallocyanin–chromalum stains the DNA double helix of sperm heads of all three species, and that the helices are aligned and tightly packed, which agrees with the physical findings in the preceding paragraph. This study will be published separately when it has been completed; it is clear now that the results validate the methods used in this volume.

Further identification and discrimination of the type of nucleic acid which is stained is achieved by the use of RNase and DNase prior to staining with GCA. There is no doubt that when specimens are observed by light microscopy, the effect of RNase is to prevent the subsequent staining of nucleoli and ergastoplasm by GCA. This failure of RNA to stain might be caused by combination of the enzyme with nucleic acids in such a way that enzymic hydrolysis is blocked as is the attachment of the dye, or the failure to stain may be caused by the hydrolysis of nucleic acids. The fact that the UV absorption at the nucleic acid absorption peak is decreased markedly after enzyme treatment *(19)* eliminates the first possibility and firmly establishes the second.

The results of enzymic hydrolysis are never as clear with the electron microscope, even when ultrathin sections of the same specimens are observed. It is true that there is a great loss of contrast by the nucleolus, but the nucleolus remains recognizable in electron micrographs. Similarly, in most cells which have been studied, there is a marked decrease in contrast of ribosomes. The residual contrast after RNase treatment is probably attributed to the combination of hydrolytic products of RNAs with basic groups of proteins, and to the combination of acidic groups of proteins with the enzymes. The effects of DNase are similarly not as dramatic when observed with the electron microscope as when viewed with the light microscope, and probably for similar reasons. In addition to these complexities is a certain amount of distortion of submicroscopic structures which always accompanies immersion of specimens, even after extensive cross-linking, in the water of the enzyme solution. Despite these difficulties, the results of enzyme treatment are helpful and sometimes surprisingly convincing, as is seen in nerve and cartilage cells (see this volume, Chapter 12 and Vol. II, Chapter 5).

The dye molecules attached to the phosphate groups extend outward from the backbone(s) of the nucleic acids as stacks. As can be seen from a space-filling model, the result is that the diameter of the double helix of DNA is approximately doubled. Even slight deviations from an end-on view of the axis of DNA and GCA cause the DNA–dye complex to appear in projections like a thread. GCA achieves two effects: (1) amplification of the mass, so that the DNA is denser and appears in electronmicrographs as threads, and (2) a twofold amplification of the apparent diameter. The stain is so nearly specific for nucleic acids in freeze-dried material that the background is virtually nil with the light microscope and only slightly greater with the electron microscope, with the exception of mitochondria (see Vol. II, Chapter 3).

Each DNA double helix would appear as a thread about 45 A in diameter. Most commonly, the threads are arranged as second-order coils whose diameter and pitch are nearly always about 400 A. Observations of models of second-order coils of DNA–GCA were made in various planes of sectioning and projected as shadowgraphs. The most common appearance, in sections about 500 A thick, is of a series of circles and ellipses. These have been identified in chromatin of all cell types studied, and also in chromosomes, where the dimensions of the second-order coil may differ. When such coils are juxtaposed in various combinations, they may give the appearance of coarse granules attached to the finer threads of DNA–GCA. These and other aspects of second-order coils of DNA are illustrated in Fig. 3.1.

Some sections of GCA-stained specimens were observed directly with the light microscope, while others were stained with toluidine blue, which favors optical viewing and photography.

Controls

Although soluble RNAs (sRNAs) are retained following postfixation in alcohol (after freezing and drying), they are probably largely extracted by the water of the staining solution. This statement is based on the observation that alcohol-treated sRNA, after drying, goes into solution in GCA. Specimens postfixed in alcohol serve as a baseline for the others treated with FFDNB or the mercury chelate, where sRNAs were made partly insoluble and are stained by GCA.

Ribonuclease or deoxyribonuclease was applied to freeze-dried specimens after postfixation, but prior to staining with GCA; the specimens were then embedded and sectioned. As a control on enzyme activity, specimens were treated with cross-linking reagents and immersed in water alone for comparable times and then stained with GCA. Some ultrathin sections of vapor postfixed specimens already stained with gallocyanin–chromalum were treated with ribonuclease, washed, and dried for viewing, as an additional control on the identification of sRNA in electron micrographs. Retention of the plastic during enzyme treatment of the section reduces markedly, without completely abolishing, morphological distortions.

Because mercury has a particular affinity for SH groups, an attempt was made to see whether their removal by oxidation influenced staining by the mercury chelate.

PROCEDURES

Specimens were cut in a moist box to a thickness of 0.1 to 0.2 mm, placed on thin aluminum foil, and frozen ultrarapidly. They were dried at −40°C and infiltrated with 95% alcohol, or treated *in vacuo* with a cross-linking vapor, and then infiltrated with 95% alcohol. Both steps were described in Chapter 2, this volume.

1. 95% Alcohol. Some of the specimens remain in 95% alcohol while others pass successively during 4 hours through 80, 70, 60, 50, and 40% alcohol to

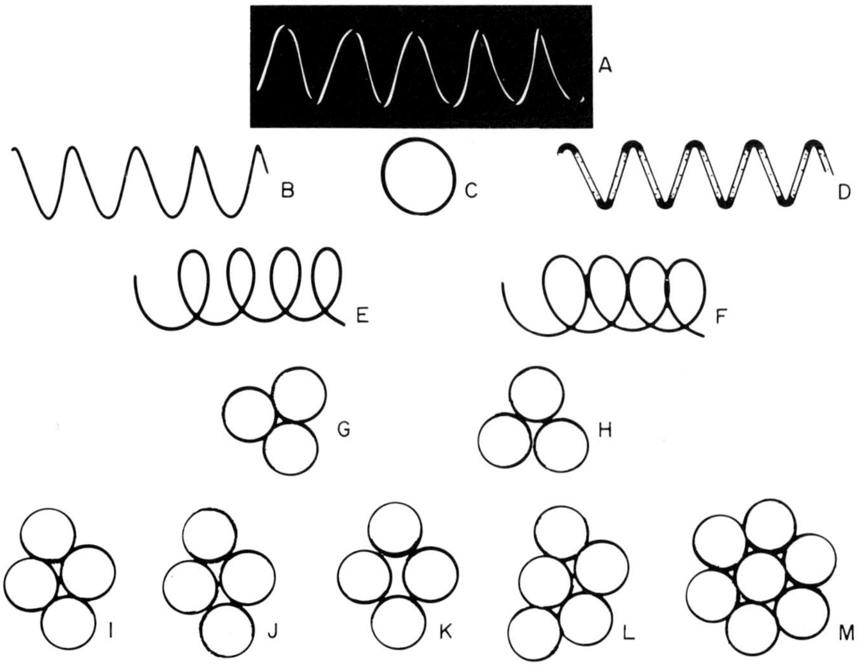

FIG. 3.1. The appearance of a second-order DNA coil when viewed in various planes and combinations as shadowgraphs. The wire represents the first-order coil (the DNA double helix) plus the attached molecules of gallocyanin–chromalum. The period and diameter of the second-order coil is about 400 A. The coil is viewed in three dimensions in (A). In (B) and (C), the shadows are shown as they would appear when the coil is viewed at right angles and parallel to the axis. (D) is a shadow drawing of a similar coil made of partly transparent material to emphasize density differences. (E) and (F) are shadows of the second-order coil at angles intermediate between those represented in (B) and (C). The central clear (triangular) area which appears in (G) and (H) when three second-order coils are viewed in their axis, would appear as a line or elongated dot when the angle is altered. When four (I, J, K), five (L), and seven (M) parallel second-order DNA coils are viewed in a similar way, the dimensions of the central line or dot vary somewhat. All figures are photographs of models, with the exception of (D), which is a drawing.

GCA where they remain for 20 hours. They are then washed in water, and dehydrated by successive steps during 4 hours, and remain overnight in 95% alcohol. At this point, all the specimens stained and unstained are embedded in water-insoluble or water-soluble Durcupan (w-i or w-s Durcupan), as described in Chapter 2, this volume.

2. *FFDNB Vapor.* The dried specimens are treated with FFDNB vapors for 5 hours at 50°C. Excess vapors are removed by evacuating the tube T (this volume, Chapter 2, Fig. 2.1) at 60°C overnight. Specimens are infiltrated at room temperature with 95% alcohol admitted slowly and removed from the foil in 95% alcohol, where they remain for 1 day. They are then cut into small pieces. Some specimens are retained without staining while others are stained with GCA, and all are embedded in w-s Durcupan as in method (1) above.

3. *Mercury Chelate Vapor.* Specimens were dried in tube S (this volume, Chapter 2, Fig. 2.1) and treated with vapors at 0°C for 3 hours. Excess vapors were removed. Specimens were infiltrated with 95% alcohol at room temperature, kept in 95% alcohol for 1 day, and then cut into smaller pieces. Some remained in 95% alcohol, and others were stained with GCA as in method (1). All were embedded in w-s Durcupan.

Controls

To test for the effects of enzymes, specimens postfixed with 95% alcohol, alone or after the customary treatment with FFDNB or Hg hfac, were immersed in 0.1% aqueous solution of ribonuclease (RNase) or deoxyribonuclease (DNase) for ½ hour. They were washed briefly in water and dehydrated, after which they remained in 95% alcohol for 1 day. The specimens were stained with GCA and embedded in w-s Durcupan as in method (1), and sections about ¼ μ were cut for study with the light microscope. Other ultrathin sections were viewed with the electron microscope.

Another method for testing the effects of ribonuclease at the level of the electron microscope was to prepare ultrathin sections of specimens postfixed in Hg hfac, stained with GCA, and embedded in w-s Durcupan. Ultrathin sections were mounted on copper grids and dried. They were passed successively through three drops of the RNase solution for 1 minute each, and allowed to remain in a large drop for 30 minutes. They were drained with filter paper before immersion in the next drop. The grids were passed through six drops of water, and again drained after each drop. After the final draining, grids were dried and ready for study with the electron microscope. Sections which were passed through water alone served as controls. Both were compared with sections of comparable thickness of the same specimen not treated with enzyme or water.

An additional control was made to test whether sulfhydryls could significantly influence the staining with the mercury chelate. For this test, specimens were dried and treated with ozone at −80°C for 40 minutes as described later in Chapter 1 (Vol. II). Excess ozone was removed rapidly and the specimens were treated with vapors of the mercury chelate for 5 hours at 0°C. Excess vapor was removed by evacuating for 2 hours at 0°C and then overnight at room temperature. Specimens were embedded in w-i Durcupan.

Polymerized blocks were trimmed and sections were cut, briefly floated on water, mounted on grids, and examined with a Hitachi HU-11A electron microscope operating at an accelerating voltage of 50 kv. Electron micrograph plates were enlarged two times in printing in nearly all cases.

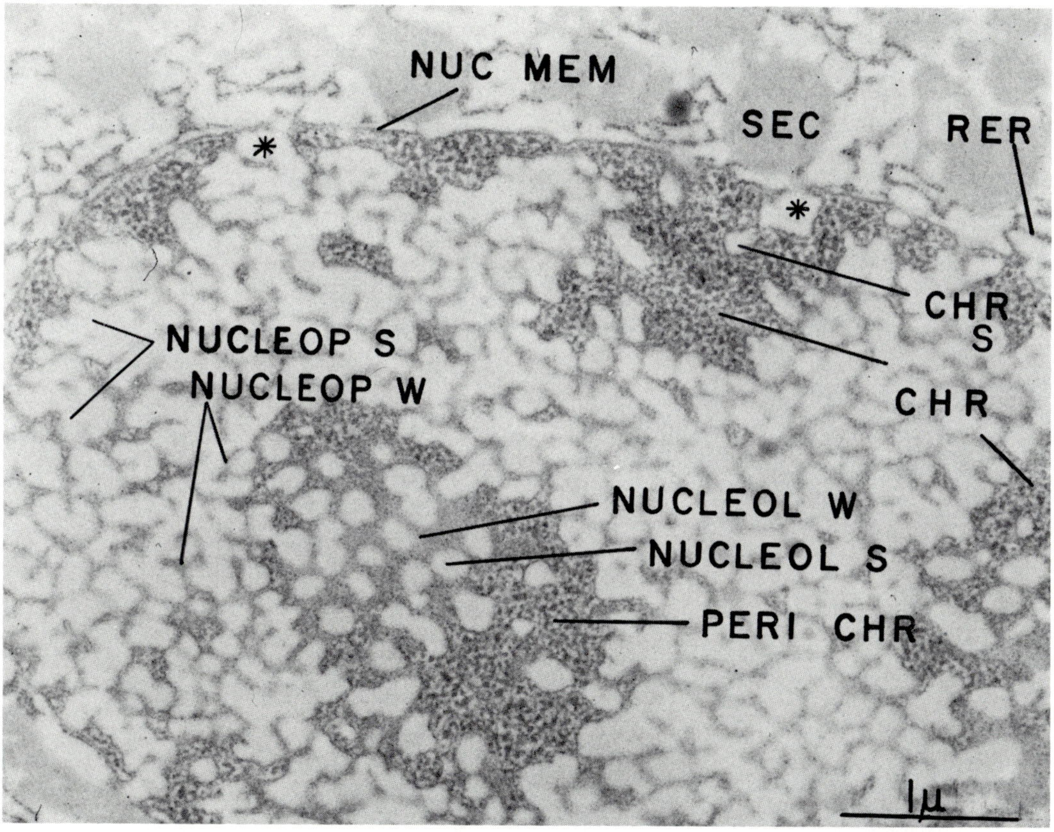

FIG. 3.2. Mouse pancreas, postfixation in 95% alcohol, staining with gallocyanin–chromalum. Orientation electron micrograph to show the spongelike nature of the nucleus with interconnecting "spaces" separated by walls. Chromatin masses (CHR, PERI CHR) are characteristically dense, nucleolar walls (NUCLEOL W) are less dense, while the nucleoplasmic walls (NUCLEOP W) are least dense. Nucleolar spaces (NUCLEOL S), chromatin spaces (CHR S), and nucleoplasmic spaces (NUCLEOP S) are interconnected. Spaces between chromatin masses (*) are separated by thin walls from the cytoplasm. The rough endoplasmic reticulum (RER) is quite dense and granular in contrast with the secretory granules (SEC). The site of the nuclear membrane (NUC MEM) is indicated. × 24,000.

Observations

In order to shorten this section, some detailed descriptions are given in the figure legends only. The term "microscopic" is used in describing structures which can be seen with the light microscope; when structures cannot be resolved with the light microscope, they are referred to as submicroscopic.

Pancreas

1. Postfixation in Alcohol and Staining with Gallocyanin–Chromalum (GCA) (Figs. 3.2–3.4). The nucleus is marked by dense small and large masses of chromatin, the larger ones corresponding with similar bodies observed with the light

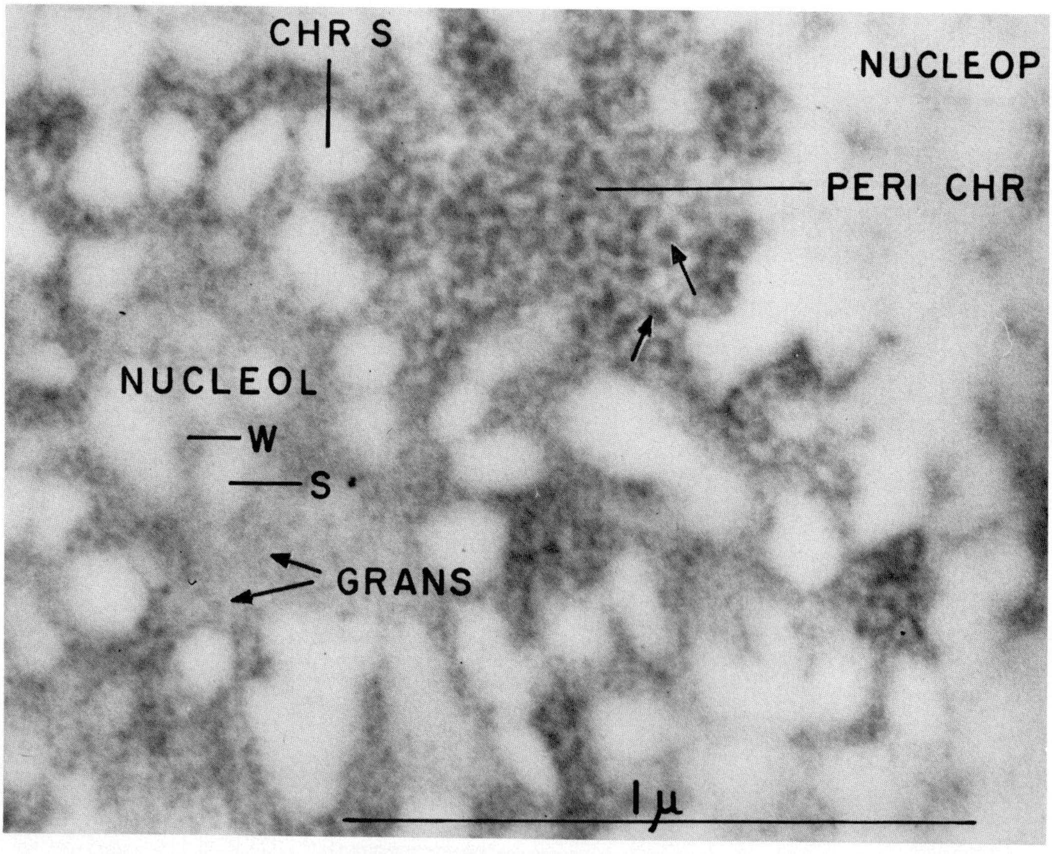

Fig. 3.3. Mouse pancreas, postfixation in 95% alcohol, staining with gallocyanin–chromalum. Nucleolus and perinucleolar chromatin. Small and large RNA granules (GRANS) in the nucleolar walls (NUCLEOL W) and in the nucleoplasmic walls are visible against a clear or pale background. Perinucleolar chromatin (PERI CHR) is present mainly as granules (arrows). Occasionally, threadlike structures are visible. All other abbreviations same as in Fig. 3.2. × 82,000.

microscope. In electron micrographs, most of them are distributed discontinuously on the periphery of the nucleus, but some appear in the nucleoplasmic portion, and others enclose the nucleolus to varying extents. All three may, in fact, be connected to each other at different levels. Some of these clumps are microscopic, while others are submicroscopic. Between the peripherally disposed chromatin masses are clear areas which are directly continuous with the spongelike "spaces" of the nucleoplasm; they are separated from the cytoplasm by the nuclear membrane. Similar "spaces" in the chromatin masses also communicate with the spongelike "spaces" of the nucleoplasm. These relations are clearer at somewhat higher magnifications, when the chromatin masses are resolvable into chiefly coarse dense granules, although some fine filaments can be observed.

The nucleolus is less dense than the perinucleolar chromatin. It is composed of granules of various sizes from about 140 A in diameter to granules at the limit of resolution. They lie in a "homogeneous" material which affords them a very slight background density. The nucleolar substance is penetrated by apparently

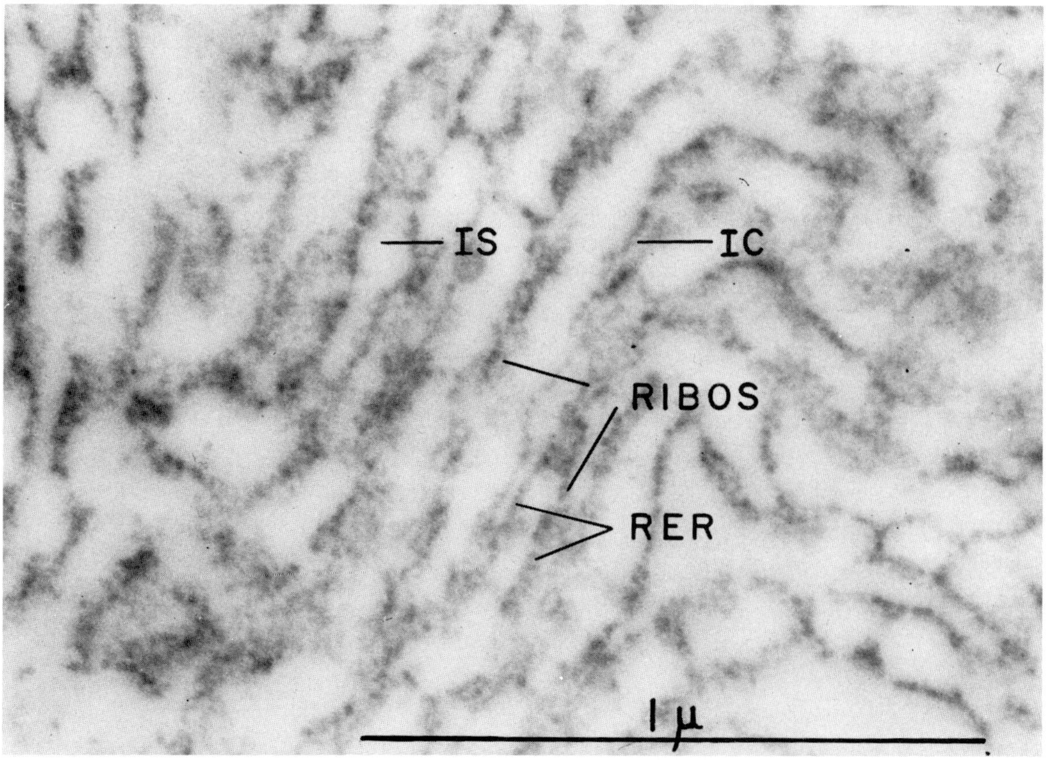

FIG. 3.4. Mouse pancreas, postfixation in 95% alcohol, staining with gallocyanin–chromalum. Ribosomes (RIBOS) of the rough endoplasmic reticulum (RER) are discrete. The intracisternal space (IC) is generally narrow; wider intercisternal spaces (IS) are clear. There are no indications of ribosomal groupings. × 82,000.

interconnected spongelike "spaces" which are continuous with similar spaces in the chromatin masses and in the nucleoplasm. It should be noted that the intranucleolar walls abut directly on the perinucleolar chromatin.

The perinucleolar chromatin (as well as other chromatin masses) extends out into the nucleoplasm, which appears as a three-dimensional, spongelike, moderately dense material which in sections appears as netlike, submicroscopic reticulations. They appear to have a faint homogeneous background within which lie granules indistinguishable from those in the nucleolus, though they are less numerous.

Outside the nucleus, separated from it by a clear gap, is an interrupted, thin dense layer which appears like the rough endoplasmic reticulum (RER) and seems to be continuous with it at some points. The RER is especially prominent

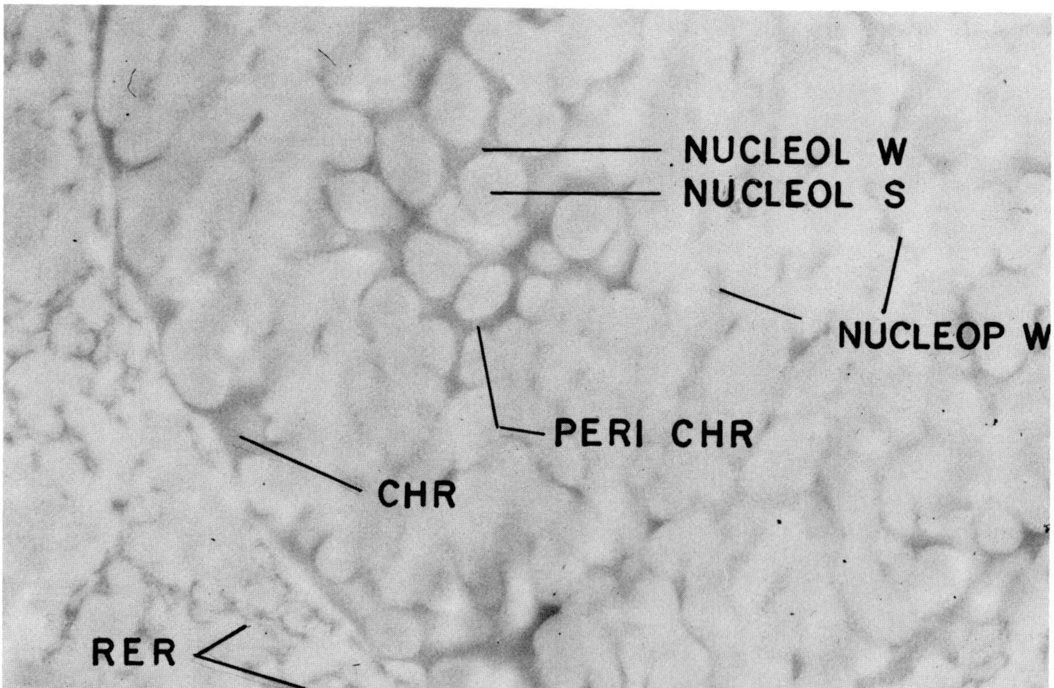

FIG. 3.5. Mouse pancreas, postfixation in 95% alcohol, no stain. No granules are detectable in the nucleus where the chromatin (CHR) is denser than the nucleolus (NUCLEOL), and the nucleoplasmic walls (NUCLEOP W) are barely indicated. In the cytoplasm, the RER is densest, and roughly approximates the chromatin in density. × 24,000.

at the base of the cell, but extends also into the remainder of the cytoplasm as apparently separated strands and figures. The RER reaches out to all surfaces of the cell. At higher magnification, the dense "lines" of the RER are resolvable into densely stained ribosomes, about 200 A in diameter, which are closely related to the lipid component described in Chapter 1 (Vol. II). The space between the paired layers of the RER, the intracisternal space, is narrow; the space between adjacent paired RER membranes, the intercisternal space, is larger and commonly less dense.

Some of the paired membranes closely abut on secretory granules. Others impinge on an unstained cytoplasmic network. This is clearly visible in electron micrographs of low magnification, and is coextensive with the Golgi structures stained more specifically with the lipid methods.

2. Postfixation in Alcohol, Unstained (Fig. 3.5). The nucleus is clearly demarcated by the rather uniformly dense masses of chromatin, which show no indications of coils or granules. The perinucleolar chromatin is readily observable and encloses the thinner walls of the nucleolar spongelike structure. The walls of the nucleoplasmic spongework are barely indicated, and are continuous with extensions of chromatin masses. In the cytoplasm, the RER is dense, and only in some regions are the ribosomes distinguishable. A comparison of these electron micrographs with those stained with GCA shows how little contrast unstained preparations have and how much contrast has been added in selective sites by GCA.

Although alcohol is a good denaturant, it is not adequate to consistently protect the submicroscopic structure against the solution, swelling, and deformations which takes place during the subsequent prolonged immersion in the water of the GCA staining solution. In addition, the alcohol cannot protect RNAs from solution by the water. To obviate these difficulties, specimens are treated with cross-linking reagents, one of which is FFDNB.

3. Postfixation and Staining in Vacuo *with Vapors of Difluorodinitrobenzene (FFDNB), Followed by Staining with GCA (Figs. 3.6–3.11).* The outstanding consequences of this method of preparation, as compared with those following postfixation with alcohol alone are: (1) the more consistent morphology, e.g., fewer areas of swelling; (2) the chromatin masses are resolvable into finer granules, circles, ovals, and other figures which are all parts of second-order DNA coils, the first-order DNA coil (the double helix) being about 45 A in diameter; (3) the occurrence of a cloud of moderate density in the nucleolus which often obscures the granules; (4) the similar increase in unresolvable (background) density in the walls of the spongelike nucleoplasm; (5) the occurrence in many small areas of the RER of a similar background density; and (6) the grouping in the RER of ribosomes in clusters.

4. Postfixation and Staining in Vacuo *with Vapors of FFDNB Alone (Figs. 3.12 and 3.13).* This should be compared with electron micrographs of specimens postfixed in alcohol alone, and of specimens postfixed with FFDNB and stained in addition with GCA. The RER is less prominent than after staining with GCA. Also, in the nucleus, the distinctions between the chromatin, nucleolus, and nucleoplasmic walls are considerably reduced. As after postfixation in alcohol, no internal structure can be discerned in the chromatin masses and the chromatin remains uniformly dense.

5. *Postfixation and Staining* in Vacuo *with Vapors of Mercury (II) Hexafluoroacetylacetonate (Hg hfac), Followed by Staining with GCA (Figs. 3.14–3.17).* The background density in the nucleolus, nucleoplasmic walls, and the RER is prominent. The appearances of second-order DNA coils in the chromatin, and of granules in the nucleolar and nucleoplasmic walls and in the RER

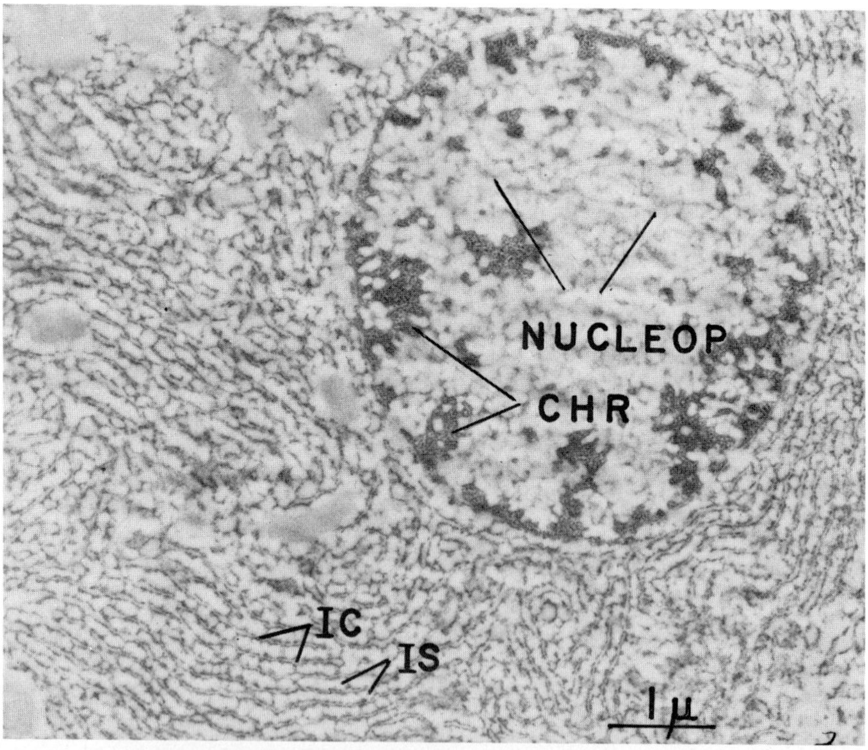

FIGS. 3.6–3.11. Mouse pancreas, postfixation and staining *in vacuo* with vapors of difluordinitrobenzene (FFDNB), followed by staining with GCA.

FIGS. 3.6 and 3.7. Figure 3.6 is of the basal portion of the cell; Fig. 3.7 of the luminal portion. The RER is better preserved than after postfixation with alcohol. The intracisternal (IC) and intercisternal (IS) spaces are more uniform. The braching, spongelike Golgi structures (GOLGI STRUC) have little stainable material. The chromatin masses (CHR) in the nucleus are denser than the nucleolus (NUCLEOL), which in turn is denser than the nucleoplasmic walls (NUCLEOP). Spaces in the nucleolus and nucleoplasm are continuous with those of the chromatin. Other labels are given previously. × 14,000.

are clearly retained, though frequently blurred by the homogeneous stainable component in which they are embedded.

6. *Postfixation and Staining* in Vacuo *with Vapors of Hg hfac Alone (Fig. 3.18).* In general, the appearance is the same as that after postfixation and staining with vapors of FFDNB alone. The chief differences are that the chromatin in the nucleus and the RER in the cytoplasm have more contrast. Specimens treated with ozone could not be distinguished from those postfixed with vapors of the mercury chelate without prior oxidation.

7. *Effects of Ribonuclease (RNase) and Deoxyribonuclease (DNase) on Staining with GCA (Figs. 3.19–3.24).* Freeze-dried specimens were postfixed *in vacuo* with vapors of Hg hfac and by subsequent immersion in alcohol. Some specimens were then stained with GCA. The remainder were then transferred through a graded series of alcohol to water. Some were then treated with RNase (Worthington, twice recrystallized), 1 mg/ml in distilled water for 6 hours at 28°C and others with DNase (Worthington, DCL-11S), 1 mg/ml in distilled water for 6 hours at 28°C with sixteen changes of solution. Of these, some were immersed in water

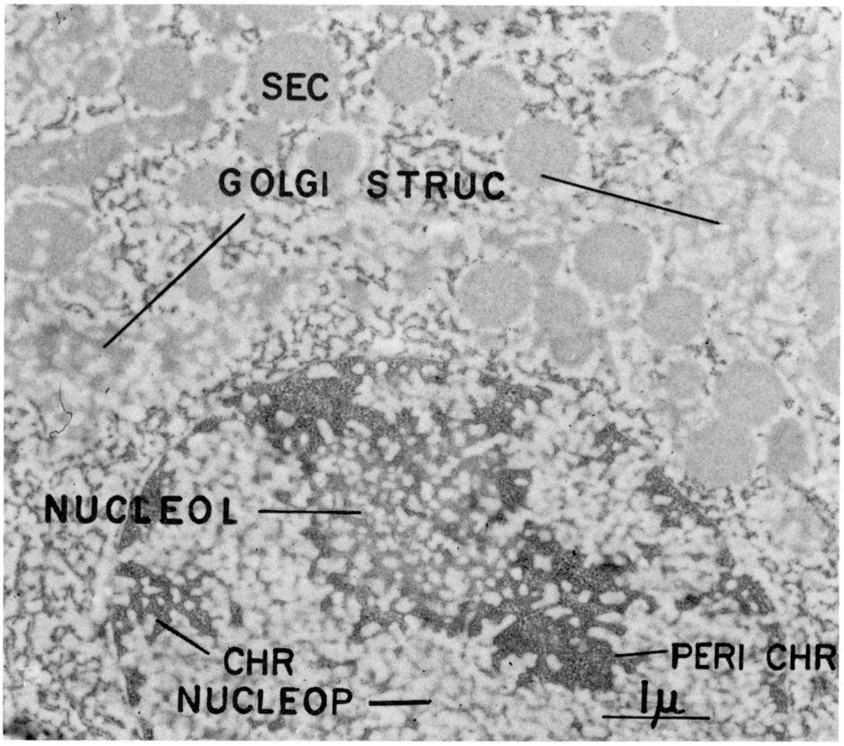

Fig. 3.7.

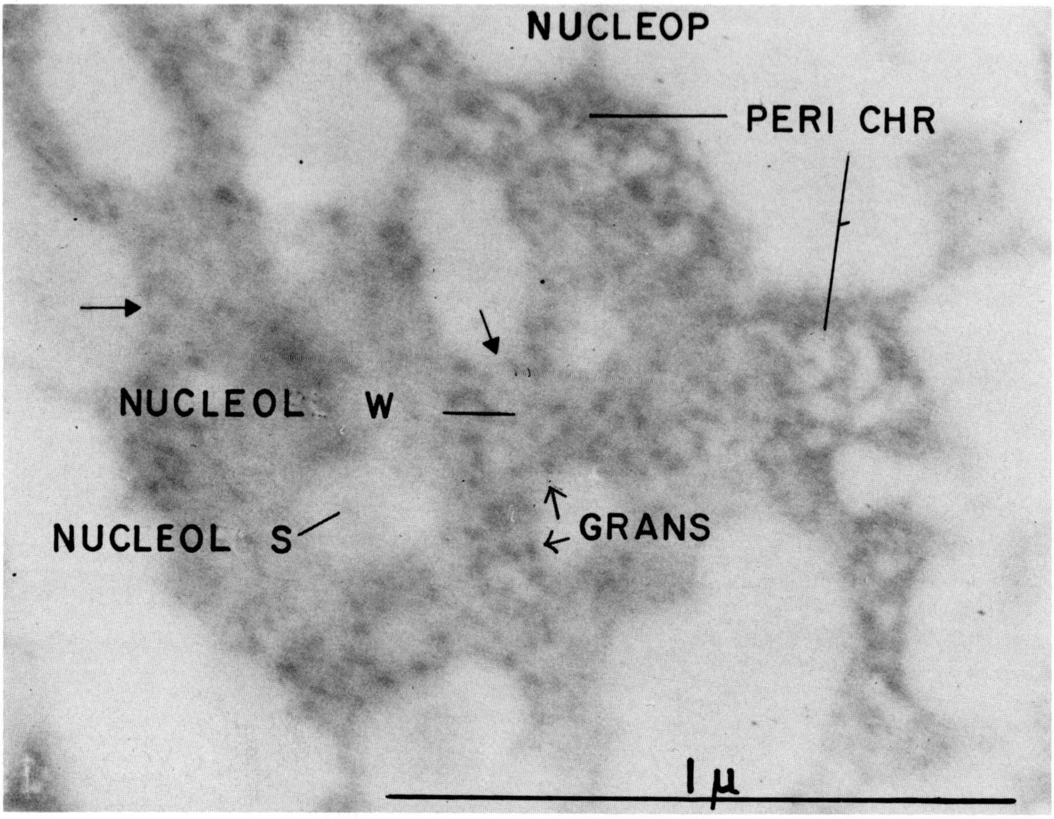

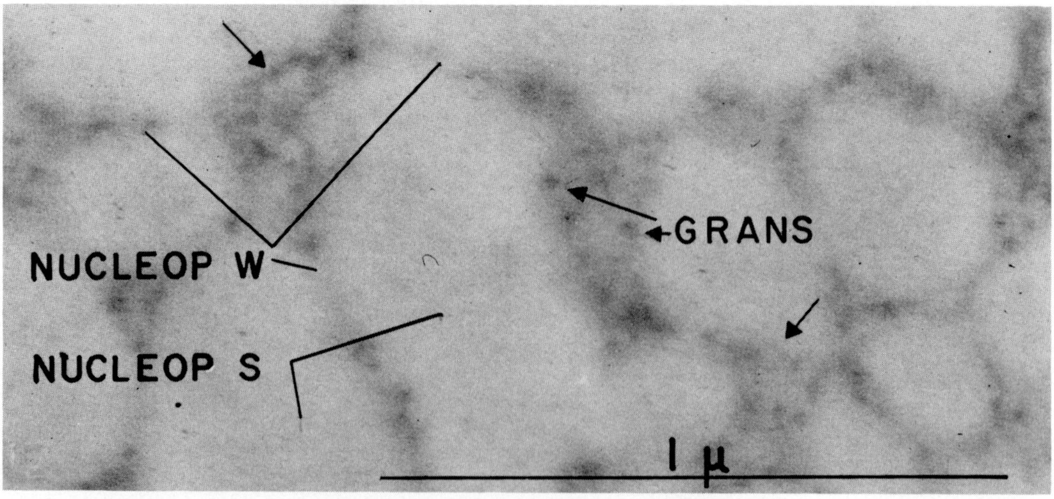

FIGS. 3.8 (TOP) AND 3.9 (BOTTOM). Nucleolus, perinucleolar chromatin (PERI CHR), and nucleoplasm. Granules in the walls of the nucleolus and nucleoplasm are resolved with difficulty because of the background density, and only the larger granules (GRANS) can be seen. The perinuclear chromatin (PERI CHR) comprises granules and coils cut in various planes. Arrows with solid heads point to coils. × 82,000.

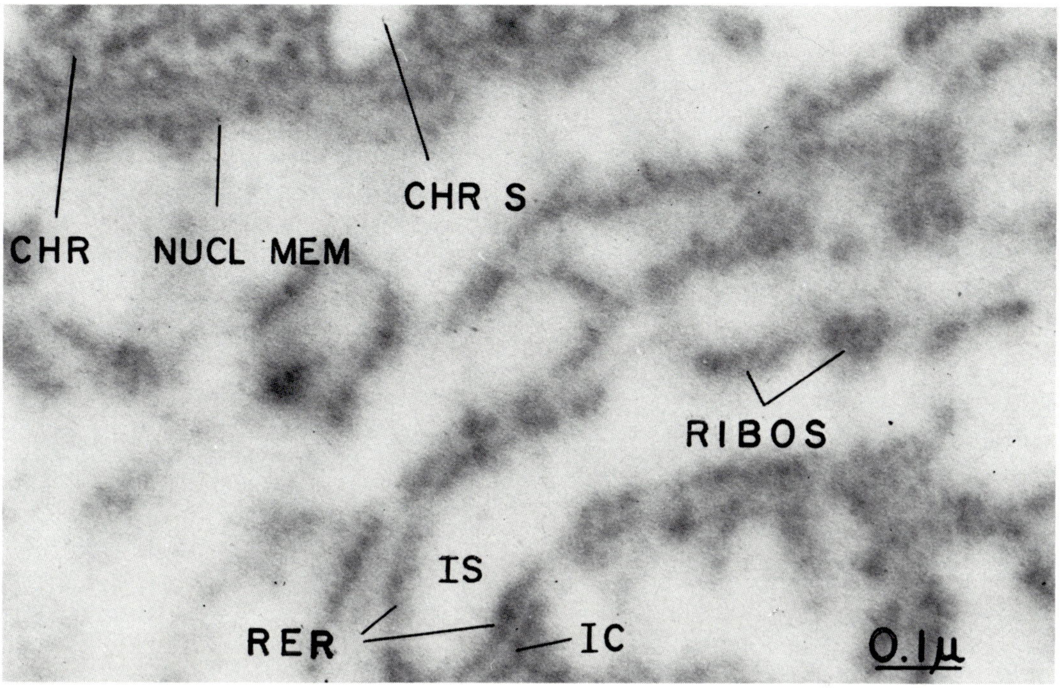

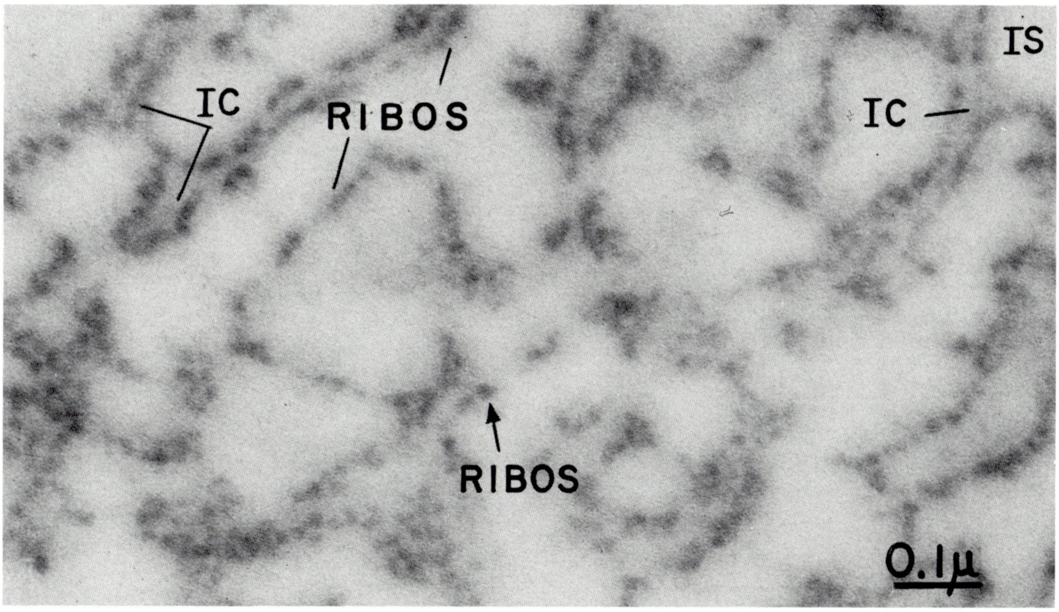

FIGS. 3.10 (TOP) AND 3.11 (BOTTOM). Ribosomes (RIBOS) of the rough endoplasmic reticulum (RER) are clustered in groups. In some parts they are not readily resolvable because of the moderately or equally dense background. Even where clearly resolvable, they lie in a homogeneous background. Granules of the outer nuclear membrane (NUC MEM) are dense. Sectioned coils of chromatin (CHR) and chromatin spaces (CHR S) are also visible. The intracisternal space (IC) is uniformly narrower than the intercisternal space (IS). × 118,000.

alone, as zero enzyme controls. All were washed briefly, dehydrated gently, and hardened for 1 day in 95% alcohol. These were then stained in the usual manner with GCA and embedded in w-s Durcupan. Ultrathin sections were examined with the electron microscope, and thicker sections (about 0.25 μ) were mounted on glass slides and stained with toluidine blue. As indicated earlier, the morphology seems to be retained better after these procedures than after postfixation in alcohol alone, probably because of the preliminary cross-linking, but is still not altogether satisfactory.

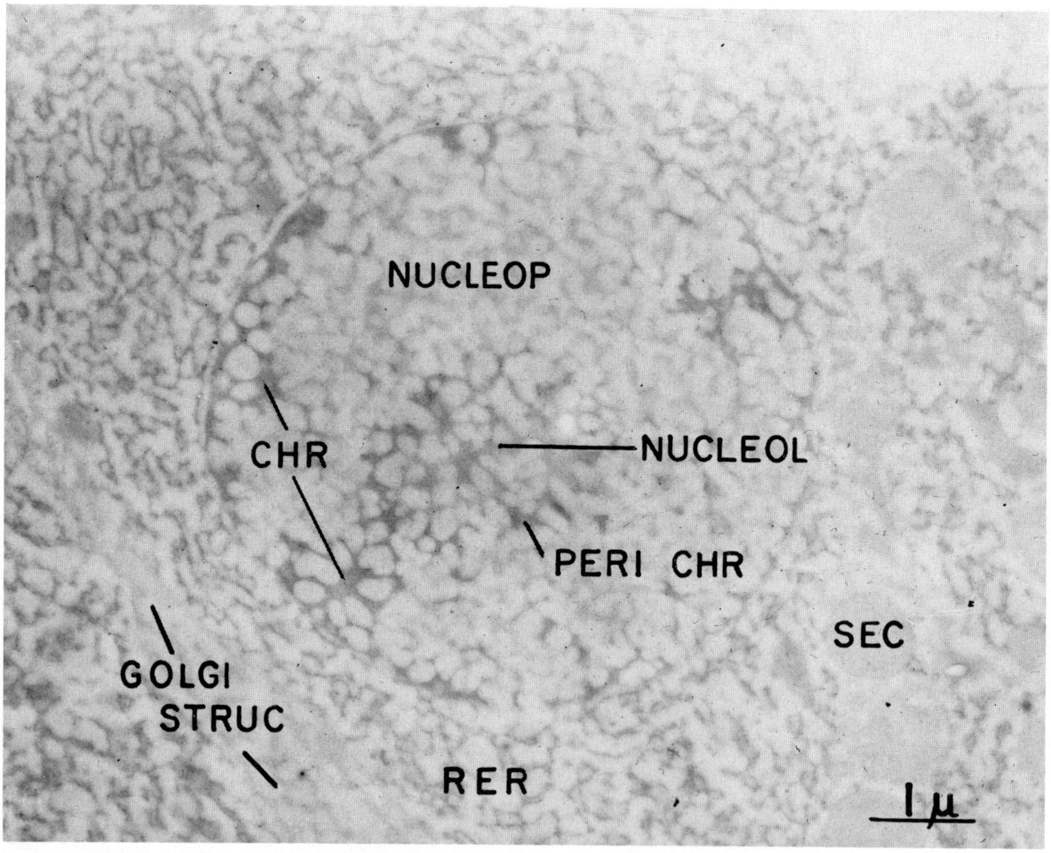

FIGS. 3.12 and 3.13. Postfixation and staining *in vacuo* with vapors of FFDNB alone; no GCA. As compared with specimens postfixed with alcohol, and not subsequently stained, the density of all protoplasmic components has increased, but not selectively. The rough endoplasmic reticulum (RER) is not clearly distinguished by density from the Golgi apparatus (GOLGI STRUC) and from the secretory granules (SEC). The intracisternal space (IC) is rather uniform. Ribosomes (RIBOS) are not clearly defined, and are grouped. Differences in density in the walls of nuclear structures are not as clear as after alcohol postfixation. The chromatin (CHR, PERI CHR) of the nucleus appears uniformly dense. Other abbreviations as given previously. Figure 3.12, × 14,000; Fig. 3.13, × 82,000.

As compared with zero enzyme control preparations, which are fairly satisfactory (Fig. 3.19), the effects of RNase as observed with the electron microscope are to reduce contrast markedly in the nucleolus though not in the RER; DNA in the chromatin masses is largely retained (Fig. 3.20). This is in contradistinction with the complete removal of nucleolar and ergastoplasmic stain when enzyme-treated (Fig. 3.23) and water control (Fig. 3.22) specimens are compared with the light microscope. The effects of DNase are to remove the DNA of the chromatin masses, while retaining much of the nucleolar and ribosomal density, as observed with the electron microscope (Fig. 3.21). This also is in contradistinction with the apparently complete retention of nucleolar and ergastoplasmic stain, while the chromatin (DNA) staining is abolished (Fig. 3.24). The discrepancy between the observations of sections of the same specimen block observed with the light and electron microscope have been discussed on p. 55.

An interesting set of observations was made on ultrathin sections mounted on grids and then treated with an RNase solution. The sections were of pancreas postfixed with Hg hfac and subsequently stained with GCA and embedded in water-soluble Durcupan. After treatment of the sections for 30 minutes to 1 hour, the background density in the walls of the nucleolus and nucleoplasm, as well as in the RER, was significantly reduced, so that the granules in all three sites were more clearly detectable than in equivalent sites of control sections treated for the same times in water only. The diffuse RNA component fixed, even after vapor fixation with cross-linking reagents and staining with GCA, is selectively digested by the action of RNase. DNase is as ineffective in removing the diffuse RNA component as is water. This sensitivity of RNA to RNase even after treatment with

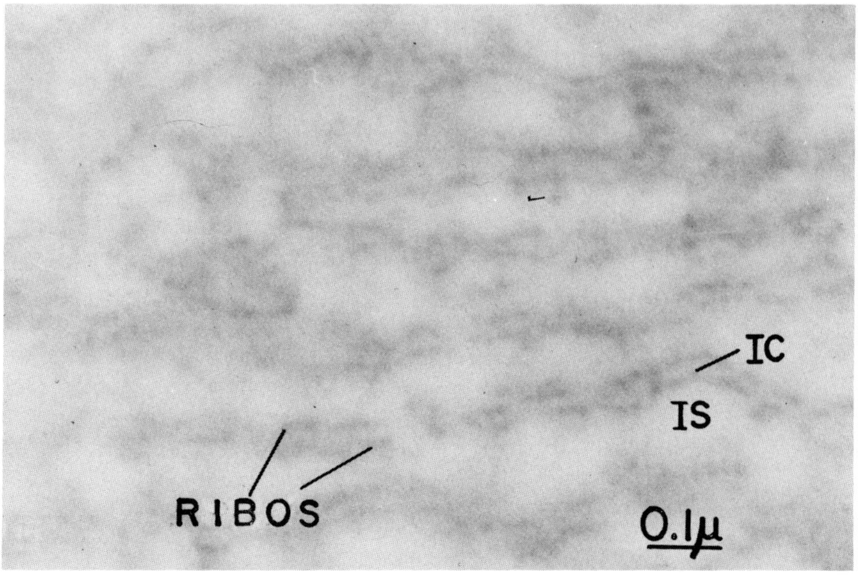

Fig. 3.13.

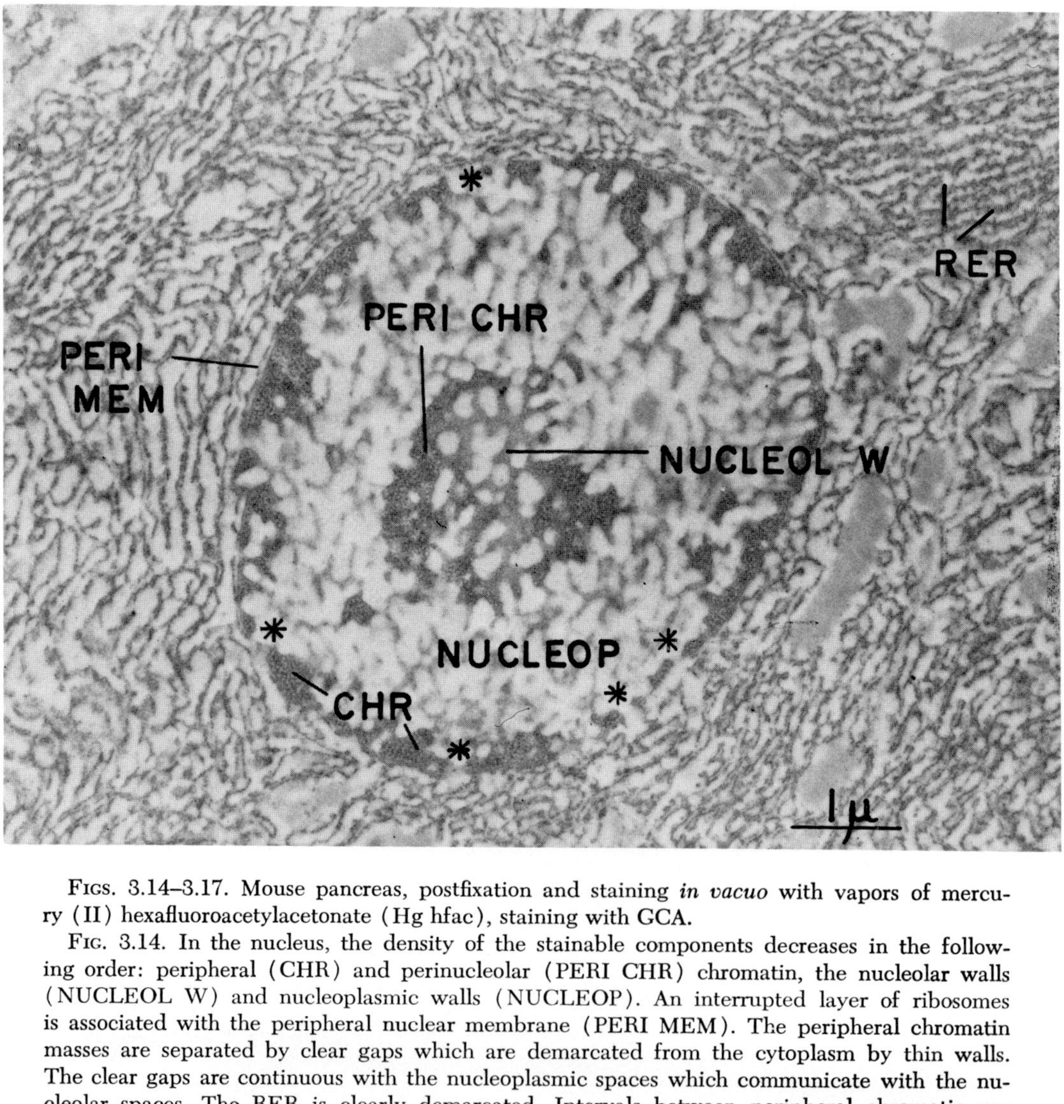

Figs. 3.14–3.17. Mouse pancreas, postfixation and staining *in vacuo* with vapors of mercury (II) hexafluoroacetylacetonate (Hg hfac), staining with GCA.

Fig. 3.14. In the nucleus, the density of the stainable components decreases in the following order: peripheral (CHR) and perinucleolar (PERI CHR) chromatin, the nucleolar walls (NUCLEOL W) and nucleoplasmic walls (NUCLEOP). An interrupted layer of ribosomes is associated with the peripheral nuclear membrane (PERI MEM). The peripheral chromatin masses are separated by clear gaps which are demarcated from the cytoplasm by thin walls. The clear gaps are continuous with the nucleoplasmic spaces which communicate with the nucleolar spaces. The RER is clearly demarcated. Intervals between peripheral chromatin are marked by an asterisk (*). × 14,000.

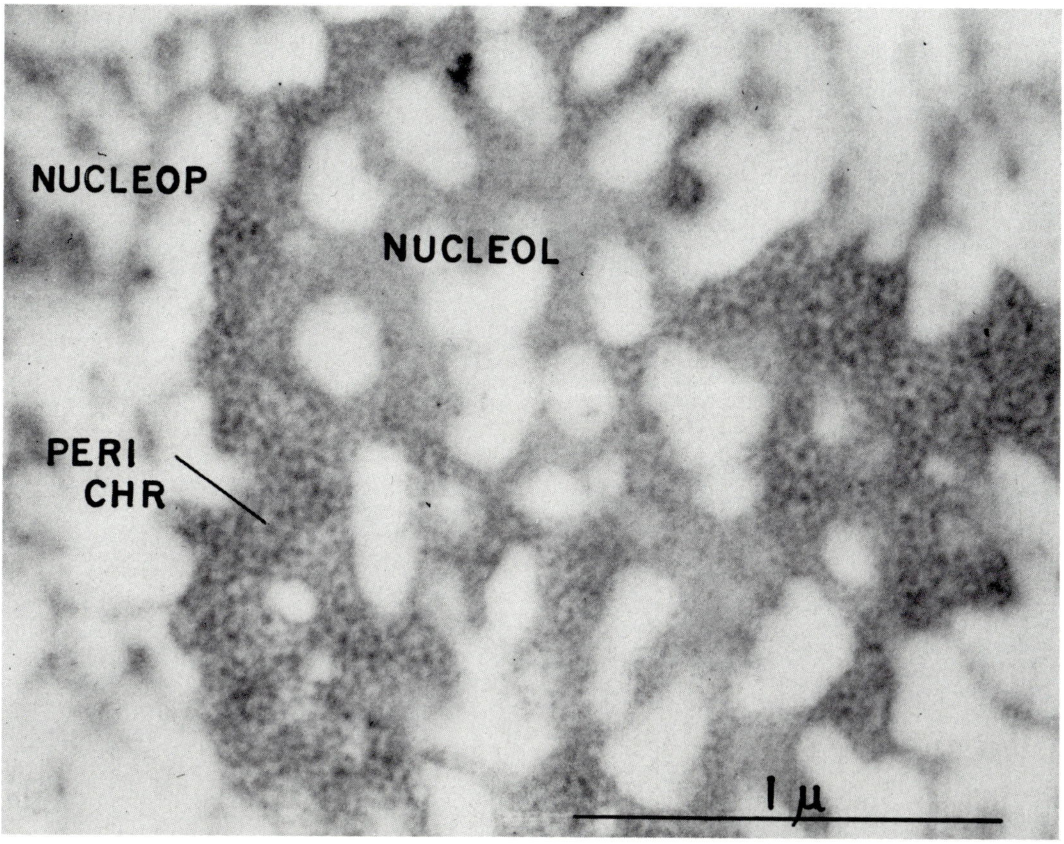

FIG. 3.15. Nucleolus (NUCLEOL), perinucleolar chromatin (PERI CHR), and nucleoplasm (NUCLEOP). The background density in the walls of the nucleolus and nucleoplasm makes it difficult to resolve the finest granules. In the chromatin, the DNA occurs as coils in various planes and angles. × 56,000.

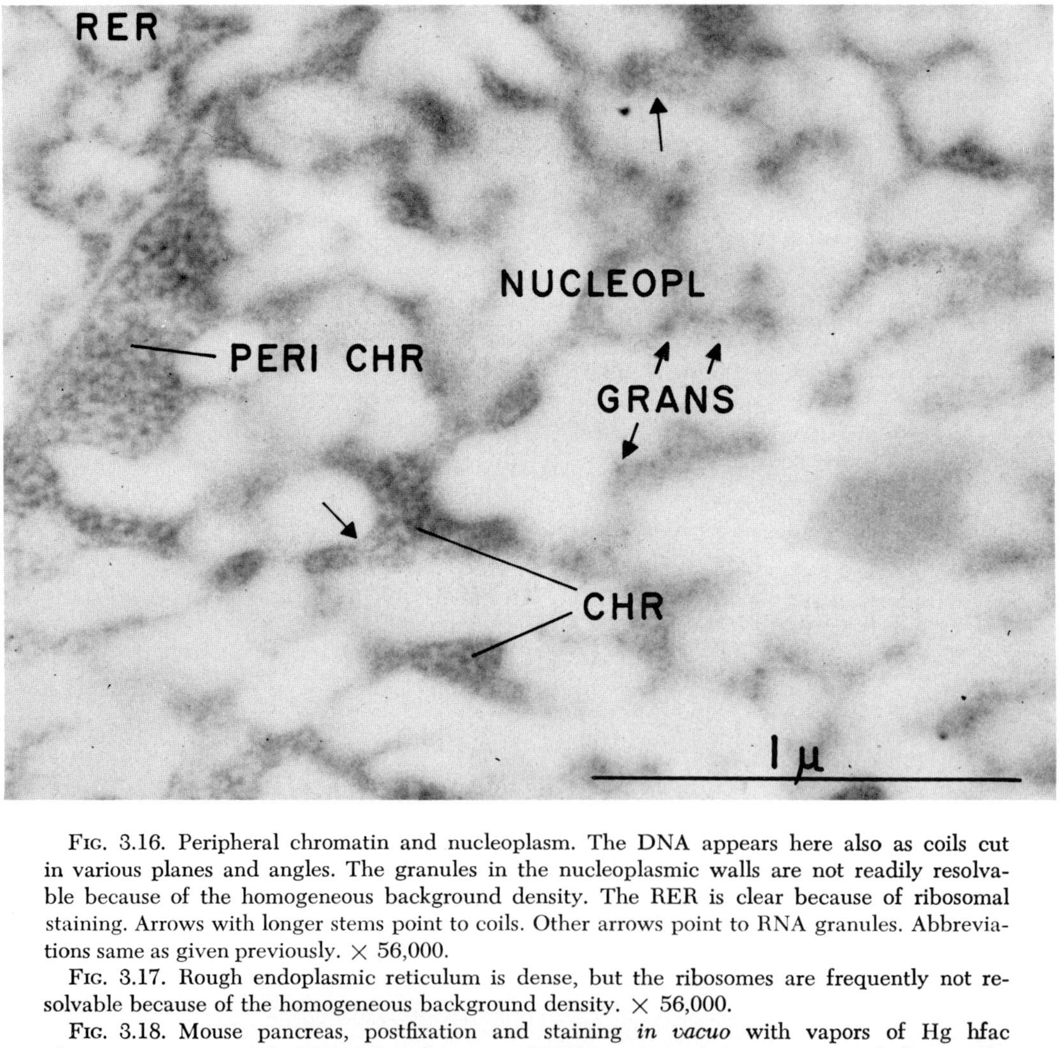

FIG. 3.16. Peripheral chromatin and nucleoplasm. The DNA appears here also as coils cut in various planes and angles. The granules in the nucleoplasmic walls are not readily resolvable because of the homogeneous background density. The RER is clear because of ribosomal staining. Arrows with longer stems point to coils. Other arrows point to RNA granules. Abbreviations same as given previously. × 56,000.

FIG. 3.17. Rough endoplasmic reticulum is dense, but the ribosomes are frequently not resolvable because of the homogeneous background density. × 56,000.

FIG. 3.18. Mouse pancreas, postfixation and staining *in vacuo* with vapors of Hg hfac alone; no staining with GCA. The chromatin (CHR) appears homogeneous and dense. The walls of the nucleoplasm (NUCLEOP W) are also uniform, though less dense. The RER is rather dense. PERI MEM, peripheral nuclear membrane. × 56,000.

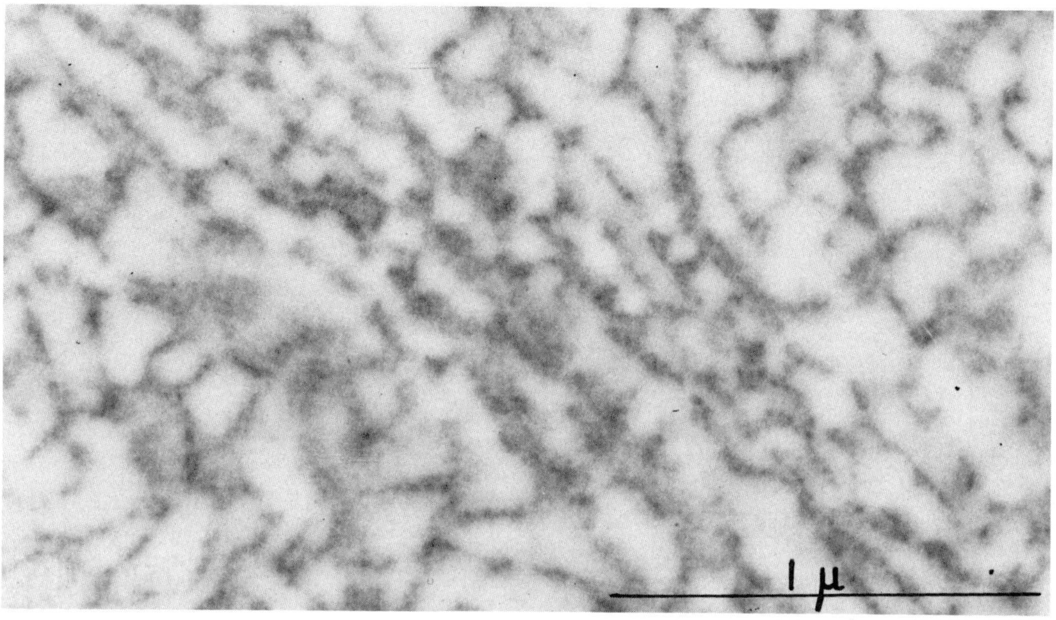

FIG. 3.17.

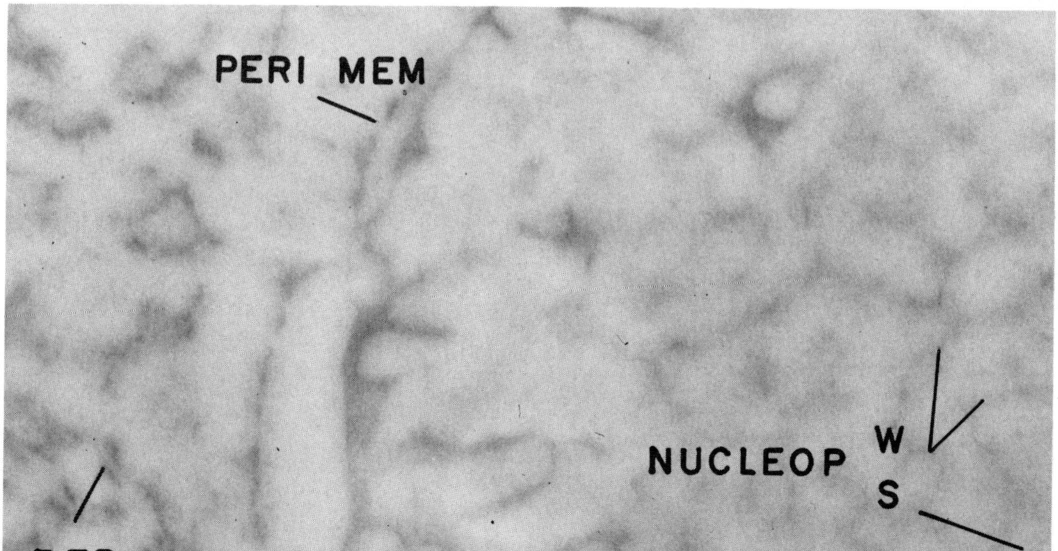

FIG. 3.18.

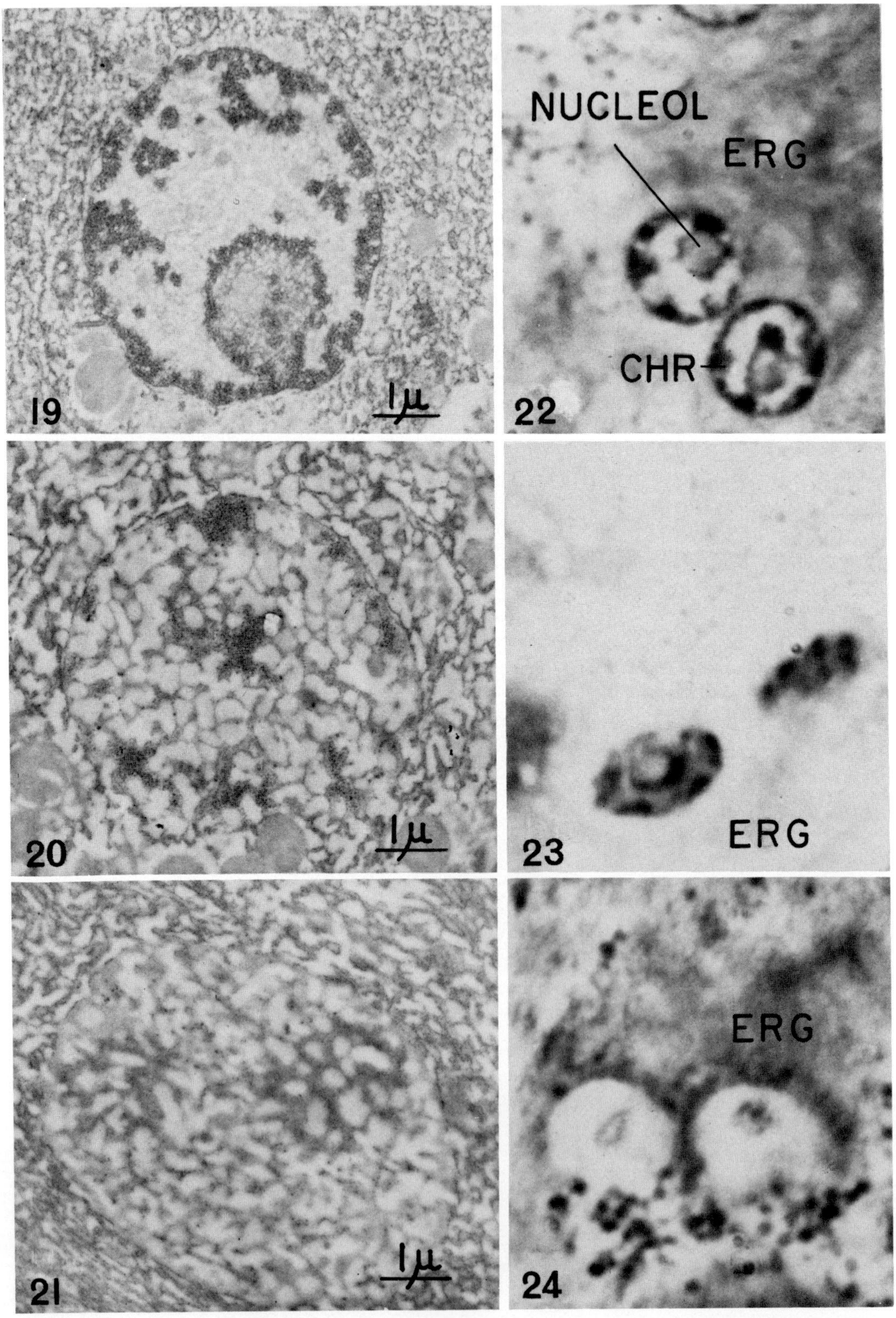
19
1μ
20
1μ
21
1μ
NUCLEOL
ERG
CHR
22
23
ERG
ERG
24

cross-linking vapors and gallocyamin-chromalum is not really contrary to the findings of Leduc and Bernhard *(31)*, as their specimens probably were devoid of sRNA after the action of the fluid fixative. Further observations on tRNA and mRNA are described in the Appendix to this chapter.

The pertinent differences of the various methods of postfixation and staining of the pancreas are summarized in Table 3.1, p. 84.

Liver

1. *Postfixation in Alcohol, Staining with GCA (Figs. 3.25–3.27).* The nuclear chromatin is distributed in small masses throughout the nucleus, especially at the periphery of the nucleus and nucleolus. The nucleolus stains less deeply. Still paler are the netlike cords which appear in electron micrographs as a reticulum, but in three dimensions, which probably represent interconnected columns of material which is spongelike. The spaces between the walls are clear, and communicate with similar spaces which penetrate the nucleolus and larger chromatin masses. These less dense spaces and their more dense walls extend to the nuclear membrane. The chromatin masses consist chiefly of numerous dense granules about 200 A or more in diameter, though some filaments may be seen.

The walls of the nucleoli are packed with numerous small granules which range from about 100 A in diameter to about the limit of resolution. Similar granules are present also in the walls of the nucleoplasm, where they are not as numerous. The background (unresolved) density is minimal in both regions. In the cytoplasm, the ribosomes are clearly stained, and these extend continuously along the RER, especially in the region of mitochondria.

2. *Postfixation in Alcohol, Unstained (Fig. 3.28).* Cell structures have less contrast than in stained preparations. Throughout the nucleus, the denser walls

Figs. 3.19–3.21. Mouse pancreas. Electron micrographs of sections of specimens treated with enzymes as in Figs. 3.22–3.24. × 11,000.

Fig. 3.19. Water control, no enzyme treatment. The nucleolus, nucleoplasm, and chromatin of the nucleus as well as the RER are clearly distinguishable.

Fig. 3.20. Ribonuclease treatment. The density of the nucleolus is reduced; the chromatin masses are almost as dense as in the control specimens.

Fig. 3.21. Deoxyribonuclease treatment. The density of the RER is perhaps slightly reduced, as compared with the controls, but the density of the nucleolus has not been reduced appreciably. The chromatin masses are not stained.

Figs. 3.22–3.24. Mouse pancreas. Photomicrographs of sections of specimens fixed *in vacuo* with vapors of Hg hfac, treated with DNase and RNase, stained with GCA, and embedded in Durcupan. Sections were mounted on slides and subsequently stained with toluidine blue. × 3200.

Fig. 3.22. Water control, no enzyme treatment. Nucleolus (NUCLEOL), chromatin (CHR), and ergastoplasm (ERG) are stained.

Fig. 3.23. Ribonuclease treatment. Nucleolar and cytoplasmic basophilia are notably reduced, while the chromatin is deeply stained.

Fig. 3.24. Deoxyribonuclease treatment. Nucleolar and cytoplasmic basophilia persist, while the staining of the chromatin is abolished.

are clearly distinguished from the less dense channels. In the chromatin masses and nucleoli, the walls are thicker and denser than in the nucleoplasm. No granules can be discerned in any of these structures. In the cytoplasm, the mitochondria are enclosed by the denser RER which also extends as columns between them. Throughout the RER, ribosomes are clearly resolvable.

3. *Postfixation and Staining* in Vacuo *with Vapors of Difluorodinitrobenzene (FFDNB), Followed by Staining with GCA (Figs. 3.29–3.31).* The distribution

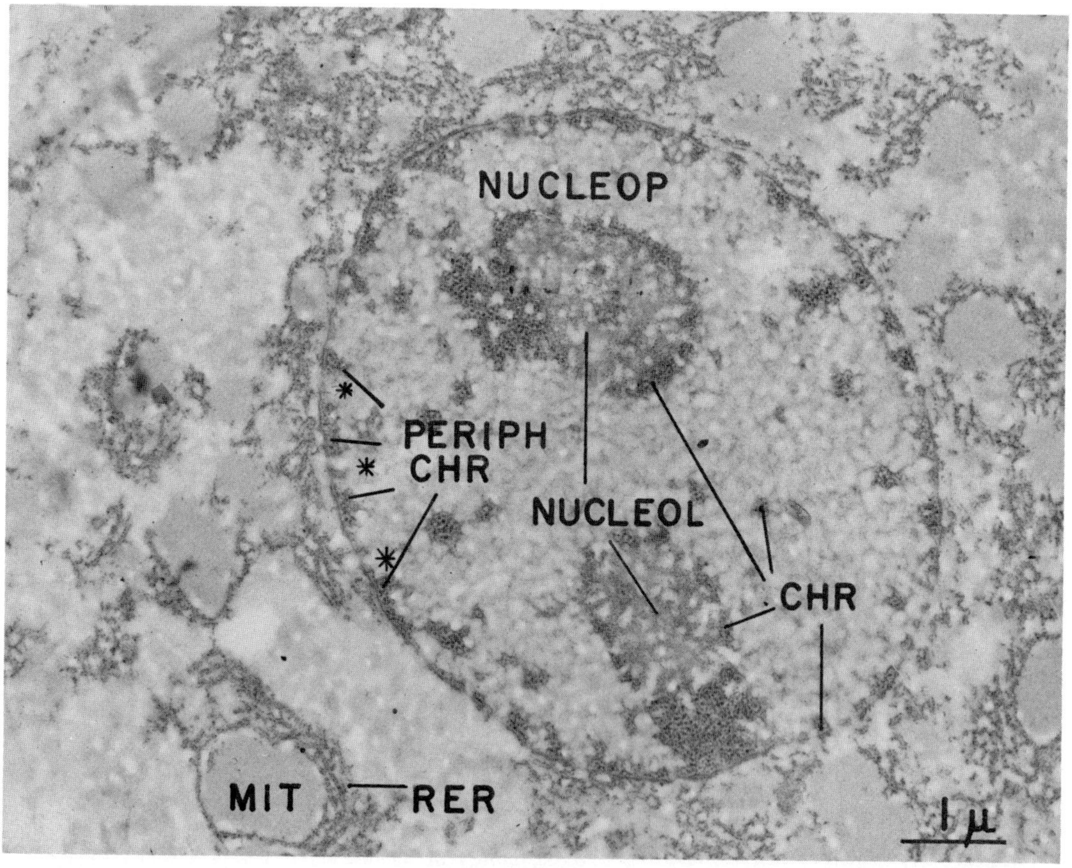

FIGS. 3.25–3.27. Mouse liver, postfixation in 95% alcohol, staining with GCA. The sponge-like structure of the nucleus is reflected by the denser walls which lie between the less dense channels in the nucleoplasm (NUCLEOP), nucleolus (NUCLEOL), and the chromatin (CHR). The "spaces" extend to the surface of the nucleus, between the peripheral chromatin masses (PERI CHR); the spaces are marked with an asterisk (*). Granules containing RNA are numerous in the nucleolus, and less numerous in the nucleoplasmic walls (GRANS). The DNA appears chiefly as very dense granules although some filaments are also present. In the rough endoplasmic reticulum, which is especially rich around the mitochondria (MIT), the ribosomes are discrete. All the granules have high contrast because of the absence of background density. Figure 3.25, × 14,000; Figs. 3.26 and 3.27, × 56,000.

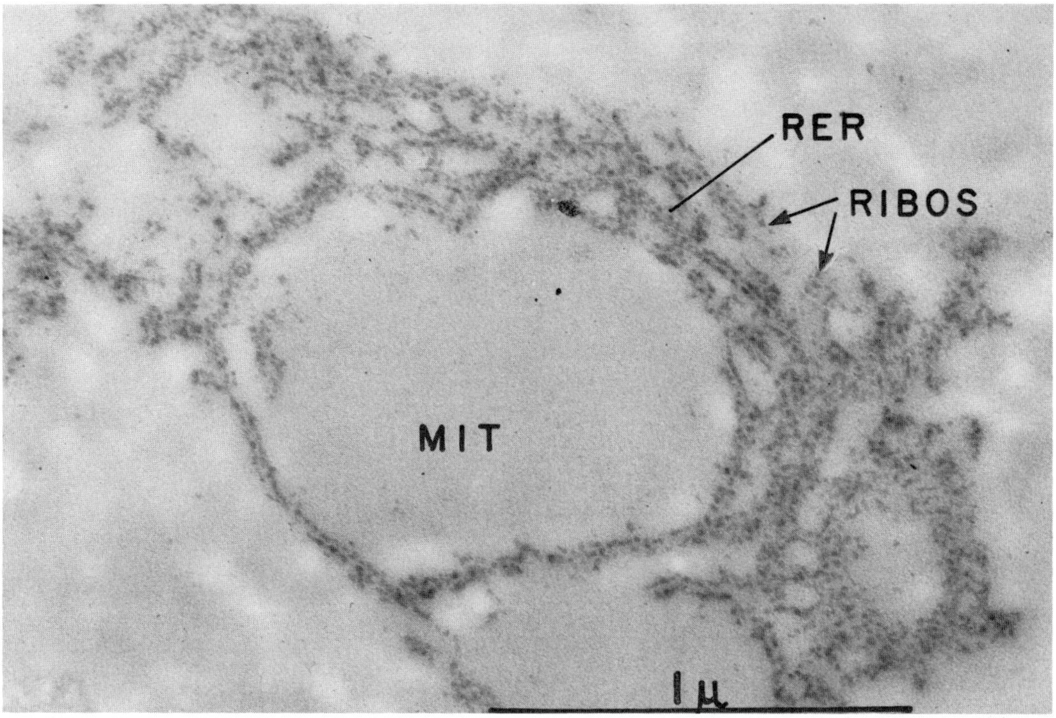

FIG. 3.26. ▲ ▼ FIG. 3.27.

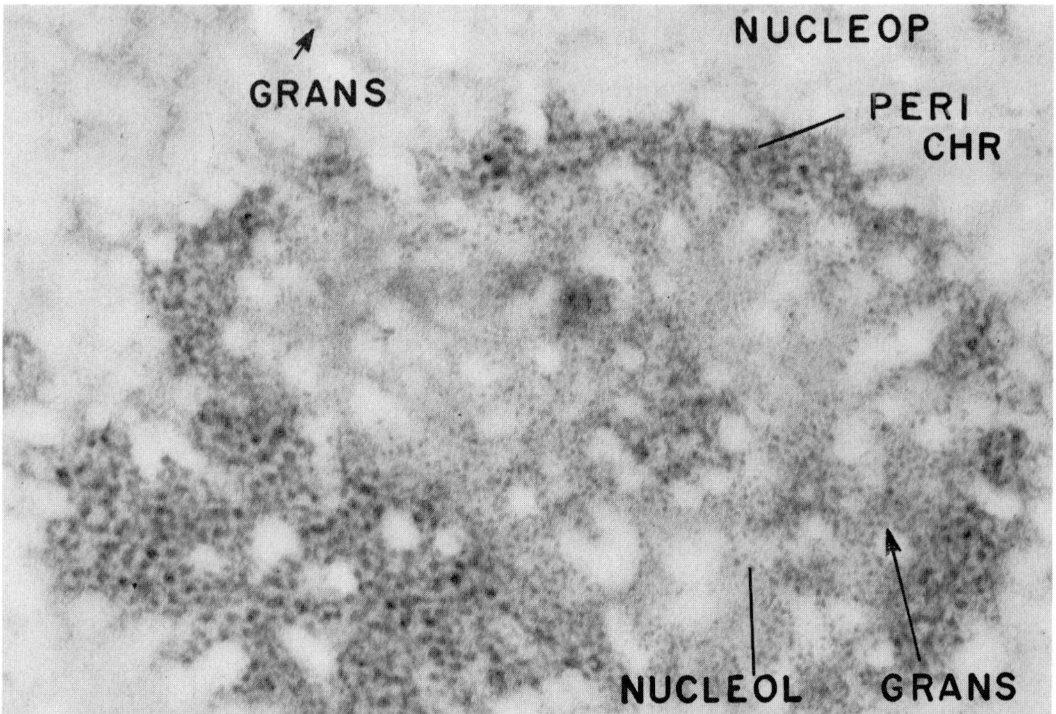

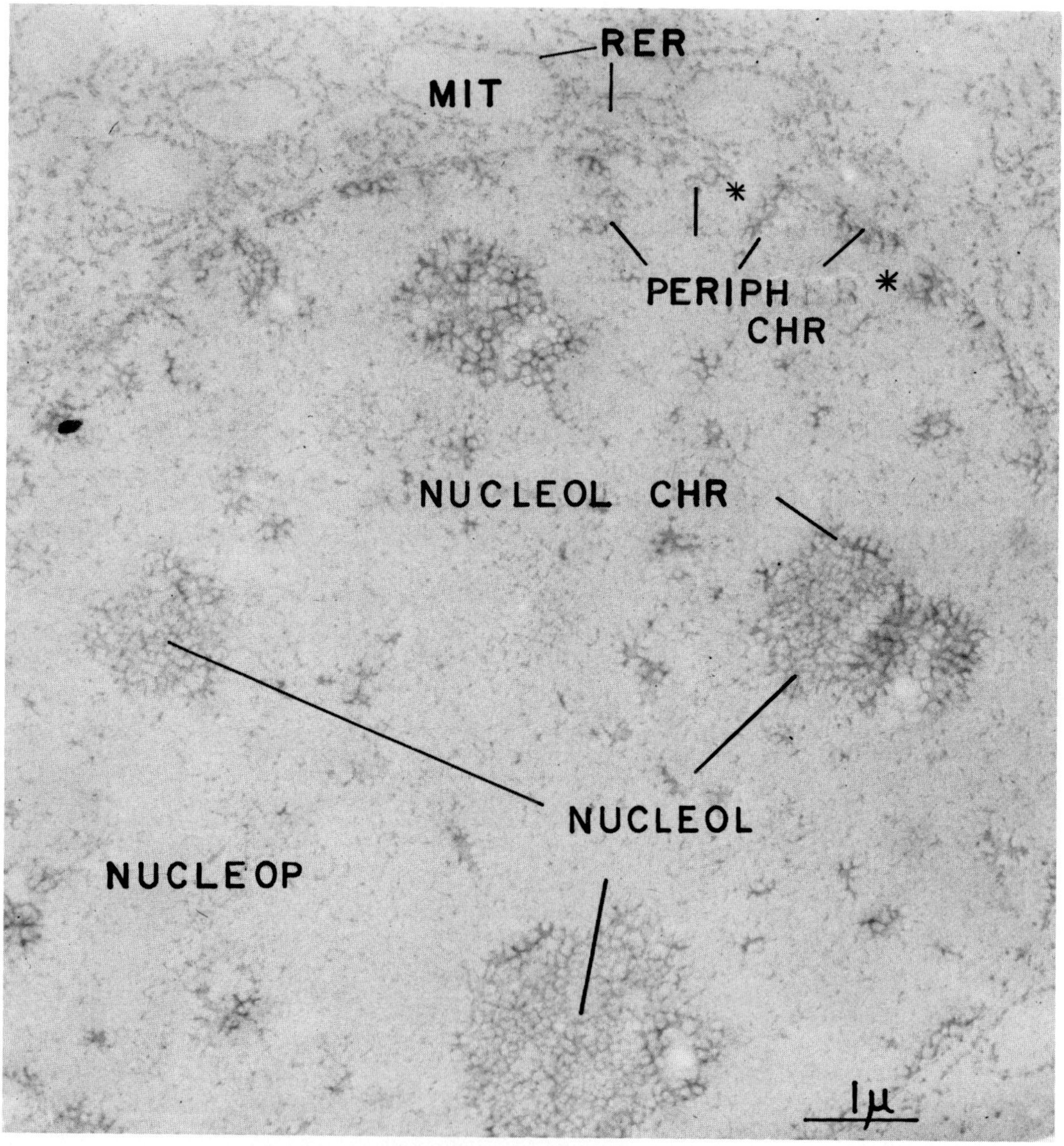

FIG. 3.28. Mouse liver, post fixation in 95% alcohol, no stain. As compared with Figs. 3.25–3.27, there is less contrast between the various structures. The dense walls and less dense channels of the spongelike structure of the nucleoplasm (NUCLEOP) are clear and continuous with equivalent structures of the nucleolus (NUCLEOL) and the prominent walls of chromatin masses (PERIPH CHR, NUCLEOL CHR). The nucleoplasm extends between peripheral chromatin masses (PERIPH CHR) to the nuclear membrane. Neither granules nor coils are detectable in the nucleus. In the RER, the walls are denser than those in other portions of the cytoplasm; but ribosomes are not readily distinguished. Mitochondria (MIT) are less dense. × 14,000.

of dense material superficially resembles that of specimens postfixed in alcohol without subsequent staining, with the difference that there seems to be more contrast (though with a general diffuse appearance) than after alcohol alone. In the nucleoplasm, granules of about 100 A or less in diameter are present in the denser cords, which have a diffuse unresolvable background density. The small granules are crowded together in the nucleolus, which also has a background density. There are relatively few granules in the chromatin, as most of the circular, coiled,

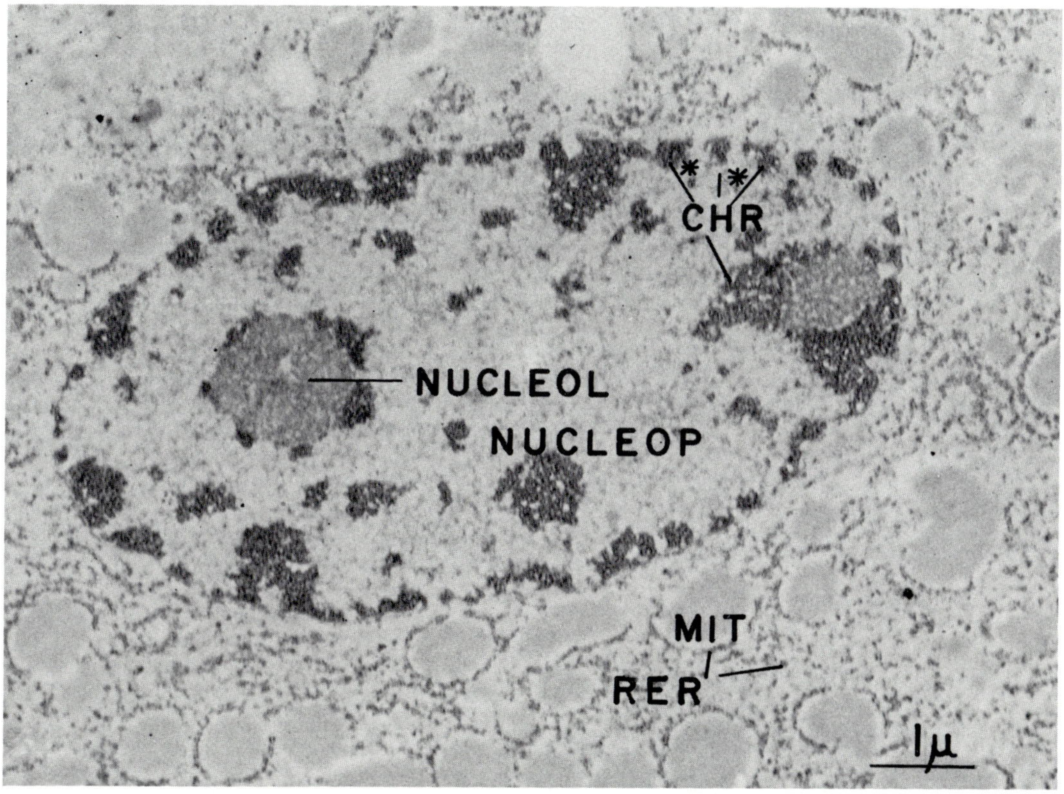

FIG. 3.29.

FIGS. 3.29–3.31. Mouse liver, postfixation and staining *in vacuo* with vapors of FFDNB, followed by staining with GCA. The nucleoli and chromatin masses are denser than the walls of the nucleoplasm. The walls of the nucleoplasm contain numerous small granules embedded in a homogeneous background density. The granules are more numerous in the nucleolus than in the nucleolar walls. The chromatin masses are resolvable into numerous coils and figures which are based on fibrils about 45 A in diameter which form secondary coils of a diameter of about 400 A. Some granules are also present, but their interpretation is uncertain. In the cytoplasm, the ribosomes of the RER are grouped in clusters and are not as clearly resolved as in Figs. 3.25–3.27. This is attributed to the presence of a background or unresolvable density which surrounds the ribosomes. The unresolvable density is attributable to soluble RNAs which have been preserved for subsequent staining by the cross-linking agent. Arrows with solid heads point to coils. All abbreviations as given earlier. Figure 3.29, × 14,000; Figs. 3.30 and 3.31, × 56,000.

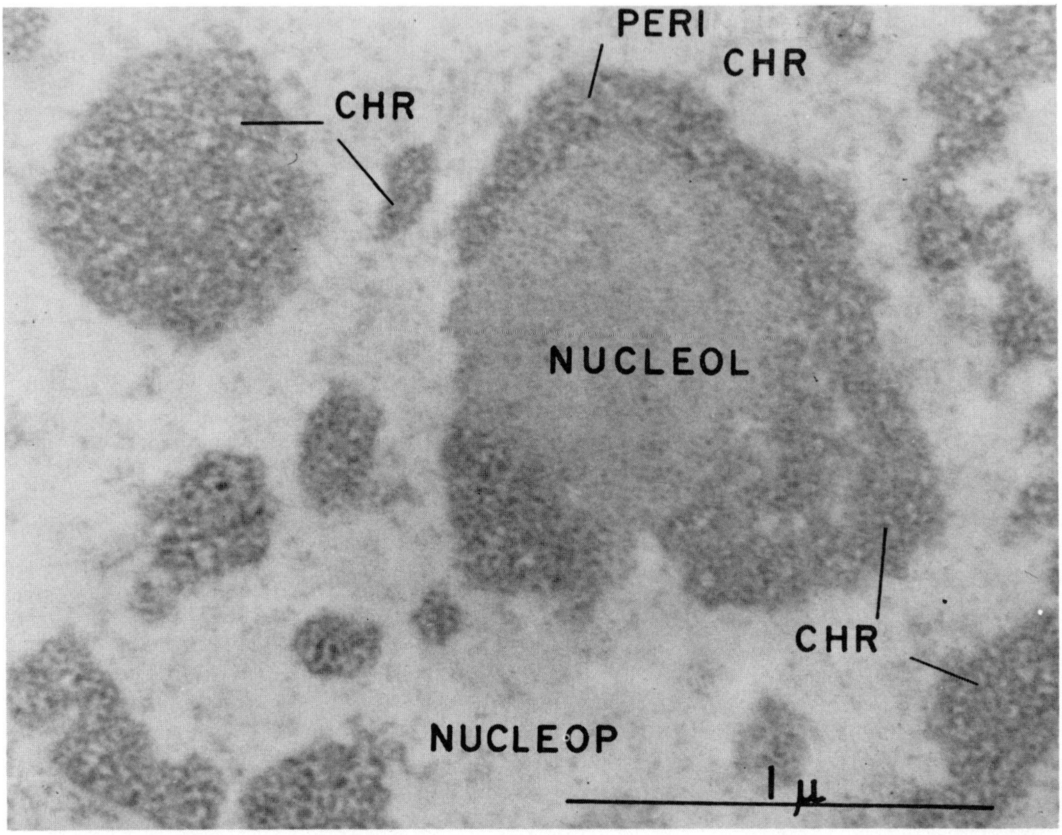

FIG. 3.30. ▲

▼ FIG. 3.31.

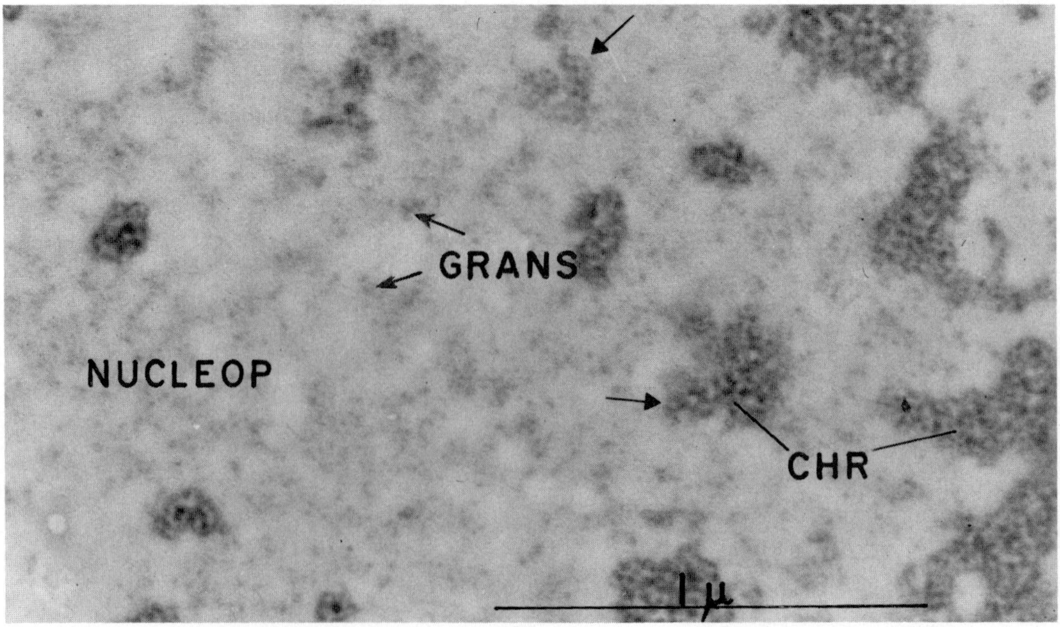

and finely granular structures can be interpreted in terms of second-order coils with a periodicity of about 400 A. The clear spaces of the nucleoplasm, nucleoli, and chromatin are continuous and extend to the surface of the nucleus. The ribosomes of the RER are frequently connected by a generally increased and unresolvable dense material, so that they are sometimes not as sharply delineated as after postfixation in alcohol.

4. *Postfixation and Staining* in Vacuo *with Vapors of FFDNB Alone (Fig. 3.32).* Specimens have less contrast than those fixed in alcohol alone. The density of the chromatin masses and nucleoli does not much exceed that of the nucleoplasm. In the cytoplasm, the density of the mitochondria and cytoplasm between the cisternae of the RER is not markedly different from the density of the ribosomes of the RER. This relatively uniform density of all parts of the protoplasm

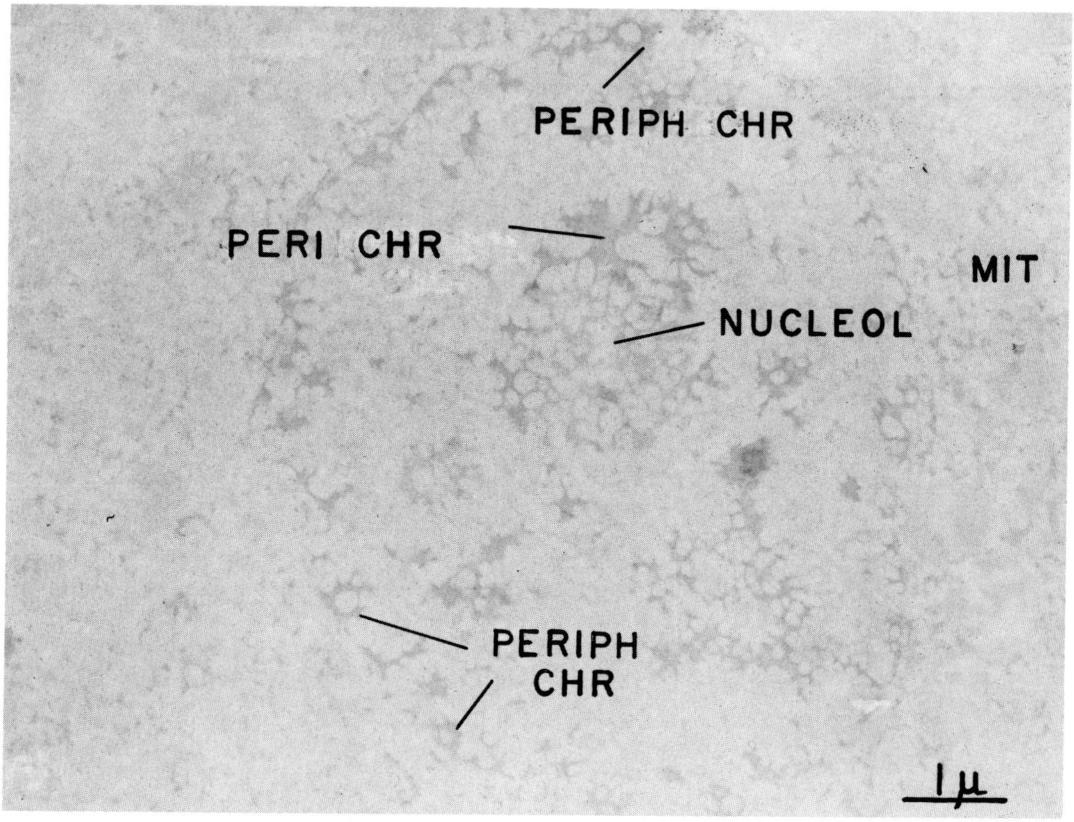

FIG. 3.32. Mouse liver, postfixation and staining *in vacuo* with vapors of FFDNB; no staining with GCA. The normally higher, intrinsic contrast of structures containing nucleic acids is reduced by the enhanced contrast of protein-rich structures which have reacted with the FFDNB. Thus the chromatin, nucleoli, and ribosomal portion of the RER are not clearly detectable. Abbreviations as given previously. × 14,000.

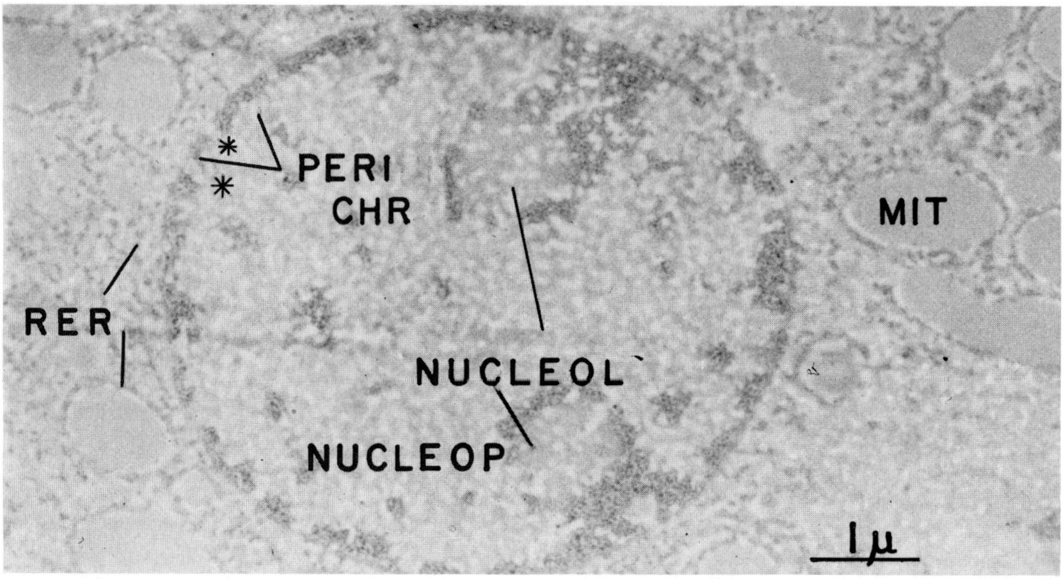

FIG. 3.33. ▲ ▼ FIG. 3.34.

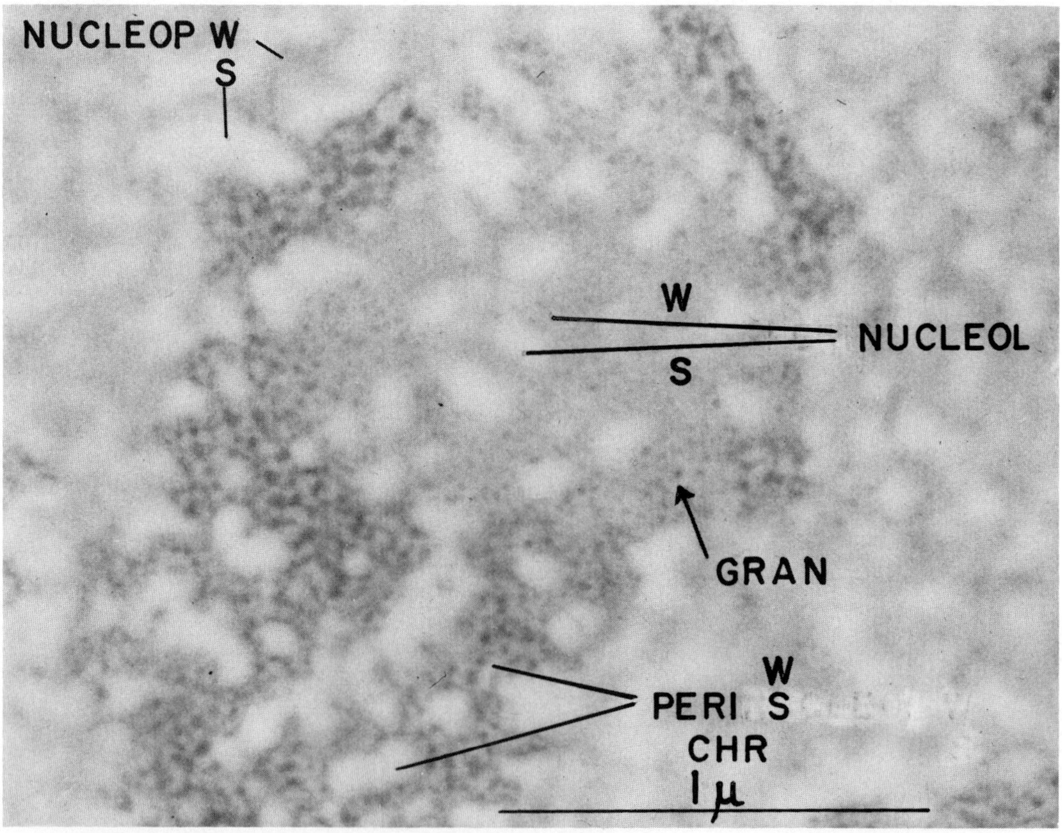

FIGS. 3.33–3.35. Mouse liver, postfixation and staining *in vacuo* with vapors of Hg **hfac**, followed by staining with GCA. These electron micrographs closely resemble those of specimens treated with FFDNB and also stained with GCA, in Figs. 3.29–3.31. There is some shrinkage in the cytoplasm of Fig. 3.33, especially notable around mitochondria (MIT). Abbreviations as indicated previously. Figure 3.33, × 14,000; Figs. 3.34 and 3.35, × 56,000.

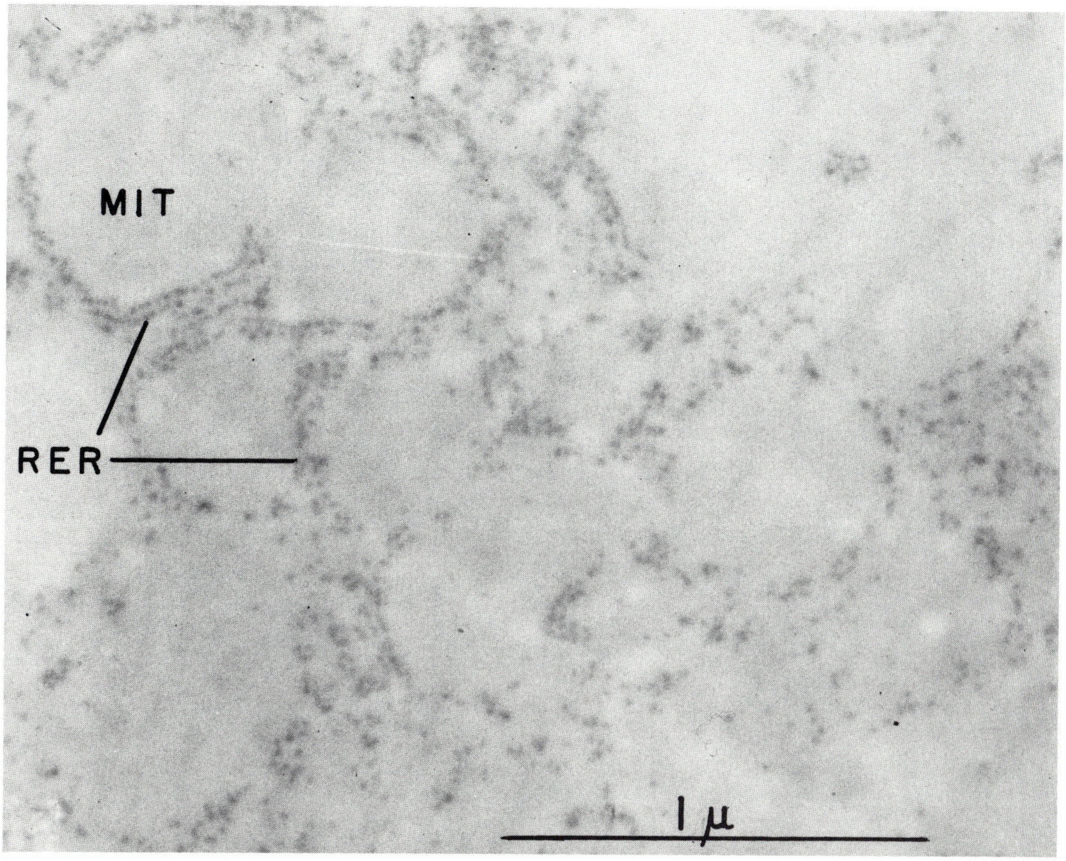

FIG. 3.35.

is probably caused by the added mass of FFDNB which reacts primarily with amino acids, polypeptides, and proteins, and to the very slight reaction, if any, with nucleic acids.

5. *Postfixation and Staining* in Vacuo *with Vapors of Mercury (II) Hexafluoroacetylacetonate, Followed by Staining with GCA (Figs. 3.33–3.35).* In general, the staining is quite comparable with the preparations treated with vapors of FFDNB and stained with GCA. The most prominent features are the increased background density in nucleoplasm and nucleolus, as well as in the RER, and in the aggregation of ribosomes. This results in (apparently) somewhat fuzzy electron micrographs which lack the crispness of those specimens postfixed in alcohol and stained with GCA. As after treatment with vapors of FFDNB, the nonresolvable, stained material is probably sRNA.

6. *Postfixation and Staining* in Vacuo *with Vapors of Hg hfac Alone (Figs. 3.36 and 3.37).* The denser walls of nucleoplasm, nucleolus, and chromatin appear darker than the less dense channels between them. The nucleolar walls are denser and thicker. The walls of the chromatin are still more so. No granules or

TABLE 3.1

Summary of Staining of Nucleic Acids in Pancreatic Acinar Cells by Gallocyanin–Chromalum after Using Various Postfixations

Postfixation and staining	Nucleus			Cytoplasm		
	DNA	Background density in nucleolus and nucleoplasm	Granules in nucleolus and nucleoplasm	Ribosomes	Ribosomal aggregates	Background density in rough endoplasmic reticulum
Alcohol + GCA	Chiefly granules	Weak	Clear	Clear	None	None or trace
Alcohol alone	Obscured by protein	Not apparent	Not visible	Barely detectable	None	Not apparent
FFDNB + GCA	Chiefly first- and second-order coils	Present	Partly obscured	Clear, sometimes obscured	Present	Present
FFDNB alone	Obscured by protein	Present	Mostly obscured	Barely detectable	—	Present
Hg hfac + GCA	Chiefly first- and second-order coils	Marked	Obscured	Clear, frequently obscured	Present	Marked
Hg hfac alone	Obscured by protein	Present	Mostly obscured	Low contrast	—	Present

coils are clearly demarcated in the nucleus. In the cytoplasm, the ribosomes of the RER are clear, but are not sharply demarcated. By comparison with preparations treated only with vapors of FFDNB, the contrast of structures containing nucleic acid has become somewhat more marked than in the specimens postfixed in alcohol. This is probably caused by the reaction of the mercury atoms of the vapor with nucleotides, to match the added density caused by the reaction with protein components.

The differences of the various methods of postfixation and staining of the liver resemble those summarized for the pancreas in Table 3.I, and will not be repeated here.

Discussion

Ribosomes are associated with the lipid–protein layer of the RER; around them are varying amounts of RNA. The former are stained by GCA after postfixation in alcohol alone, or after postfixation with cross-linking vapors, such as FFDNB and the mercury chelate. The latter are also stained by GCA, and are visible only after the use of cross-linking vapor reagents. After cross-linking, the homogeneously stainable component is removed by RNase, even from stained sections, though not by DNase or by water treatment alone. The homogeneous component corresponds with sRNA. The ribosomes are frequently grouped in clusters, which may correspond with polysomal units. The lipid–protein membranes of opposite surfaces of the cisternae are usually separated by a rather uniform distance. One of the important functions of the lipid components of the RER may be the maintenance of a molecular organization of the protein and RNA constituents. Such a molecular orientation could be of the greatest importance in concentrating and orienting transfer RNA (tRNA) and in maintaining intrapolysomal structural integrity. The intercisternal space may extend for relatively long distances, of the order of microns, after which it may impinge on some Golgi surface.

In Chapter 2 (this volume), the protein component of the Golgi net is described. In Chapter 1 (Vol. II), the lipid component of the Golgi complex is described as a highly pleomorphic reticulum whose lipid content is, in addition, very variable. While most of the three-dimensional mesh appears solid, some of the cords appear to have a lumen, and some minor tubelike connections are made with the RER. The most pronounced feature, however, is the enormity of the surface of the mesh, much of which is in contact with intercisternal spaces, where ribosomes and sRNA are concentrated.

In the nucleus, certain structures, especially those whose contrast is selectively enhanced by GCA, contain insoluble nucleic acid. This chemical identification is strengthened by the results of treatment of specimens with RNase and DNase. In the chromatin, the linear, circular, and oval figures whose structural element is a thread about 45 A thick actually represent the DNA double helix, whose diameter

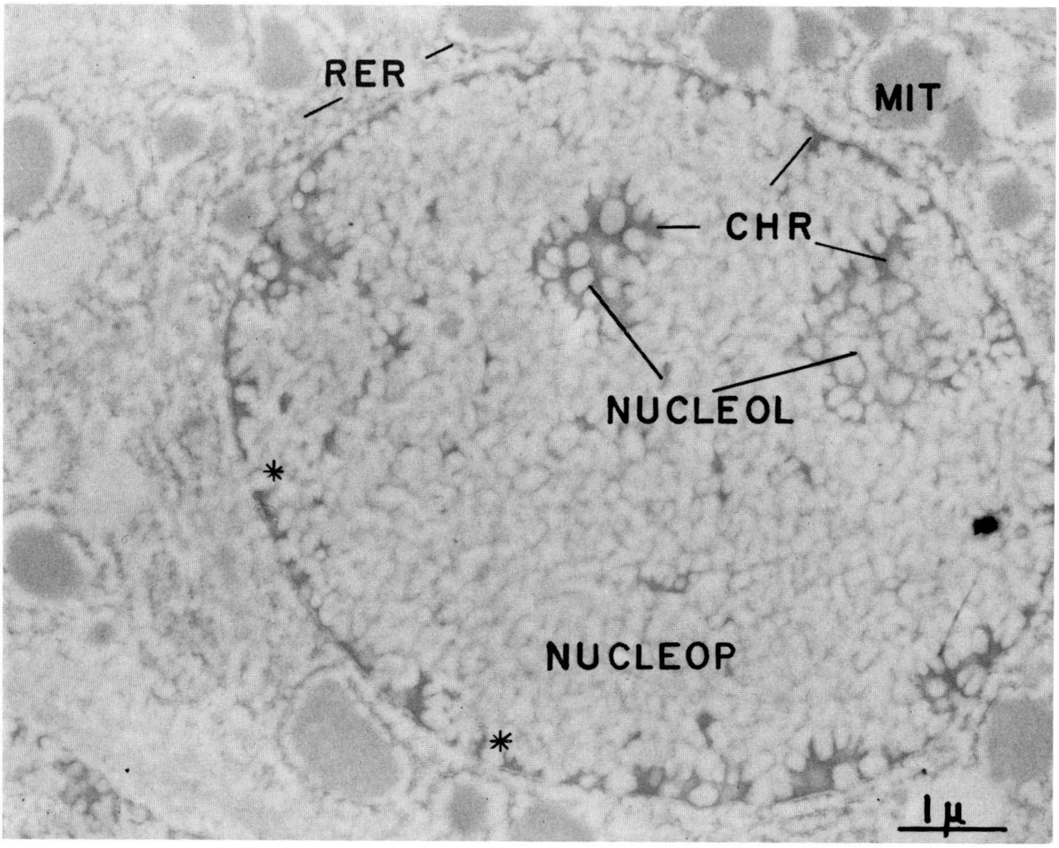

FIGS. 3.36 and 3.37. Mouse liver, postfixation and staining with vapors *in vacuo* of Hg hfac alone; no staining with GCA. Because the reagent probably reacts with nucleotides, the sites of nucleic acid are not obscured by the higher density of protein components which also react with the reagent. No granules or coils are detectable in the walls of the nucleolus, chromatin masses, or nucleoplasm. In the cytoplasm, ribosomes of the RER are grouped in clusters, in which the ribosomes are not readily resolved because of the background density of the material in which they are embedded. There is some shrinkage in the cytoplasm of Fig. 3.36. Nuclear gaps between peripheral chromatin are marked by an asterisk (*). See abbreviations as given in previous legends. Figure 3.36, × 14,000; Fig. 3.37, × 56,000.

is approximately doubled by the GCA molecule. GCA is attached nearly stoichiometrically to the phosphate on the outside of the double helix. This may be referred to as the molecular or first-order coil whose detailed structure is not resolvable by the method used in this report. This has sometimes been observed to be a straight line for some distance, but almost always it appears as some form or figure of a larger coil cut in various ways and lying in various planes in the section. This larger, or second-order coil has a diameter of about 400 A and a period

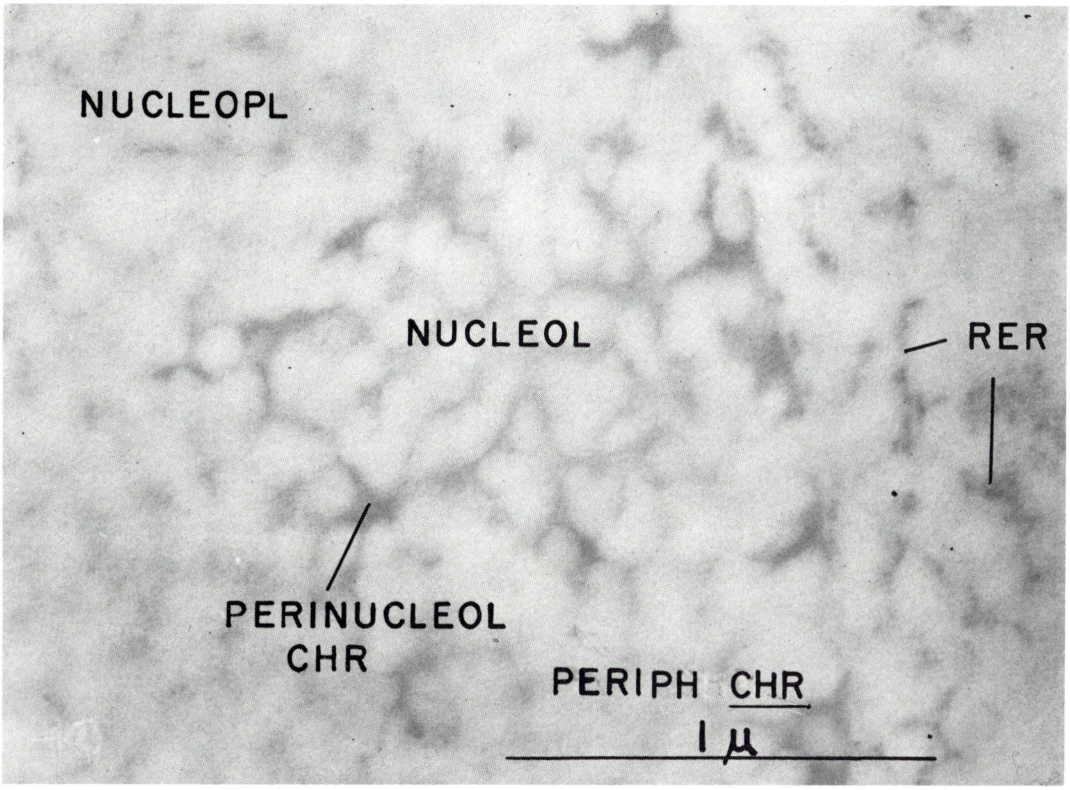

FIG. 3.37.

of about the same dimension. Both orders of DNA coil could no longer be stained selectively after treatment of the specimen with DNase, but resisted digestion by RNase. In these preparations, the morphology alone should suffice to identify the DNA coils in the nucleus, though, of course, supplementary confirmation is desirable.

It should be noted that although the vapor method of preparation results in a very high proportion of DNA figures which can be interpreted as first- and second-order DNA coils, nevertheless some granules or thick rods may also be observed. The proportion of such DNA figures and their size is smaller than when vapors are not used (as after postfixation in alcohol). It is possible that coarse, granular, or rodlike DNA particles observed in vapor-treated specimens may then represent a minority of DNA coils which are not well fixed; or they may represent particular groupings of secondary DNA coils.

It should also be noted that the fibrillar component noted in nucleoli after fixation in aqueous fluids *(4, 10)* was not observed in the material examined. This may be interpreted in several ways: (1) Fibrils of RNA may be present in the nucleolus, but may be obscured by the diffusely stained sRNA. (2) Fibrils of

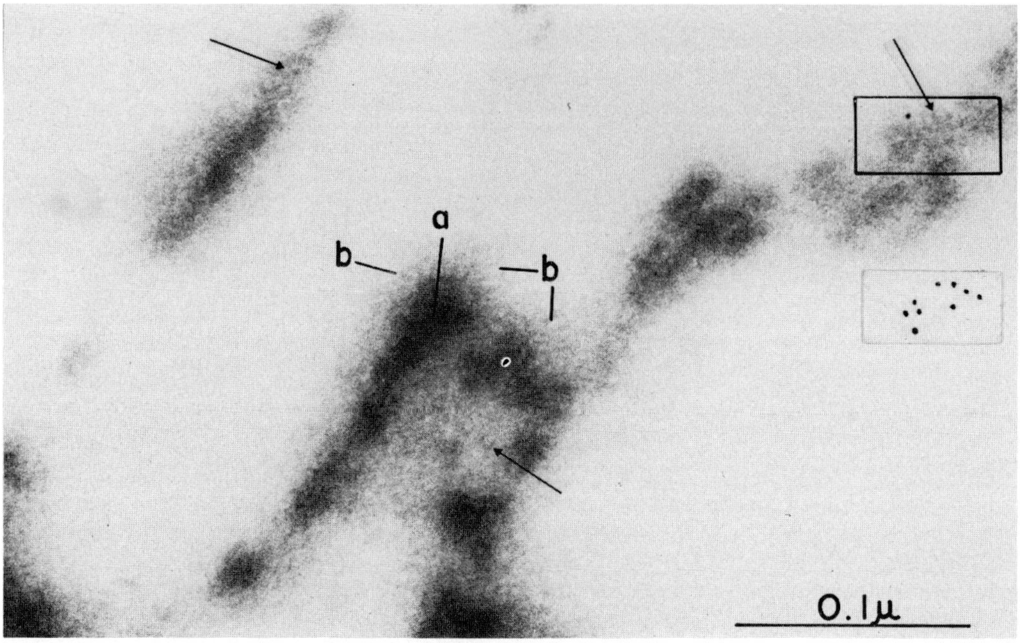

FIG. 3.38.

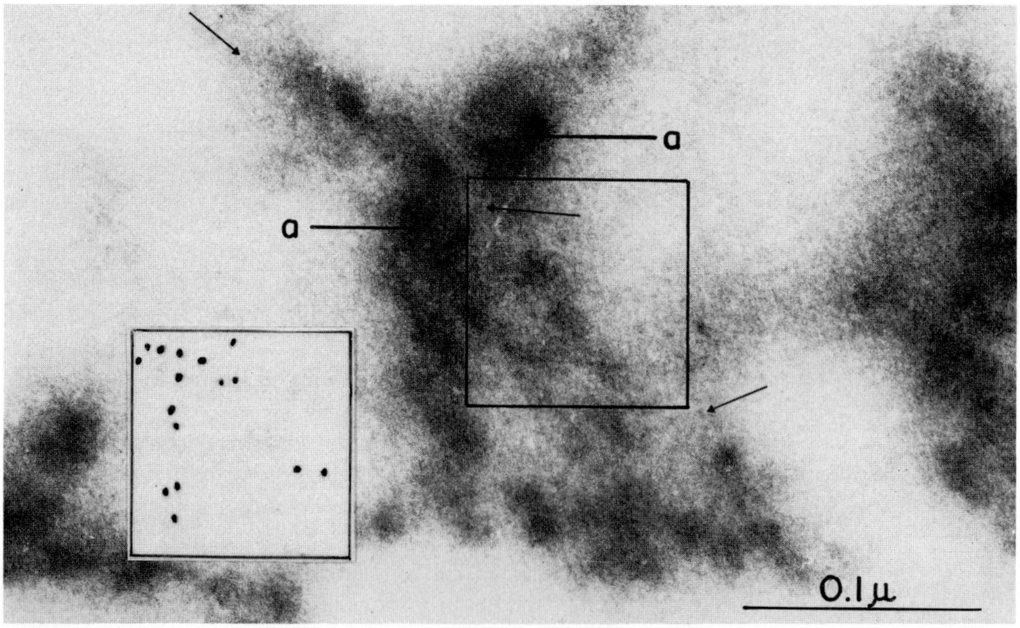

FIG. 3.39.

FIGS. 3.38–3.41. Pancreas, freeze-dried, postfixed, and stained with vapors of Hg hfac and stained *en bloc* with GCA for nucleic acids. Minute particles of nucleic acid marked by arrows are present around more densely stained ribosomes. Some of these particles are indicated in the inserts. Where they are very numerous, they cannot be clearly resolved, as at (a). Sometimes their arrangement in rows may be observed as if they radiate from the ribosome, as at (b). Nucleic acid within ribosomes seems irregularly disposed, but actually conforms to a pattern of organization described on p. 94. × 305,000. G2 print.

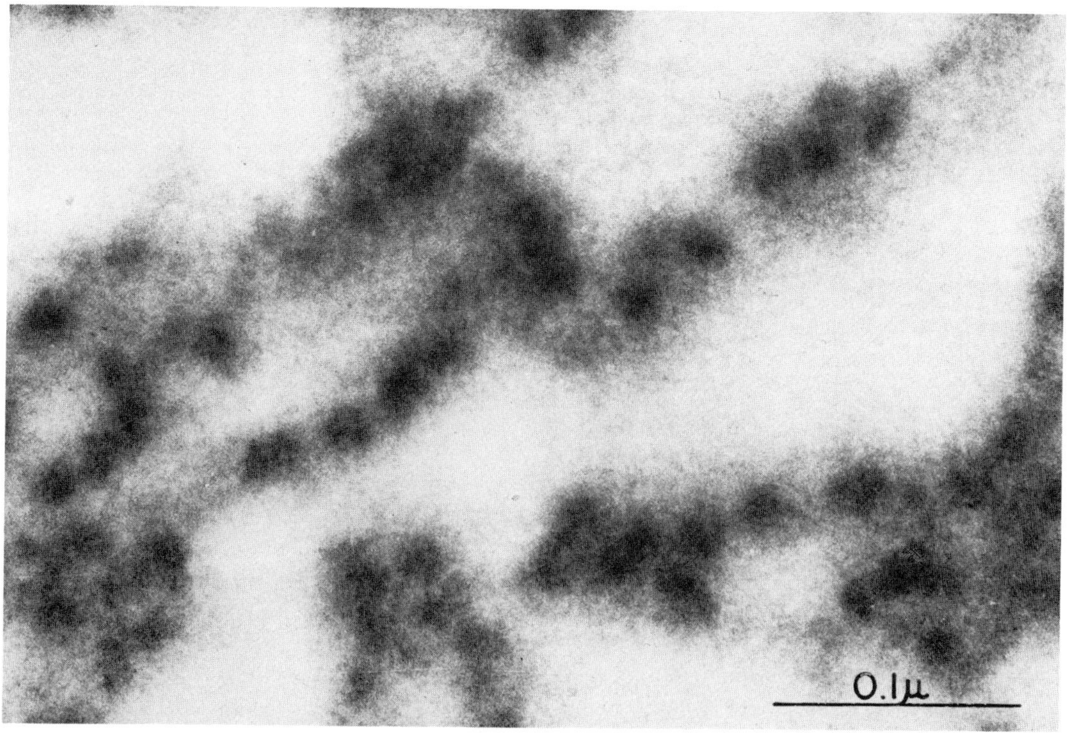

FIG. 3.40. ▲

▼ FIG. 3.41.

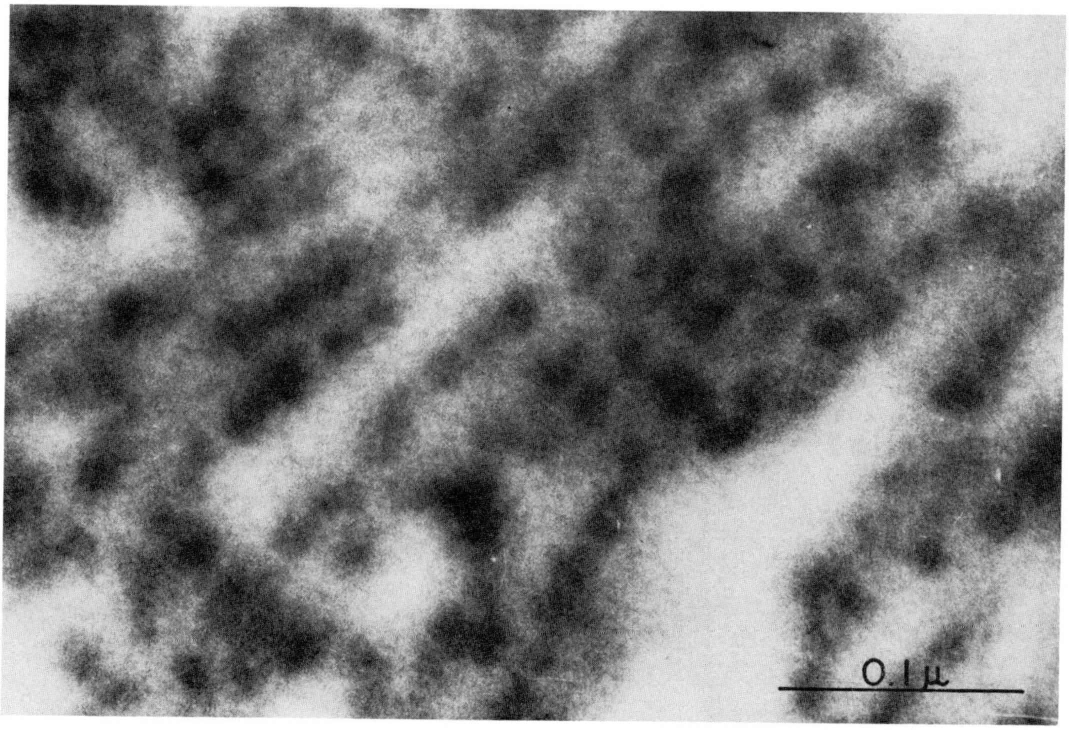

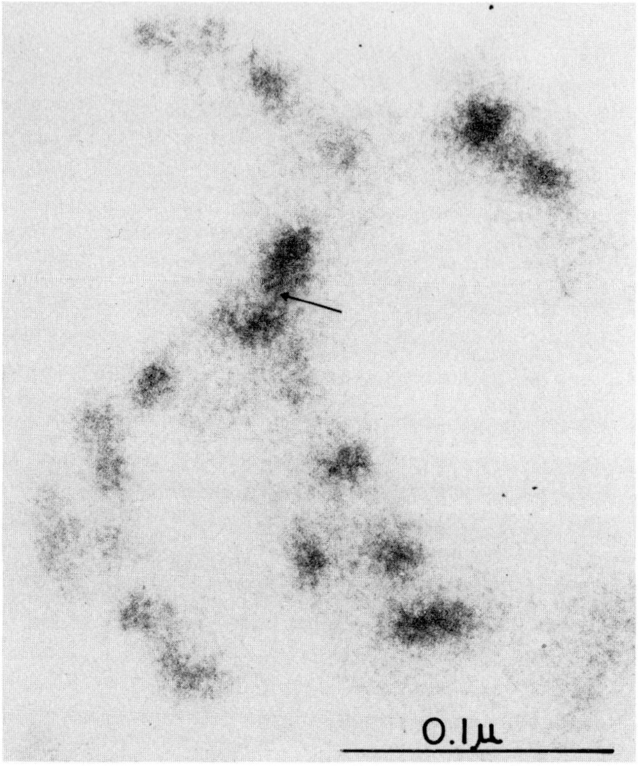

Fig. 3.42. Liver, as prepared in Figs. 3.38–3.41. A straight filament, which is probably mRNA, joins two ribosomes (arrow). Internal structure of some ribosomes is displayed through the distribution of RNA, which is densely stained. × 350,000. G2 print.

protein may be present in the nucleolus, but may be obscured by the homogeneous protein matrix which infiltrates them and is of the same density. (3) On the other hand, it is possible that fibrils which are observed after the use of aqueous fixatives may be the result of precipitation during fixation of one of among many protein components, most of which are dissolved, or may be the result of denaturation of RNA granules which have uncoiled as a consequence of fixation. Further comments on nucleolar fibrils are made in Chapter 5 (Vol. II).

The insoluble granules of the nucleus vary in size, and range from about 100 A to the limit of resolution. They are not removed by DNase, or are they removed entirely by RNase. The granules are presumably digested with difficulty by this enzyme under the conditions of the methods employed. They may represent a form of RNA not digestable with RNase *(7)*. These granules lie in an unresolvable background of sRNA which itself is removed from specimens and even from stained sections by RNase, but which is not affected markedly by DNase or by a suitable water control. As in the cytoplasm, the homogeneous component is not preserved for staining except through prior cross-linking; after postfixation with alcohol, for example, the background staining is largely absent.

The nucleus is organized somewhat like a sponge, whose walls are more or less continuous with each other, even though they may differ chemically, and whose spaces are interconnected. The "spaces" are in fact originally filled with protein which can be observed after more extensive denaturation than is customary. The differential loss of stainability of the material in the "spaces" indicates that the protein present in these spaces differs from that in the walls (see discussion, p. 47 for an analysis of this concept). The nucleoplasm and nucleolus seem to be very similar, in that the walls of both contain protein, RNA granules, and sRNA. The granules are more numerous in the nucleolus, and the walls in which they lie are thicker. The smaller granules seem to be relatively more numerous in the walls of the nucleolus than in those of the nucleoplasm. As in the cytoplasm, the channels extend to the surface of the nucleus. They may appear to be continuous with walls in the cytoplasm which also contains "spaces." It is possible that the denser walls of the spongelike structure of the nucleus represent the paths along which transport of nucleotides and polynucleotides take place in the nucleus. In this case, one would further postulate that the transport system is directly connected, on the one hand, with the chromatin and the nucleolus, where the products of DNA activity are derived and/or stored, and, on the other hand, with the bare patches of the nuclear membrane between chromatin masses where one may postulate that the nuclear "pores" are located. The mechanism underlying this suggested transport system is not known.

A feature common to the nuclei of pancreatic acinar and hepatic cells, as well as all other cell types studied thus far, is the close relation of chromatin masses to the nuclear membrane as well as the occurrence of gaps, chiefly submicroscopic in size, between them. The latter may be related to the transport system, as suggested above. The possible significance of the former is not clear. The relative constancy of the chromatin pattern in each cell type studied thus far suggests that the relation of chromatin masses and nuclear membrane is in some way essential for the expression of cell differentiation and function and for orderly cell divisions during mitosis. A vast amount of literature bearing on the peripheral disposition of chromatin was systematized by Vanderlyn *(55)*, whose analysis antedated the Watson-Crick model. A reinterpretation of some of the same and more recently acquired data is presented in Chapter 14 (this volume).

The development of a reliable method for the demonstration of the DNA double helix *in situ* has made possible some exploratory studies of the nature of the systems which control genetic activity in highly developed metazoan cells. The work of Jacob and Monod *(28)* led to the concept of the operon system as a means of genetic control of cell metabolism in *Escherichia coli.* How widespread this type of regulation may be in higher plants and animals is unknown; it may be universal. On the other hand, there is no doubt that multicellular organisms present a much greater complexity in development, differentiation, and metabolic processes than is found in bacteria. In fact, it has been suggested that more complex systems of genetic control of metabolic systems may be present in higher organisms, and that these controlling factors function at the chromosomal (chromatin) level, rather than at the genic level alone *(3, 8)*. McClintock's studies *(35)* on corn have revealed the presence of such regulatory systems which, when disturbed, result in a variegation in the plant tissues. She has given reasons for believing that the controlling elements are associated with the heterochromatic parts of the chromosomes. Variegations, in which the heterochromatic regions of chromosomes play a part, also occur in *Drosophila (49)* and mouse *(46)*. There are many indications that the genes in heterochromatic regions are inactive *(24, 32)*, and it has been suggested that heterochromatin and euchromatin are interconvertible states *(9, 12)*.

The methods described in this report are being used in an effort to define the patterns of organization of the DNA double helix in different cell types, in different states of activity of several cell types, during differentiation, and during cell death. The DNA molecular pattern has been studied in the nuclei of over forty different cell types in the mouse, and is different in each (Chapter 8). It has been studied in the euchromatin, heterochromatin, and puffs of the giant chromosomes in *Drosophila* salivary gland (Chapter 4). In all instances, the DNA double helix appears in electron micrographs of stained specimens as a filament about 45 A in diameter. With few exceptions this is arranged as a second-order coil with a period and diameter of about 400 A. The secondary coil is the same in heterochromatin and euchromatin in all cells studied. It is only their arrangement which may be different, the secondary coils being single or in pairs in euchromatin and rather tightly crowded in heterochromatin. Other patterns which affect the accessibility of DNA to nucleoplasmic exchange have also been observed and will be described in later chapters.

The occurrence of nucleoprotein coils in chromosomes has been amply reviewed by DuPraw *(15)*. In all those which he and others have described, the identification of the DNA component is at best vague, and lacking a positive identification through selective staining of either the nucleic acid or the protein component. Moreover, the possibilities of solution, reprecipitation, and rearrangement on the surface of the water trough have been largely ignored *(62)*. In our studies, neither the first- nor second-order of DNA coils seems to have a well defined sheath. Instead, the coils are embedded in a homogeneous protein environment of very nearly the same density and contrast (see Chapters 2 and 5, Vol. I). It is possible, of course, that in our studies, the DNA coils are enclosed in a protein sheath of uniform thickness, but that these nucleoprotein units are in turn bathed in a protein matrix of the same density and contrast. If such nucleoprotein coils exist, we suppose they may become visible after solution of the matrix by the method of preparation of the specimens on the water trough, when the nuclei are ruptured and the chromosomes are exposed directly to the water.

A common feature of the protoplasm in all cells, as reported from the senior author's laboratories, is the occurrence of less dense submicroscopic vacuoles enclosed by their denser walls. The submicroscopic vacuoles may more accurately be referred to as pseudovacuoles, since their contents have approximately the same density as their walls, as can be seen in preparations where the protein contents are more completely denatured by suitable postfixatives. When less completely denatured, the contents of the pseudovacuoles are not stained by the reagents used and the vacuoles appear less dense than their walls. The less dense and more dense parts of protoplasm have been regarded as components of a minimally two-phase system in equilibrium (Chapter 2, this volume). The properties of such a two-phase system were regarded as significant in maintaining cellular homeostasis. The operation of this important property need not be considered abrogated by the fact that the vacuoles are pseudovacuoles, i.e., that the density of protein in the vacuoles is of the same order as that in the walls. If it is assumed that some (even nearly all) of the protein extracted from the pseudovacuoles differs from that of the remainder of the protoplasm, and that some of this protein is the same as that in the walls, but less concentrated, then the elements of the two-phase system are present in protoplasm. The reinterpretation of the submicroscopic vacuolar structure of protoplasm does not modify significantly the theoretical importance of the pseudovacuoles in the maintenance of homeostasis with respect to small molecules or atoms. It should be stated that the expression "two-phase system" is the simplest one, and that if the concept is at all applicable, it represents a series of two- or multiphase systems.

Summary

A method is described for the fixation and staining of nucleic acids in cells. Freeze-dried preparations free of detectable ice crystals are postfixed *in vacuo* with vapors which cross-link reactive groups of certain soluble and insoluble cell components and thus tend to make them insoluble in subsequent steps. These include staining with gallocyanin–chromalum. In addition to making nucleic acids insoluble, the method retains DNA in its extended (coiled) state, its diameter is increased by the stain molecules, and its contrast is intensified. In the cytoplasm of the cells studied, ribosomes are clearly delineated. In addition, ribosomes are frequently enclosed in a cloud of sRNA.

The outstanding quality of the nucleus (as of the cytoplasm) is its spongelike appearance. The walls, and the spaces between them, are continuous through nucleoplasm, nucleolus, and chromatin. The submicroscopic pseudovacuoles contain protein(s) which remain unstained in most procedures, giving rise to the "spaces" observed in electron micrographs. The density and thickness of the walls increase in the following order: chromatin > nucleolus > nucleoplasm. The chromatin comprises chiefly first-order DNA coils (the double helix) organized as second-order DNA coils with a period and diameter of about 400 A. In the preceding chapter, it was shown that the DNA helices are embedded in a homogeneous matrix which is largely protein in nature. Both insoluble (granular) and soluble (diffuse) RNA are described in the nucleoplasm and nucleolus.

These observations are discussed briefly in relation to concepts of euchromatin and heterochromatin, genetic control in metazoan cells, differentiation, cellular homeostasis, and an intranuclear transport system.

Appendix

A few electron micrographs of exceptional quality were made of unusually thin (transparent) sections of specimens postfixed with vapors of Hg hfac and stained with GCA. The most striking aspect was the presence of small, dense granules about 50 A in size. When present in large numbers, they were unresolvable (Figs. 3.38–3.41). They are crowded around and between ribosomes. In certain regions, the ribosomes are surrounded by dense regions which seem to radiate from the ribosomes (Figs. 3.38–3.41), appearing in electron micrographs like lines radiating from ribosomes. In favorable sites, the radiations are resolvable into 50 A particles. Both groups of small granules, which are stainable with GCA, are perhaps tRNA particles, namly aminoacyl-tRNA. If so, their preservation is probably due to the use of the cross-linking agent and the subsequent unmasking by the more or less selective staining with GCA. Further proof of the identity of these particles is necessary. The significance of the distribution and relations of these RNA particulates is discussed in the last chapter of Volume 2.

The internal structure of ribosomes seems in electron micrographs to be extremely variable. This is so probably because of the great variation in the positioning of an unsymmetrical structure (the ribosome) in relation to the electron beam. A densely stained central sheet or pillar runs through the ribosome, and fans out from the core toward the periphery at several levels (Figs. 3.38–3.41). The spaces between these minute shelflike regions are less dense. From these photographs, the internal structure is as follows: There is a central core of RNA, with several sheets extending from it toward the periphery of the ribosome. Between these sheets are the globular protein molecules described in Chapter 2 (Vol. I). It is as if each protein molecule is enclosed or wrapped around by an RNA fabric in the form of very thin sheets. The net result is an exceedingly intimate relation between protein and RNA. The relation of these sheets of RNA to the folds and coils described by Spirin and Gavrilova is unknown *(64)*.

On rare occasions a thread of uniform dimensions can be seen between ribosomes (Fig. 3.42). This is probably the RNA component of a messenger, whose protein component was described in this volume, Chapter 2 (Figs. 2.31, 2.36, and 2.37).

References

1. Albersheim, P., and Killias, U. (1963). The use of bismuth as an electron stain for nucleic acids. *J. Cell Biol.* **17**, 93–103.
2. Barnicot, N. A., and Huxley, H. E. (1965). Electron microscope observations on mitotic chromosomes. *Quart. J. Microsc. Sci.* **106**, 197–214.
3. Becker, H. J. (1964). Die genetischen Grundlagen der Zelldifferenzierung. *Naturwissenschaften* **51**, 205–211.

4. Bernhard, W. (1965). Ultrastructural aspects of the normal and pathological nucleolus in mammalian cells. *Natl. Cancer Inst. Monogr.* **23**, 13–38.
5. Bernhard, W. (1969). A new staining procedure for electron microscopical cytology. *J. Ultrastruct. Res.* **27**, 250–265.
6. Bernard, W., and Granboulan, N. (1968). Electron microscopy of the nucleolus in vertebrate cells. *In* "The Nucleus" (A. J. Dalton and F. Haguenau, eds.), pp. 81–149. Academic Press, New York.
7. Bonner, J., and Ts'o, P. (Eds). (1964). "The Nucleohistones" Holden-Day, San Francisco, California.
8. Brink, R. A. (1964). Genetic repression in multicellular organisms. *Amer. Natur.* **98**, 193–211.
9. Brown, S. W. (1966). Heterochromatin. *Science* **151**, 417–425.
10. Busch, H., and Smetana, K. (1970). "The Nucleolus." Academic Press, New York.
11. Caspersson, T. (1940). Nukleinsäureketten und Genvermehrung. *Chromosoma,* **1**, 605–619.
12. Cooper, K. W. (1959). Cytogenetic analysis of major heterochromatic elements (especially Xh and Y) in *Drosophila melanogaster,* and the theory of "heterochromatin." *Chromosoma* **10**, 535–588.
13. Darzynkiewicz, D., Gledhill, B. L., and Ringertz, N. R. (1969). Changes in deoxyribonucleoprotein during spermeogenesis in the bull. *Exp. Cell Res.* **58**, 435–438.
14. De Robertis, E. (1964). Advances in the ultrastructure of the nucleus and chromosomes. *Nat. Cancer Inst. Monogr.* **14**, 33–55.
15. DuPraw, E. J. (1968). "Cell and Molecular Biology." Academic Press, New York.
16. Einarson, L. (1951). On the theory of gallocyanin-chromalum staining and its application for quantitative estimation of basophilia. A selective staining of exquisite progressivity. *Acta Pathol. Microbiol. Scand.* **28**, 82–102.
17. Finck, H. (1958). An electron microscope study of basophile substances of frozen-dried rat liver. *J. Biophys. Biochem. Cytol.* **4**, 291–300.
18. Gersh, I. (1959). Fixation and staining. *In* "The Cell" (J. Bracket and A. E. Mursky, eds.), Vol. 1, pp. 21–66. Academic Press, New York.
19. Gersh, I., and Bodian, D. (1943). Some chemical mechanisms in chromatolysis. *J. Cell. Comp. Physiol.* **21**, 253–279.
20. Gersh, I., Vergara, J., and Rossi, G. L. (1960). Use of anhydrous vapors in post-fixation and in staining of reactive groups of proteins in frozen-dried specimens for electron microscopic studies. *Anat. Rec.* **138**, 445–460.
21. Gledhill, B. L., Gledhill, M. P., Rigler, R., and Ringertz, N. R. (1966). Changes in deoxyribonucleoprotein during spermeogenesis in the bull. *Exp. Cell Res.* **41**, 652–665.
22. Hall, C. E., and Litt, M. (1958). Morphological features of DNA macromolecules as seen with the electron microscope. *J. Biophys. Biochem. Cytol.* **4**, 1–4.
23. Hay, E. D. (1968). Structure and function of the nucleolus in developing cells. *In* "The Nucleus" (A. J. Dalton and F. Haguenau, eds.) pp. 1–79. Academic Press, New York.
24. Heitz, E. (1933). Die somatische Heteropyknose bei *Drosophila melanogaster* und ihre genetische Bedeutung. *Z. Zellforsch. Mikrosk. Anat.* **20**, 237–287.
25. Horobin, R. W., and Murgatroyd, L. B. (1968). The composition and properties of gallocyanin-chrome alum stains. *Histochem. J.* **1**, 36–54.
26. Huxley, H. E., and Zubay, G. (1961). Preferential staining of nucleic acid-containing structures for electron microscopy. *J. Biophys. Biochem. Cytol.* **11**, 273–296.
27. Isenberg, I. (1957). The use of organic dyes in electron microscopy. *Bull. Math. Biophys.* **19**, 279–292.
28. Jacob, F., and Monod, J. (1961). On the regulation of gene activity. *Cold Spring Harbor Symp. Quant. Biol.* **26**, 193–211.
29. Kiefer, G. (1970). Recent developments in gallocyanin-chrome alum staining. *In* "Introduction to Quantitative Cytochemistry" (C. L. Wied and G. F. Bahr, eds.), Vol. 2, pp. 199–208, Academic Press, New York.
30. Lafontaine, J.-G. (1968). Structural components of the nucleus in mitotic plant cells. *In* "The Nucleus" (A. J. Dalton and F. Haguenau, eds), pp. 151–196. Academic Press, New York.

31. Leduc, E. H. and Bernhard, W. (1961). Ultrastructural cytochemistry. Enzyme and acid hydrolysis of nucleic acids and protein. *J. Biophys. Biochem. Cytol.* **10**, 437–455.

32. Lyon, M. F. (1961). Gene action in the X-chromosome of the mouse (Mus musculus L.) *Nature (London)* **190**, 372–373.

33. Macinnes, J. W., and Uretz, R. B. (1968). DNA organization in the mature sperm of several orthoptera by the method of polarized fluorescence microscopy. *J. Cell Biol.* **38**, 426–436.

34. Marinozzi, V. (1963). The role of fixation in electron staining. *J. Roy. Microsc. Soc.* **81**, 141–154.

35. McClintock, B. (1951). Chromosome organization and genic expression. *Cold Spring Harbor Symp. Quant. Biol.* **16**, 13–47.

36. Monneron, A., and Bernhard, W. (1969). Fine structural organization of the interphase nucleus in some mammalian cells. *J. Ultrastruct. Res.* **27**, 266–288.

37. Moses, M. J., and Coleman, J. R. (1965). Structural patterns and the functional organization of chromosomes. *In* "The Role of Chromosomes in Development" (M. Locke, ed.), pp. 11–49. Academic Press, New York.

38. Mundkur, B. (1961). Electron microscopical studies of frozen-dried yeast. II. The nature of basophile particles and vesicular nuclei in Saccharomyces. *Exp. Cell Res.* **25**, 1–23.

39. Mundkur, B. (1964). Submicroscopic cytochemical organization of interphase nuclei revealed by protein reagents and gallocyanin-chromalum. A study of Ehrlich ascites cells. *Z. Zellforsch. Mikrosk. Anat.* **63**, 52–80.

40. Mundkur, B. (1965). Bifunctional cross-linking reagents and the fine structure of protein in Ehrlich ascites tumor cell nuclei. *Cancer Res.* **25**, 1035–1051.

41. Pattri, H. O. E. (1932). Über die Doppelbrechung der Spermien. *Z. Zellforsch. Mikrosk. Anat.* **16**, 723–744.

42. Porter, K. R. (1960). Problems in the study of nuclear fine structure. *4th Int. Conf. Electron Microsc.*, Vol. 2, pp. 186–199. Springer-Verlag, Berlin.

43. Ris, H. (1966). Fine structure of chromosomes. *Proc. Roy. Soc. London Ser.* **B***164*. 246–257.

44. Ris, H., and Kubai, D. F. (1970). Chromosome structure. *Annu. Rev. Genet.*, **4**, 263–294.

45. Ruch, F. (1966). Birefringence and dichroism of cells and tissues. *In* "Physical Techniques in Biological Research" (A. W. Pollister, ed.), 2nd ed. Vol. IIIA., pp. 57–86. Academic Press, New York.

46. Russell, L. B., and Bangham, J. W. (1961). Variegated-type position effects in the mouse. *Genetics* **46**, 509–525.

47. Sandritter, W., Kiefer, G., and Rick, W. (1966). Gallocyanin Chrome Alum. *In* "Introduction to Quantitative Cytochemistry" (G. L. Wied, ed.), pp. 295–326. Academic Press, New York.

48. Schmidt, W. J. (1937). "Die Doppelbrechung von Karyoplasma, Zytoplasma und Metaplasma." Gebrüder Borntraeger, Berlin.

49. Schultz, J. (1936). Variegation in *Drosophila* and the inert chromosome regions. *Proc. Nat. Acad. Sci. U.S.* **22**, 27–33.

50. Stoeckenius, W. (1961). Electron microscopy of DNA molecules "stained" with heavy metal salts. *J. Biophys. Biochem. Cytol.* **11**, 297–310.

51. Swift, H. (1962). Nucleoprotein localization in electron micrographs: metal binding and radioautography. *In* "The Interpretation of Ultrastructure" (R. J. C. Harris, ed.), pp. 213–232. Academic Press, New York.

52. Swift, H. (1962). Nucleic acids and cell morphology in dipteran salivary glands. *In* "The Molecular Control of Cellular Activity" (J. M. Allen, ed.), pp. 73–125. McGraw-Hill, New York.

53. Swift, H., and Adams, B. J. (1962). Simple methods for nucleic acid and acid mucopolysaccharide localization in electron micrographs. *5th Int. Congr. Electron Microsc.*, Vol. 2, pp. 0–4. Academic Press, New York.

54. Terner, J. Y., and Clark, G. (1960). Gallocyanin-chrome alum. I. Technique and specificity. *Stain Technol.* **35**, 167–177.

55. Vanderlyn, L. (1949). The heterochromatin problem in cytogenetics as related to other branches of investigation. *Bot. Rev.* **15**, 507–582.

56. Watson, M. L., and Aldridge, W. G. (1961). Methods for the use of indium as an electron stain for nucleic acids. *J. Biophys. Biochem. Cytol.* **11**, 257–272.
57. Watson, M. L., and Aldridge, W. G. (1962). Considerations of nucleic acid morphology in fixed tissues. *5th Int. Congr. Electron Microsc.*, Vol. 2, pp. 0–5. Academic Press, New York.
58. Wilkins, M. H. F. (1956). Physical studies of the molecular structure of deoxyribose nucleic acid and nucleoprotein. *Cold Spring Harbor Symp. Quant. Biol.* **21**, 75–88.
59. Wilkins, M. H. F. and Battaglia, B. (1953). Note on the preparation of specimens of oriented sperm heads for X-ray diffraction and infrared absorption studies on some pseudomolecular behavior of sperm. *Biophys. Biochim. Acta* **11**, 412–415.
60. Wilkins, M. H. F., and Randall, J. T. (1953). Crystallinity in sperm heads: molecular structure of nucleoprotein *in vivo*. *Biophys. Biochim. Acta* **10**, 192.
61. Wolfe, S. L., Beer, M., and Zobel. C. R. (1962). The selective staining of nucleic acids in a model system and in tissue. *5th Int. Congr. Electron Micros.* Vol. 2, pp. 0–6. Academic Press, New York.
62. Zirkin, B. R., and Wolfe, S. L. (1970). The chemical composition of nuclei and chromosomes isolated by the Langmuir trough technique. *Chromosoma* **32**, 162–170.
63. Zobel, C. R., and Beer, M. (1961). Electron stains. I. Chemical studies on the interaction of DNA with uranyl salts. *J. Biophys. Biochem. Cytol.* **10**, 335–346.
64. Spirin, A. S., and Gavrilova, L. P. (1969) . "The Ribosome." Springer-Verlag, New York.

4

Cytochemical Study of Nucleic Acids in Salivary Gland Chromosomes of Drosophila Larvae

Eileen S. Gersh and Isidore Gersh

Several reports have appeared on the submicroscopic structure of chromosomes of salivary glands *(3, 6, 7, 10, 29, 32, 39, 47, 49, 53, 59)*. Although these reports have emphasized the structural complexity of the polytene chromosomes, they have failed to illuminate critically or offer acceptable solutions to what seem now to be a most important series of problems for the interpretation of gene action: the relations, at the supramolecular level, of nucleic acids and proteins to the bands, interband regions, puffs, and heterochromatic regions, as well as to the remaining extrachromosomal structures of the nucleus. Studies at the molecular level by very elegant and sensitive methods may be extended to higher levels of organization only speculatively because of the insufficient morphological resolution of the methods employed *(14, 15, 21–23, 48, 69)*. Other difficulties could be attributed, in large part, to the fact that the genetic materials of chromosomes have not been adequately preserved by the fixatives employed, and in part to the lack of specificity of the methods for staining nucleic acids. In addition, proteins are probably significantly displaced, distorted, or extracted during fixation. These difficulties are not limited to studies of polytene chromosomes, but are in large part responsible for the mournful reviews of chromosomal structure in general *(58, 62, 73)*. The relevance of many earlier submicroscopic studies is open to question essentially because the disorder created by such methods of fixation and staining is so great as to obscure the fundamentally strictly ordered genetic system.

We believe these difficulties are largely overcome in the work reported in this chapter and in Chapter 5 (this volume) by the method of fixation and staining employed. This involves ultrarapid freezing and drying followed by the use of cross-linking agents in the vapor phase. This results in a kind of extensive insolubilization of otherwise soluble proteins of cells, and makes possible a subsequent selective staining of nucleic acids with gallocyanin–chromalum while retaining their morphological relationships (this volume, Chapter 3; Vol. II, Chapter 5). The method results in the staining of the DNA double helix (the first-order coil). With only two exceptions up to this time (chromatolytic nerve cells, this volume, Chapter 12, and at certain points in the mitotic cycle of developing *Drosophila* eggs, this volume, Chapter 13) this first-order coil is organized as a second-order coil. In all somatic cell nuclei studied thus far, with the exception of *Drosophila* eggs, the diameter of the second-order coil is about 400 A. This applies to about forty cell types of the mouse, some cell types in other mammals, the imaginal disk of *Drosophila*, and one sample of plant material. The same method also stains RNA, which appears as granules in the nucleolus, nucleoplasm, and cytoplasm, and as a background density unresolvable in routine electron micrographs because the magnification is too low. The former probably corresponds with rRNA (at least in the cytoplasm) and the latter with at least a fraction of soluble RNA.

This chapter is limited to a consideration of nucleic acids in salivary gland chromosomes. Protein in the same nuclei is treated later in this volume (Chapter 5).

Material and Methods

Salivary glands of late third instar larvae of wild-type *Drosophila melanogaster* were dissected in *Drosophila* Ringer's solution *(26)* and transferred to a small square of aluminum foil. Excess solution was removed by blotting, a process which is accompanied by flattening of the gland. This is technically advantageous in that the chances of ice crystal formation during freezing are reduced. The specimen was frozen ultrarapidly by immersion in Freon 13 at about −175°C, and dried *in vacuo* at a temperature below −40°C. The vacuum tube was transferred to ice. Vapors of mercury hexafluoroacetylacetonate and of water were introduced into the vacuum tube without interrupting the vacuum. After 6 hours, the reagent tube was removed and excess vapors were removed by evacuation for 2 hours at 0°C and then overnight at room temperature. Specimens were infiltrated *in vacuo* with 95% alcohol. When the specimens were covered with alcohol, the vacuum was broken for the first time. Specimens were removed and hardened at ambient temperature and atmospheric pressure for 1 day. Then they were stained with gallocyanin–chromalum at pH 1.9, washed, dehydrated, hardened for 1 day in 95% alcohol, and infiltrated with, and embedded in, water-soluble Durcupan. Sections were cut with the ultramicrotome and mounted on Formvar-coated grids for viewing with the electron microscope at magnifications of 7100 and 28,000. Prints were made with a magnification of two times. The method was described in greater detail in this volume (Chapter 3).

The staining procedure does not discriminate between DNA and RNA. Nucleases were used on other salivary glands to distinguish these components. Specimens were frozen ultrarapidly as described above, and postfixed for 5 hours at 50°C *in vacuo* with vapors of difluorodinitrobenzene in the presence of water vapor. The cross-linking prevents or reduces the solution, swelling, and subsequent reprecipitation which may take place during and after immersion in the aqueous enzyme solutions. Excess vapors were evacuated from the specimens at 60°C overnight. On cooling, specimens were infiltrated *in vacuo* with 95% alcohol, where they were allowed to harden by immersion for 1 day at ambient temperature and atmospheric pressure. The glands were then hydrated and treated with many changes of solutions containing RNase or DNase. Many changes were used to reduce the possibly toxic action toward the enzymes of unreacted or partially

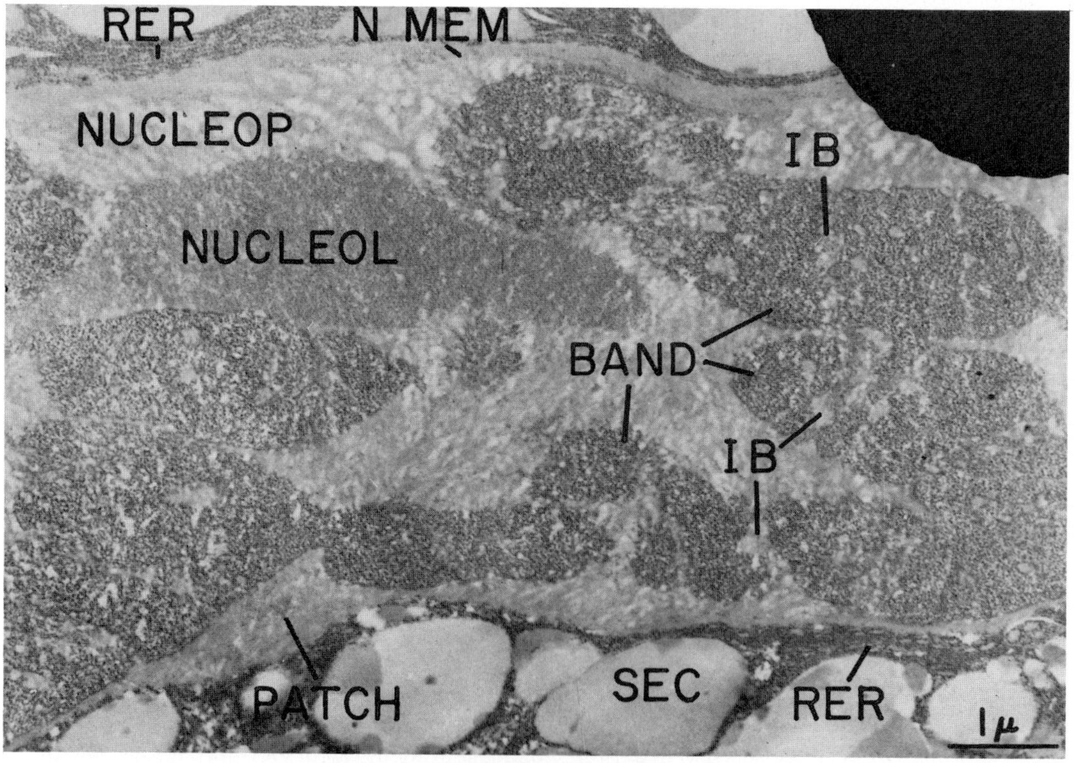

FIGS. 4.1–4.4. Figures 4.1 and 4.3 show the cytoplasm which contains secretory granules (SEC) surrounded by the rough endoplasmic reticulum (RER). The spongelike nucleoplasm (NUCLEOP) is contained within the unstained nuclear membrane (N MEM). Denser patches in the nucleoplasm comprise RNA granules (PATCH). The nucleolus (NUCLEOL) is similarly structured but the walls of its spongelike structure are richer in RNA granules. The chromosomes contain dense and less dense areas which correspond with band (BAND) and interband (IB) regions. Parts of the nucleolar organizing chromosome (N CHROM) are attached to the surface of and lie within the nucleolus, × 14,200. Figures 4.2 and 4.4 are parts of the same fields at higher magnification, to show the RNA granules (R GRANS) in the nucleolus and nucleolar organizing chromosome (N CHROM). In Fig. 4.4, dense circles and ovals are second-order DNA (DNA II) coils cut in various planes. Fig. 2, × 50,500; Fig. 4, × 56,000.

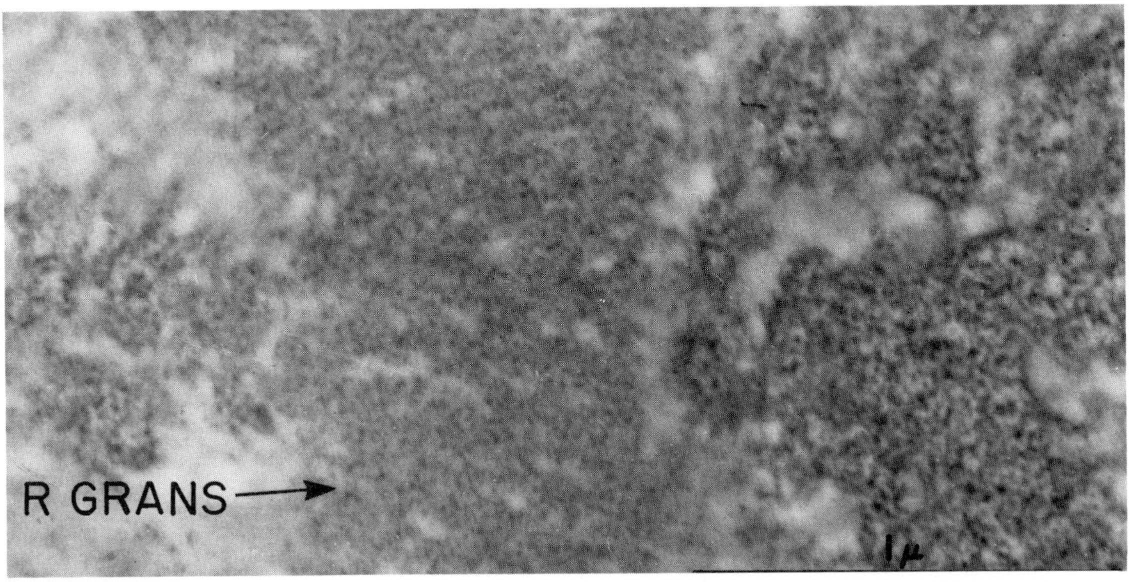

FIG. 4.2. ▲

▼ FIG. 4.3.

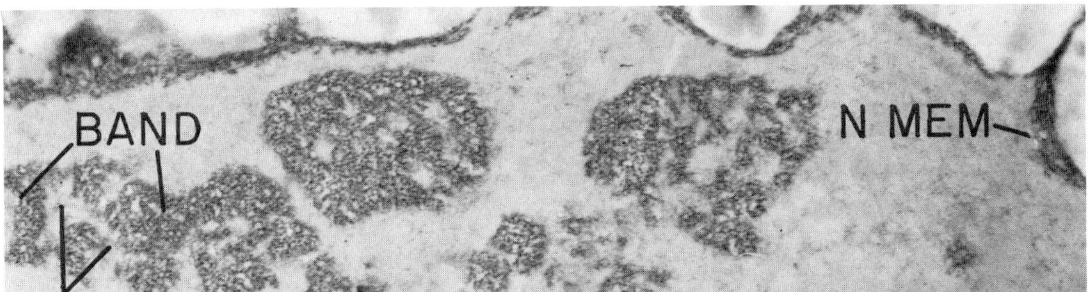

FIG. 4.3. See p. 100 for complete legend.

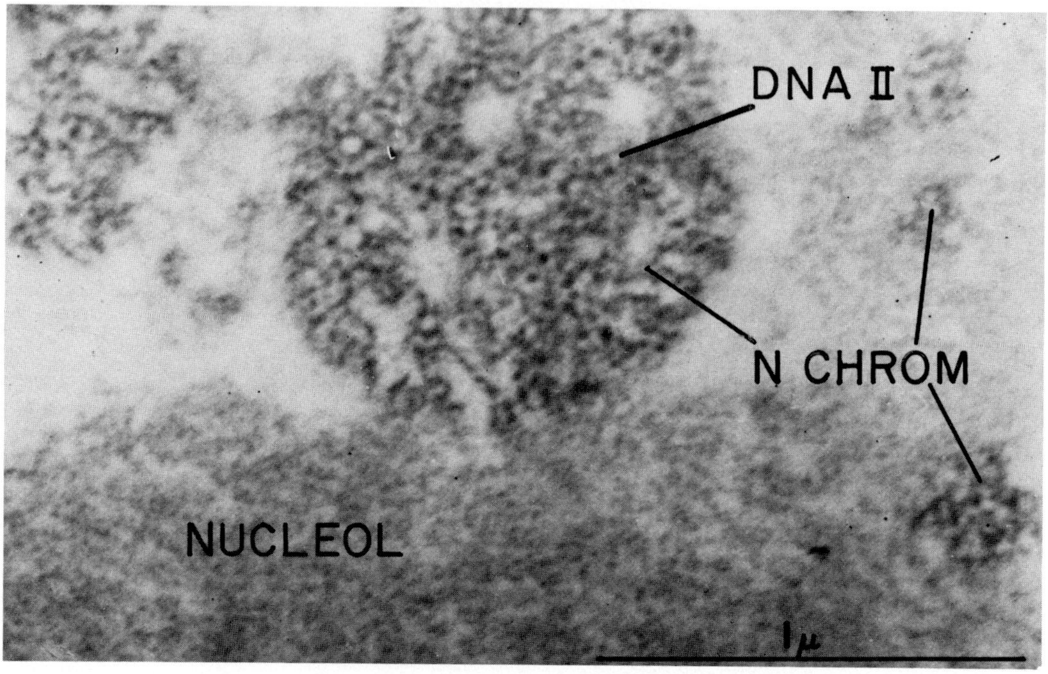

FIG. 4.4. See p. 100 for complete legend.

reacted or released cross-linking reagent. Specimens were then washed with water, dehydrated, hardened, and stained with gallocyanin–chromalum. After this they were hardened again and embedded in water-soluble Durcupan as above. The blocks were sectioned; sections were examined and photographed in the electron microscope under the same conditions as those used in Chapter 3, this volume.

Ribonuclease removes all or most of the RNA so that the remaining substance of high contrast consists mainly of DNA; deoxyribonuclease removes all or most of the DNA and the remaining substance represents mostly RNA. Both enzymes used successively remove all or most of both types of nucleic acids, and the remaining material represents the intrinsic, nonspecific density attributable mainly to proteins, and to enzyme and hydrolytic products adsorbed by them.

The use of enzymes is marked by several uncertainties. For example, the rate of hydrolysis of some forms of DNA and RNA may differ, and some may resist digestion altogether *(7)*. Thus, it is possible that in our material some nucleic acid remaining after the action of ribonuclease may nevertheless be RNA, and not all nucleic acid remaining after the action of deoxyribonuclease need be RNA. In some instances in this work, and in several instances in the literature, the sites of nucleic acids in cells are identified after use of nucleases by holes or empty spaces. In some instances, one can be sure that not only nucleic acid but also protein had been dissolved by water in enzyme solution. We have avoided this, for the most part, by prior cross-linkage of proteins in a nonfluid medium. It is probable that despite the insolubilization of proteins by prior cross-linking, minor displacements could not be prevented. It is clear, in addition, that DNA and granular RNA were both swollen as a consequence of the exposure to the water of the enzyme solution. Despite these uncertainties, the nucleases are very useful in many instances in discriminating, with a high degree of probability, between DNA and RNA.

It was thought to be important to select certain parts of chromosomes to study the fine structure of such regions and the morphological state and amount of nucleic acids in relation to chromosomes in different states of activity. The regions selected were several parts of the heterochromatic chromocenter, several puffs,

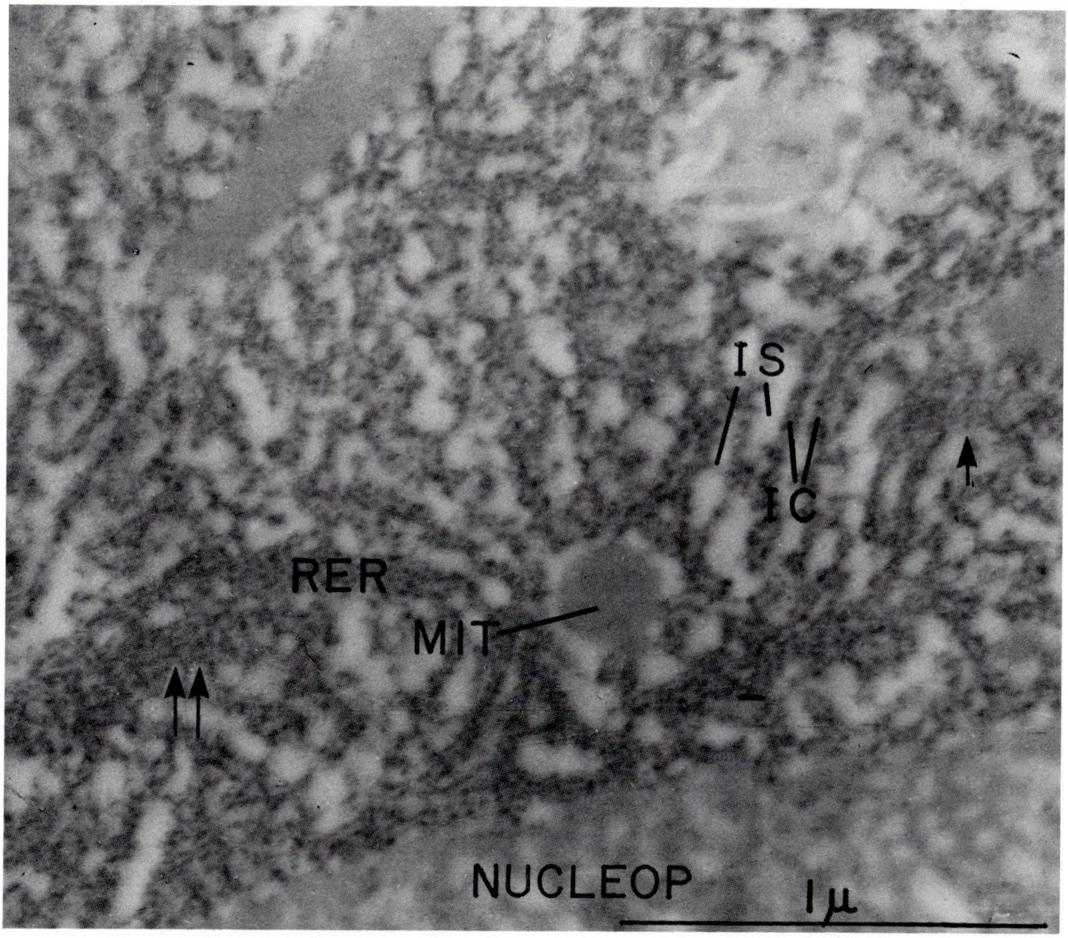

FIG. 4.5. The spongelike structure of the nucleoplasm (NUCLEOP) is apparent. In some parts of the rough endoplasmic reticulum (RER) the ribosomes are discrete (↑). In others they lie in a uniform (unreasonable) density of RNA (↑ ↑). The intracisternal spaces (IC) are uniformly narrow, while the intercisternal spaces (IS) are wider. The small granules are probably mitochondria (MIT). × 56,000.

region 3 of the X chromosome, the tip of the X chromosome, and the base of chromosome 3L. For this purpose specimens were fixed, stained, and embedded in the usual way. Sections were cut 6-μ thick with a sliding microtome and a slanted hard-steel knife, and mounted on separate slides in a nonpolymerizing mixture of Durcupan components ABD. The sections were examined and those nuclei were selected for further study in which some known cytological component could be recognized and lay in a possibly favorable plane. The nuclei were photographed or drawn to aid in subsequent identification. These 6-μ sections were reembedded in water-soluble Durcupan, and sliced with an ultramicrotome. The sections were mounted at random, and all sectioned nuclei recognized on the fluorescent screen were photographed at $\times$ 4000. A photomontage was made of the $\times$ 2 enlarged prints, and the selected chromosome was photographed again, this time at a magnification of $\times$ 28,000, with customary enlargement of the prints. The resulting photomontage was, of course, too large for reproduction. In order

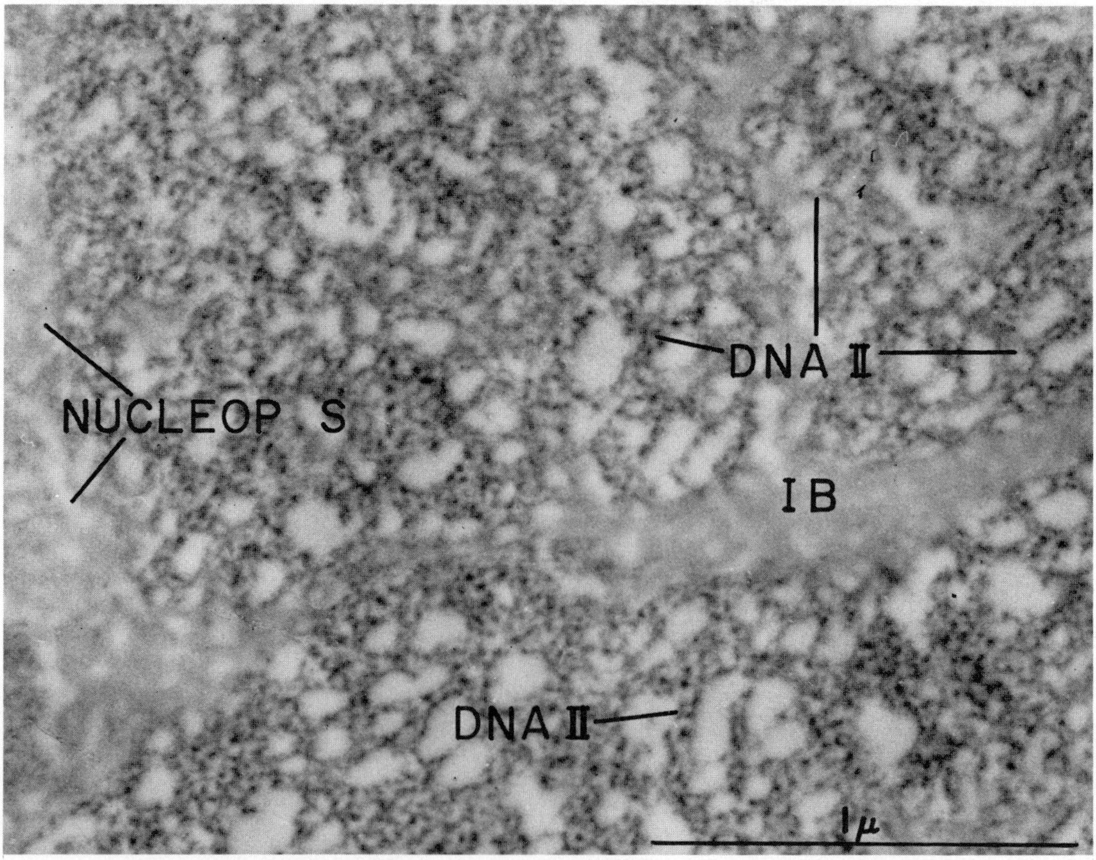

FIG. 4.6. Second-order DNA coils (DNA II) are arranged particularly around larger spaces some of which communicate with the nucleoplasmic (NUCLEOP S) and interband (IB) spaces. $\times$ 56,000.

to present the results at all, tracings were made of the enlarged photomontage, and the various parts were coded according to their content of RNA and second-order coils of DNA. These tracings were then reduced for publication. These coded tracings with a few electron micrographs on which the tracings were based are presented to illustrate the structures.

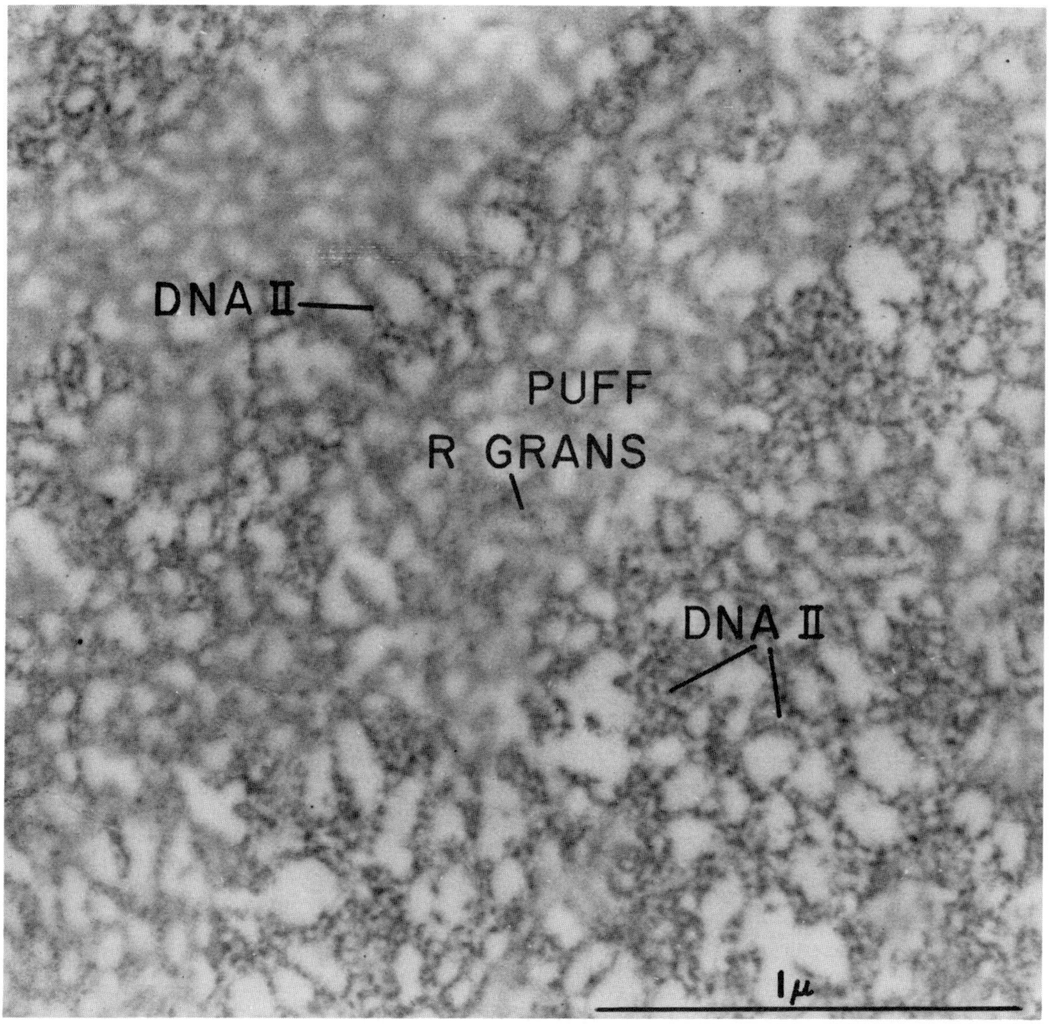

FIG. 4.7.

FIGS. 4.7 and 4.8. Sections probably through chromosomal puffs. In Fig. 4.7, the puff (PUFF) is small. The spongelike walls of the chromosome contain numerous RNA granules (R GRANS) between a thick and thin band of DNA arranged in second-order coils (DNA II). In addition, there is much unresolvable RNA. Figure 4.8 is a somewhat larger puff, and the RNA granules (R GRANS) and background density are also prominent. Lying in the midst of this region are finely distributed second-order DNA coils (DNA II). × 56,000.

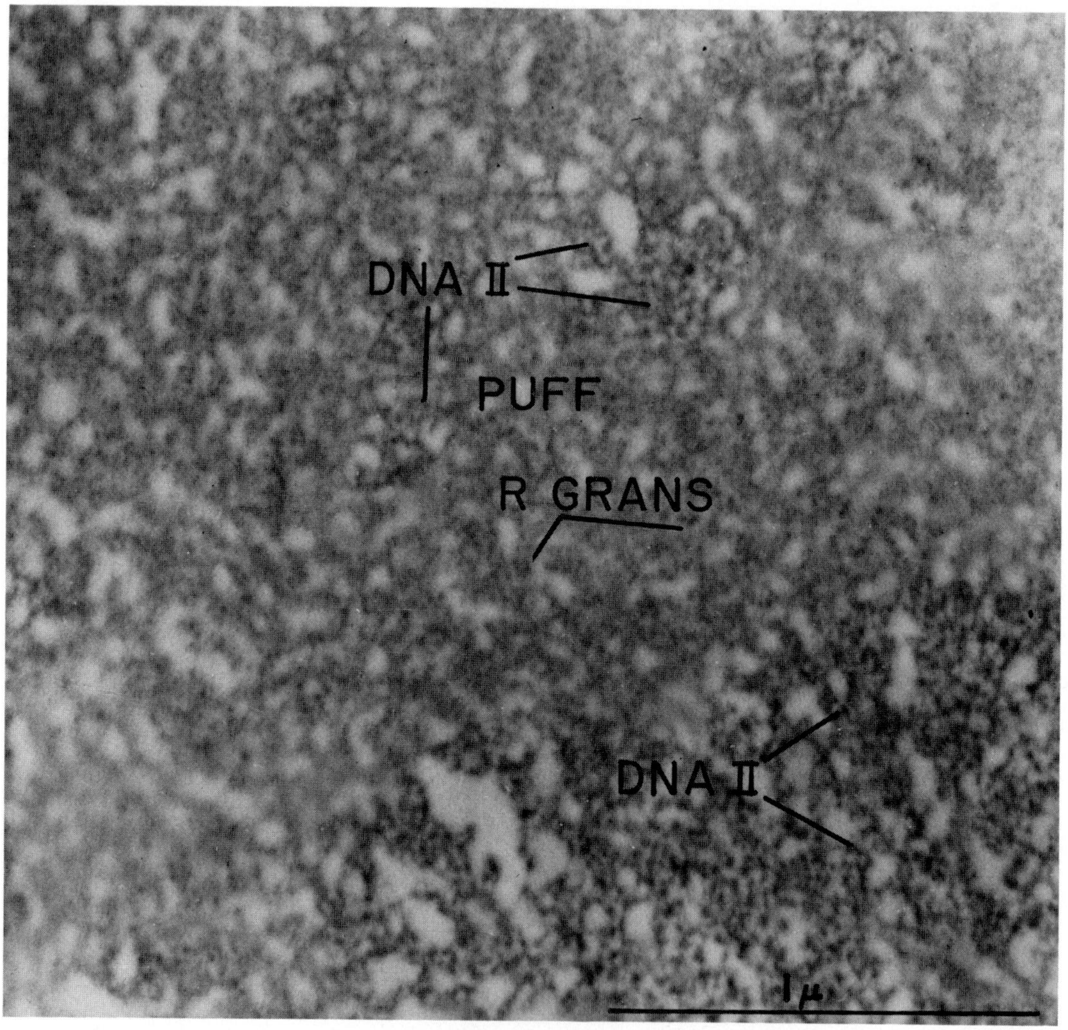

FIG. 4.8. See p. 105 for complete legend.

Observations

The cytoplasm of the salivary gland cells is rich in rough endoplasmic reticulum (RER), whose course is indicated by the alignment of dense ribosomes around the nuclear membrane and the secretory granules (Figs. 4.1, 4.3, and 4.5.) While some ribosomes are discrete, most seem to be enveloped in a cloud of moderate density of unresolved material, which represents, in part transfer RNA (tRNA). The nuclear membrane is unstained and appears structureless (Figs. 4.1 and 4.3). The nucleoplasm consists of numerous submicroscopic dark (denser) and light (less dense) regions. The former are the walls of the spongelike structure of the nucleoplasm, while the latter are the "pores" or channels of the spongelike structure. The walls in certain regions are free of dense granules, while in others, the walls contain numerous dense granules frequently embedded in an unresolvable cloud of moderate density. In small regions, these granules are concentrated as patches which are more obvious in thick sections (Fig. 4.3). At low magnification, it can be seen that the less dense spaces are continuous with those of the nucleolus and chromosomes as gaps and slits of low density. In the bands, they are narrow and irregular (Figs. 4.3 and 4.6), while in the interband regions they are wider. There is no evidence of a chromosomal or nucleolar capsule.

The nucleolus consists of numerous granules ranging from about the size of ribosomes to the limit of resolution. It is possible that the larger granules represent

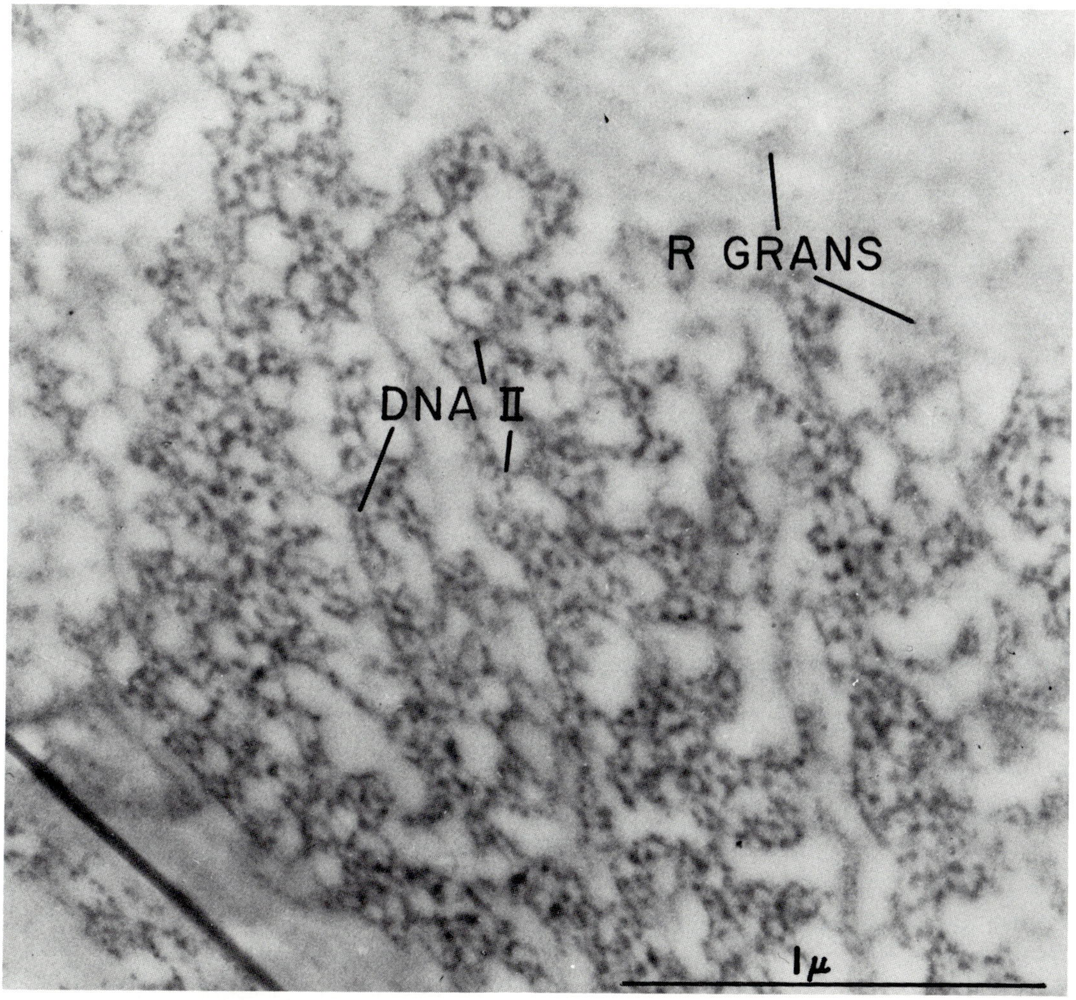

Fig. 4.9.

Figs. 4.9–4.11. In some parts of the chromosome, the second-order DNA coils (DNA II) are loosely packed compared to those in Figure 4.6. RNA granules (R GRANS) are present in some of the spongelike walls, but are less numerous than those seen in Figs. 4.7 and 4.8. × 56,000.

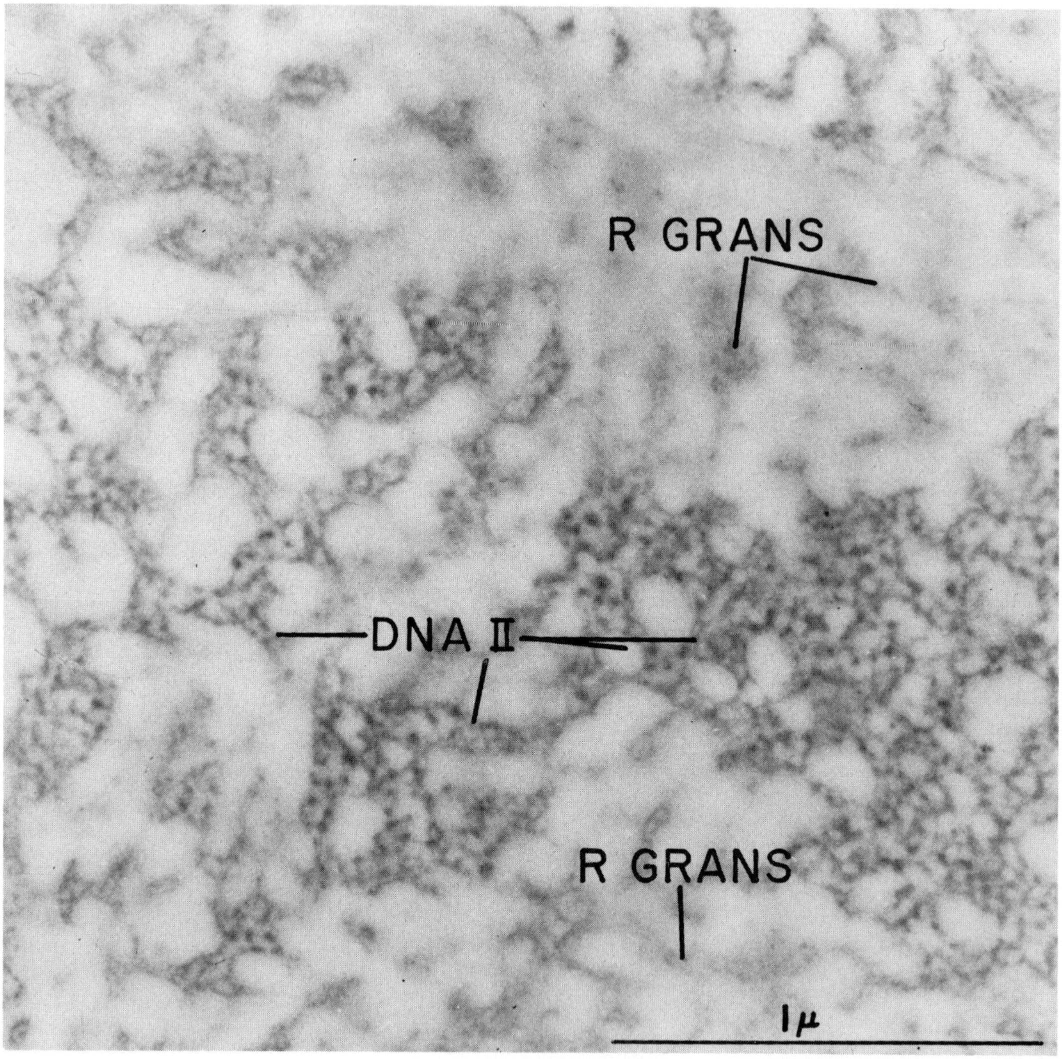

FIG. 4.10. See p. 107 for complete legend.

fortuitous patterns of over- or underlay of smaller granules. These granules are rich in RNA as will be shown later. They are contained within regions continuous with the nucleoplasmic walls, and the less dense spaces are continuous with similar spaces in the nucleoplasm. Small regions occur in some parts on the surface and internally which are composed of secondary coils of DNA, and which are probably portions of the nucleolar organizing chromosomes (Figs. 4.3, 4.4, 4.12, 4.16, and 4.20). The evidence for this identification is given below.

Most of the dark bands of chromosomes appear as compound structures made up of lamellae (Figs. 4.12, 4.14, and 4.16). Some of them are rich in DNA, which appears in electron micrographs as circular or oval lines which represent second-order coils. Granules rich in RNA are present between the coils. These lamellae

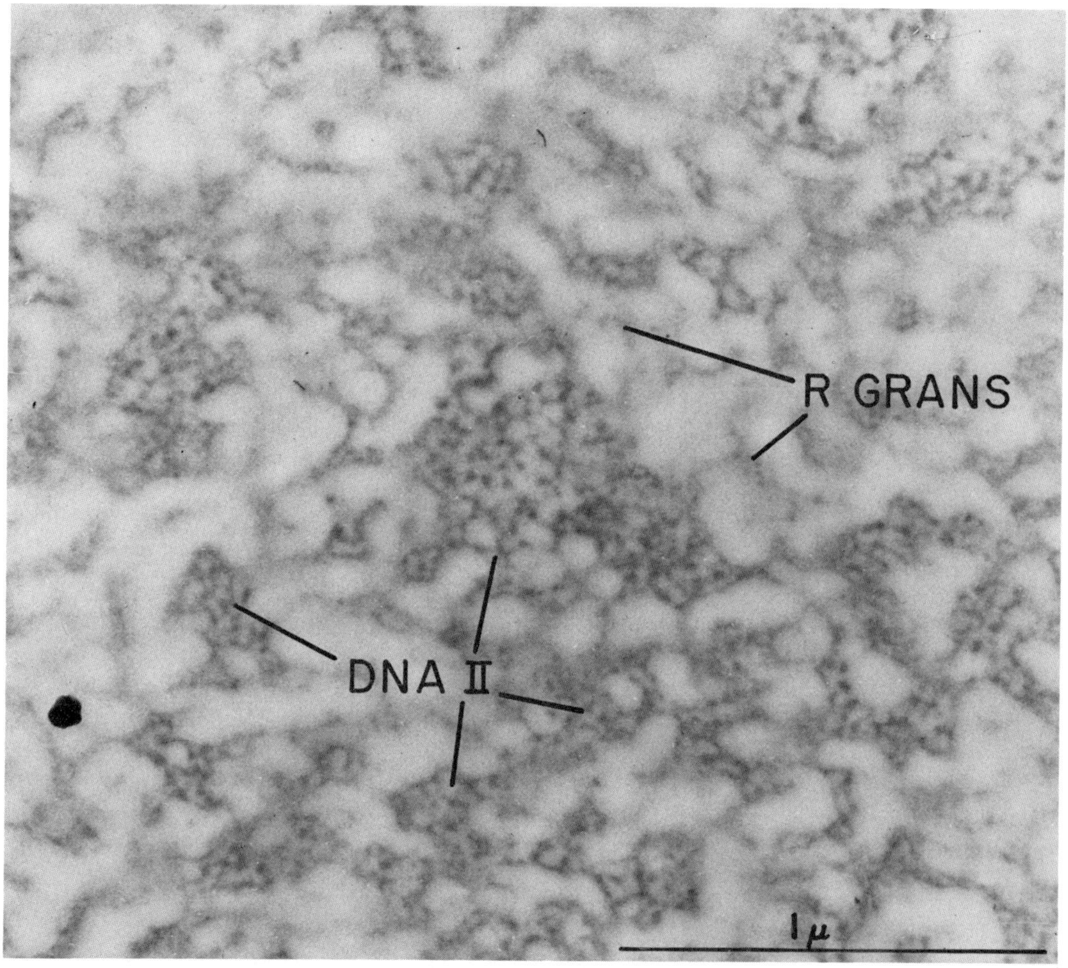

FIG. 4.11. See p. 107 for complete legend.

alternate with others where the RNA granules are very numerous and the DNA less prominent. Together, the two types of lamellae constitute the dark band. The interbands resemble the DNA-poor lamellae, but have fewer RNA granules. There are also less dense spaces within the DNA-rich lamellae which continue into the spongelike spaces of the slits or gaps in the dark bands, the interband spaces, and the nucleoplasm (Figs. 4.3, and 4.6–4.11). Granules rich in RNA may be seen in the spongelike walls both in the chromosome and in the nearby nucleoplasm, and they are especially numerous in chromosome puffs. The evidence for these statements will be given below.

Characterization of Nucleic Acids from Enzyme-Tested Preparations

Effects of Ribonuclease. In the cytoplasm, nucleoplasm, and nucleolus, the contrast is enormously decreased, with a consequent increase in contrast of the second-order DNA coils, especially those associated with the nucleolus. Regions between the dark bands and between lamellae of chromosomes are also reduced in contrast. As a result, the second-order DNA coils which pass from band to band become more prominent (Figs. 4.12–4.19). These are especially notable in Figs. 4.18 and 4.19, which are more probably tangentially oriented with respect to the plane of sectioning through the bands.

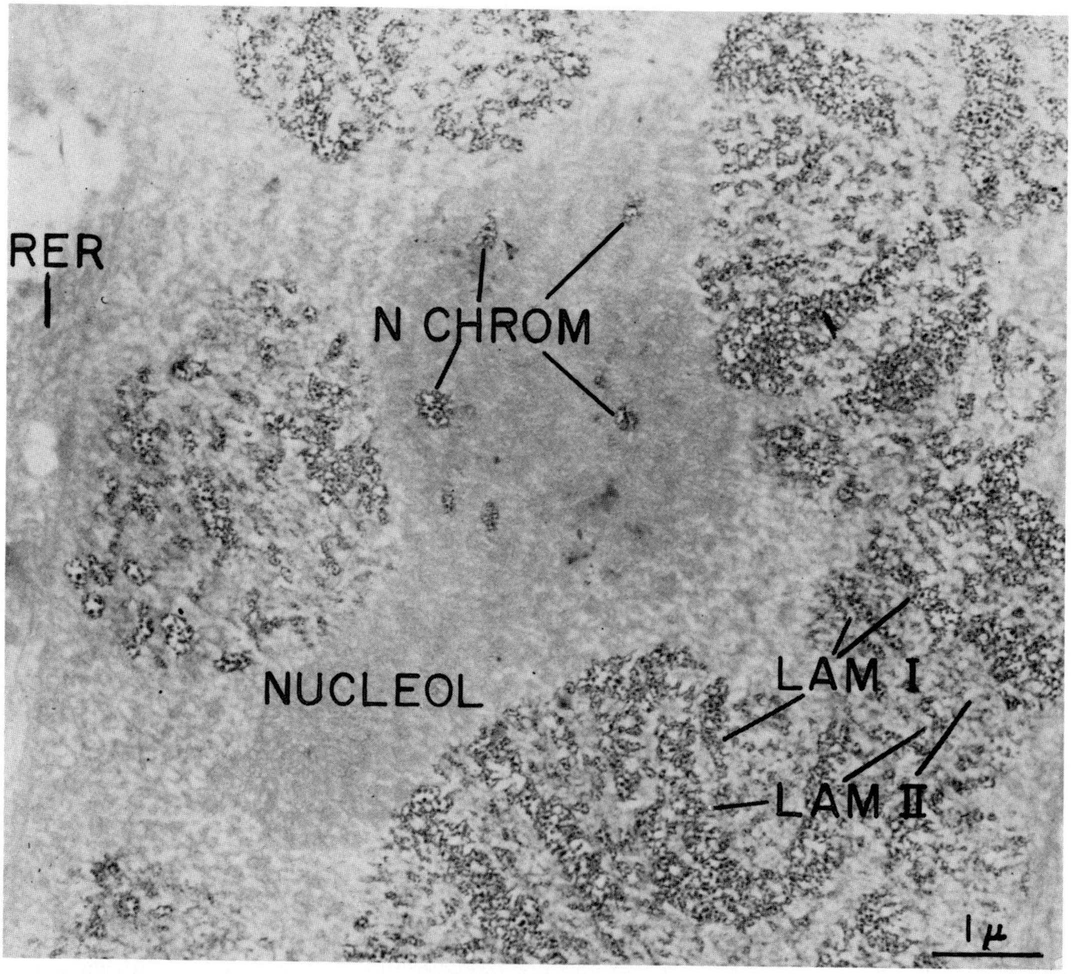

FIG. 4.12.

FIGS. 4.12–4.17. Figures 4.12 and 4.14 are low power electron micrographs showing the results of treating cross-linked salivary glands with ribonuclease before staining with gallocyanin-chromalum for nucleic acids. The low density areas are the RER, the nucleolus (NUCLEOL), and certain regions between lamellae (LAM II) from which most or all of the RNA has been removed by enzymic action. The patches of RNA granules in the nucleoplasm are no longer recognizable. All or most of the structures of high density have DNA. Chromosomal DNA (N CHROM) is present within the nucleolus (NUCLEOL). Chromosomes are composed of lamellae (LAM I) separated by pale regions from which RNA has been removed (LAM II). × 14,200.

Also clarified in these enzyme-treated specimens are enlarged spaces in the bands and lamellae around which are oriented the customary second-order DNA coils (Figs. 4.14–4.19). These have been interpreted as sites where the second-order coils are organized as relational coils. The term relational coil refers to two strands or two parts of the same strand which are twisted around one another. In this instance, the strands are second-order helices which seem to be folded back on themselves and twisted. This same arrangement has been suggested by Crick *(19)*. There seem to be large variations in thickness of the lamellae, but these may be only apparent, depending on how the plane of section passes through single lamellae, or the degree of overlap of adjacent lamellae which deviate from the strictly transverse.

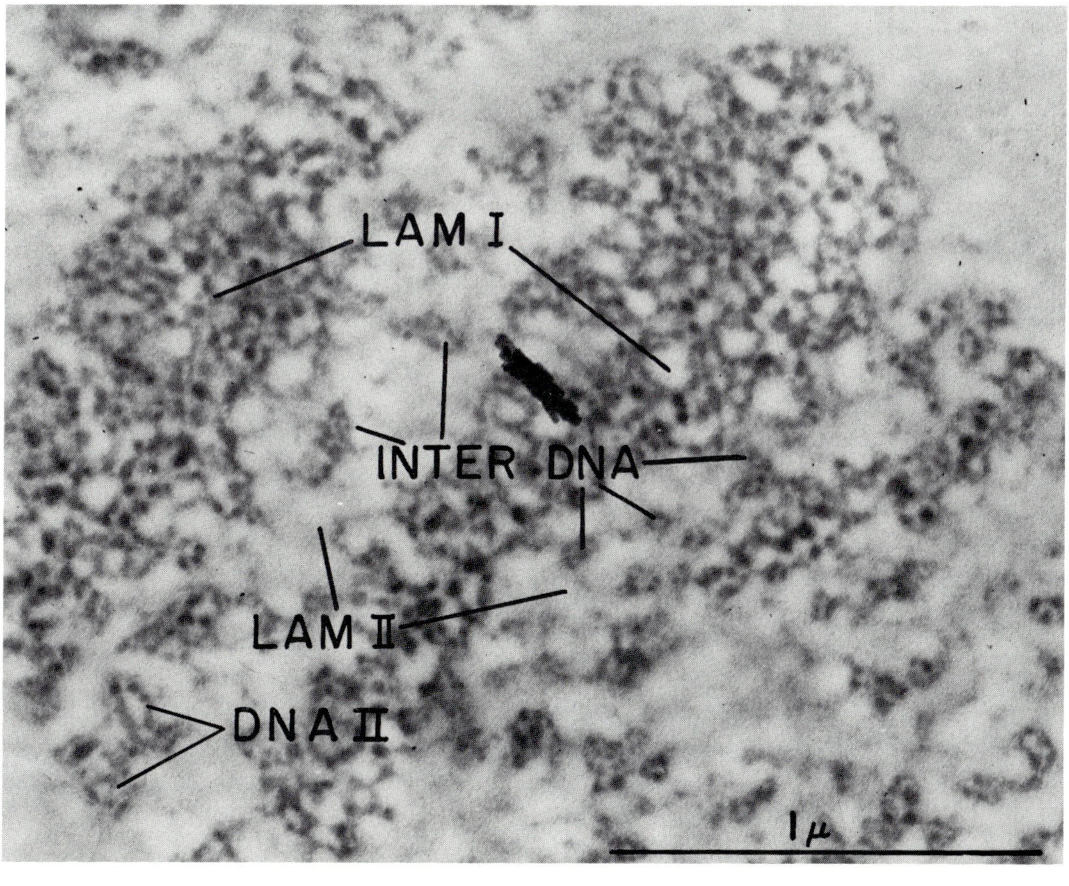

FIG. 4.13.

Figures 4.13, 4.15, 4.16, and 4.17 are higher magnifications. Seen here are second-order DNA coils (DNA II) grouped in lamellae (LAM I) around rather large spaces. The lamellae are not uniform, but seem to be interconnected with others or interrupted in many places. Between these, in the lamellae cleared of RNA (LAM II), are portions of isolated second-order DNA coils (INTER DNA) which obviously join the second-order coils of adjacent lamellae. × 56,000.

Effects of Deoxyribonuclease. Contrast of the cytoplasmic, nucleoplasmic, and nucleolar granules is somewhat reduced, as compared with normal photographs (Figs. 4.20 and 4.21). The customary high contrast of second-order DNA coils of intranucleolar chromosome regions is nearly abolished. When digestion is incomplete, residual DNA occurs as large granules. The smaller RNA granules occur in columns which correspond with lamellae II. The evidence is clear that the regions (lamellae) rich in RNA granules alternate with regions (lamellae) rich in DNA. RNA granules are also present in the interband regions, nucleoli, nucleoplasm, and, of course, in the ribosomes of the RER.

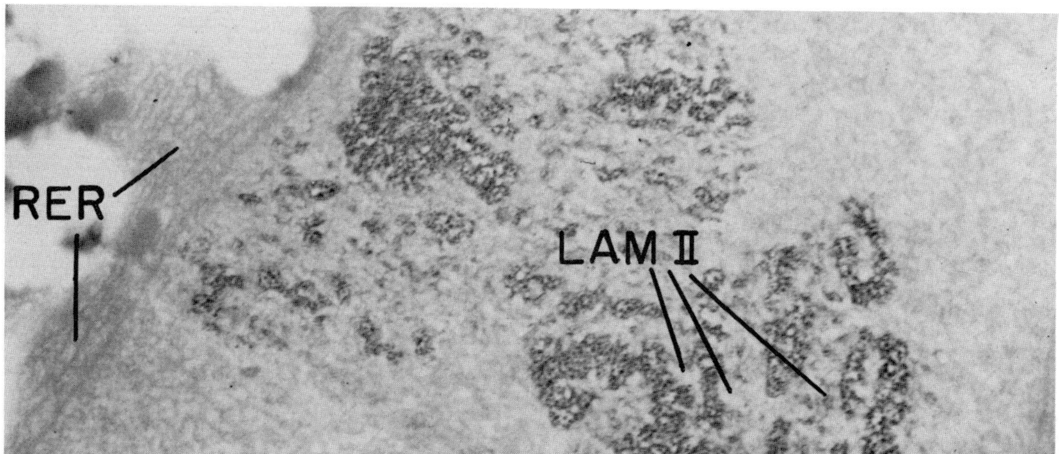

FIG. 4.14. See p. 110 for complete legend.

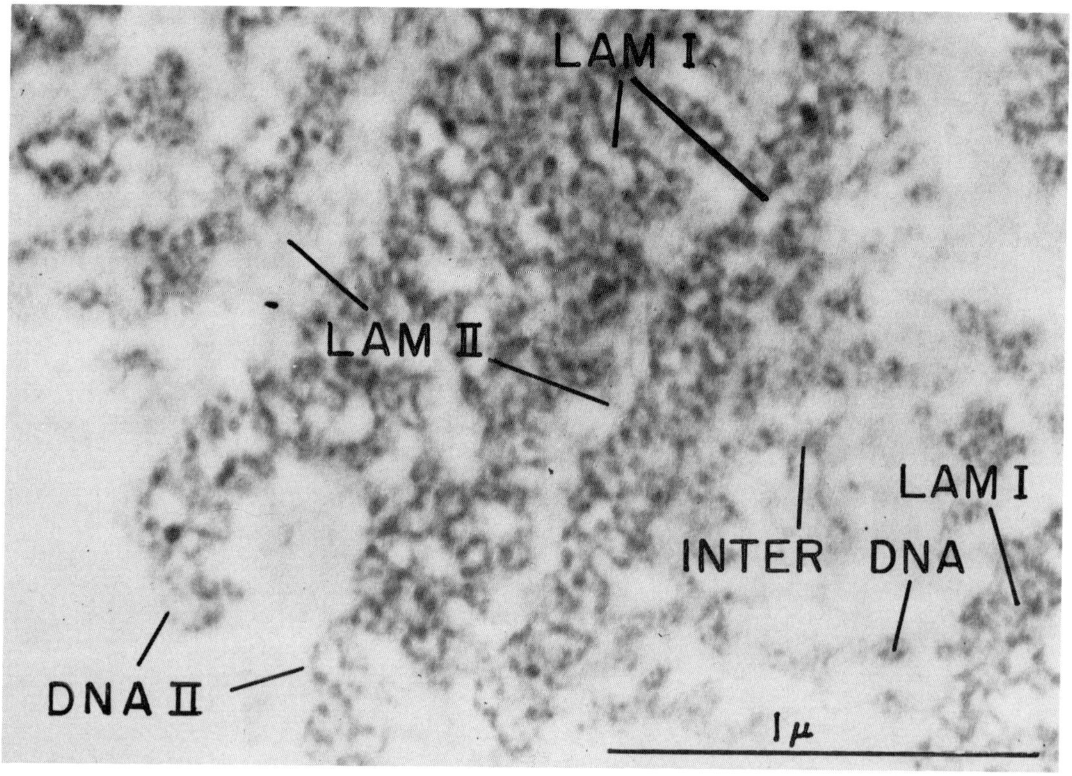

Fig. 4.15. See pp. 110–111 for complete legend.

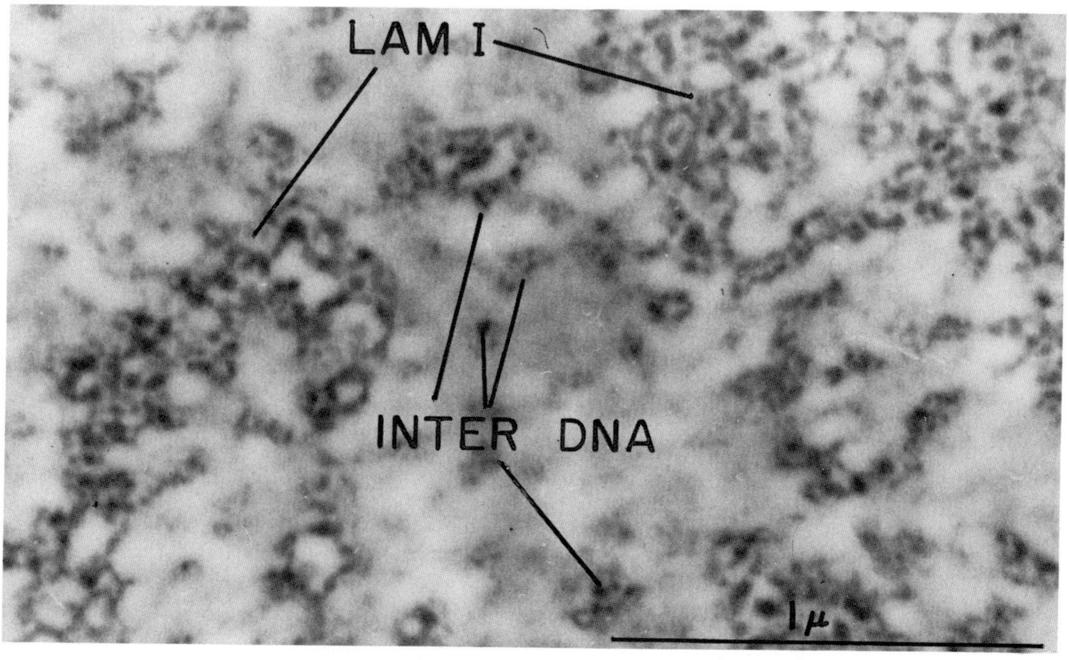

Fig. 4.16. See pp. 110–111 for complete legend.

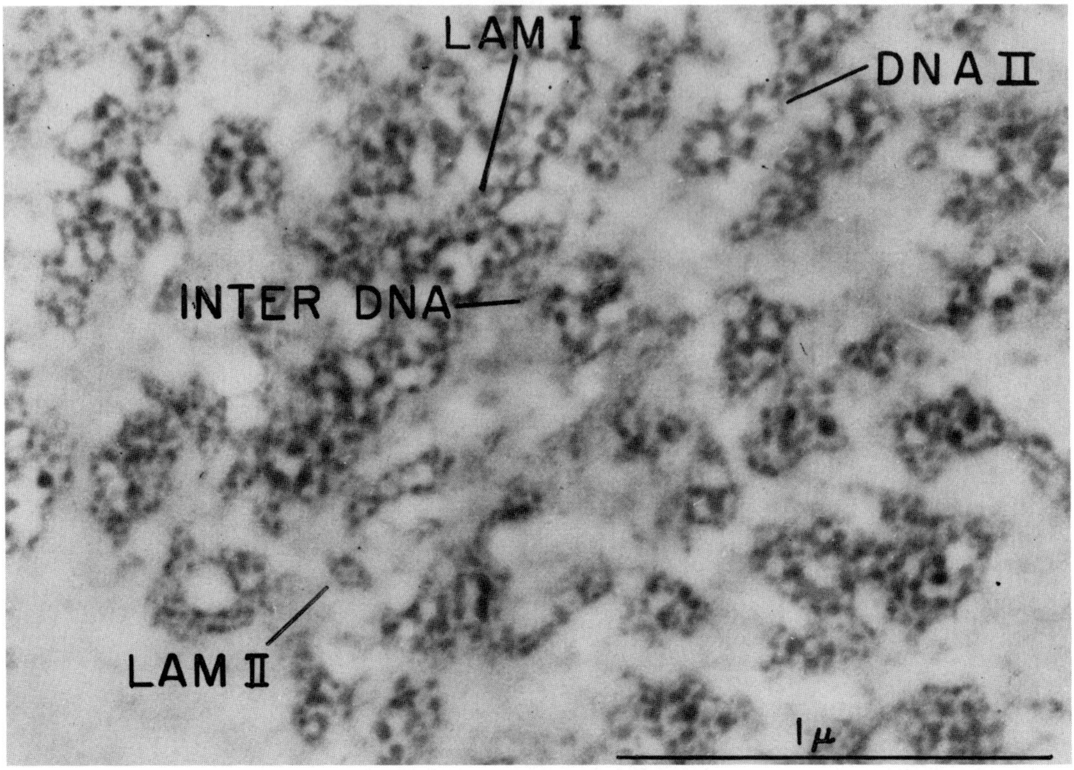

FIG. 4.17. See p. 111 for complete legend.

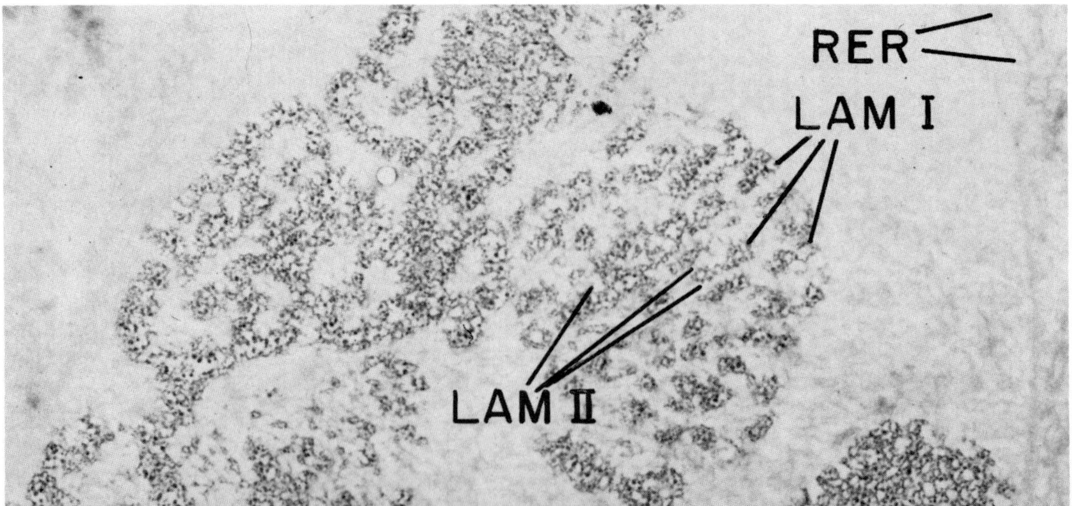

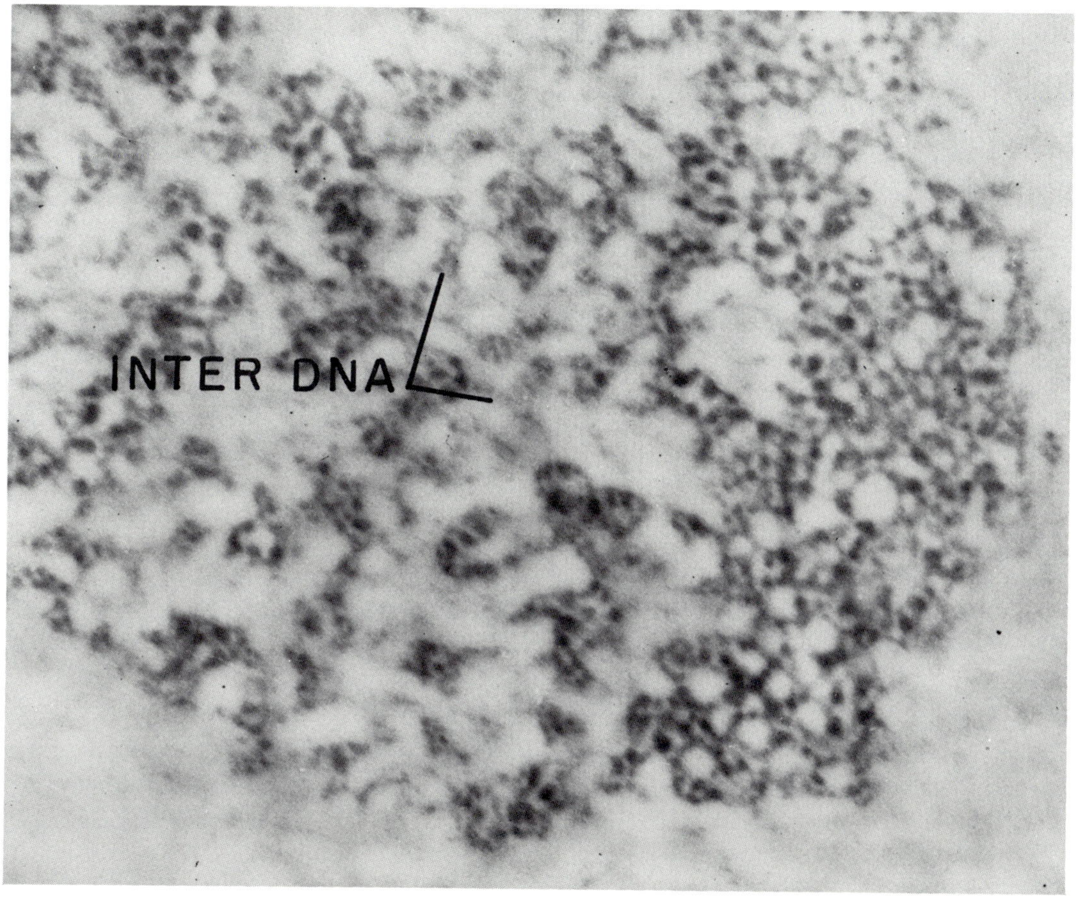

FIG. 4.19.

FIGS. 4.18 and 4.19. Electron micrographs of probably tangential sections of chromosomes to show portions of second-order DNA coils coursing in the interval (LAM II) between the DNA-rich lamellae (LAM I). Abbreviations as given previously. × 56,000.

◄ FIG. 4.18.

Effects of Successive Treatment of Salivary Glands with Deoxyribonuclease and Ribonuclease

Contrast is markedly reduced throughout the nucleus and cytoplasm to a more or less flat uniformity (Fig. 4.22).

Descriptions of Identified Parts of Chromosomes

1. *Known Base of 3L (Fig. 4.23, A–E).* This was identified in a 6-μ section by the band pattern (A), which is identical with that in a squash preparation of another salivary gland nucleus stained with orcein (B). The chromosome is clearly recognizable in a photomontage of electron micrographs of a section about 600 A thick through the same nucleus (C). The dark bands are not so clearly distinguished as in (A) or (B) because the dark bands now appear divided into numerous interconnected subbands or lamellae, these being of such small dimensions and so irregular that details of the band pattern vary greatly from (ultramicro) section to section. At this particular level of sectioning, the bands seem identifiable as indicated in the diagram (D). This diagram was made from a photomontage of electron micrographs printed at a magnification of × 56,000, and then reduced for publication. A constituent frame appears in (E). Here can be seen the second-order DNA coils with some granules in adjacent regions outlined by spaces which are continuous with similar spaces in the nucleoplasm. The walls of these spaces are also continuous with those of the spongelike nucleoplasm, and may (as noted above) contain dense granules.

2. *Known Tip of X Chromosome (Fig. 4.24, A–E).* This was identified in a 6-μ section (A) from the similarity in band pattern with that in a squash preparation of another gland stained with orcein (B). When this section was reembedded and, in turn, sectioned, the same tip of the chromosome was identified in a photomontage of electron micrographs of low magnification (C). Here again the dark bands are subdivided into interdigitating subdivisions, but the correspondence in band pattern is clear, and is indicated in the diagram in (D). A frame from the photomontage of electron micrographs at higher magnification is shown in (E). The distribution of nucleic acids is shown in (D). In the solid black regions, the second-order DNA coils are dominant. In the spongelike walls between these regions are pale staining DNA coils and RNA in the form of granules embedded in a diffuse density (cross-hatched and coarsely stippled). The nucleoplasmic walls in the vicinity also contain this diffuse form of RNA (finely stippled).

3. *Known Bands of X Chromosome, Region 3 (Fig. 4.25, A–E).* This was identified in a 6-μ section (A) by the similarity in band pattern with that in a squash preparation of another gland stained with orcein (B). The fine structure of this same region was described by Berendes *(9)*. When the same section was sectioned by ultramicrotomy, the nucleus appears in electron micrographs at low magnification as in the photomontage (C). The dark bands of this chromosome region comprise branching walls containing characteristic second-or-

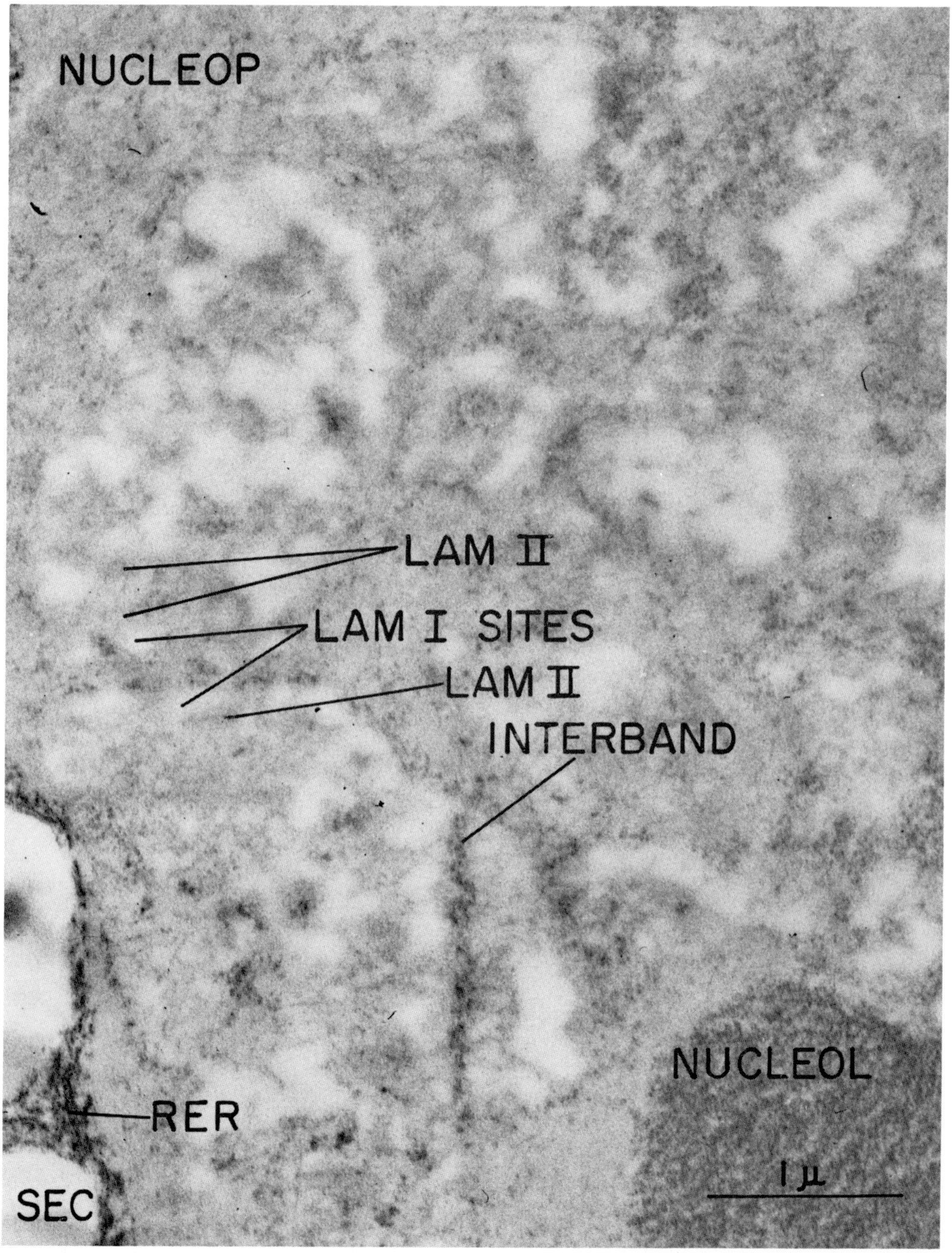

FIG. 4.20. Electron micrograph to show the effects of treating cross-linked salivary glands with deoxyribonuclease before staining with gallocyanin–chromalum for nucleic acids. The DNA coils have been largely removed by enzymic action, the spaces formerly occupied by them being mostly blank (LAM I sites). RNA granules are visible in lamellae (LAM II), interbands, nucleolus (NUCLEOL), nucleoplasm (NUCLEOP) and in the cytoplasmic ribosomes of the rough endoplasmic reticulum (RER). The latter surround secretory granules (SEC). Some granules in the nucleoplasm and interbands are aggregated. × 28,000.

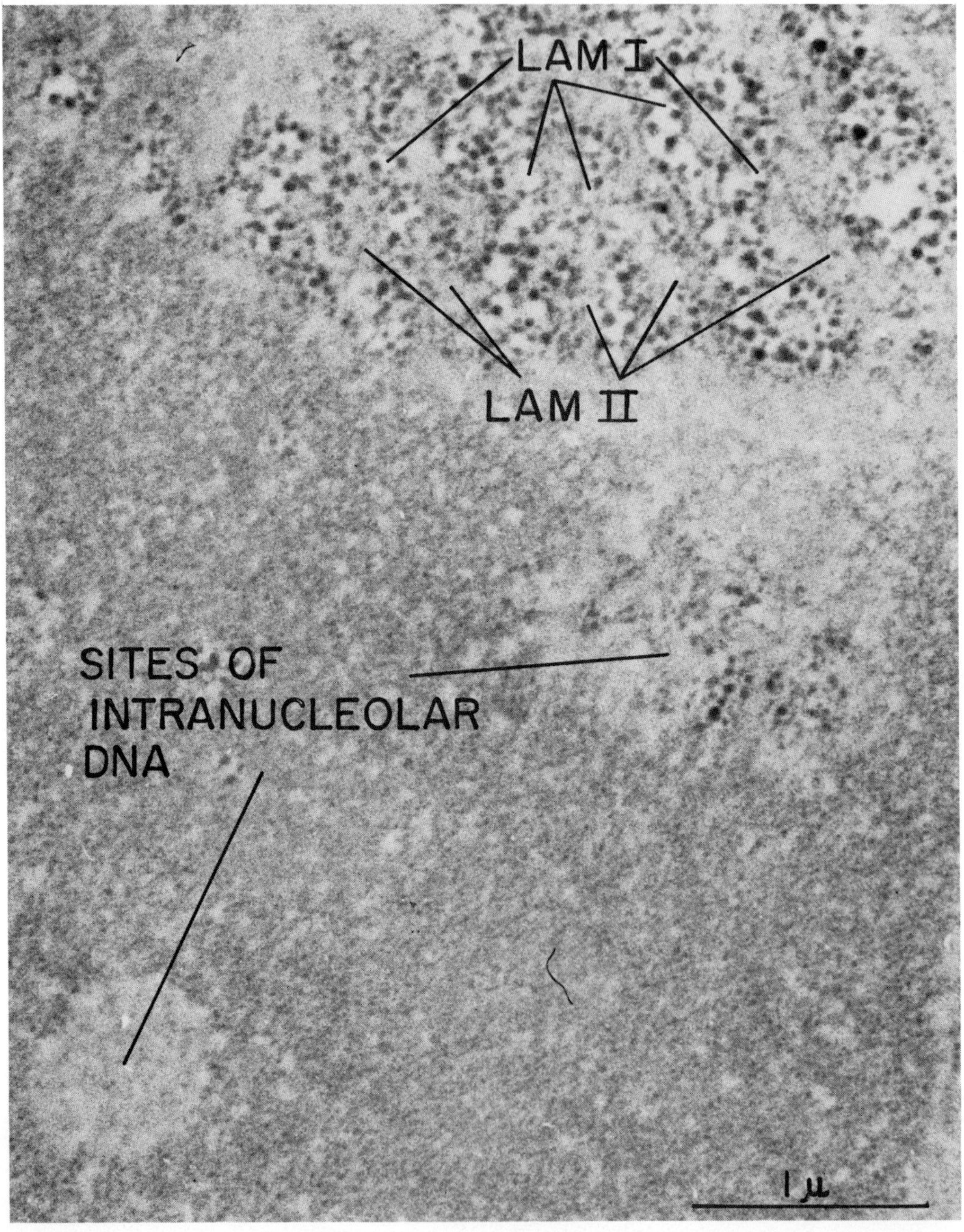

Fig. 4.21. Electron micrograph of a preparation treated with DNase as in Fig. 4.20. The clear spaces in the nucleolus (NUCLEOL) were occupied by intranucleolar second-order coils of DNA. This was almost completely removed by the action of the enzyme. RNA granules are recognizable in the same sites as in Fig. 4.20. and, in addition, in the sites where intranucleolar DNA had been present before treatment with DNase. × 32,000.

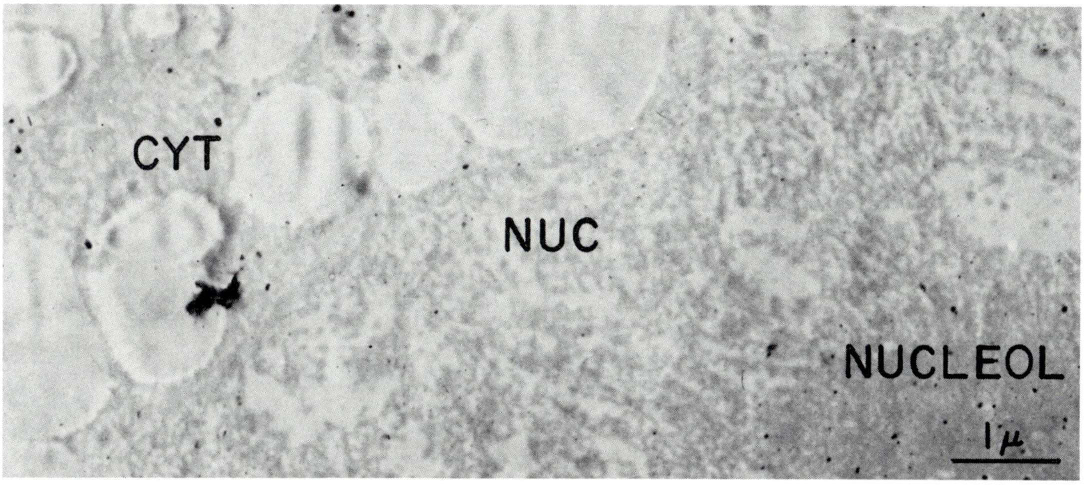

FIG. 4.22. Contrast is drastically reduced when salivary glands are treated first with deoxyribonuclease and then with ribonuclease prior to staining with gallocyanin–chromalum for nucleic acids. All parts of the RER, nucleolus (NUCLEOL), nucleoplasm, and chromosomes have the same minimal density because both types of nucleic acid have been removed from both nucleus (NUC) and cytoplasm (CYT). The process of digestion in aqueous solutions is accompanied by solution, displacement, and reprecipitation of residual protein components despite the prior action of the cross-linking reagent. × 14,200.

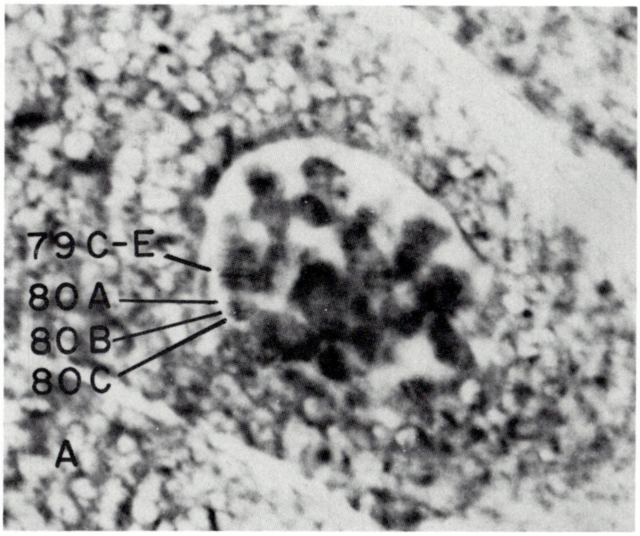

FIG. 4.23. Identification and analysis of a portion of the base of chromosome 3L. Bands are numbered following Bridges *(11)*. (A) Photomicrograph of a 6-μ section through a salivary gland nucleus to show banding characteristic of the base of 3L. × 1800.

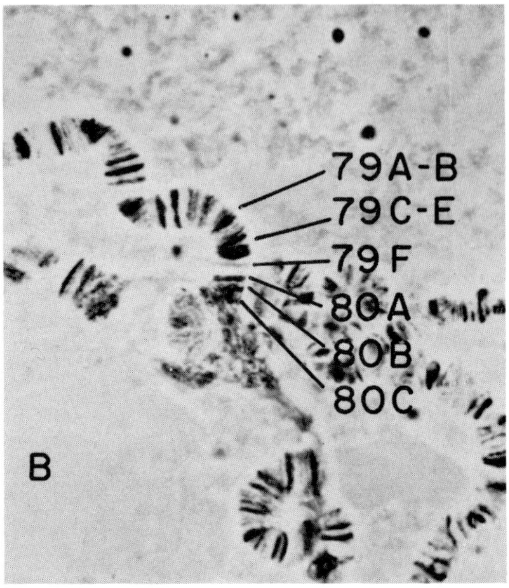
79A-B
79C-E
79F
80A
80B
80C
B

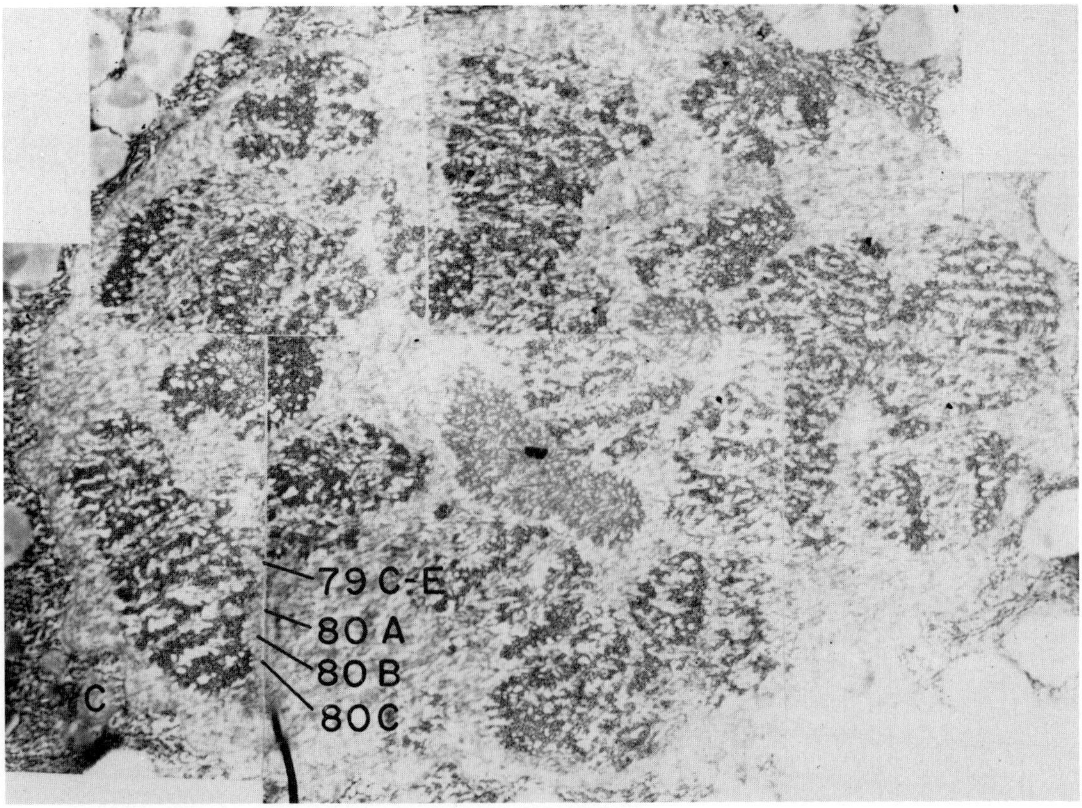
79C-E
80A
80B
80C
C

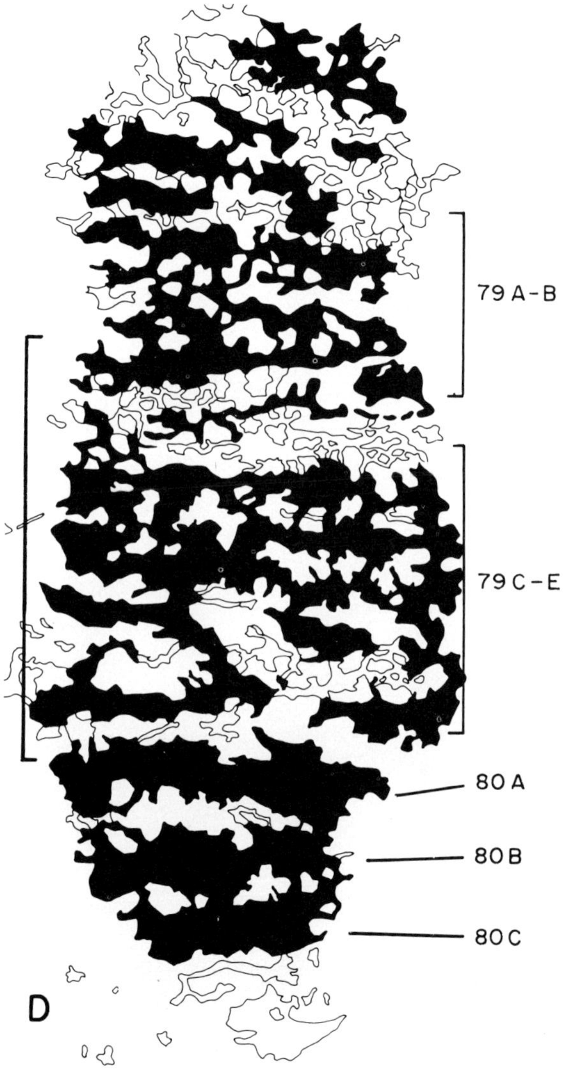

FIG. 4.23(B–E). (B) Photomicrograph of a salivary gland squash preparation to show the identity in banding at the base of 3L. × 1200. (C) Photomontage of electron micrographs through the same part of the nucleus shown in (A), in a plane of section favorable for illustration of the same bands as in (A) and (B) above. × 5300. (D) Tracing of the banding pattern of the identified chromosome based on electron micrographs taken at high magnification, and subsequently reduced for publication. Most of the bands and lamellae (black) contain dark staining sections of DNA coils. The remainder contain pale staining sections of DNA coils, together with some RNA granules. The bracket encloses the part of the chromosome which was traced from the electron micrograph in (E). (E) Typical electron micrographs from which the tracings were made. × 56,000. See text p. 116 for description.

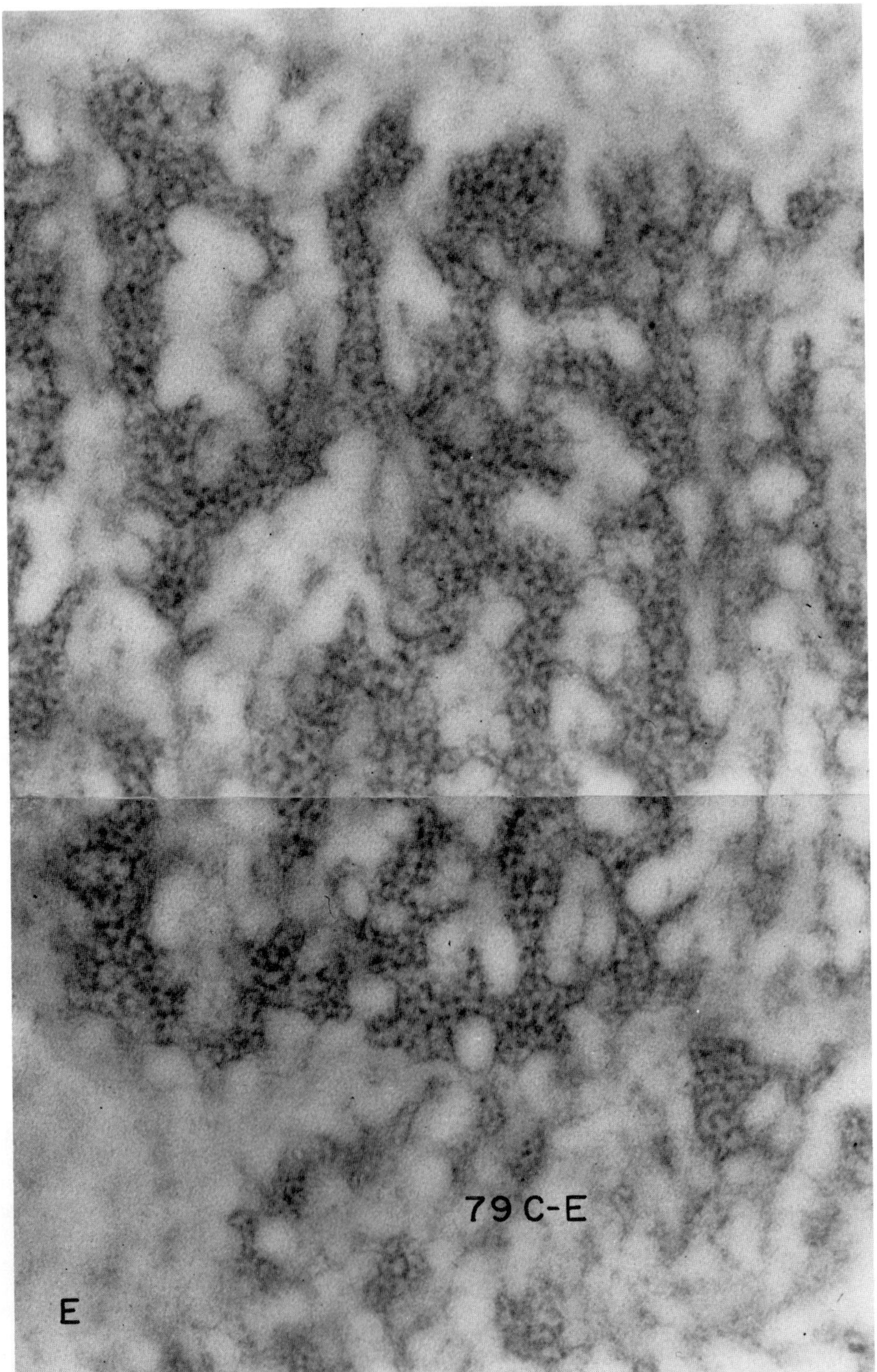

FIG. 4.23E.

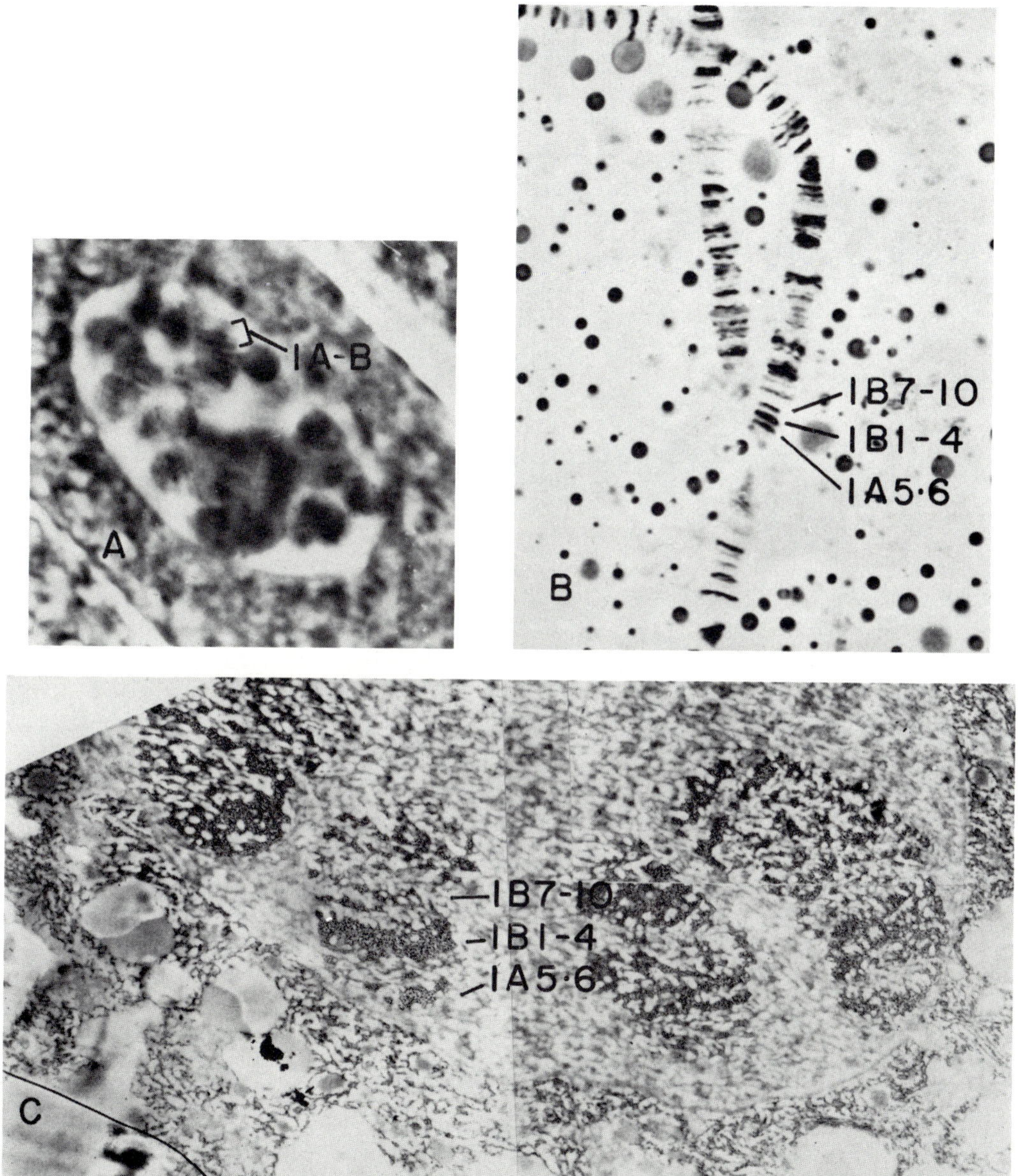

FIG. 4.24. Identification and analysis of the tip of the X chromosome. Bands are numbered following Bridges *(12)*. (A) Photomicrograph of a 6-μ section through a salivary gland nucleus showing banding characteristic of the tip of the X chromosome. $\times$ 1800. (B) Photomicrograph of a salivary gland squash preparation to show the identical banding at the tip of the X chromosome. $\times$ 1200. (C) Photomontage of electron micrographs of an ultrathin section of the same nucleus shown in (A). $\times$ 5300. (D) Tracing of the banding pattern of this chromosome tip based on tracings of electron micrographs at high magnification, and subsequently reduced for publication. Black patches contain chiefly dark staining sections of DNA coils. Stippled and cross-hatched regions contain pale staining sections of DNA coils and RNA granules. Finely stippled regions are nucleoplasm and chromosomal clefts. The bracket encloses the part of the chromosome region which was traced from the electron micrograph in (E). (E). Typical electron micrograph from which tracings were made. $\times$ 56,000. See text p. 116, for description.

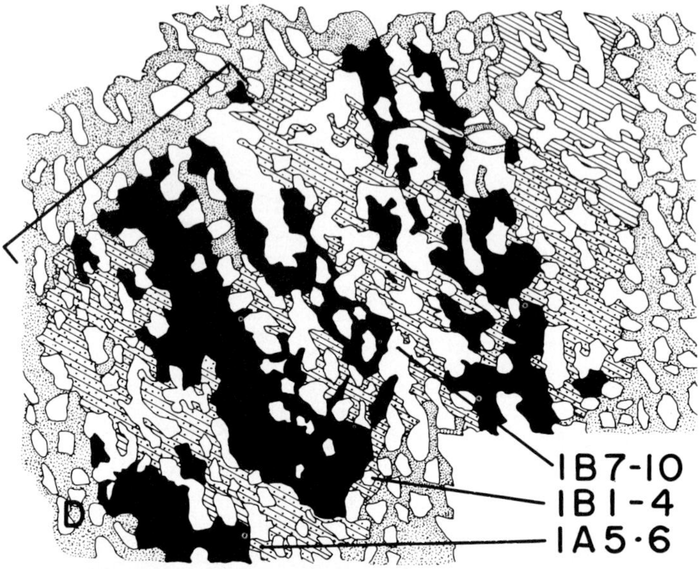

FIG. 4.24D.

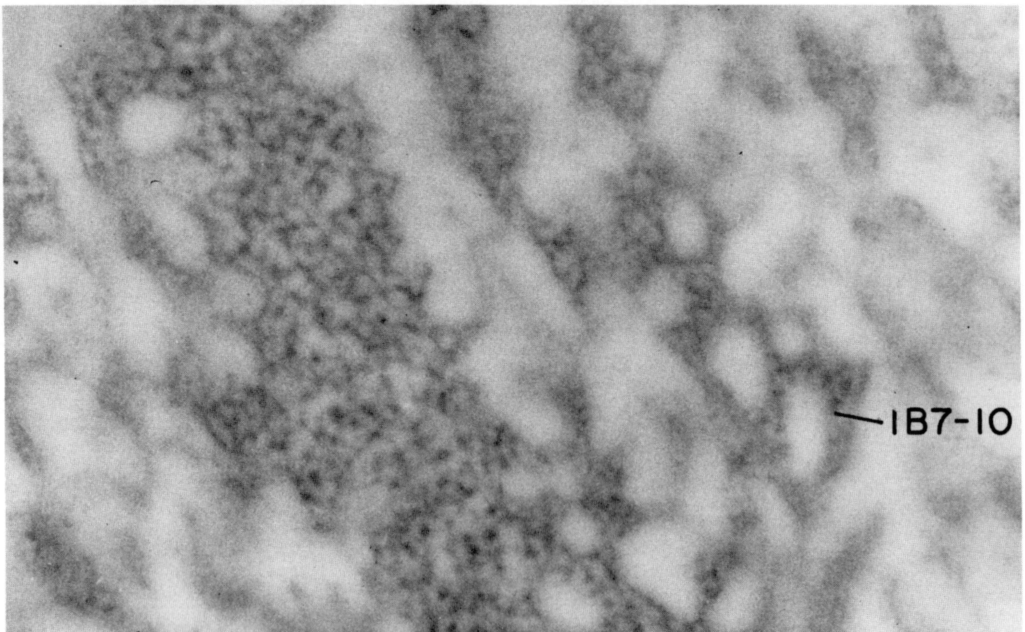

FIG. 4.24E.

der DNA coils, as shown in the diagram prepared from a tracing of a photomontage made from electron micrographs at high magnification (D). The code in the diagram is the same as that used in the diagram of the known tip of the X chromosome. An original electron micrograph on which the diagram is based appears as (E).

4. Known Puff Region (Fig. 4.26, A–E). The puffs identified in 6-μ sections (A) are probably those in region 74–75 of the left arm of chromosome 3 shown in the squash preparation of another gland stained with orcein (B). The same nucleus was further sectioned by ultramicrotomy and the puff was identified in a photomontage of electron micrographs at low magnification (C). The same region was reconstructed from a photomontage of electron micrographs at higher magnification and the resulting coded diagram was reduced for printing (D). Two frames from the original photomontage are presented in (E). In addition to the spongelike dense areas representing the second-order DNA coils, are the swollen regions characterized by pale staining secondary DNA coils, with numerous RNA granules embedded in a uniform background density of unresolvable RNA. Both DNA coils and the RNA occupy the spongelike walls which are separated by less dense spaces. The walls of the chromosome and nucleoplasm are connected, as are the spaces.

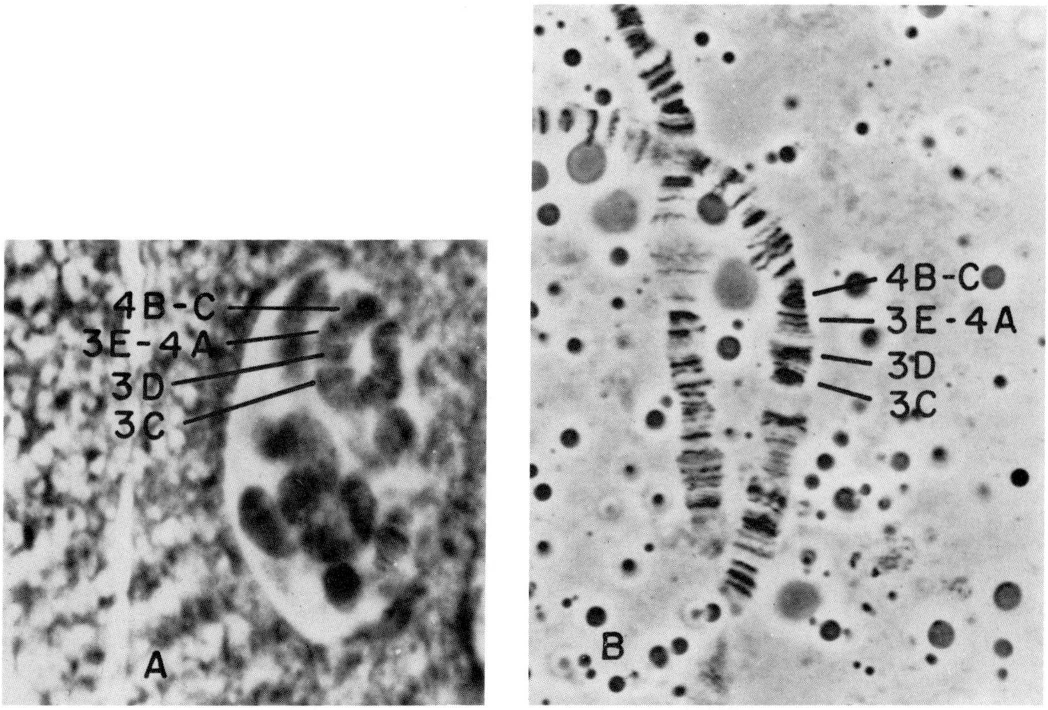

FIG. 4.25. Identification and analysis of region 3 of the X chromosome. Bands are numbered following Bridges *(12)*. (A) Photomicrograph of a 6-μ section through a salivary gland nucleus to show banding characteristic of this region. × 1800. (B) Photomicrograph of a salivary gland squash preparation to show the identical banding of the same region. × 1200.

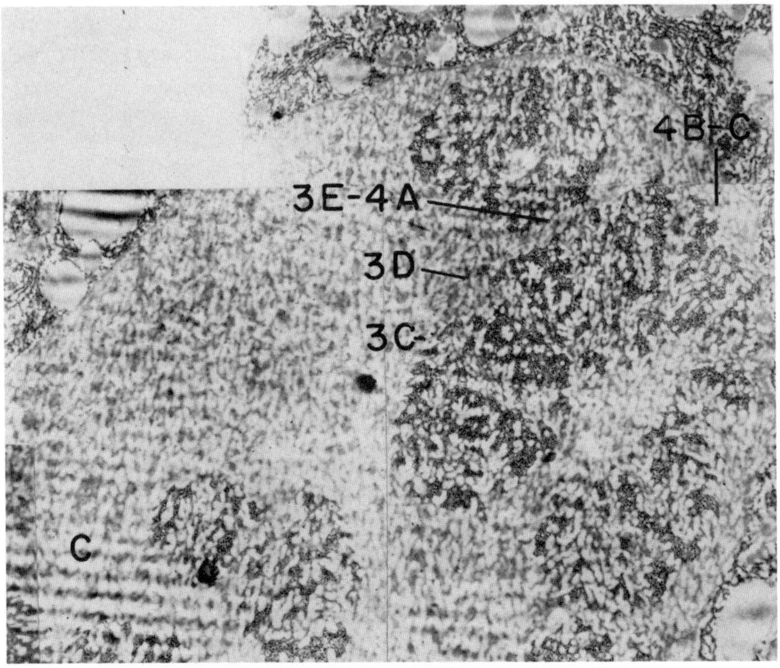
4B-C
3E-4A
3D
3C
C

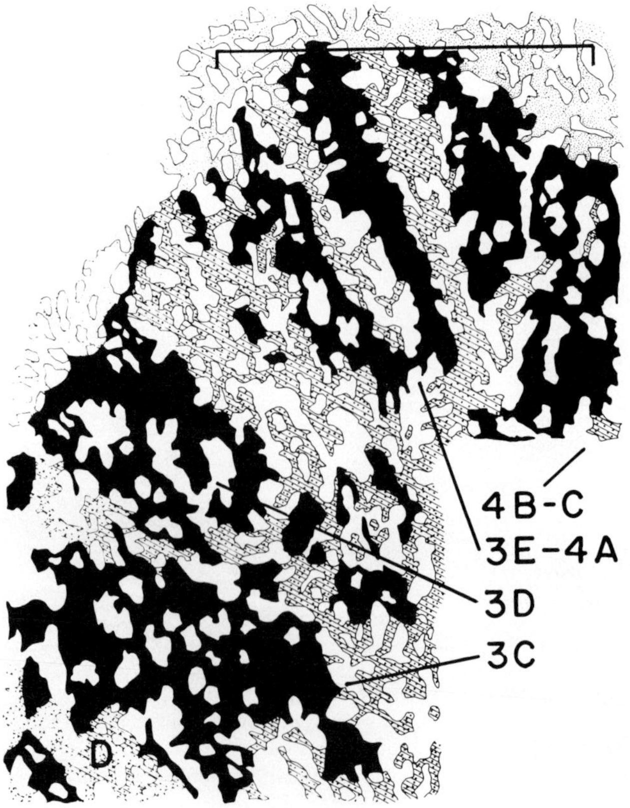
4B-C
3E-4A
3D
3C
D

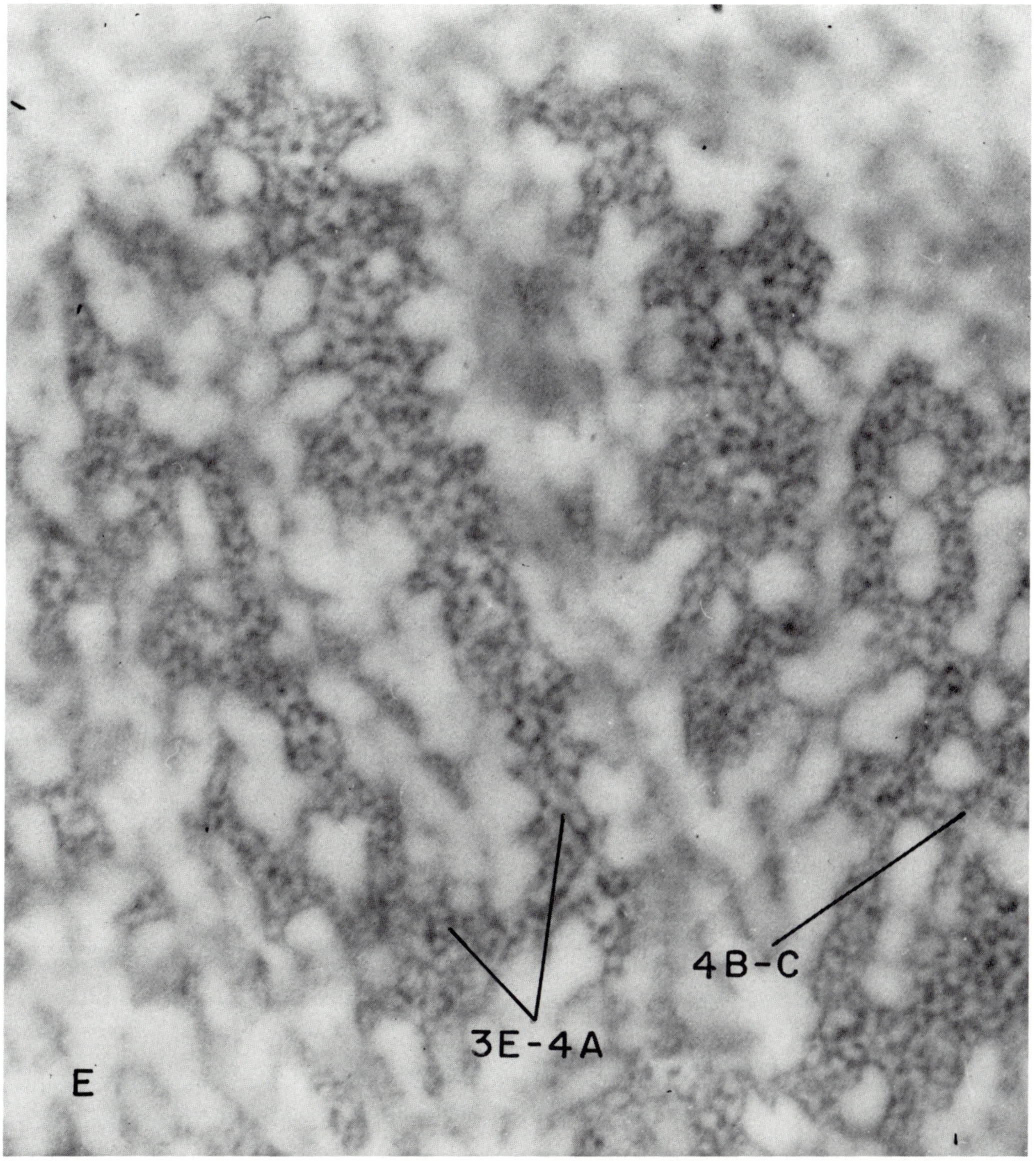

FIG. 4.25 C–E. (C) Photomontage of electron micrographs of the same nucleus shown in (A), with region 3 of the X chromosome marked. × 5300. (D) Tracing of the banding pattern of the same region based on tracings of electron micrographs at high magnification, and subsequently reduced for publication. Black patches contain chiefly dark staining sections of DNA coils. Stippled and cross-hatched regions contain pale staining sections of DNA coils and RNA granules. Finely stippled regions are nucleoplasm. The bracket encloses the part of the chromosome region which was traced from the electron micrograph in (E). (E) Typical electron micrograph from which the tracings were made. × 56,000. See text, p. 116 for description.

5. Known Heterochromatic Region (Fig. 4.27, A–D). This was identified in 6-μ sections (A) and the position relative to the nearby chromosomes was identified in a photomontage of low power electron micrographs of an ultrathin section through the same nucleus (B). A photomontage of the heterochromatic region was then prepared from higher power electron micrographs and a tracing of this was coded as above (C). A sample electron micrograph on which the dia-

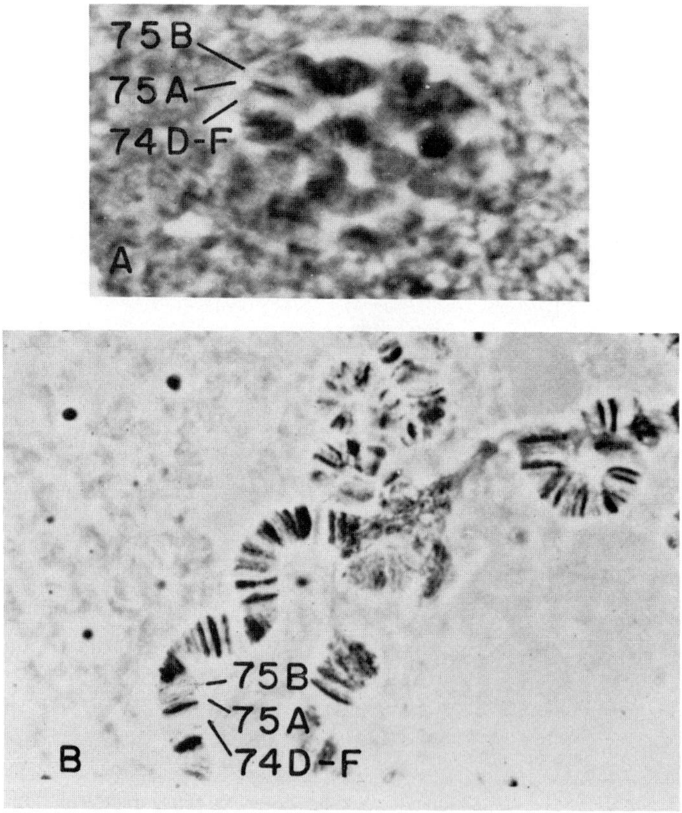

FIG. 4.26. Identification and analysis of a known puff region. (A) Photomicrograph of a 6-μ section through a salivary gland nucleus to show the puffs which were studied. Bands are numbered following Bridges *(11)*. × 1800. (B) Photomicrograph of a salivary gland squash preparation to show the location of the same puffs with their characteristic appearance and relationships. × 1200. (C) Photomontage of electron micrographs of the same nucleus shown in (A) with the puff region marked. × 5300. (D) Tracing of the banding pattern of this region based on electron micrographs at high magnification, and subsequently reduced for publication. Black patches contain chiefly dark staining sections of DNA coils. Stippled and cross-hatched regions contain pale staining sections of DNA coils and RNA granules. Small patches of cross-hatching alone are poor in DNA and RNA. Finely stippled regions are nucleoplasm and chromosomal clefts. The bracket encloses the part of the chromosome region which was traced from the electron micrograph in (E). (E) Typical electron micrographs from which the tracings were made. × 56,000. See text p. 125 for description.

gram was based appears as (D). In general, the heterochromatic region is characterized by a maximum of second-order DNA coils, with a minimum of RNA in the spongelike walls between the DNA-rich regions.

Discussion

The basic structure of the nucleus is spongelike. This structure is common to chromosomes, nucleolus, and nucleoplasm. The walls constitute the denser phase of the sponge; the cavities are less dense. The walls of nucleoplasm, nucleolus, and chromosomes are continuous, as are the cavities.

In the chromosomes, identifiable euchromatic bands are resolved into several smaller lamellae. (See Fig. 4.28 for diagrams to assist in elucidation of the description of the relation of the DNA double helix and RNA to chromosomal structure.) These contain DNA and RNA, both concentrated in interconnected

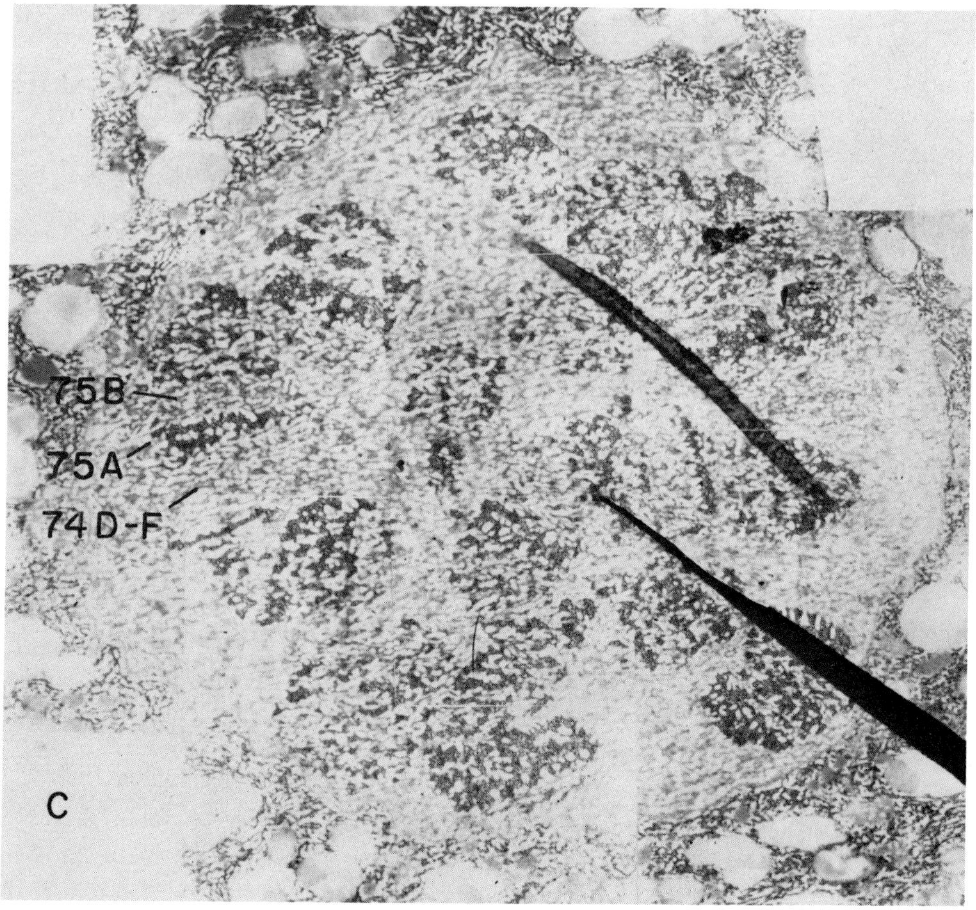

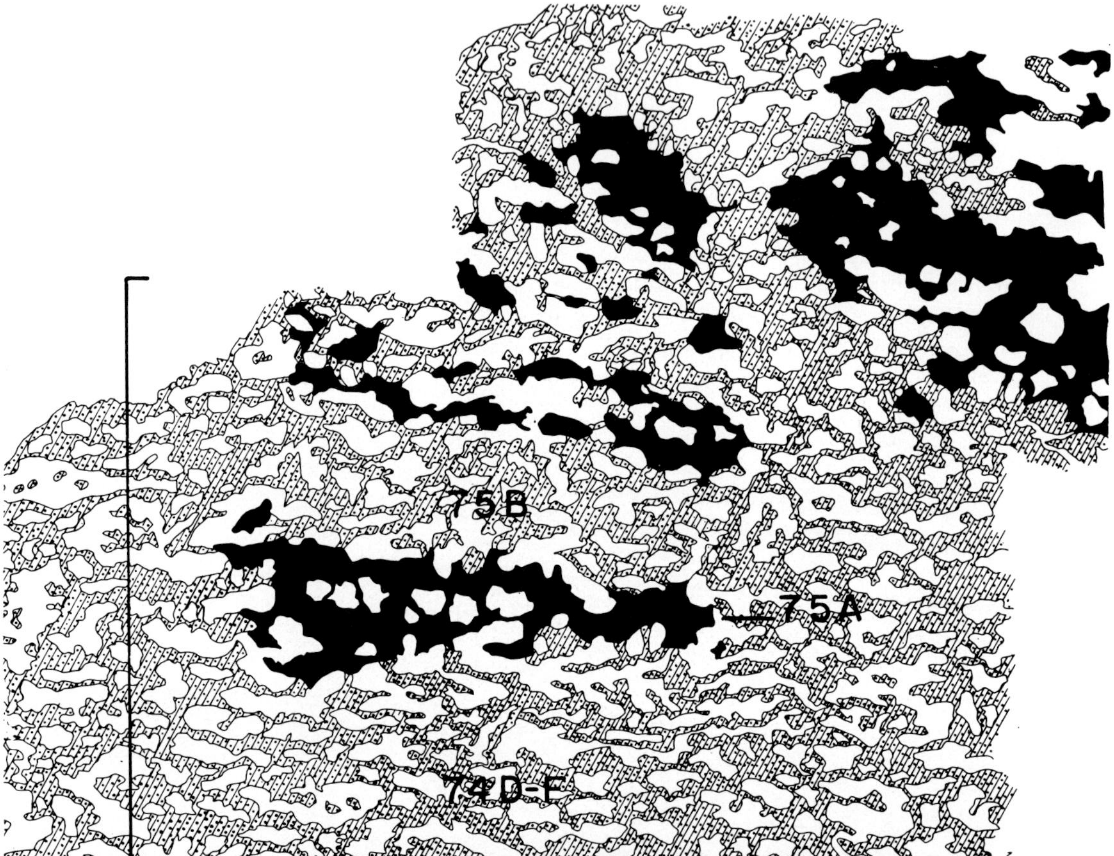

FIG. 4.26D. Opposite: FIG. 4.26E.

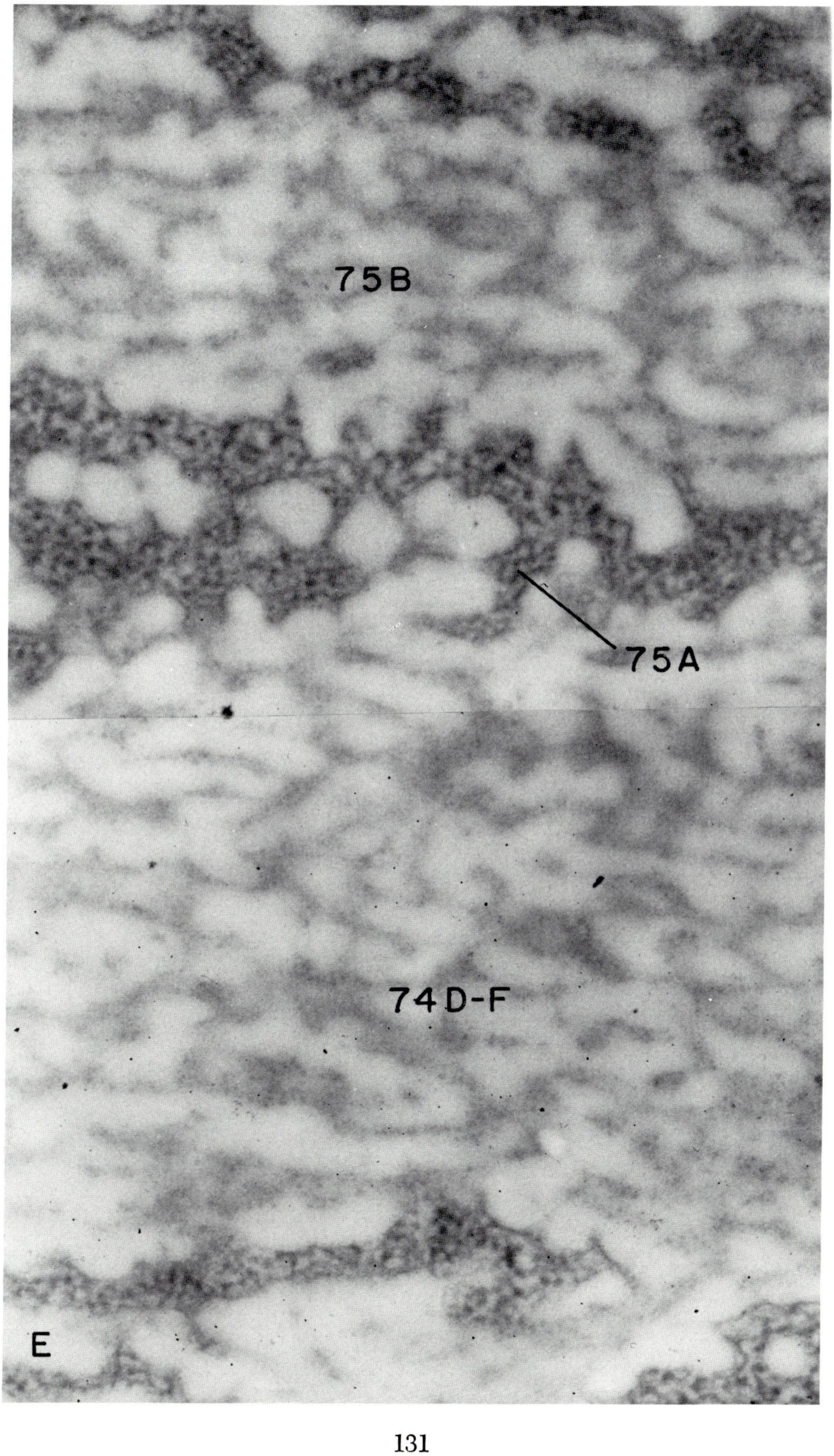
75B
75A
74D-F
E

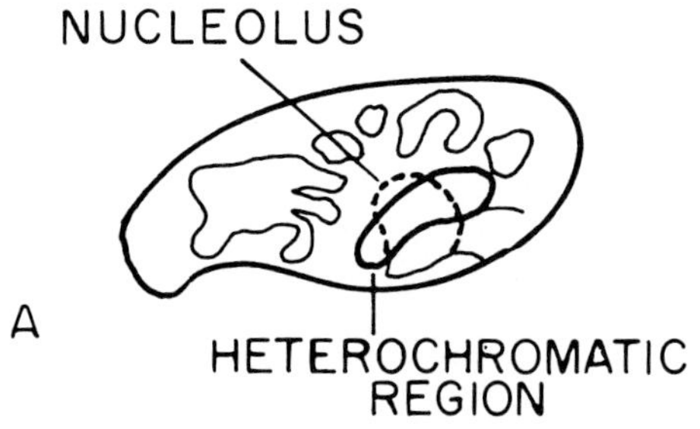

FIG. 4.27. Identification and analysis of a chromosome region known to be heterochromatic. (A) Drawing of a nucleus with the heterochromatic region marked. (B) Photomontage of lower power electron micrographs of the same nucleus with the same regions marked. × 5300. (C) Tracing of the banding pattern of the identified heterochromatic region based on electron micrographs at high magnification, and subsequently reduced for publication rendered as a mirror image. Black patches contain dark staining sections of DNA coils. Small patches of cross-hatching alone are poor in DNA and RNA. Coarsely stippled regions contain pale staining sections of DNA coils. Finely stippled regions are nucleoplasm and chromosomal clefts. The bracket encloses the part of the chromosome region which was traced from the electron micrograph as in D. (D) Typical electron micrographs from which the tracings were made. × 56,000. See text p. 128 for description.

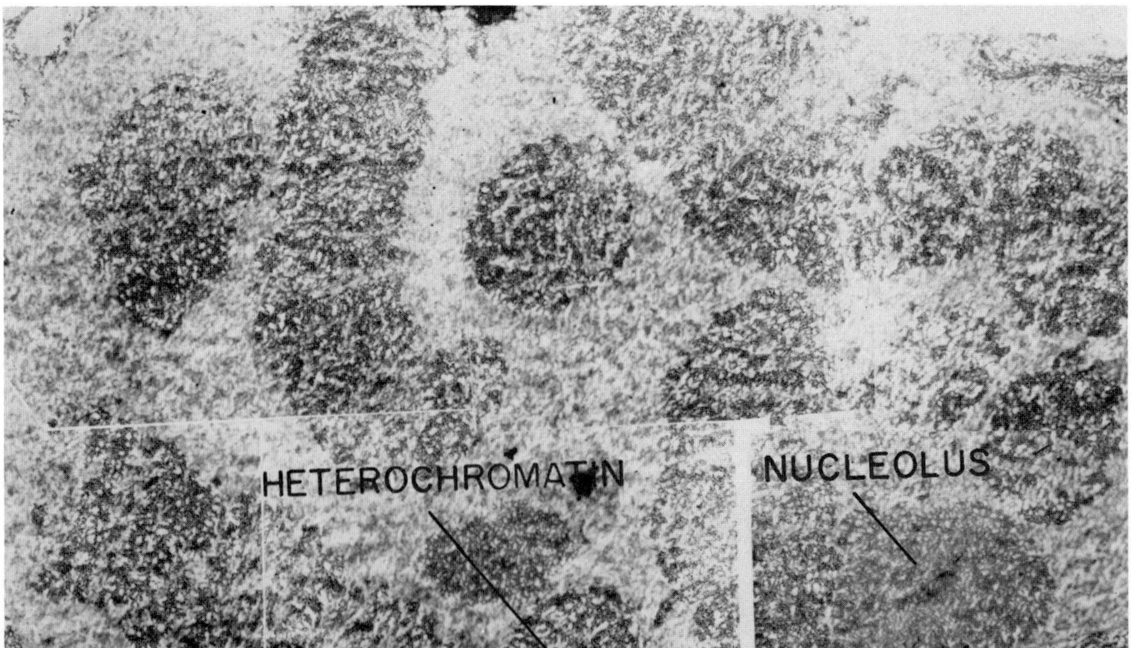

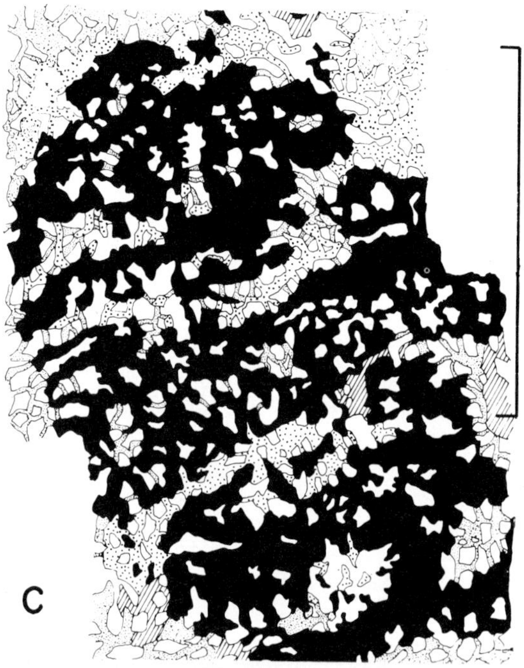

C

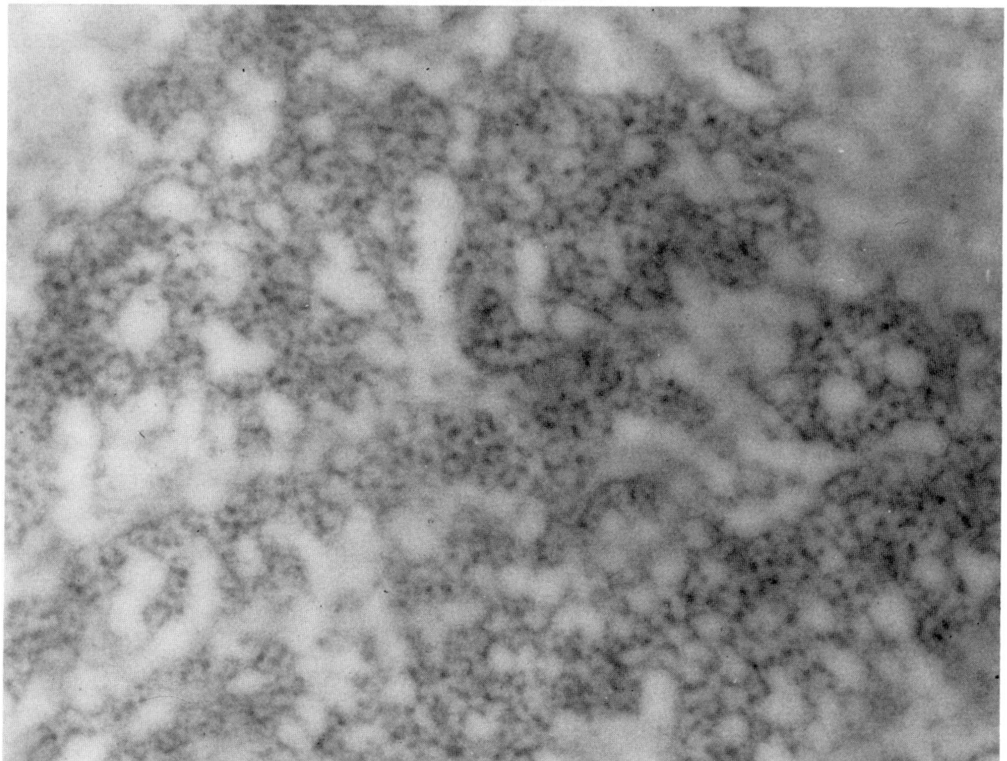

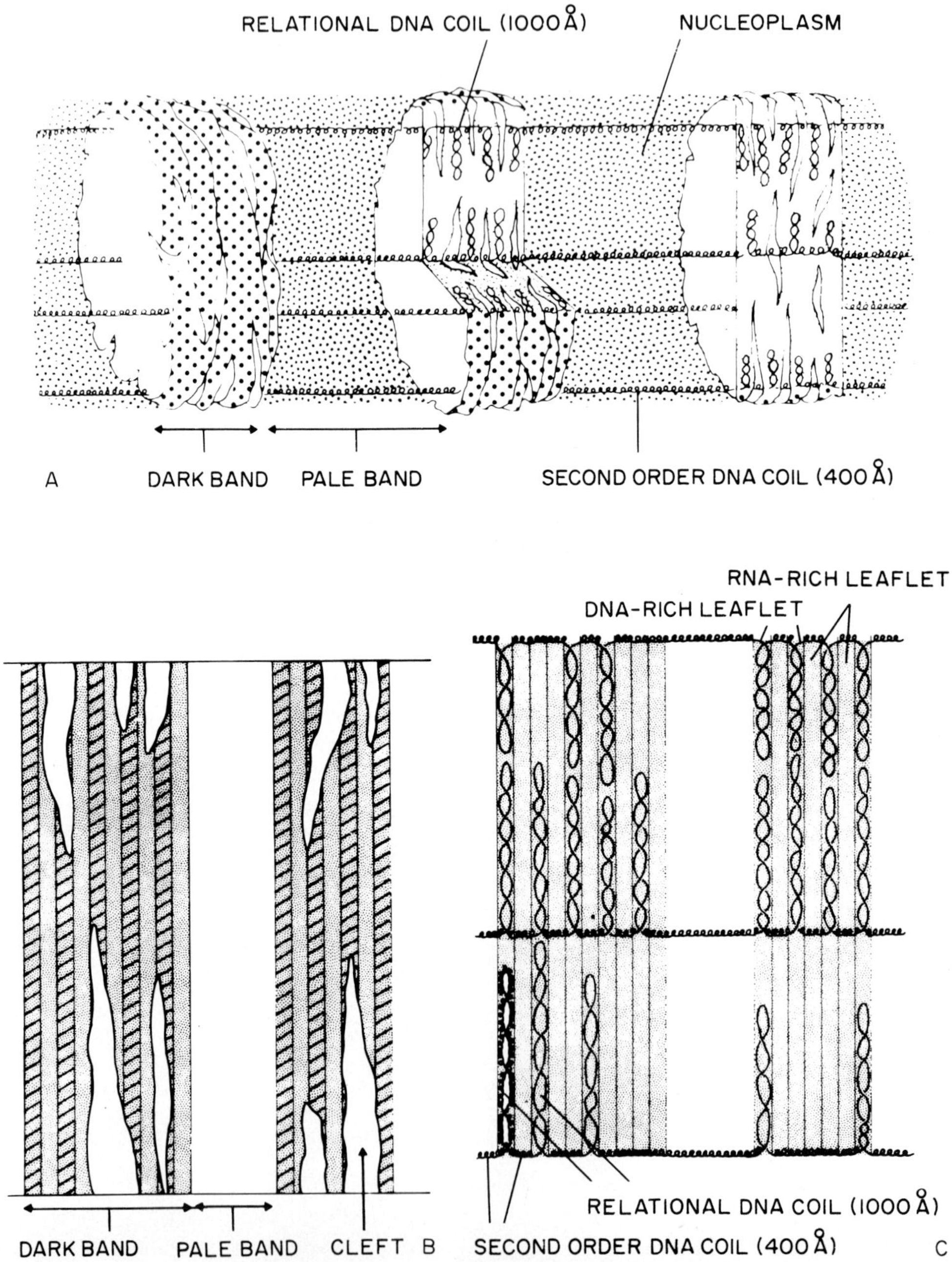

FIG. 4.28. Model of distribution of DNA and RNA in salivary gland chromosomes based on submicroscopic cytochemical studies. Bands and interbands are traversed by second-order coils (about 400 A in diameter) of the first-order DNA double helix (about 20 A in diameter). The surface of the bands is irregular and is characterized by clefts which penetrate internally (B). Within the dark bands the second-order coils form relational coils about 1000 A in diameter. The relational coils are not distributed uniformly in the dark bands, but are confined in lamellae, which alternate with other lamellae rich in RNA granules. RNA is also present between the relational coils. These relations of clefts and coils to the bands and interbands are illustrated in (A), (B), and (C), where it is indicated that leaflets may be incomplete. The irregular fusions of leaflets of the same kind are not indicated. The model is highly idealized.

lamellae of irregular thickness, about 1000 A thick. These lamellae may be incomplete and may not extend completely across the chromosome. In the RNA-rich lamellae, the first-order DNA coil (the DNA double helix) is arranged as second-order DNA coils lying within the walls of the spongy structure. Also within the walls, between the coiled strands, are some RNA granules. Within the DNA-rich lamellae, the second-order DNA coils seem to form relational coils (see p. 134). These are crowded together in the lamellae and impart to chromomeres of polytene chromosomes their characteristically dark appearance. The coils in adjacent DNA-rich lamellae are connected by second-order coils which run through the RNA-rich lamellae between them. Thus the second-order DNA coils are continuous, and course in DNA-rich bands and lamellae and through interbands and RNA-rich lamellae. In puffs, where the second-order DNA coiled strands are more widely separated than in other parts of the chromosome, RNA granules are unusually abundant. By contrast, in heterochromatic regions of the chromosome the second-order DNA coils may be more aggregated than in euchromatic regions and there is an absence or paucity of RNA.

The nucleolus of salivary gland cells of *Drosophila* is similar in its spongelike structure to that of hepatic and exocrine pancreatic cells. The walls contain RNA granules which range in size from that of ribosomes to the limit of resolution in routine electron micrographs. The nucleolus is associated with (and penetrated by) DNA of the nucleolus organizing regions in the form of 400 A coils.

In the nucleoplasm, the walls may contain RNA granules distributed generally, or concentrated in small patches, and a diffuse RNA which is unresolvable in routine electron micrographs. In the region of puffs, both forms of RNA are abundant; in the region of heterochromatin, the RNA granules are scarce or absent. It was suggested in Chapter 3 (this volume) that the walls of the nucleoplasm constitute part of the transport system in the somatic cell nucleus, and the observations in salivary gland cells are consistent with this interpretation.

Packing of the DNA Genonema in the Salivary Gland Chromosome

One question to be considered is whether the second-order helix and the relational coil of DNA visualized by our techniques are sufficient to provide for the packing of the first-order helices within the dimensions of the salivary gland chromosomes.

The salivary gland nucleus contains approximately 1000 times as much DNA as a *Drosophila* sperm nucleus. If it is assumed that each mitotic chromosome consists during the G_1 period of a single DNA strand or chromonema *(43)* and that a ninefold geometric replication occurs in the development of the mature salivary chromosome in *Drosophila melanogaster* larvae *(73)*, then there are 1024 strands in a salivary chromosome pair, each one equivalent to a strand in the mitotic chromosome. The total length of DNA double helix in a mature salivary gland nucleus is about 91 meters *(22)*. The total length of DNA in a haploid set of single strands can then be arrived at by dividing this figure by 1024. The length of DNA per haploid set thus calculated is approximately 9×10^4 μ.

If the G_1 chromosome has more than one DNA strand, the number of strands in the salivary chromosome would be increased by this factor and the total length of each strand would be correspondingly reduced. We shall use the value of 9 × 10^4 μ in order to find the maximum degree of packing required, given the above assumptions, to accommodate the DNA strand within the length of a single set of salivary chromosomes.

Bridges *(11)* gave the total length of the lax, unstretched chromosomes as seventy times that of the gonidial chromosomes (7.5 μ), which would be about 500 μ. In order to accommodate the 9 × 10^4 μ of DNA within a length of 500 μ, an overall reduction in length by 180-fold must be brought about by packing. The packing factor is not the same all along the DNA strand, for, as we have seen, the DNA of bands is very compact while that of interbands is simply coiled into a second-order helix. The interbands make up about two-fifths of the whole length of the euchromatic chromosome parts, i.e., 200 μ, and the degree of packing achieved in the second-order helix (as in all helices whose diameter and pitch are equal) is 3.3, so that 660 μ of strand are included in the interbands. Disregarding for the time being the proximal heterochromatic regions, of whose structure we are uncertain, much of the remaining 89,340 μ must be packed into the 300 μ of length occupied by bands, calling for a packing factor of perhaps 300 times—that is, an additional packing of 90 times over and above the 3.3 times packing attained in the second-order helix. We shall see that the relationally coiled loops described above could account for such a packing factor.

Since the diameter of the approximately circular relational loops is about 1000 A, each circle comprises about 3140 A of the second-order helix. A relational coil of ten such loops would consist of 31,400 A of second-order helix and, if all circles lay in a plane perpendicular to the chromosome axis, it would occupy a length of only 350 A along the chromosome. This would give a packing factor of about 90 times.

The cross sections of the paired chromosomes, each about 2.5 μ in diameter, would, however, only encompass about one thousand such loops. This layer of circles would represent the relational coils of only one-tenth of the 1024 strands. Those of other strands would have to be layered and, in all, ten such layers, each 350 A deep, are called for in an average band. The combined depth of these layers would be 0.35 μ and together with the RNA-rich lamellae they might correspond to a band, visible with the light microscope, having a thickness of 0.70 μ. This would represent a sizable band, as the bands visible with the light microscope range from 0.2 to 1.0 μ in thickness, with a majority of rather fine bands.

As already mentioned, these calculations do not account for the heterochromatic regions. In these regions, the DNA is particularly densely packed and, making allowances for the length of DNA strand included there, this would reduce the packing factor required in the bands of the so-called euchromatic regions.

On the other hand, the assumption that the chromonemata of the salivary chromosome are not equivalent to those of mitotic chromosomes but differ in having a linear redundancy of DNA sequences *(5)*, would aggravate the packing problem. If the increased length of the salivary chromosomes is considered to be due to a linear reduplication of some or all genes rather than to extension of the densely packed mitotic chromonema, then the 1024-fold increase in DNA content would not be simply the result of an eight-fold lateral replication, but would presumably result from a combination of lateral and linear reduplications. The number of chromonemata would then have to be less than 1024 in order to give a total DNA 1024 times that of a single-stranded haploid set. Each strand would be longer than the 9×10^4 μ arrived at on the assumption of 1024 strands. Since the total length of salivary chromosomes remains unchanged, an even greater packing factor would be required to accommodate these chromonemata. This consideration diminishes the attractiveness of the idea that the salivary chromosomes increase in length by intercalation of redundant DNA sequences. This hypothesis raises an additional problem as to the mechanism by which each linear reduplication is brought about. Under the circumstances, there is justification for the alternative hypothesis, that the linear growth of salivary chromosomes is simply due to extension of a highly coiled mitotic chromonema.

Gene Activity in Puffs

Granular RNA is found in the nucleolus and also in puffs and interband regions of the chromosomes. Within the bands, it is present in the RNA-rich lamellae and to a minor extent in the DNA-rich lamellae. It is also present in the nucleoplasm.

While it seems safe to assume, following the work of Ritossa *et al.* *(64)*, that a high proportion of the nucleolar granules contain ribosomal RNA (rRNA), it is hardly likely that the granules in the other locations are the same. The presence of rRNA at these nonnucleolar sites would indicate either that it was being synthesized in these sites or that the granules were ribosomes or ribosome precursors which would participate in the synthesis of protein within the chromosomes. It is clear, both from the work cited above, and from hybridization of labeled rRNA with DNA of salivary chromosomes *(54)*, that little if any rRNA is synthesized outside of the nucleolus organizing region, and this precludes the possibility that a substantial part of the RNA granules found elsewhere are rRNA formed *in situ*. There are more of these granules in the puffs than elsewhere. Although, in some cases, puffing is accompanied by an increase in protein content, there is little evidence of protein synthesis in the puffs *(66)*. Thus, we are led to the conclusion that the RNA in the puffs probably is not rRNA, even though preribosomal RNA has been reported as moving into some (unspecified) regions of the chromosomes from the nucleolus *(57, 61)*.

It is possible to present an argument in favor of considering puffs as sites of transfer RNA (tRNA) synthesis. Thus, it has been suggested *(65)* that formation of tRNA might be a function of the *Minute* loci. There is evidence of considerable coincidence between sites of *Minute* loci and sites of ectopic pairing (nonhomologous attachments) of salivary chromosomes *(34)*. More recently similar data have been presented relating ectopic pairing with puffing regions *(40)*. It would be difficult, however, to reconcile the sporadic appearance of some puffs with the requirements of tRNA synthesis. It has been shown *(30)* that synthesis of nucleolar (presumably ribosomal) RNA continues in the third instar of *Chi-*

ronomus, when the nucleolus is already large. One would expect that tRNA, being another part of the machinery of protein synthesis, would follow a similar steady course. Apart from this, there is an insufficiency of puffing regions to account for the required number of tRNA loci (see below).

The instances in which the presence of known puffs coincide with the presence of specific cell product *(2, 3, 33)* suggest that at least in some cases, the puffs synthesize messenger RNA (mRNA), which serves as template for synthesis of specific polypeptides. In view of Clever's evidence that some of the mRNA of the salivary gland cells is stabilized and long-lived *(16, 17),* it would not be surprising to find it in granular form, possibly as ribonucleoprotein particles. There is some evidence for this occurrence in rat liver nuclei *(35),* and it has also been shown that RNA particles can be reversibly denatured to a threadlike form *in vitro (68, 72).* Beermann has suggested that the protein in some puffs, which is apparently produced elsewhere and simply accumulates in the puffs, is used to package RNA *(5).* Our results show the presence of RNA, both soluble and particulate, in the nucleoplasm surrounding puffs as well as in the puffs themselves, as would be expected of gene products finding their way out to the cytoplasm. Thus there is reason to believe that the intrachromsomal particulate RNA, as well as a part of that in the nucleoplasm, is mRNA.

Chromosome Structure and Gene Activity

The idea that genes are present in the interband spaces has always been kept open *(24, 42).* Nevertheless, there has been a tendency to identify the bands as the seat of the genetic material *(51, 52)* even to the extent of formulating a "one gene–one band" hypothesis *(4, 45).* This tendency probably derives in part from Belling's concept of the identity of genes with the "ultimate chromomeres" *(8)* and may even go back to earlier work such as that of Correns *(18).* The many instances in which deficiency for a single band is accompanied by a genetic effect at a particular locus have confirmed this tendency to identify genes with bands or chromomeres. Intensive studies of induced recessive lethals in restricted chromosome regions have failed to produce more than a one to one ratio of susceptible loci and bands *(1, 36, 37, 45, 60, 71).*

Such data are not critical, however. The deficiencies in question could include interband DNA. There might be deficiencies for interband DNA alone which would be cytologically undetectable. As for the lethals, it is known that some genetic loci can be lost without a lethal effect on the organism *(50),* so that a one to one relationship between lethals and bands would actually indicate a greater than one to one relationship between genes and bands.

On the other hand, there is much evidence to support the assumption that there are genes in the interband regions as well as in the bands; that the genes which are active in the salivary gland nucleus are in the extended second-order coils of DNA in the interbands and puffs, and perhaps also in the RNA-rich lamellae within bands; and that there are many other genes, in the bands as well as in the proximal heterochromatic regions, which perform no function in the salivary gland nuclei, though they are active in nuclei of various other cell types.

The arguments supporting this thesis are summarized below. For a fuller understanding, the reader is referred to the extensive discussion published elsewhere *(31)*.

1. Morphologically, the interbands resemble puffed regions, while the bands resemble heterochromatic regions in the disposition of their DNA coils and in the relative amount of RNA present. Likewise, RNA synthesis, as indicated by incorporation of [^{3}H]-uridine, occurs in puffs but not in heterochromatic regions, and the autoradiographs can reasonably be interpreted as showing RNA synthesis in interbands rather than within bands *(28, 55, 56)*.

2. In the development of the salivary gland nucleus, there is a progressive increase in the total number of bands and the relative length of chromosome occupied by bands as opposed to interbands *(13, 27)*. This parallels the development of many maturing cells, in which the amount of dense chromatin increases as development and differentiation proceed. Our interpretation of the change is that many gene systems, including those involved in cell division and growth, are turned off and sequestered in the form of heterochromatic masses. The behavior of the salivary chromosomes is consistent with this picture if we assume that the DNA of interbands and puffs constitutes the active part of the genome.

3. The assumption that there are active genes in interbands and puffs, and more numerous inactive genes in bands and proximal heterochromatin, is more consonant with contemporary estimates of the total number of genes in the *Drosophila* genome *(44)* than is the assumption of a one gene–one band relationship. Taking into consideration the redundancy of certain fundamentally important genes, of which those for rRNA *(63)*, tRNA *(65)*, 4S and 5S RNA, and histones *(41, 74)* are probably only a part, a one gene–one band relationship, giving a total of some five thousand genes per haploid chromosome set, is quite inadequate. As explained elsewhere *(31)* one would expect something like this figure for the number of genes which are active in any given cell type, and a number one or two orders of magnitude larger for the total genome.

The amount of DNA in the bands is easily large enough to represent a total of the order of 50,000 to 500,000 genes. There is evidence that some of the redundant genes may be in the banded regions rather than in heterchromatic regions *(25)*. The amount of DNA in an average salivary chromosome band visible with the light microscope can be spun out to at least a 10-μ length of DNA double helix *(45)*. This is enough DNA to code for several genes. The band must, therefore, comprise a considerable amount of nonsense DNA, or include several genes—either repetitions of the same gene, as suggested by Beerman *(5)*, or a sequence of different genes. We suggest that the actual situation is a synthesis of the last two alternatives, brought about in the evolutionary process following the initial duplication in tandem of individual genes. The duplicated members, initially identical, evolve in the course of time both structurally and functionally. At a given time such as the present, some bands may contain identical tandem repeats, recently formed, while others contain repeats at various stages of evolution, including pseudoalleles and completely different, nonallelic, loci. This concept is close to that formulated by Lewis *(46)* and differs from that of Beermann *(5)* in ascribing a more orthodox function, as well as a greater degree of differentiation, to the units within bands.

Ellenhorn, Prokofyeva, and Muller *(24)* considered the possibility that genes are present in both bands and interbands. Our conclusions are in accord with this suggestion. In part, they follow Koltzoff *(42)*, who assumed that all the bands are heterochromatic, and hence contain no genes; he concluded that all genes are in the interband threads. Our proposal is that the interbands contain active genes, and that the bands as well as the heterochromatic regions contain genes which remain inactive except when a banded region is converted into a puff. In regions which are genetically active, the DNA molecules are rather separate, while in genetically inactive regions they are crowded together, and these conditions are interconvertible.

Even if the bands contain inactive genes, it is possible that the extended DNA of the RNA-rich lamellae is genetically active. Another possibility is that this RNA is a structural element of the bands which combines specifically with some nucleotide sequences of the DNA and, either alone or in conjunction with protein molecules, holds the second-order DNA helices together in compacted form *(7, 20, 38)*. In addition to its structural function, this RNA would be part of a control system involving groups of adjacent genes (see Chapter 8).

It should be noted, in conclusion, that if our interpretation is correct, it would require a revision of the terminology applied to the salivary gland chromosome. At present, the only regions considered heterochromatic are the proximal regions adjacent to the centromeres, and numerous short intercalary regions at certain positions along the chromosome arms. The clearly banded arms are considered to be mainly euchromatic. According to our interpretation, the euchromatic regions would be the interbands and puffs and the heterochromatic regions would include not only proximal and intercalary heterchromatin *(34)*, but also all bands. It might, in fact, be more sensible to discard the terms heterochromatic and euchromatic, and refer instead to the dense or compacted (inactive) and less dense (active) DNA coils. The term "compaction" has already been suggested to replace the earlier "heterochromatization" *(67, 70)*.

Summary

Using the method for nucleic acids described in Chapter 3 (this volume), their distribution was studied in salivary gland chromosomes of late third instar larvae of *Drosophila*. As in other cells, the basic structure of the nucleus is spongelike, with the denser walls of nucleoplasm, nucleolus, and chromosomes all continuous with each other; the pseudovacuoles are similarly continuous with each other. The walls in the nucleoplasm and nucleolus contain a variable number of granules rich in RNA. These are present also in chromosomes. Chromosomes consist of bands and interband regions. The DNA double helix occurs in both regions as a second-order coil with a periodicity and diameter of about 400 A. In the bands, lamellae (I) rich in DNA (with some RNA) alternate with lamellae (II) rich in

RNA (with some DNA). The DNA traverses lamellae I and II as well as the interband region as second-order helices; in addition, in lamellae rich in DNA (I), the second-order helical coils form relational coils of larger dimensions. Several regions of the chromosomes were identified according to Bridge's maps and the distribution of nucleic acids in them was studied. Of particular interest, was the contrast between regions known to be heterochromatic and those known as puffs. In the former, the second-order coils of DNA were tightly packed and there were few RNA granules. In the latter, the second-order coils were widely separated, and there were numerous RNA granules between them, spreading out into the adjacent nucleoplasm. The cytochemical observations are discussed extensively in relation to three main topics: the packing of DNA in polytene chromosomes, gene activity in puffs, and gene activity in relation to structure.

References

1. Alikhanian, S. I. (1937). A study of the lethal mutations in left end of the sex-chromosome in *Drosophila melanogaster*. *Zool. Zh.* **16**, 247–279. (Russian with English summary.)
2. Baudisch, W., and Panitz, R. (1968). Controlle eines biochemischen Merkmal in den Speicheldrüsen von *Acricotopus lucidus* durch ein Balbiani-Ring. *Exp. Cell Res.* **49**, 470–476.
3. Beermann, W. (1961). Ein Balbiani-ring als Locus einer Speicheldrüsen-mutation. *Chromosoma* **12**, 1–25.
4. Beerman, W. (1962). Riesenchromosomen. *Protoplasmatologia*, **6D**, 1–16.
5. Beermann, W. (1967). Gene action at the level of the chromosome. *In* "Heritage From Mendel" (R. A. Brink, ed.), pp. 179–201. University of Wisconsin Press, Madison, Wisconsin.
6. Beermann, W., and Bahr, G. F. (1954). The submicroscopic structure of the Balbiani-ring. *Exp. Cell Res.* **6**, 195–201.
7. Bekhor, I, Bonner, J., and Dahmus, G. K. (1969). Hybridization of chromosomal RNA to native DNA. *Proc. Nat. Acad. Sci. U. S.* **62**, 271–277.
8. Belling, J. (1928). The ultimate chromomeres of *Lilium* and *Aloe* with regard to the numbers of genes. *Univ. Calif. Publ. Bot.* **14**, 307–318.
9. Berendes, H. D. (1970). Polytene chromosome structure at the submicroscopic level. I. A map of region X, 1-4E of *Drosophila melanogaster*. *Chromosoma* **29**, 118–130.
10. Berendes, H. D., and Meyer, G. F. (1968). A specific chromosome element, the telomere of *Drosophila* polytene chromosomes. *Chromosoma* **25**, 184–197.
11. Bridges, C. B. (1935). Salivary chromosome maps. *J. Hered*, **26**, 60–64.
12. Bridges, C. B. (1938). A revised map of the salivary gland X-chromosome of *Drosophila melanogaster*. *J. Hered.* **29**, 11–13.
13. Buck, J. B. (1937). Growth and development of the salivary gland chromosomes in *Sciara*. *Proc. Nat. Acad. Sci. U. S.* **23**, 423–428.
14. Buck, J. B., and Melland, A. M. (1942). Methods for isolating, collecting and orienting salivary gland chromosomes for diffraction analysis. *J. Hered.* **33**, 173–184.
15. Caspersson, T. (1939). Nukleinsäureketten und Genvermehrung. *Chromosoma* **1**, 605–619.
16. Clever, U. (1969). Chromosome activity and cell function in polytenic cells. II. The formation of secretion in the salivary glands of *Chironomus*. *Exp. Cell Res.* **55**, 317–322.
17. Clever, U., Storbeck, I., and Romball, C. G. (1969). Chromosome activity and cell function in polytenic cells. I. Protein synthesis at various stages of larval development. *Exp. Cell Res.* **55**, 306–316.
18. Correns, C. (1902). Uber den Modus und den Zeitpunkt der Spaltung der Anlagen bei den Bastarden von Erbsen-typus. *Bot. Z.* **60**, 65–82.

19. Crick, F. (1971). General model for the chromosomes of higher organisms. *Nature (London)* **234**, 25–27.
20. Darnell, J. E., Jr. (1968). Ribonucleic acids from animal cells. *Bacteriol. Rev.* **32**, 262–290.
21. DuPraw, E. J., and Rae, P. M. M. (1966). Polytene chromosome structure in relation to the "folded fibre" concept. *Nature (London)* **212**, 598–600.
22. DuPraw, E. J., (1968). "Cell and Molecular Biology." Academic Press, New York.
23. Edström, J.-E. (1964). Chromosomal RNA and other nuclear RNA fractions. *In* "The Role of Chromosomes in Development" (M. Locke, ed.), pp. 137–152. Academic Press, New York.
24. Ellenhorn, J., Prokofyeva, A. A., and Muller, H. J. (1935). The optical dissociation of *Drosophila* chromomeres by means of ultraviolet light. *Dokl. Akad. Nauk. SSSR* **1**, [N.S.], 234–241.
25. Entingh, T. D. (1970). DNA hybridization in the genus *Drosophila. Genetics* **66**, 55–68.
26. Ephrussi, B., and Beadle, G. W. (1936). A technique of transplantation for *Drosophila. Amer. Natur.* **70**, 218–225.
27. Frolova, S. L. (1938). Development of the giant nuclei of salivary gland type in different organs of *Drosophila,* and their comparison with "metabolic" nuclei of larvae and imago. *Biol. Zh.* **7**, 703–736.
28. Fujita, S. (1965), Chromosomal organization as a genetic basis of cytodifferentiation in multicellular organisms. *Nature (London)* **206**, 742–744.
29. Gay, H. (1956). Chromosome-nuclear membrane-cytoplasmic interrelations in *Drosophila. J. Biophys. Biochem. Cytol. Suppl.* **2**, 407–414.
30. Gay, H. (1964). Chromosome Structure and Function. *Carnegie Inst. Wash. Yearb.* **63**, 605–614.
31. Gersh, E. S. Studies on the *Sal locus in Drosophila pseudoobscura.* III. Puffing in polytene chromosomes, its relation to position effect, and its interpretation in terms of fine structure and the molecular pattern of DNA in active and inactive chromosome regions. In preparation.
32. Goodman, R. H., and Spiro, D. (1962). Electron microscopic studies of giant salivary gland chromosomes. *Exp. Cell Res.* **27**, 359–363.
33. Grossbach, U. (1969) . Chromosomen-aktivität und biochemische Zelldifferenzierung in den Speicheldrüsen von *Camptochironomus. Chromosoma* **28**, 136–187.
34. Hannah, A. M. (1951). Localization and function of heterochromatin in *Drosophila melanogaster. Advan. Genet.* **4**, 87–125.
35. Henshaw, E. C. (1968). Messenger RNA in rat liver polyribosomes: evidence that it exists as ribonucleoprotein particles. *J. Mol. Biol.* **36**, 401–411.
36. Hochman, B. (1971). Analysis of chromosome 4 in *Drosophila melanogaster.* II. Ethyl methane-sulfonate induced lethals. *Genetics* **67**, 235–252.
37. Hochman, B. (1972). The detection of four more vital loci on chromosome 4 in *Drosophila melanogaster. Genetics* **71**, s.71.
38. Holmes, D. S., Mayfield, J. E., Sander, G., and Bonner, J. (1972). Chromosomal RNA: its properties. *Science* **177**, 72–74.
39. Jacob, J., and Sirlin, J. L. (1963). Electron microscope studies on salivary gland cells. I. The nucleus of *Bradysia mycorum* Frey (Sciaridae), with special reference to the nucleolus. *J. Cell Biol.* **17**, 153–165.
40. Kaufmann, B. P., And Iddles, M. K. (1963). Ectopic pairing in salivary-gland chromosomes of *Drosophila melanogaster.* I. Distributional patterns in relation to puffing. *Port. Acta Biol. Ser.* **A7**, 225–248.
41. Kedes, L. H., and Birnstiel, M. L. (1971). Reiteration and clustering of DNA sequences complementary to histone messenger RNA. *Nature New Biol.* **230**, 165–169.
42. Koltzoff, N. (1934). The structure of the chromosomes in the salivary glands of *Drosophila. Science* **80**, 312–313.
43. Laird, C. D. (1971). Chromatid structure: relationship between DNA content and nucleotide sequence diversity. *Chromosoma* **32**, 378–406.
44. Laird, C. D., and McCarthy, B. J. (1969). Molecular characterization of the *Drosophila* genome. *Genetics* **63**, 865–882.

45. Lefevre, G., Jr. (1971). Salivary chromosome bands and the frequency of crossing over in *Drosophila melanogaster*. *Genetics* **67**, 497–513.
46. Lewis, E. B. (1967). Genes and gene complexes, *In* "Heritage from Mendel" (R. A. Brink, ed.), pp. 17–47. Univ. Wisconsin Press, Madison, Wisconsin.
47. Lowman, F. G. (1956). Electron microscope studies of *Drosophila* salivary-gland chromosomes. *Chromosoma* **8**, 30–52.
48. MacInnes, J. W., and Uretz, R. B. (1966). Organization of DNA in dipteran polytene chromosomes as indicated by polarized fluorescence microscopy. *Science* **151**, 689–691.
49. Moses, M. J., and Coleman, J. R. (1965). Structural patterns and the functional organization of chromosomes. *In* "The Role of Chromosomes in Development" (M. Locke, ed.), pp. 11–49. Academic Press, New York.
50. Muller, H. J. (1935). A viable two-gene deficiency phaenotypically resembling the corresponding hypomorphic mutations. *J. Hered.* **26**, 469–478.
51. Muller, H. J., and Prokofyeva, A. A. (1934). Continuity and discontinuity of the hereditary material. *Dokl. Akad. Nauk. SSSR* **4**, [N.S.], 74–83.
52. Painter, T. S., and Griffin, A. B. (1937). The structure and the development of the salivary gland chromosomes of *Simulium*. *Genetics* **22**, 612–633.
53. Palay, S. L., and Claude, A. (1949). An electron microscope study of salivary gland chromosomes by the replica method. *J. Exp. Med.* **89**, 431–438.
54. Pardue, M. L., Gerbi, S. A., Eckhardt, R. A., and Gall, J. G. (1970). Cytological localization of DNA complementary to ribosomal RNA in polytene chromosomes of *Diptera*. *Chromosoma* **29**, 268–290.
55. Pavan, C. (1965). Nucleic acid metabolism in polytene chromosomes and the problem of differentiation. *Brookhaven Symp. Biol.* **18**, 222–241.
56. Pelling, C. (1964). Ribonukleinsäure-Synthese der Riesenchromosomen: Autoradiographische Untersuchungen an *Chironomus tentans*. *Chromosoma* **15**, 71–122.
57. Pelling, C. (1970). Puff RNA in polytene chromosomes. *Cold Spring Harbor Symp. Quant. Biol.* **35**, 521–531.
58. Porter, K. R. (1960). Problems in the study of nuclear fine structure. *4th Int. Conf. Electron Microsc.* Vol. 2, pp. 186–199. Springer-Verlag, Berlin.
59. Rae, P. M. M. (1971). Whole mount electron microscopy of *Drosophila* salivary chromosomes. *Nature (London)* **122**, 139–142.
60. Rayle, R. E. (1972). Genetic analysis of a short X-chromosomal region in *Drosophila melanogaster*. *Genetics* **71**, s.50.
61. Ringborg, U., Daneholt, B., Edström, J. E., Egyházi, E., and Rydlander, L. (1970). Evidence for transport of preribosomal RNA from the nucleolus to the chromosomes in *Chironomus tentans* salivary gland cells. *J. Mol. Biol.* **51**, 679–686.
62. Ris, H. (1966). Fine structure of chromosomes. *Proc. Roy. Soc. London* **B164**, 246–257.
63. Ritossa, F. M., Atwood, K. C., Lindsley, D. L., and Spiegelman, S. (1966). On the chromosomal distribution of DNA complementary to ribosomal and soluble RNA. *Nat. Cancer Inst. Monogr.* **23**, 449–472.
64. Ritossa, F. M., and Spiegelman, S. (1965). Localization of DNA complementary to ribosomal RNA in the nucleolus organizer region of *Drosophila melanogaster*. *Proc. Nat. Acad. Sci. U. S.* **53**, 737–745.
65. Ritossa, F. M., Atwood, K. C., and Spiegelman, S. (1966). On the redundancy of DNA complementary to amino acid transfer RNA and its absence from the nucleolar organizer region of *Drosophila melanogaster*. *Genetics* **54**, 663–676.
66. Rudkin, G. T. (1964). The proteins of polytene chromosomes. *In* "The Nucleohistones" (J. Bonner and P. Ts'o, eds.), pp. 184–192. Holden Day, San Francisco, California.
67. Radkin, G. T. (1965). The structure and function of heterochromatin. *Genet. Today Proc. 11th Int. Cong. Genet.* Vol. 2, pp. 359–374. Macmillan, New York.
68. Samarina, O. P., Molnar, J., Lukanidin, E. M., Bruskov, V. I, Krichevaskaya, A. A., and Georgiev, G. P. (1967). Reversible dissociation of nuclear ribonucleoprotein particles containing mRNA into RNA and protein. *J. Mol. Biol.* **27**, 187–191.
69. Schmidt, W. J. (1939). Herstellung von Praparaten zur polarisationsoptischen Untersuchung

der Chromosomen in den Speicheldrüsenkernen der Chironomuslarven. *Z. Wiss. Mikrosk.* **56,** 1–7.
70. Schultz, J. (1963). The genetics of somatic cells. *Proc. 16th Int. Congr. Zool.* Vol. 3, pp. 216–221.
71. Shannon, M. P., Kaufman, T. C., and Judd, B. H. (1970). Lethality patterns of mutations in the zest white region of *Drosophila melanogaster. Genetics* **64,** (Suppl. 2), 58. (Abstr.)
72. Spirin, A. S. (1964). "Macromolecular Structure of Ribonucleic Acid." Reinhold, New York.
73. Swift, H. (1962). Nucleic acids and cell morphology in dipteran salivary glands. *In* "Molecular Control of Cellular Activity" (J. M. Allen, ed.), pp. 73–125. McGraw-Hill, New York.
74. Tartof, K. D., and Perry, R. P. (1970). The 5s RNA genes of *Drosophila melanogaster. J. Mol. Biol.* **51,** 171–183.

5

Distribution of Protein in Nuclei of Salivary Glands of Drosophila Larvae

Isidore Gersh and Eileen S. Gersh

From the seemingly monotonous reaction pattern of nuclear proteins of hepatic and exocrine pancreatic cells, it appeared unlikely that meaningful patterns could be picked out by the methods employed (see Chapter 2, this volume). Yet we hoped that the distribution of protein in the nuclei of salivary glands of *Drosophila* larvae might be more heterogeneous, to match the diversity in nucleic acid patterns described earlier. (See this volume, Chapter 3 and 4; Chapter 5, Vol. II.) This hope was realized to a certain extent, but the meaning of the various protein patterns is not yet clear.

Methods

The salivary glands of third instar larvae were frozen ultrarapidly as described in Chapter 2 (this volume), and dried in a vacuum at a low temperature. They were treated *in vacuo* with the vapors of protein reagents. The reagents used in this study were selected from a longer listing of vapor reagents in Chapter 2, where specific conditions of vapor treatment and subsequent procedure were given. Excess reagent is removed by evacuation and specimens are infiltrated with 95% alcohol, and then embedded in w-s Durcupan. These reagents react with certain active sites of proteins. When these sites are at the correct distance from each other, they are cross-linked. The proteins thus treated tend to become denatured and, therefore, insoluble or to have altered staining properties. In addition, their mass is increased. The consequences for cytochemistry are that there tends to be less diffusion, extraction, and displacement of preexisting polypeptides, and the latter show increased contrast on electron micrographs. It is also possible that reactivity to uranyl acetate is altered. Some specimens postfixed in 95% alcohol are subsequently treated *en bloc* with a solution of platinum tetrabromide in 95% alcohol. (Details are given in this volume, Chapter 2.) Sections were studied with the electron microscope directly, or after staining with uranyl acetate.

Observations

On the basis of the appearance of the protein components of chromosomes, observations were classified into five categories:

1. The chromosomes comprise two rather homogeneous regions, the dark bands and the paler interband regions. The latter are identified because they are

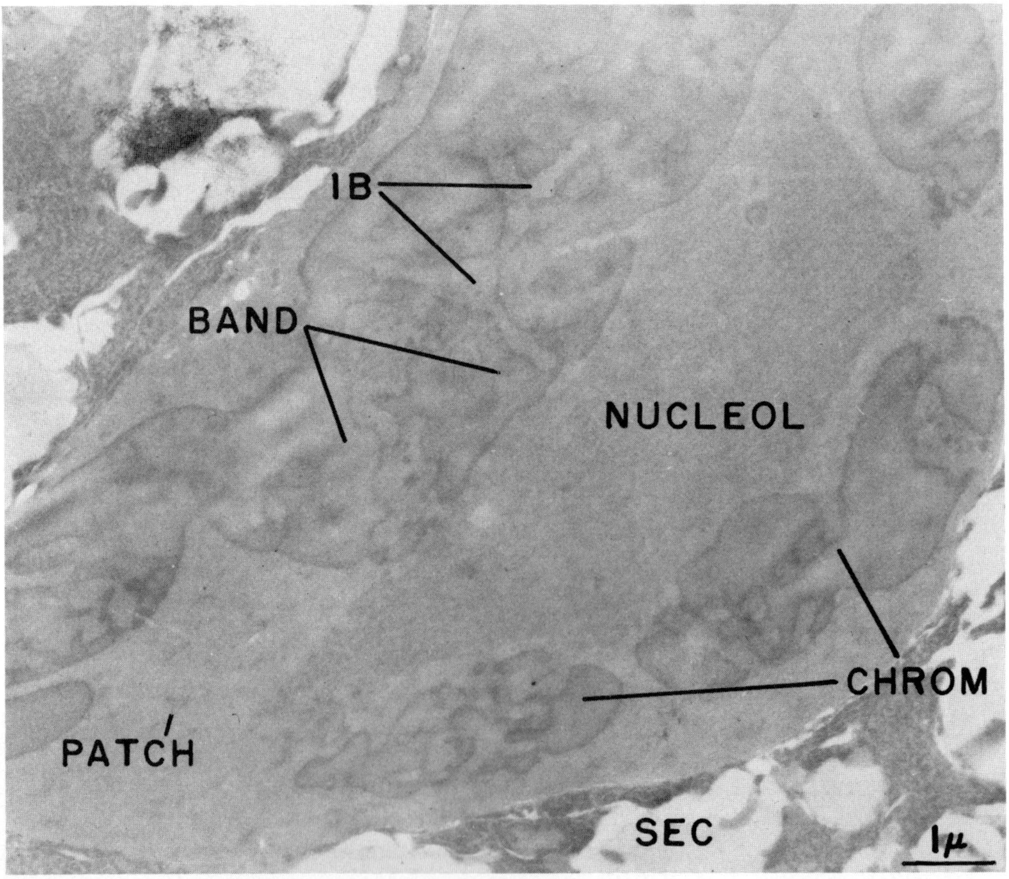

FIG. 5.1. Nucleus postfixed in alcohol, unstained. Chromosomes (CHROM) consist of dark bands (BAND) and paler interband spaces (IB). Nucleoplasm uniform, with darker patches (PATCH) containing RNA. Nucleolus (NUCLEOL) of intermediate density. Secretory granules (SEC) in cytoplasm. × 12,000.

continuous with the nucleoplasm, as in preparations stained for nucleic acids (see this volume, Chapter 4). The density of the interband regions is the same as that of the nucleoplasm except for the somewhat darker RNA patches in it. The density of the nucleolus is uniform and intermediate. This group includes specimens treated with the following methods of postfixation and/or staining: alcohol alone (no bulk stain) (Figs. 5.1 and 5.2), difluorodinitrodiphenylsulfone (FFSulfone) (Fig. 5.3), barium acetylacetonate (Ba acac), rhodium (III) hexafluoroacetylacetonate (Rh hfac), and zinc hexafluoroacetylacetonate (Zn hfac) (Fig. 5.4). After sections are stained with uranyl acetate, chromosomes have the same coarse pattern as before. The nucleoplasm appears denser, with markedly darker patches, after treatment with Ba acac, RH hfac, and Zn hfac. Nucleoli are also uniformly darker after Ba acac and Zn hfac treatment (Fig. 5.5), when the nucleolar walls appear very fine meshed. The nuclear membrane is not delineated.

2. The chromosomes are made up of very delicate stripes of differing densi-

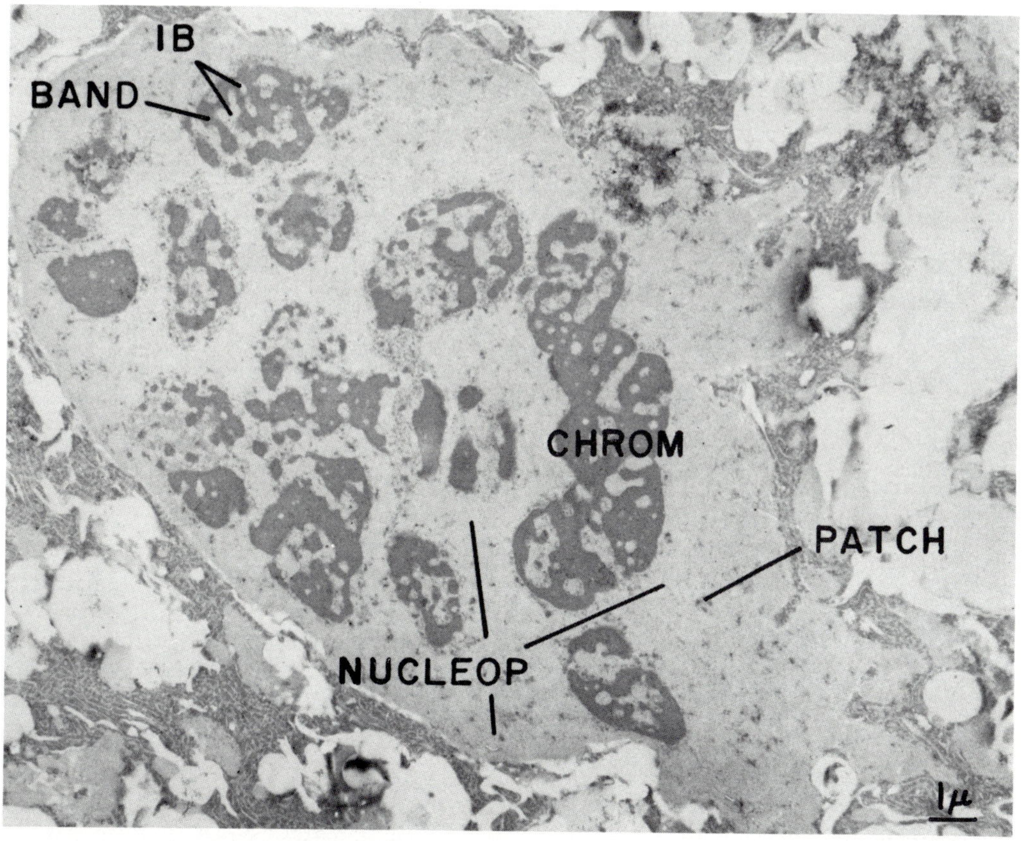

FIG. 5.2. Nucleus postfixed in alcohol, section stained with uranyl acetate. Compare with Fig. 5.1; features are all darker than unstained section. CHROM, chromosome; BAND, dark bands; IB, interband spaces; NUCLEOP, nucleoplasm; PATCH, local aggregate of RNA granules. × 6000.

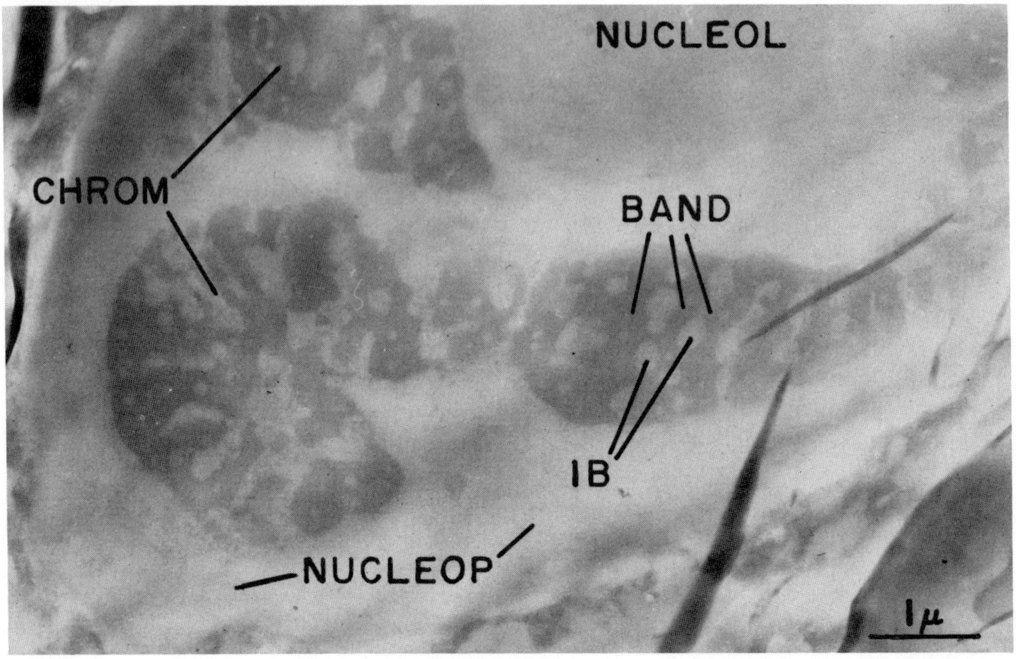

FIG. 5.3. Nucleus postfixed and stained with FFSulfone. As in Fig. 5.1, chromosomes (CHROM) consist of dark bands (BAND) and paler interband spaces (IB). Nucleoplasm (NUCLEOP) uniform. NUCLEOL, nucleolus. × 14,200.

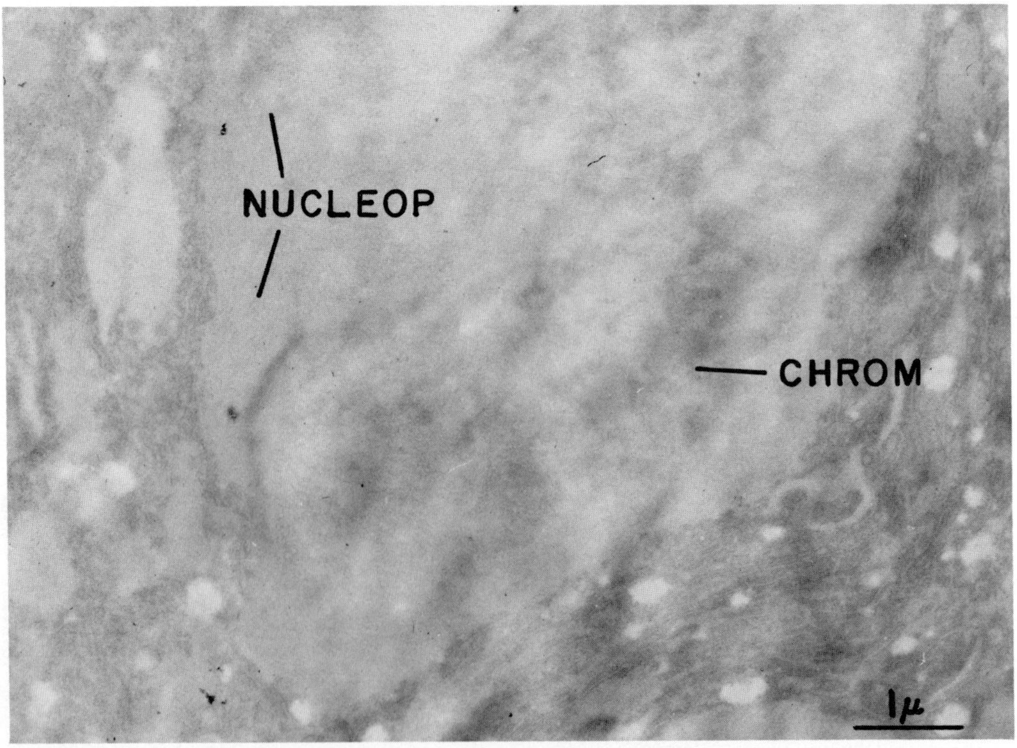

FIG. 5.4. Nucleus postfixed and stained with Zn hfac. Chromosome (CHROM) barely detectable against the stained nucleoplasmic (NUCLEOP) background. × 14,200.

ty but higher contrast, and paler intervals. The stripes appear as arborizations whose branches cross the paler intervals and join with arborizations from adjacent stripes. Whether these correspond with DNA- or RNA-rich lamellae is not clear. The nucleolus appears as a fine spongework of moderate density. The nucleoplasm is virtually unstained and is represented as a very coarse net. This group includes only specimens postfixed in alcohol and stained with an alcoholic solution of platinum tetrabromide ($PtBr_4$) (Figs. 5.6 and 5.7). After sections are stained with uranyl acetate, the chromosomes are nearly lost to view because the delicate stripes are almost completely obscured in a delicate spongework, which intervenes between them and between the chromosomes and the nucleoplasm (Fig. 5.8). The nuclear membrane is double before uranyl staining and is only slightly more prominent after staining.

3. The chromosomal stripes are somewhat coarser than above, and the arborizations are more prominent. Thus more chromosomal protein is stained. This group includes specimens postfixed and stained with vapors of difluorodinitrobenzene (FFDNB) (Figs. 5.9–5.12), lanthanum acetylacetonate (La acac), chromium (III) hexafluoroacetylacetonate (Cr hfac) (Figs. 5.13 and 5.14),

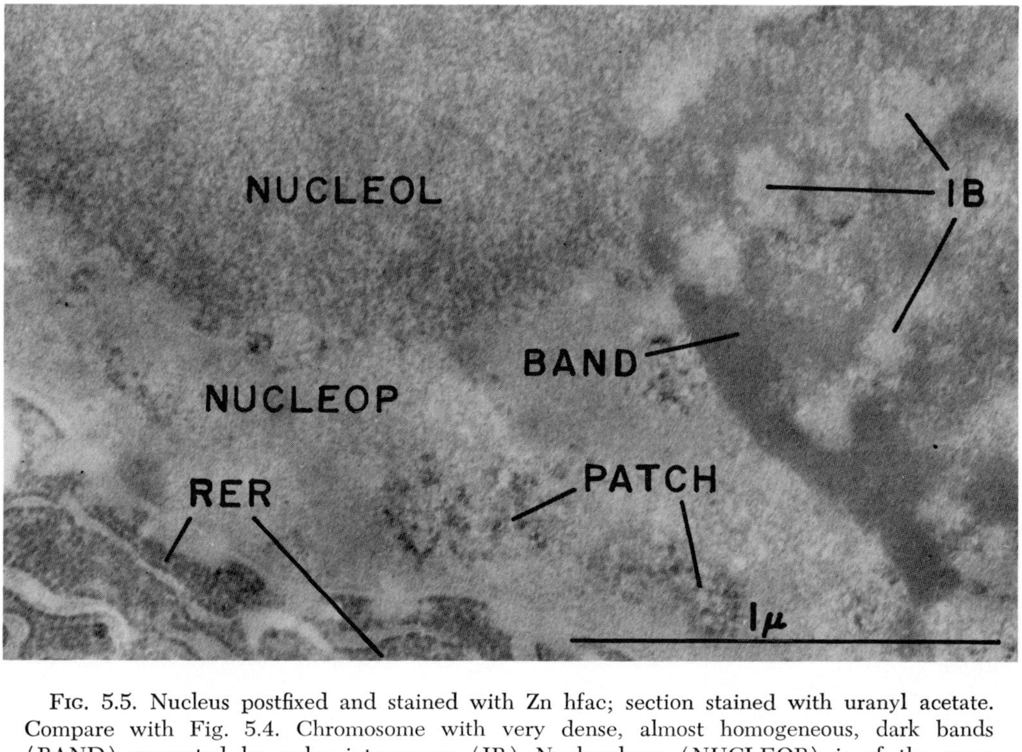

FIG. 5.5. Nucleus postfixed and stained with Zn hfac; section stained with uranyl acetate. Compare with Fig. 5.4. Chromosome with very dense, almost homogeneous, dark bands (BAND) separated by paler interspaces (IB). Nucleoplasm (NUCLEOP) is of the same density as the interband spaces and contains local aggregates of RNA (PATCH). Nucleolus (NUCLEOL) of intermediate density, with a very fine netlike texture. In the cytoplasm the RER is well stained. × 56,000.

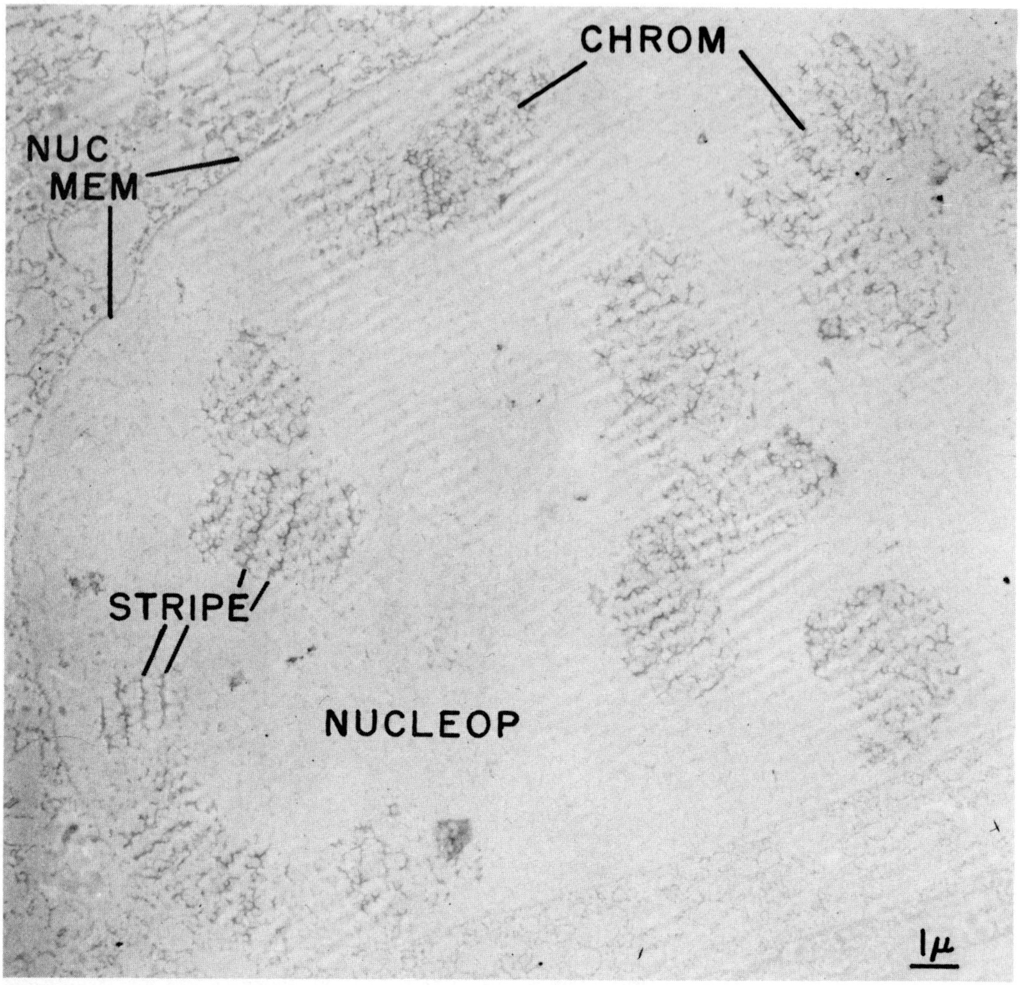

FIG. 5.6. ▲

▼ FIG. 5.7.

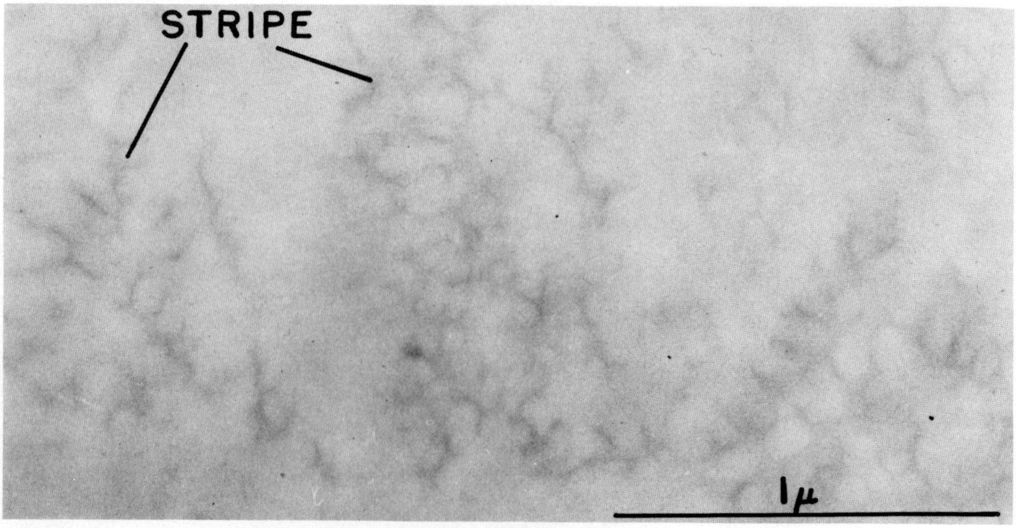

thorium (IV) hexafluoracetylacetonate (Th hfac), iron (III) acetylacetonate (Fe acac), and zirconium hexafluoracetylacetonate (Zr hfac) (Figs. 5.15 and 5.16). After sections are stained with uranyl acetate, the spongelike distribution of additionally stained protein in the chromosomal and nucleoplasmic walls, for the most part, does not obscure the chromosomes. The nuclear membrane is not stained after primary staining with FFDNB, La acac, and Fe acac, is well stained after Zr hfac, and is barely visible after Th acac. After sections are stained with uranyl acetate, nuclear membranes cannot be detected in specimens treated with FFDNB, are poorly delineated in specimens treated with Fe acac, and are visible as double membranes in all other specimens.

4. The stripes are somewhat more prominent indicating that more chromosomal protein is stained than above. This group includes specimens postfixed with vapors of calcium acetylacetonate (Ca acac) (Figs. 5.17 and 5.18), mercury (II) hexafluoroacetylacetonate (Hg hfac) (Figs. 5.19 and 5.20), and nickel (II) hexafluoroacetylacetonate (Ni hfac) (Fig. 5.21). The nuclear membrane is not visible after Ni hfac and Ca acac and only barely detectable after Hg hfac. After sections are stained with uranyl acetate, double nuclear membranes are apparent in all specimens.

5. The chromosomal protein in stripes has very little contrast, and minimal contrast is added when sections are stained with uranyl acetate. This group includes only specimens postfixed in vapors of copper (II) acetylacetonate (Cu acac). The nuclear membrane is not delineated.

Discussion

In contrast with the seemingly monotonously uniform distribution of proteins in hepatic and exocrine pancreatic cells, proteins are obviously heterogeneously distributed in salivary gland nuclei. It is easy to see why this should be so. If, for instance, we consider a single protein, such as the enzyme DNA polymerase, whose presence in the chromosome is known, we can imagine the effect of polyteny on its visiblility when stained. This enzyme consists of a single polypeptide chain which, if folded into more or less spherical form, would have a diameter of about 65 A *(1)*. In the highly dispersed DNA of the interphase nuclei of hepatic and exocrine pancreatic cells, the enzyme would occur as isolated molecules, one to each replicon, and even if stained selectively would be hard to identify. In the condensed chromatin of these nuclei, the enzyme would be more concentrated but still randomly distributed—it might stain uniformly even if no other proteins were stained at the same time. In the polytene chromosome, on the other hand, homologous replicons of the many chromonemata would each have their enzyme, so

---

FIG. 5.6 and 5.7. Nucleus postfixed and stained with alcoholic platinum tetrabromide. Chromosomes (CHROM) appear as very fine transverse stripes (STRIPE) whose arborizations join those of adjacent stripes. Nucleoplasm (NUCLEOP) very pale, with coarse net. Nuclear membrane (NUC MEM) stained. Figure 5.6, × 6000; Fig. 5.7, × 50,000.

that the molecules could be expected to form a band of 65 A width across the chromosome. This band could appear either continuous or interrupted, depending on factors such as the diameter of the chromosome at this point, and would constitute a pattern of distribution of the enzyme.

Certain protein reagents (group 1, FFSulfone, Ba acac, Rh hfac, and Zn hfac) cause maximal retention and staining of chromosomal protein—they are the least selective stains. Subsequent staining of sections with uranyl acetate does not discriminate between proteins, but only adds mass. In other groups (2–4), uranyl acetate seems to add mass to regions already stained, and, in addition, stains different proteins in chromosomes, nucleoli, and nuclear membrane. This additional staining takes place probably because prior postfixation reagents (and possibly the embedding reagents) have denatured additional proteins, with a consequential exposure of additional reactive groups that combine with the uranyl ion.

Other postfixation reagents stain more selectively. The most selective stain is $PtBr_4$ (group 2). The stains in group 3 (FFDNB, La acac, Cr hfac, Th hfac, Fe acac, and Zr hfac) are somewhat less selective. Still less selective are stains in group 4 (Ni hfac, Hg hfac, and Ca acac). Vapors of Cu acac (group 5) are of lit-

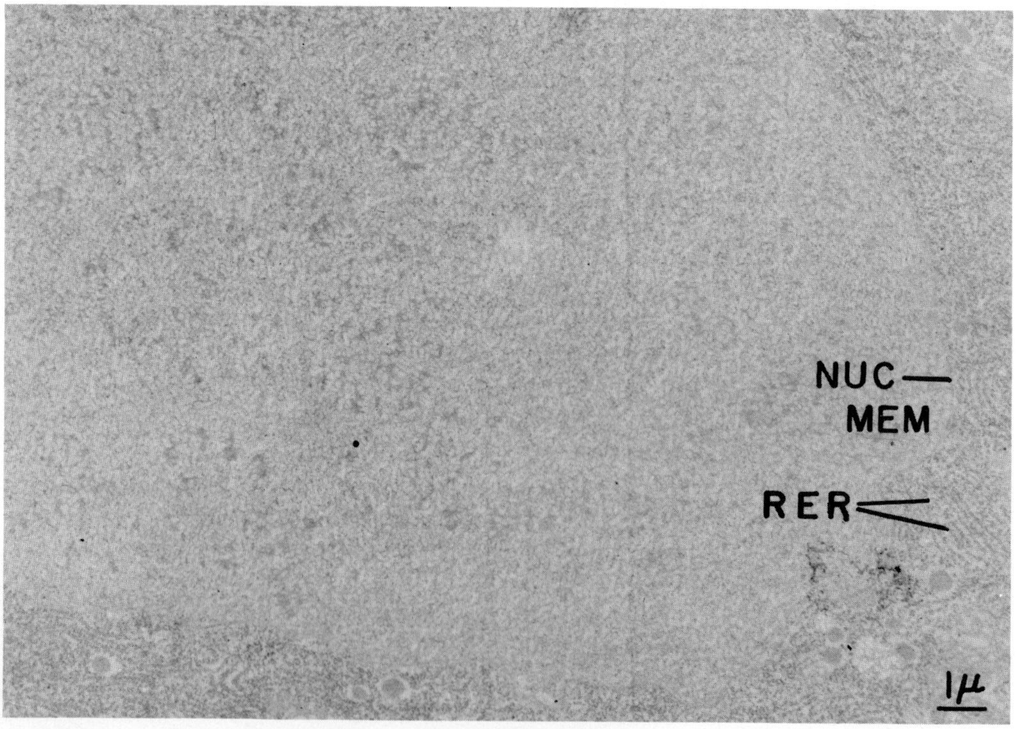

FIG. 5.8. Nucleus postfixed and stained with alcoholic platinum tetrabromide; section stained with uranyl acetate. Compare with Figs. 5.6 and 5.7. Chromosomes are obscured by the dense netlike appearance of the nucleoplasm. Nuclear membrane (NUC MEM) is delineated, and the rough endoplasmic reticulum (RER) of the cytoplasm is sharp. × 6000.

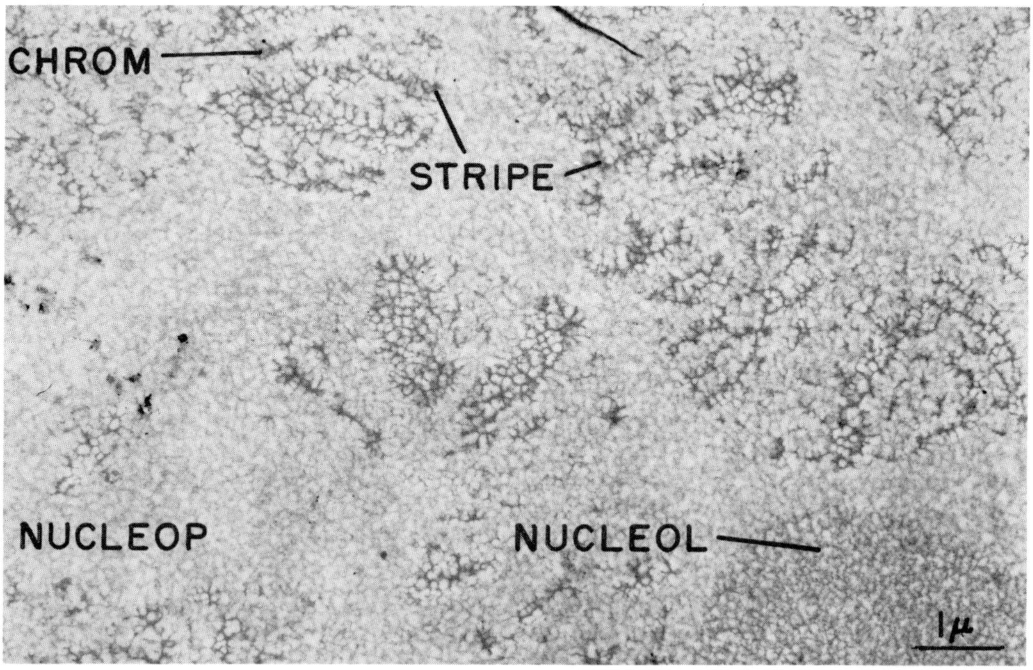

FIG. 5.9. ▲ ▼ FIG. 5.10.

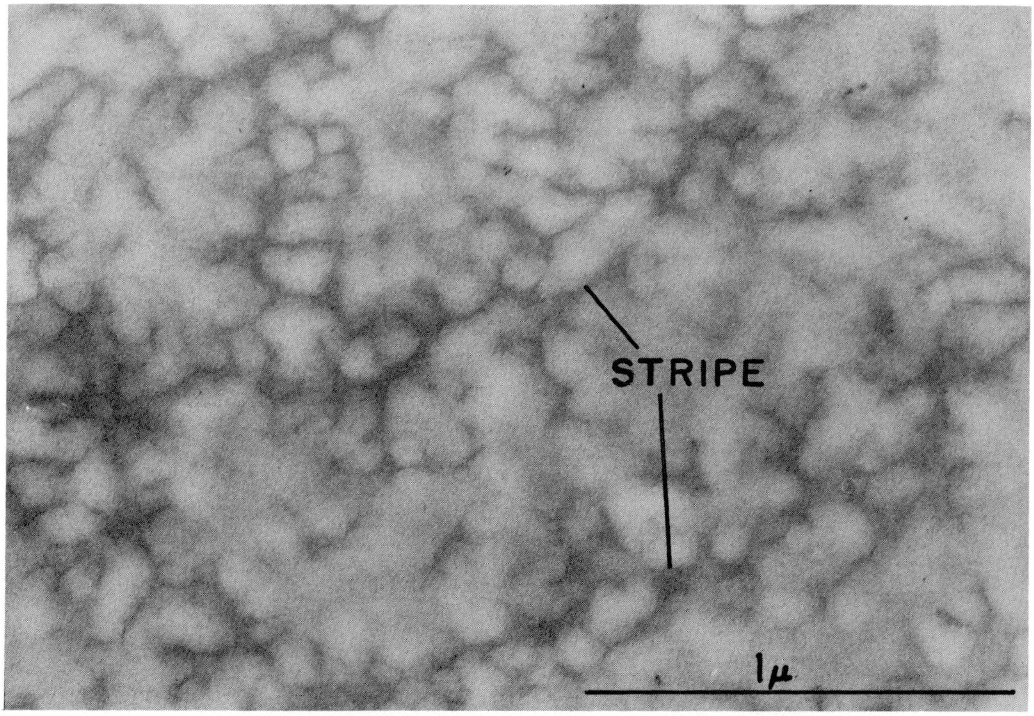

FIGS. 5.9 (TOP) and 5.10 (BOTTOM). Nucleus postfixed and stained with vapors of FFDNB. Chromosomes (CHROM) appear as fine transverse stripes (STRIPE). Nucleoplasm (NUCLEOP) appears finely reticulated. Nucleolus (NUCLEOL) is intermediate in density, and very fine meshed. Figure 5.9, × 11,200; Fig. 5.10, × 56,000.

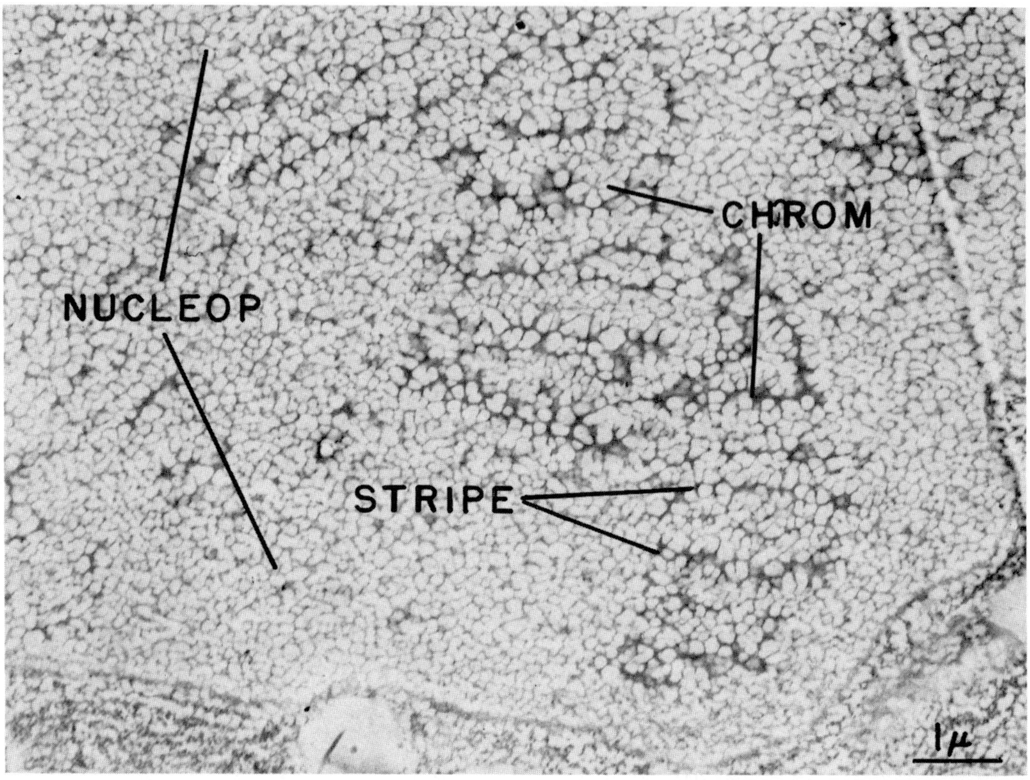

FIG. 5.11. ▲ ▼ FIG. 5.12.

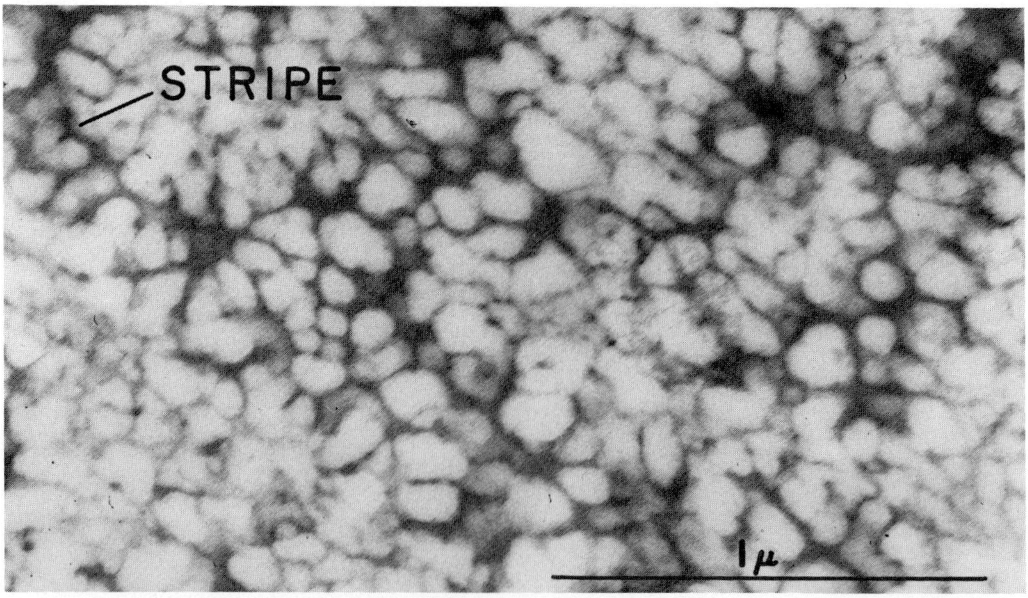

FIGS. 5.11 and 5.12. Nucleus postfixed and stained with vapors of FFDNB; section stained with uranyl acetate. Compare with Figs. 5.9 and 5.10. The dense stripes (STRIPE) of the chromosomes (CHROM) are clearly delineated, despite the enhanced contrast of the nucleoplasm (NUCLEOP) and the walls between the stripes. Figure 5.11, × 11,200; Fig. 5.12, × 56,000.

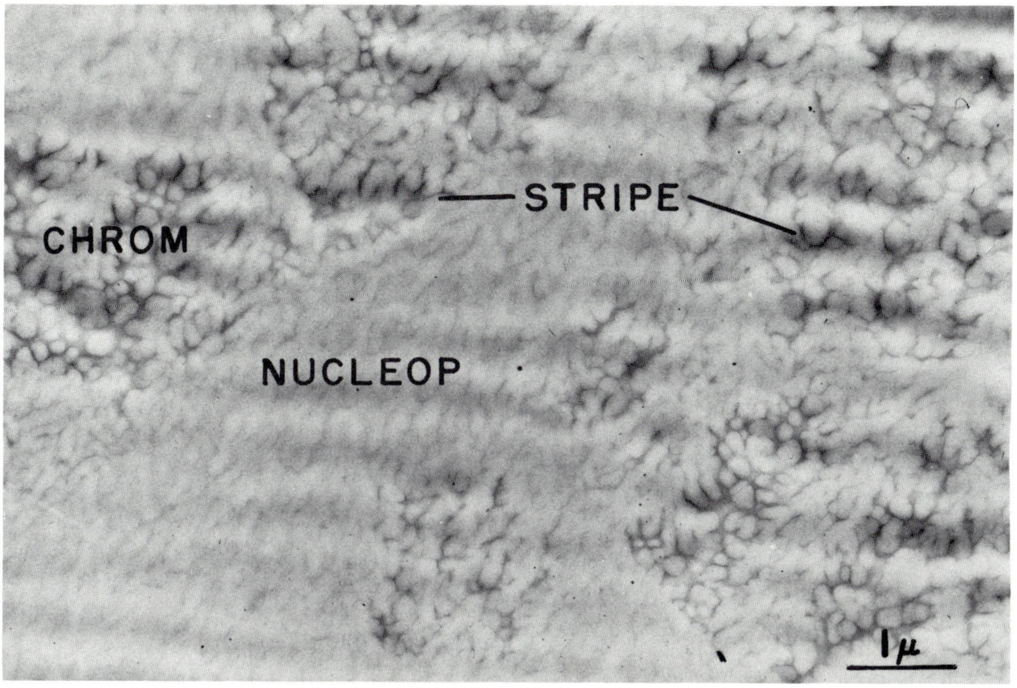

Fig. 5.13. Nucleus postfixed and stained with vapors of Cr hfac. Dense walls of stripes (STRIPE) of chromosomes are clear against the less dense walls of both chromosomes (CHROM) and nucleoplasm (NUCLEOP). × 14,200.

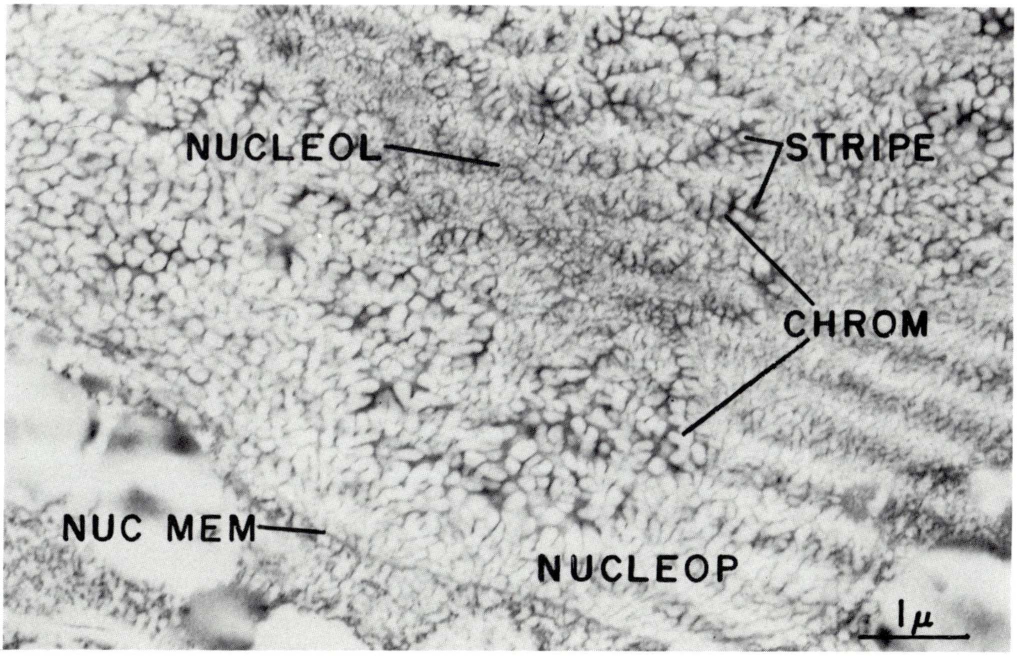

Fig. 5.14. Nucleus postfixed and stained with vapors of Cr hfac; section stained with uranyl acetate. Compare with Fig. 5.13. Chromosomes (CHROM) with their dense stripes (STRIPE) mostly retain their individuality despite the increased density of the walls between the stripes and in the nucleoplasm (NUCLEOP). Nucleolar (NUCLEOL) network is dense and close meshed. Nuclear membrane (NUC MEM) stained. × 14,200.

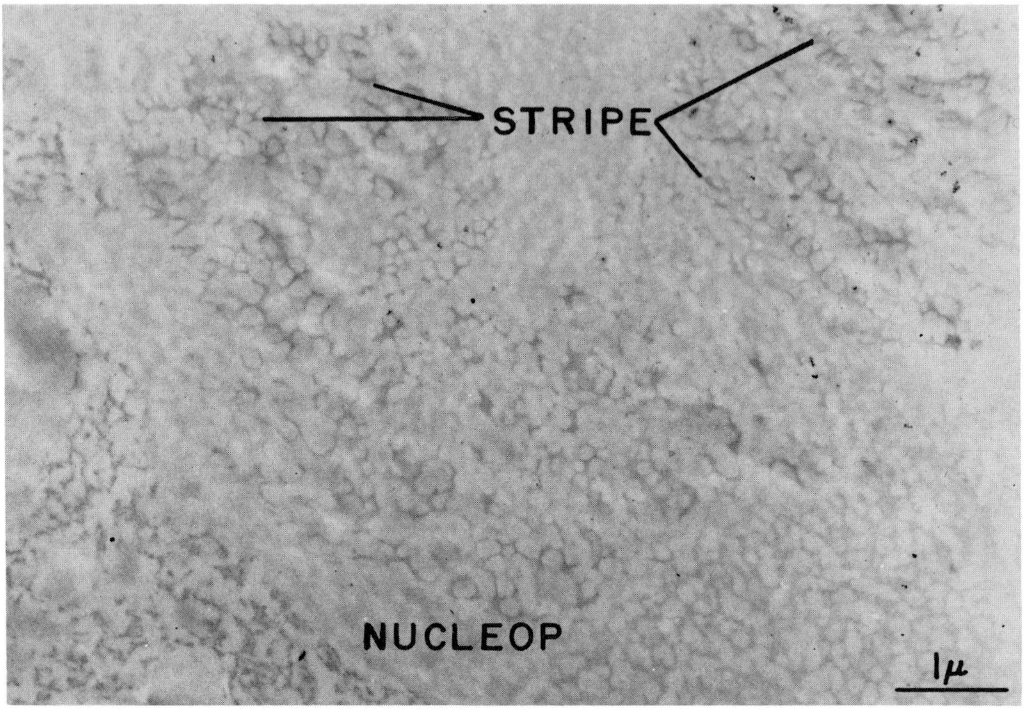

Fig. 5.15. Nucleus postfixed and stained with vapors of Zr hfac. Chromosome stripes (STRIPE) clear against pale walls of nucleoplasm (NUCLEOP). × 14,200.

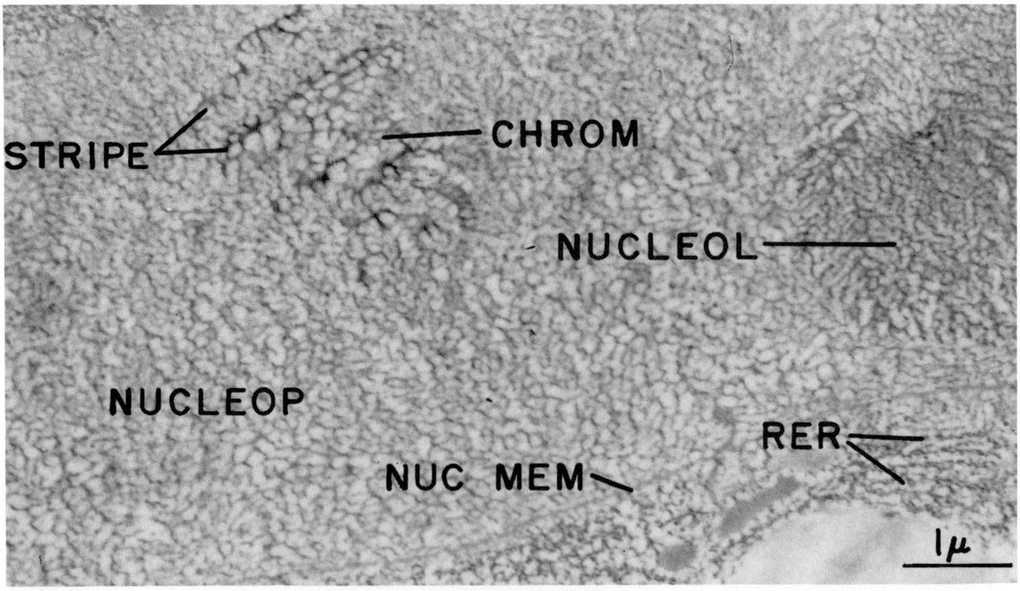

Fig. 5.16. Nucleus postfixed and stained with vapors of Zr hfac; sections stained with uranyl acetate. Compare with Fig. 5.15. Chromosome (CHROM) recognizable by its denser stripes (STRIPE) despite enhanced contrast of walls of nucleoplasm (NUCLEOP). Nuclear membrane (NUC MEM) is clear, as is also the RER in the adjacent cytoplasm. The finely meshed nucleolus (NUCLEOL) is intermediate in density. × 14,200.

tle value because minimal contrast is added. In order of increasing selectivity, primary staining by cations is listed below:

$$Ba^{2+},\ Zn^{2+} < Ni^{2+},\ Hg^{2+},\ Ca^{2+} < La^{3+},\ Cr^{3+},\ Fe^{3+},\ Th^{4+},\ Zr^{4+} < Pt^{4+}$$

(Rh^{3+} is ignored because there is little known about its reaction with proteins.) It is necessary to add that although some cations are grouped together, this does not imply that they are staining the same chemical entities in the same bands. They are grouped together only because morphologically the images of the stained material look similar even though the bands examined are quite likely from different chromosomes and almost certainly from different chromosome regions. Besides, in the case of staining of substructures within a band, even if we viewed the same band with different staining, it would be virtually impossible, in the absence of differential staining for DNA and RNA, to know whether one was looking at protein components of the DNA- or of the RNA-rich lamellae. It is

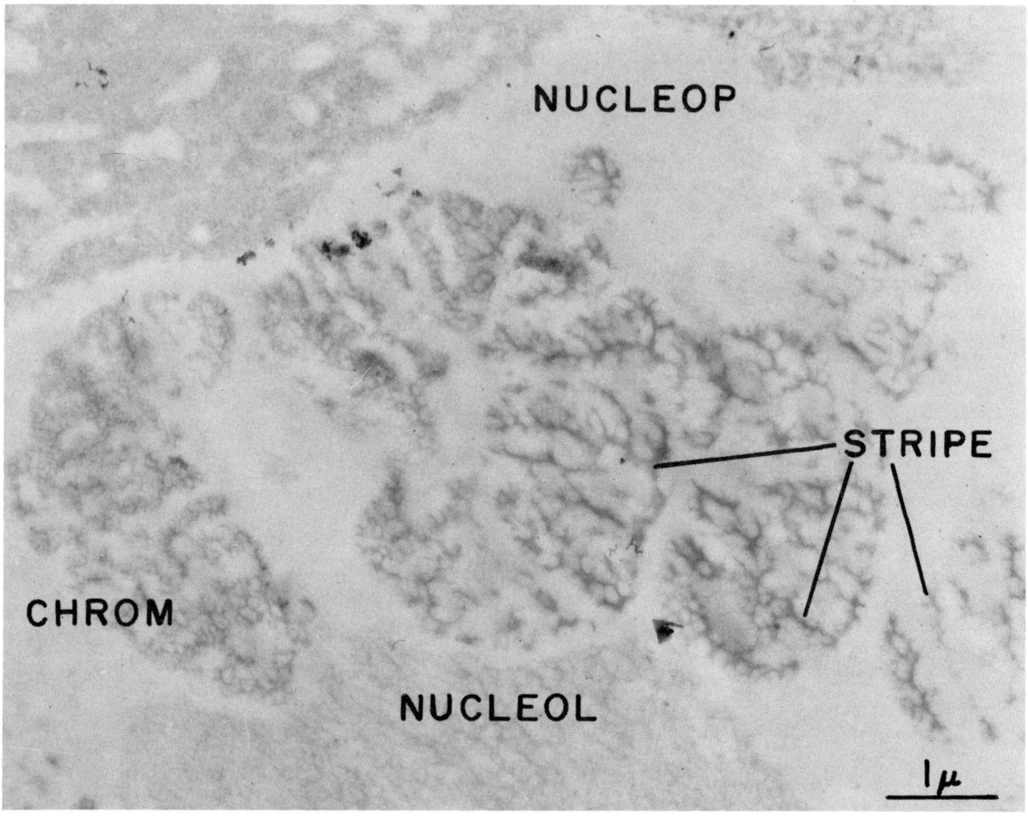

FIG. 5.17. Nucleus postfixed and stained with vapors of Ca acac. The stripes (STRIPE) of the chromosomes (CHROM) are broader and more extensively arborized than above. They stand out from the unstained nucleoplasm (NUCLEOP). The nucleolar (NUCLEOL) meshwork is intermediate in density. The RER lacks contrast. × 14,200.

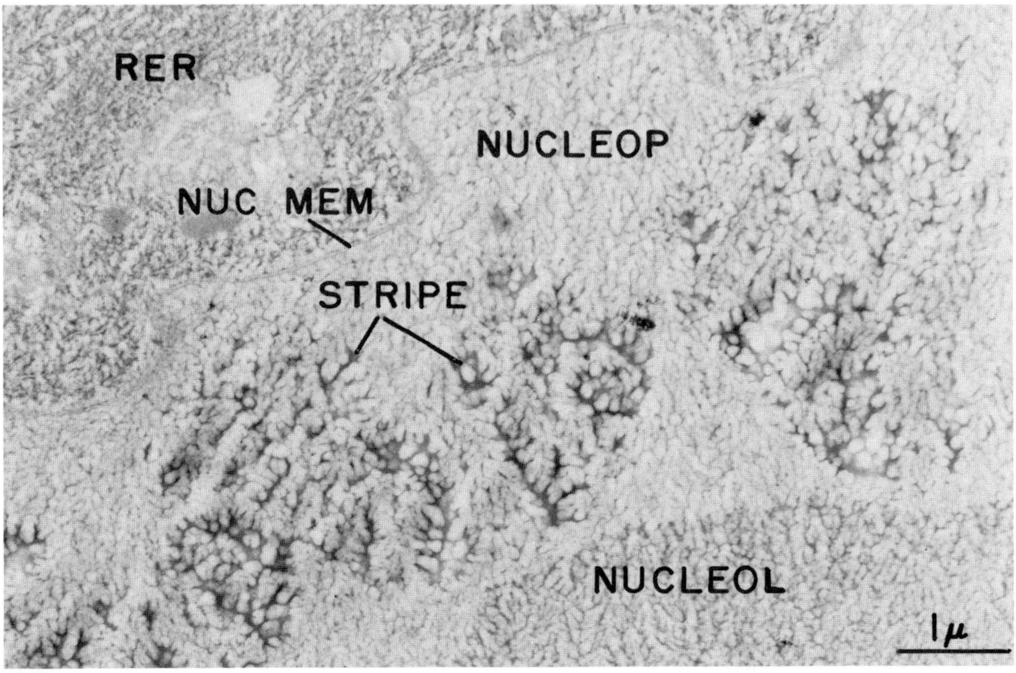

FIG. 5.18. Nucleus postfixed and stained with vapors of Ca acac; section stained with uranyl acetate. Compare with Fig. 5.17. Chromosomes (CHROM) with their very dense stripes (STRIPE) are clear despite enhanced contrast of walls of chromosome between stripes and the walls of the nucleoplasm (NUCLEOP). Nucleolar (NUCLEOL) net is of intermediate density. Nuclear membrane (NUC MEM) and RER clearly delineated. × 14,200.

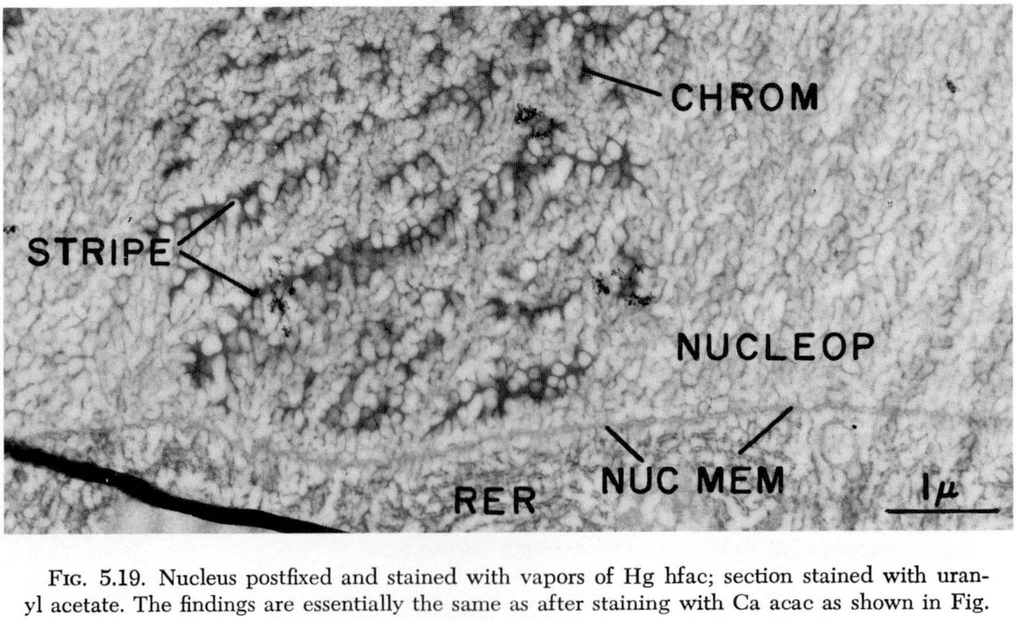

FIG. 5.19. Nucleus postfixed and stained with vapors of Hg hfac; section stained with uranyl acetate. The findings are essentially the same as after staining with Ca acac as shown in Fig. 5.18. All abbreviations given earlier. × 14,200.

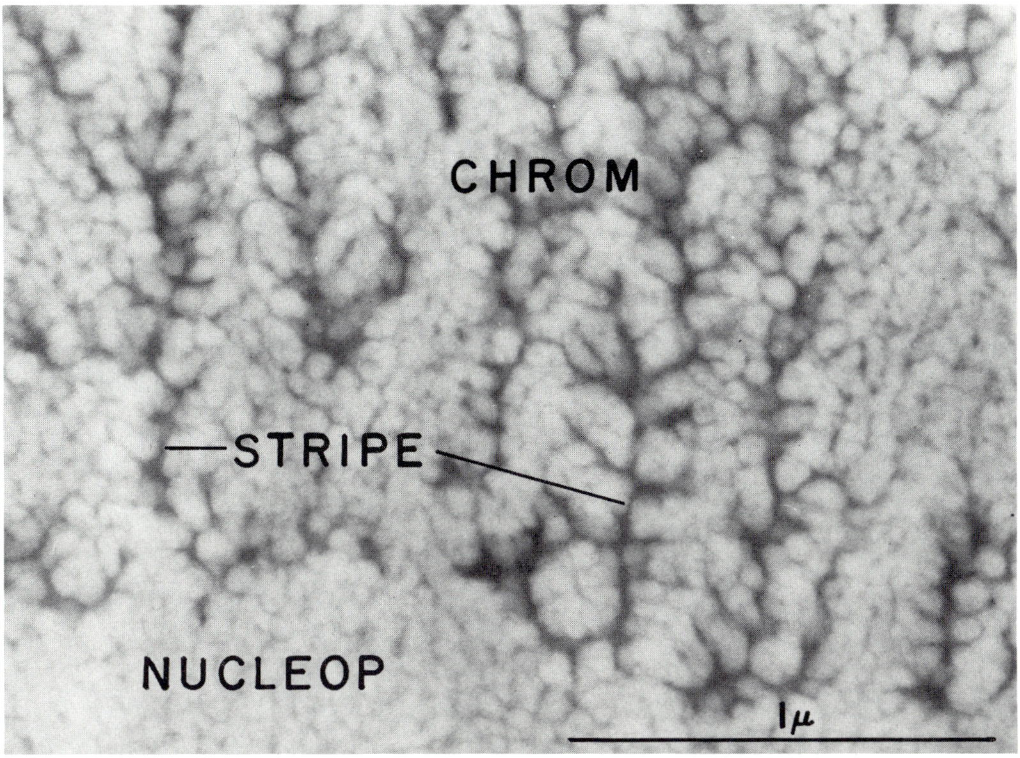

FIG. 5.20. Same nucleus and section as above to show the relation of the dense stripes (STRIPE) and their arborizations to the walls of the chromosome (CHROM) between the stripes and the walls of the nucleoplasm (NUCLEOP). × 56,000.

necessary also to consider the possibility that each group of cations may stain a class of proteins which may include a great number of proteins differing from each other in ways which are too subtle to be discriminated by the reagents employed. For instance, with the most selective stain, such as $PtBr_4$ two distinct kinds of structure are seen—fine, continuous bands across the chromosome and fragmentary ones between these bands. We cannot say whether each of these two structures is composed of a single protein or of two or more different proteins. Also it must be recalled that the cations stain only reactive parts of proteins whether these are histones or acidic proteins, and cannot be said at this time to serve as identifying markers of specific protein molecules. Some of the factors which might be involved are valence, the ease and strength of binding, the ionic radius, the reacting group(s) of the protein molecule, and the steric relations of the reacting groups. In view of the complexity of the situation, it is difficult to assign meaning to the order of increasing selectivity of primary staining noted earlier in this paragraph.

As in chromatin of hepatic and pancreatic cells, the distribution of proteins in salivary gland chromosomes is quite different from that of nucleic acids. Here,

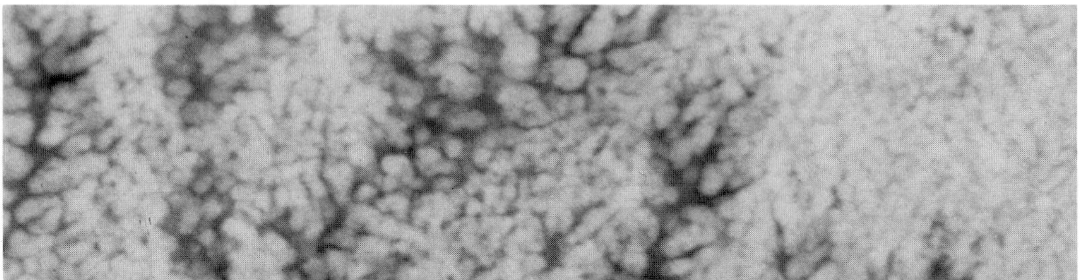

FIG. 5.21. Nucleus postfixed and stained with vapors of Ni hfac; section stained with uranyl acetate. The morphology of the chromosome and its relations are virtually identical with that in Fig. 5.20. × 56,000.

some of the protein seems to be distributed in a subband pattern such that the protein appears to serve as a kind of scaffold on which the secondary DNA coils are supported. If binding of DNA with proteins at regular intervals is responsible for the formation and retention of second-order DNA coils, and if genic action depends on the pattern of second and higher orders of coiling and their aggregation, then the role of the protein scaffolding is of critical importance. It might be that some of the protein which extends longitudinally in the chromosome, sometimes clearly arborizing from transverse protein bands, has the function of maintaining the second-order coils as in the helices which traverse the interbands and the RNA-rich lamellae described in Chapter 4 (this volume). At the same time, stains which identify proteins in masses which are probably coextensive with lamellae may reveal structural elements responsible for maintaining the dense packing of the second-order helices in the DNA-rich lamellae.

Because the protein of chromatin in hepatic and pancreatic nuclei appeared to be homogeneously distributed, it was not possible to ascertain whether or not second-order DNA coils were enclosed in sheaths. While the conclusion is still not certain, the possibility is clear that protein associated with DNA in salivary gland chromosomes lies between adjacent second-order coil helices. From the appearance of protein in nuclei fixed by cations of group 1, it is apparent that protein must also be present between the gyres.

Two general observations have an important bearing on the question whether ice crystals form during freezing and drying. The first observation has been mentioned in Chapters 1 and 2 (this volume). It is that when freeze-dried salivary glands of *Drosophila* are treated *in vacuo* with alcohol, or with certain cross-linking vapors (FFDNB, Ba acac, Rh hfac, and Zn hfac), nuclear structures appear relatively uniform, with no trace of the spaces formerly called submicroscopic vacuoles and now renamed pseudovacuoles. These do appear after the glands are treated with all other organometallic vapors. When they appear, their morphology and staining properties differ, falling into four different patterns. This speaks against the possibility that they are all products of ice crystals, for in that case, the ice crystal pattern should be the same regardless of the cross-linking agent. The second significant finding is that section staining with uranyl acetate results in alteration of the pseudovacuolar pattern, that is, of the apparent distribution of protein. For example, treatment of specimens *in vacuo* with vapors of Zr hfac results in a coarse netlike pattern; but after section staining with uranyl acetate, the nets are finer. Contrariwise, after treatment of specimens *in vacuo* with vapors of FFDNB, the protein nets are fine; but after section staining with uranyl acetate, the protein nets are coarse. These findings are best understood by regarding proteins as generally distributed; some proteins are made insoluble and stainable by certain reagents, while others are dissolved or become unstainable. Staining of the former, and absence or equal electron scattering properties of the latter generate an appearance of a less dense "space" (the pseudovacuole) enclosed in denser walls. Thus this pseudostructure of protoplasm is not formed during freezing or drying, but during subsequent treatment associated with postfixation and staining. Clearly, there is no evidence that ice crystals sufficiently large to be seen with the electron microscope are in any way related to the pseudovacuolar structure of protoplasm. The results suggest that the protein base of protoplasm is heterogeneous—one part consisting of a spongelike structure which is more readily stained, and the other part less readily stained.

The observations have a certain bearing on the nature of contrast in electron micrographs. In those instances, when staining of section with uranyl acetate resulted in further marked staining of chromosomal and/or nucleoplasmic proteins, it is clear that the vapor (postfixative) stained certain proteins, while at the same time preserving others unstained. Uranyl acetate stained these latter in many instances, as well as adding mass to the protein already stained. Both processes took place while the specimen (previously hardened during the manipulations by the effects of vapor, alcohol, and the embedding procedure) was enclosed in a cross-linked polymer (Durcupan). One could assume that no major physical displacements of solid components could take place under these circumstances. Thus it appears that in the specimens studied here, increased contrast is caused primarily or entirely by increased mass from binding of the reagent.

Summary

The distribution of protein in salivary gland nuclei was studied by methods described in Chapter 2 (this volume) involving freeze-drying with postfixation and staining *in vacuo* with cross-linking vapors. In addition, ultrathin sections were staned with uranyl acetate. Alcoholic platinum tetrabromide solutions were also used after postfixation in alcohol.

Proteins were found to be distributed throughout the nucleus, the preservation and staining of the proteins depending on the nature of the cross-linking agent used. The results indicate there is a very marked heterogeneity in the distribution of proteins in the nuclear membrane, nucleoplasm, nucleolus, and chromosome, at least as judged by the kind and degree of staining. Evidence is presented that there are submicroscopic regions, the pseudovacuoles, which appear (after certain procedures) as pale or "empty" spaces enclosed by denser staining walls. The proteins in these two regions are regarded as being different. The evidence clearly indicates that the pseudovacuoles are not the result of ice crystals which are alleged to form during freezing or drying, but are the result of selective or preferential staining during postfixation and staining after prior denaturation.

Reference

1. Englund, P. T., Deutscher, M. P., Jovin, T. M., Kelly, R. B., Cozzarelli, N. R., and Kornberg, A. (1968). Structural and functional properties of *Escherichia coli* DNA polymerase. *Cold Spring Harbor Symp. Quant. Biol.* 33, 1–9.

6

DNA Patterns of Cells of Imaginal Disks of Drosophila Larvae

Eileen S. Gersh and Isidore Gersh

It has been abundantly demonstrated *(3, 4)* that the imaginal disks of *Drosophila* are already determined during the development of the larva as wing, leg, or other structures in the adult. In addition, within each such structure, the subsequent differentiation of regions of cells, and indeed of individual cells, is more or less fixed at a time when individual cells are indistinguishable from one another in the light microscope. Since the cells are small and the resolution inadequate, it has not been possible to ascertain whether this critical state is accompanied by nuclear changes. This report is the result of an attempt to learn, by morphological methods, whether nuclear changes have taken place in determined cells. More specifically, using a cytochemical method which is sensitive enough to stain the first-order (double) DNA helix, we planned to ascertain whether changes in the DNA pattern had already taken place in determined cells.

Of the several types of specimens which could be used for this study, we selected two: the wing disk and the haltere disk. In our studies of nuclei in these disks, we first set out to determine whether, in the wing disks of late third instar wild type larvae of *Drosophila melanogaster*, the DNA molecular pattern is the same or different in the part which is determined to differentiate as wing, and the part determined as thorax. Finding that there was indeed a difference, we turned to younger (early third instar) larvae, to learn whether the DNA molecular pattern was already different in the distal and proximal parts of the wing disk, and whether the patterns resembled those in the third instar organs. We also expected that, as differentiation progresses, the DNA molecular pattern should alter and become more varied. To test this we examined cells in the developing wing disks of pupae.

Contrary to the pessimistic summary *(8)* of earlier ultrastructural analyses, our results show that there is already some nuclear differentiation in the early third instar and that several nuclear types are present in relatively early pupal stages.

Next, we made a comparison of cells in corresponding (distal) regions of the wing and haltere disks in late third instar wild-type larvae. We found again that the DNA molecular pattern is different. It then became of great interest to know what obtained in a homoeotic mutant, Haltere-mimic *(Hm)*, in which the wing approximates very closely in shape and size to a haltere. Haltere-mimic wing disks were studied from third instar larvae that were heterozygous for *Hm* (a dominant mutant) and also for Curly *(Cy)*, which also effects the wing. Another set of *Hm* wing disks was obtained from larvae heterozygous for *Hm* and Plum *(Pm*, an eye color mutant) but lacking the *Cy* mutant. Finally, the effect of *Cy* alone was studied. The group of studies of *Hm* wing disks led us to the conclusion that *Hm* does alter the DNA pattern, though the result is not identical with that found in the normal haltere.

The findings outlined above will be discussed in relation to a hypothesis concerning the nature of differentiation at the level of the genome and its morphological expression in terms of DNA molecular pattern. The problems enumerated above were the concern of others who have made ultrastructural studies of wing *(2)*, eye *(10)*, leg *(7)*, and genital disks *(9)*.

Method

The appropriate imaginal disk of larvae was dissected in *Drosophila* Ringer's solution and transferred to a small, thin square of aluminum foil. Excess fluid was drained and the specimen was frozen ultrarapidly and dried at a temperature below $-40°C$. Specimens were postfixed *in vacuo* with vapors of mercury hexafluoroacetylacetonate (Hg hfac). After hardening with 95% alcohol, specimens were stained for nucleic acids with gallocyanin–chromalum, again hardened in alcohol, and then infiltrated with water-soluble Durcupan, which was polymerized. Procedural details were given in Chapter 3 (this volume). Orientation sections $\frac{1}{4}$-μ thick were cut until the section was judged to pass lengthwise through the middle of the distal part (determined as wing) and the proximal part (determined as thorax), and the sections were mounted on slides and stained with toluidine blue. Figure 6.1 is an example of such an orientation section. At the appropriate level, ultrathin sections were cut and selected parts were studied with the electron microscope.

Observations

Wild-Type Disk, Distal Part (Figs. 6.2–6.6)

The cells are elongated and the nuclei are pale and oval in section. The cytoplasm is packed with small granules rich in RNA, except for some vacuoles. The single large nucleolus comprises numerous small granules containing RNA. The chromatin is denser than the nucleolus and comprises small irregularly shaped and interconnected masses which are not visibly

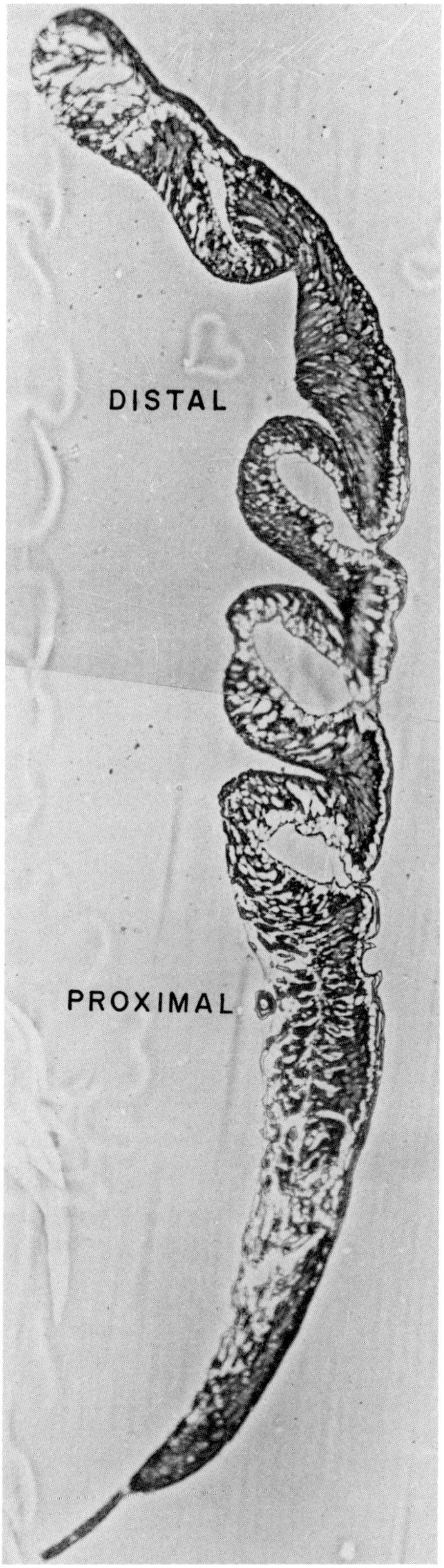

FIG. 6.1. Photomicrograph of a medial section through distal (A) and proximal (B) parts of the wing disk of late third instar wild-type and stained *en bloc* with gallocyanin–chromalum and in sections with toluidine blue. Original at × 100; × 4.5 in printing.

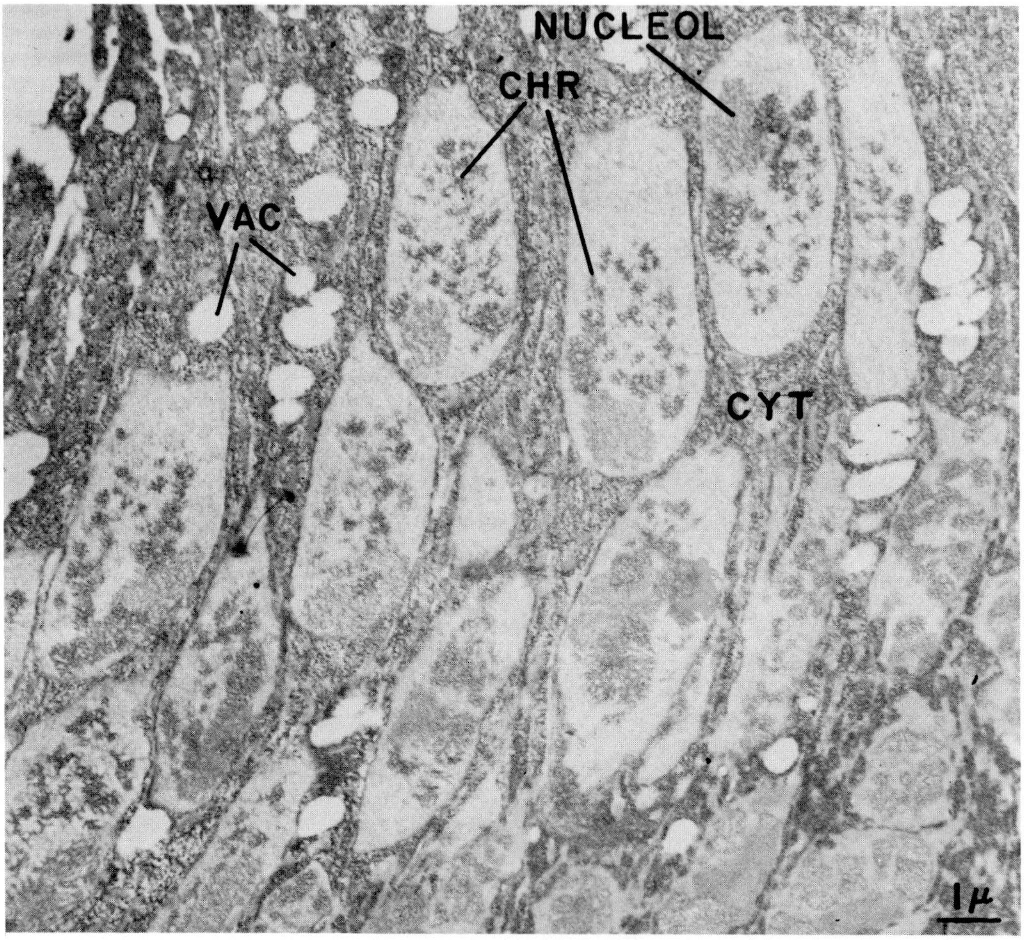

Figs. 6.2–6.6. Wild-type wing disk, late third instar, distal part. Nearly all nuclei in the distal part are like those in Figs. 6.2–6.5. A very small number are like those in Fig. 6.6 and Figs. 6.7–6.10. NUCLEOL, nucleolus; CHR, chromatin; NUCLEOP, nucleoplasm; CYT, ribosome-rich cytoplasm; VAC, vacuoles; RIBOS, ribosomes; DNA II, second-order DNA coils; GRANS, RNA granules. Figures 6.2, × 8000; Figs 6.3 and 6.6, × 25,000; Figs. 6.4 and 6.5, × 56,000.

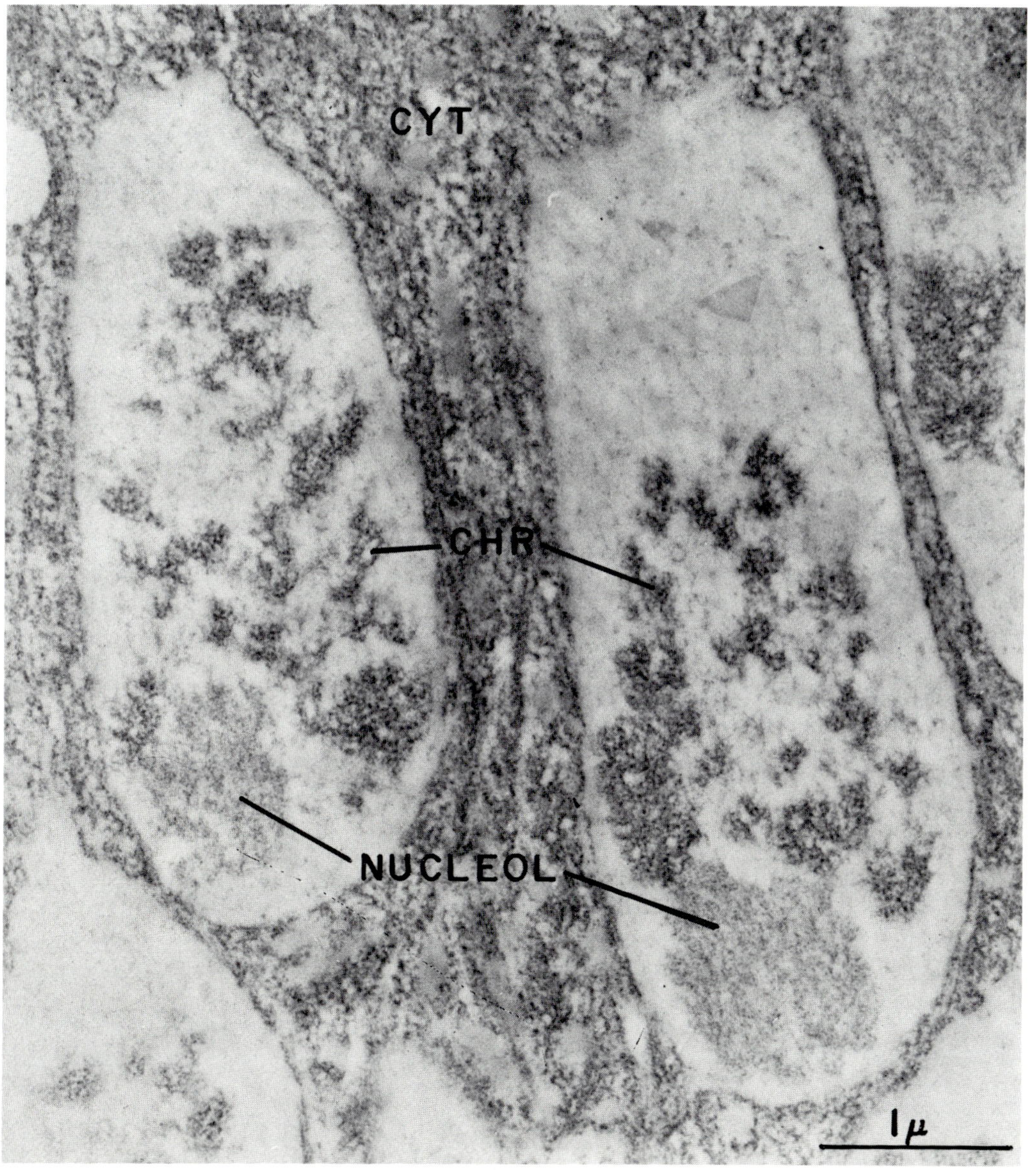

Fig. 6.3.

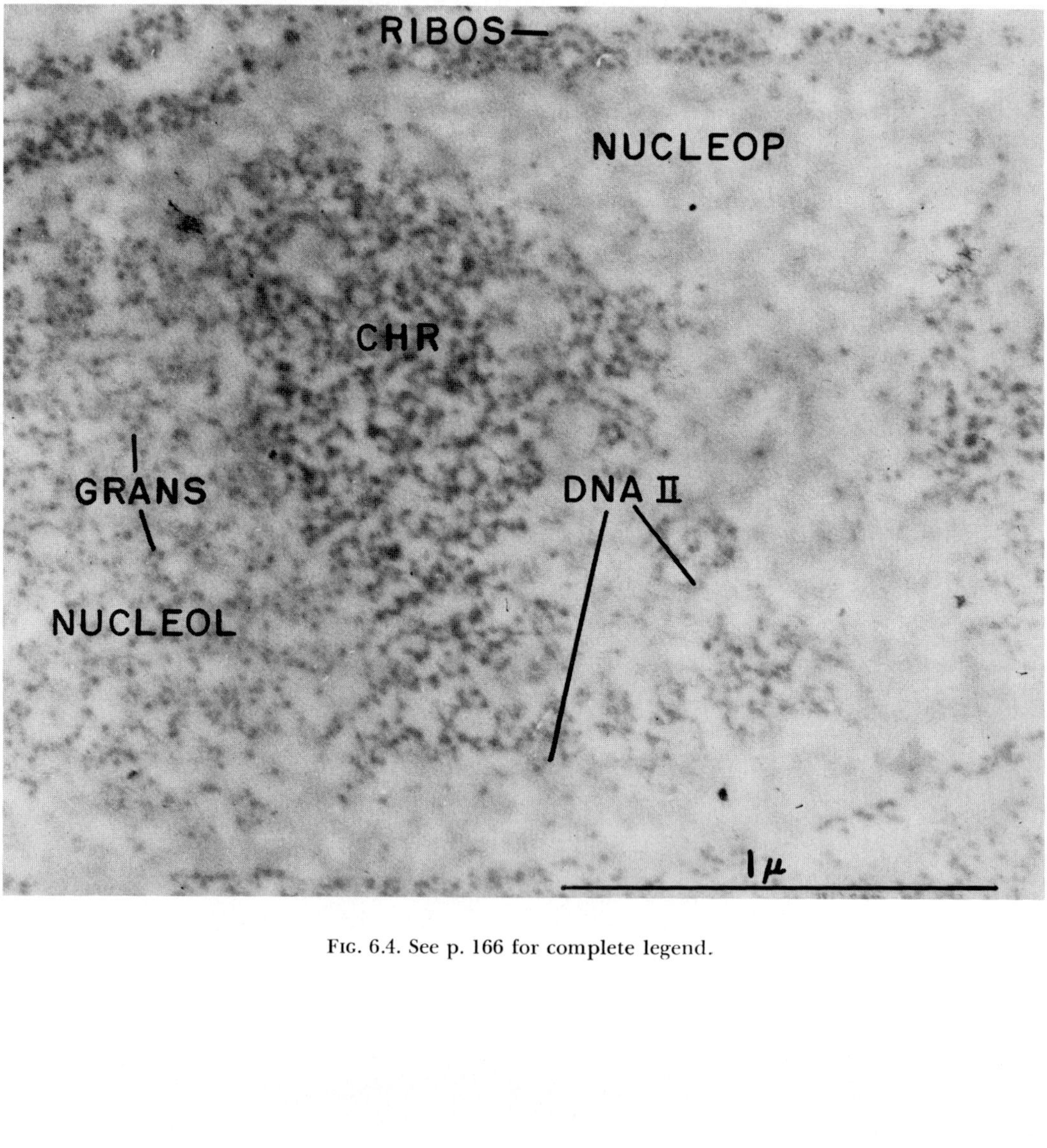

FIG. 6.4. See p. 166 for complete legend.

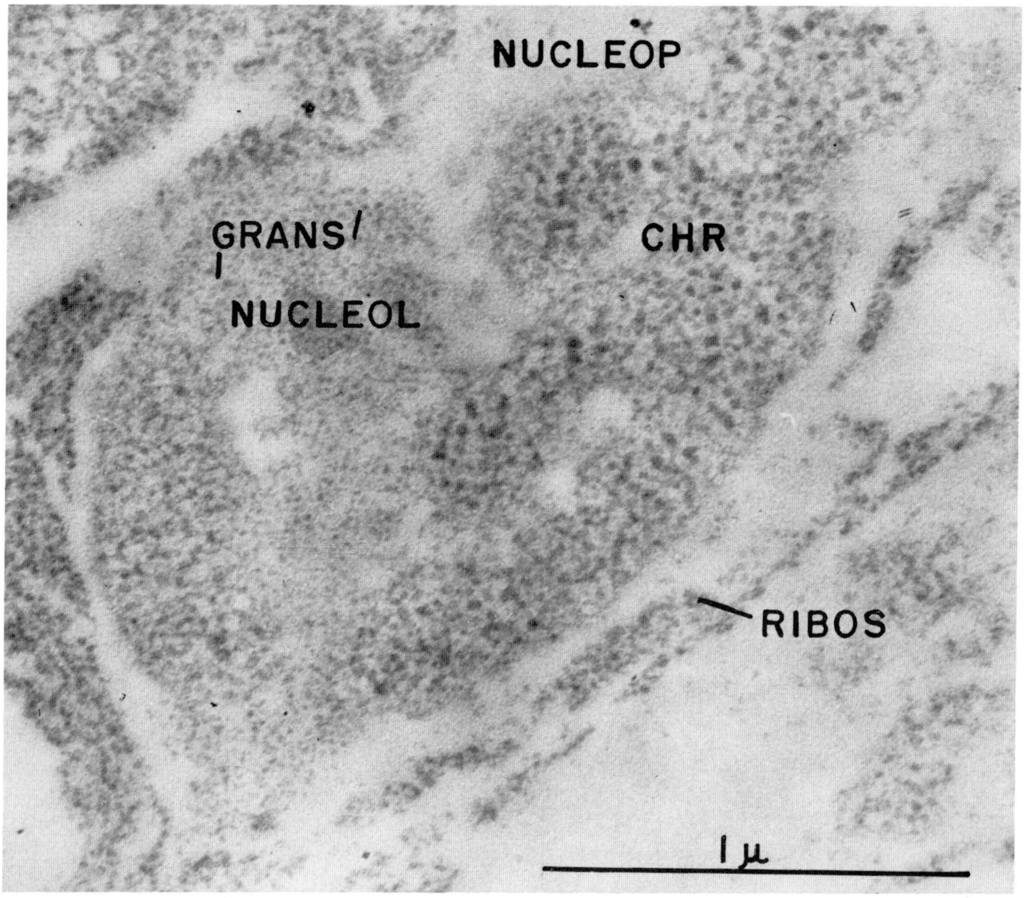

FIG. 6.5. ▲

▼ FIG. 6.6.

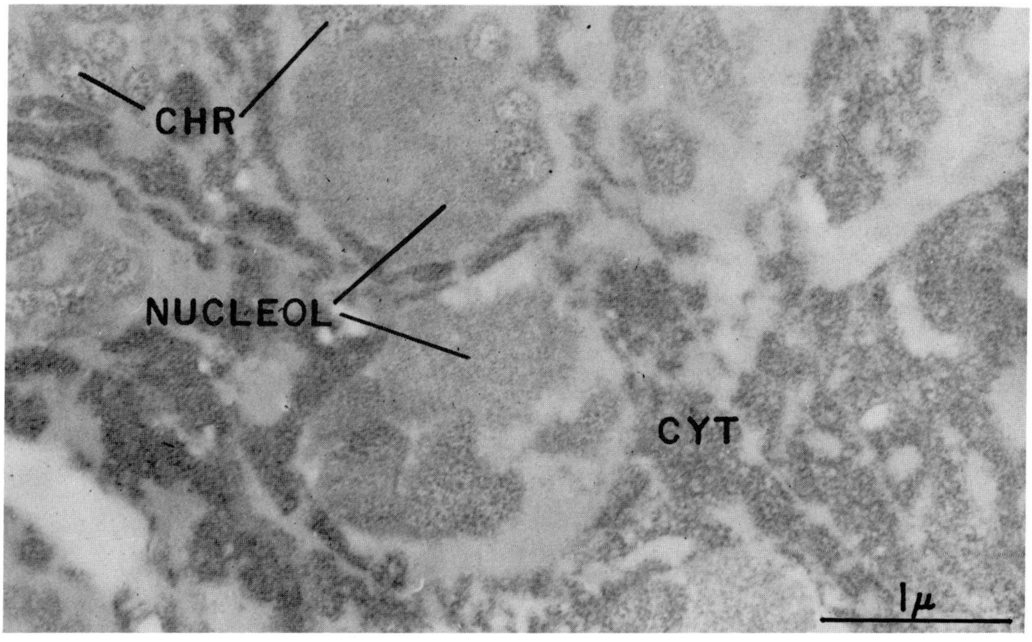

attached to the nuclear membrane. Each is made of many second-order DNA coils with a diameter of about 400 A, which are sometimes loosely arranged. The nucleoplasm consists of a spongelike barely detectable dense and less dense phase common to all nuclei studied, with few RNA granules in the denser walls of the nucleoplasm. Occasionally more spherical nuclei are found that resemble those in the proximal part (Fig. 6.6).

Wild-Type Disk, Proximal Part (Figs. 6.7 and 6.8)

Most nuclei appear less oval than in the cells of the distal part. The nucleolus appears to be more compact, the RNA granules being more tightly packed. The nucleolar granules are paler staining than most of the ribosomes. DNA is organized as circumscribed and interconnected "lobulations." The sharp outline of the lobulations is caused by fine RNA particles which surround them. RNA particles are also visible in the nucleoplasm between the lobulations, and between the latter and the nuclear membrane. The second-order DNA coils within the lobulations are rather loosely packed.

Identification of Nucleic Acids (Figs. 6.9 and 6.10)

As stated in Chapter 3 (this volume), nucleic acids in our preparations can usually be identified by solely morphological criteria. In general, DNA is in the form of first- and second-order coils, while RNA exists as granules. The identification is strengthened when other specimens can be treated with DNase or RNase prior to staining with gallocyanin–chromalum and the cell morphology is retained without too much distortion as a result of immersion in the enzyme solution. Wing disks were treated with cross-linking vapors, and then digested with ribonuclease prior to staining as described in Chapters 3 and 4 (this volume) and Chapter 5 (Vol. II). The reduction in RNA content of the cytoplasm and nucleolus results in a very prominent DNA pattern which is much the same as the DNA pattern of specimens not treated with enzyme. The DNA is partly swollen, despite the cross-linking.

Early Third Instar Wing Disk, Distal Part (Figs. 6.11 and 6.12)

The DNA pattern is less striking than at the third instar, and somewhat different in that the chromatin is less compact. In addition, numerous RNA granules are diffusely distributed in the nucleoplasm.

Figs. 6.7 and 6.8. Wild-type wing disk, third instar, proximal part. The chromatin has a characteristic "multilobulated" appearance, the "lobules" being sharply outlined by RNA granules. The rest of the nucleoplasm is relatively free of RNA granules. Abbreviations as used earlier. Figure 6.7, × 8000; Fig. 6.8, × 25,000.

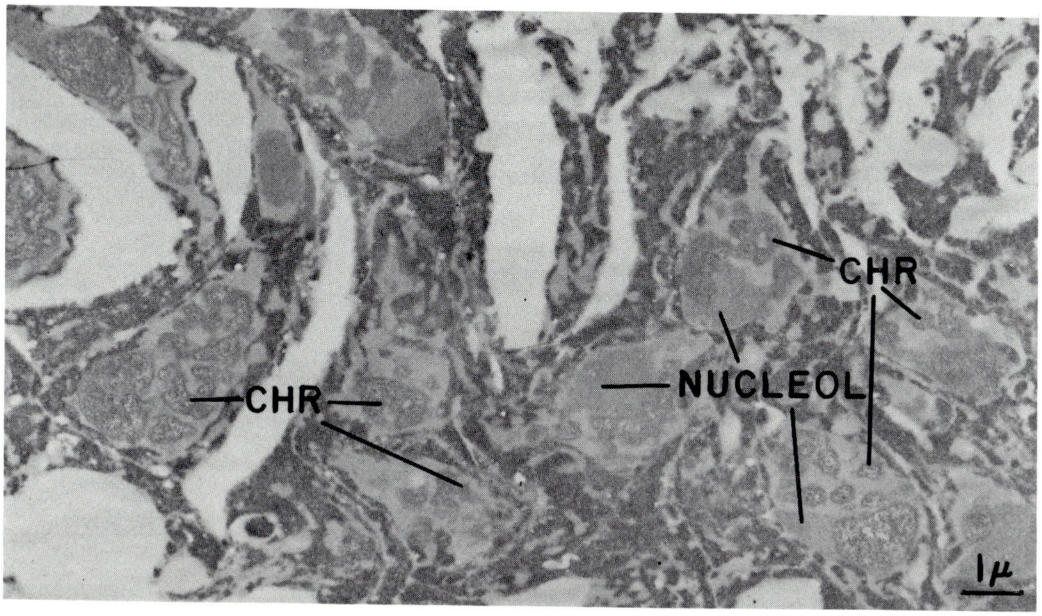

FIG. 6.7. ▲

▼ FIG. 6.8.

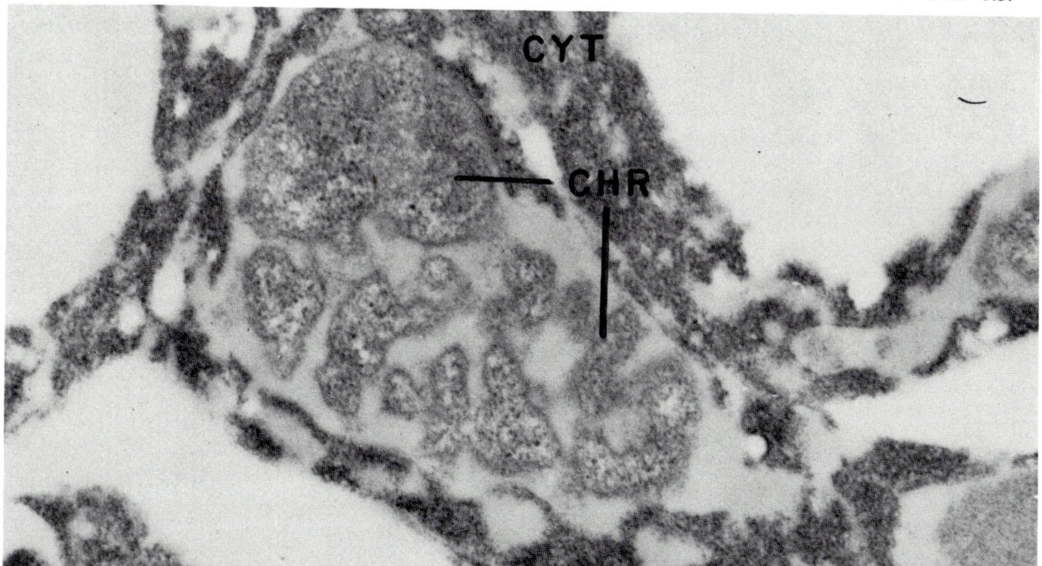

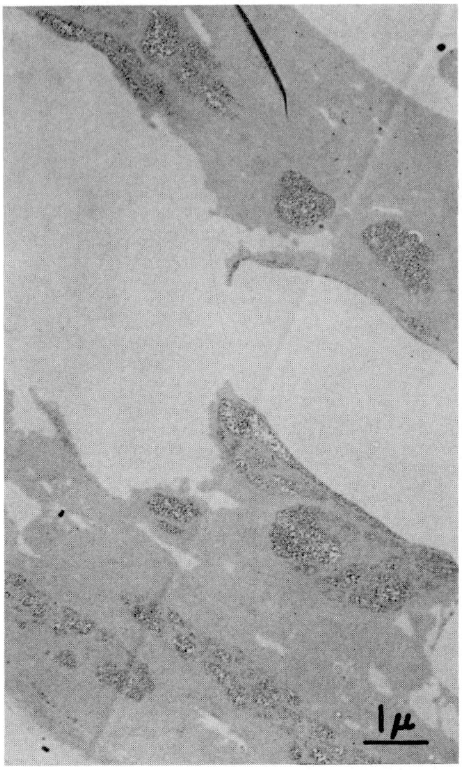

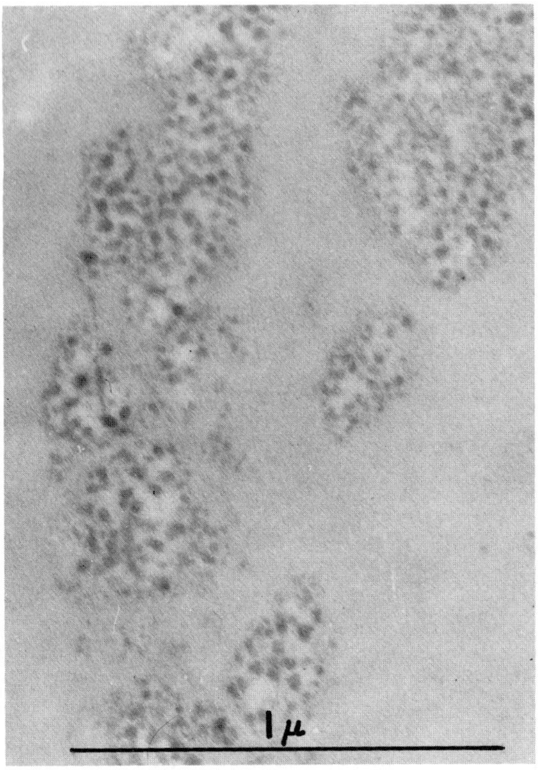

FIGS. 6.9 AND 6.10. Wild-type wing disk, late third instar, proximal part. The glands were frozen-dried, treated with a cross-linking vapor, hardened in alcohol, and then treated with ribonuclease. The specimen was then stained with gallocyanin–chromalum for nucleic acids. It is clear from these figures that ribosomes and nuclear RNA granules are of low contrast, because the RNA has been removed. The DNA remains, and shows the typical "lobulations" of chromatin. Figure 6.9, × 8000; Fig. 6.10, × 56,000.

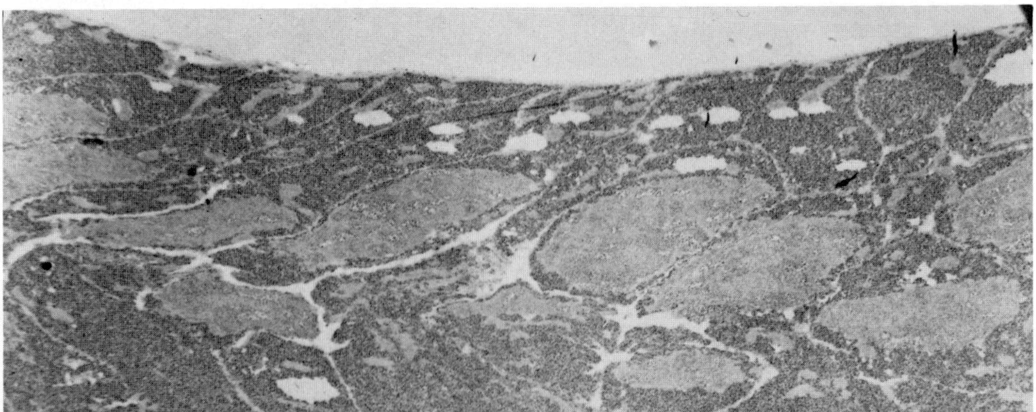

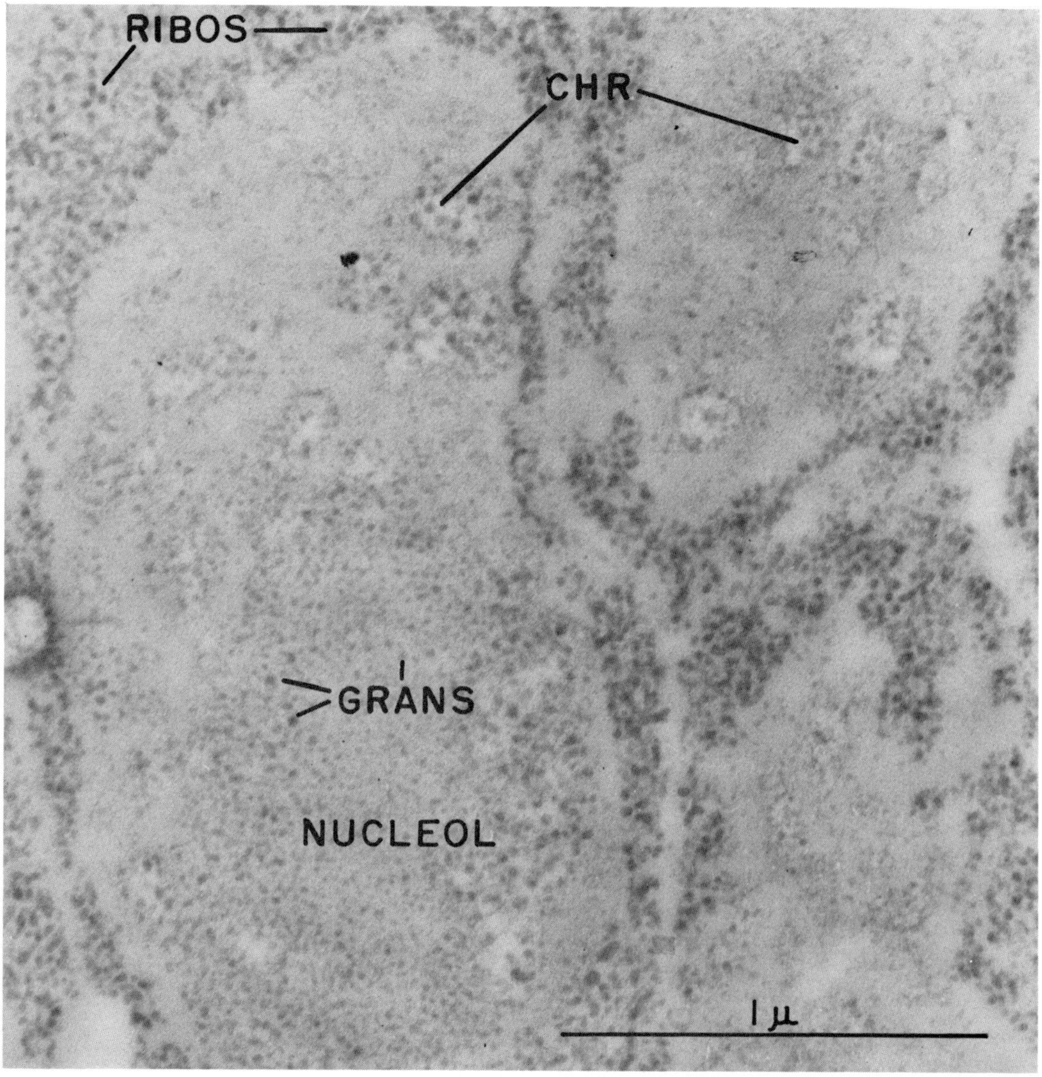

FIG. 6.12.

◄ FIG. 6.11.

FIGS. 6.11 AND 6.12. Wild-type wing disk, early third instar, distal part. Chromatin (CHR) is widely dispersed in the nucleus, and the DNA aggregates are rather small. RNA granules (GRANS) are present in the nucleoplasm, and in very loosely packed nucleoli. Other abbreviations as given earlier. Figure 6.11, × 8000; Fig. 6.12, × 44,800.

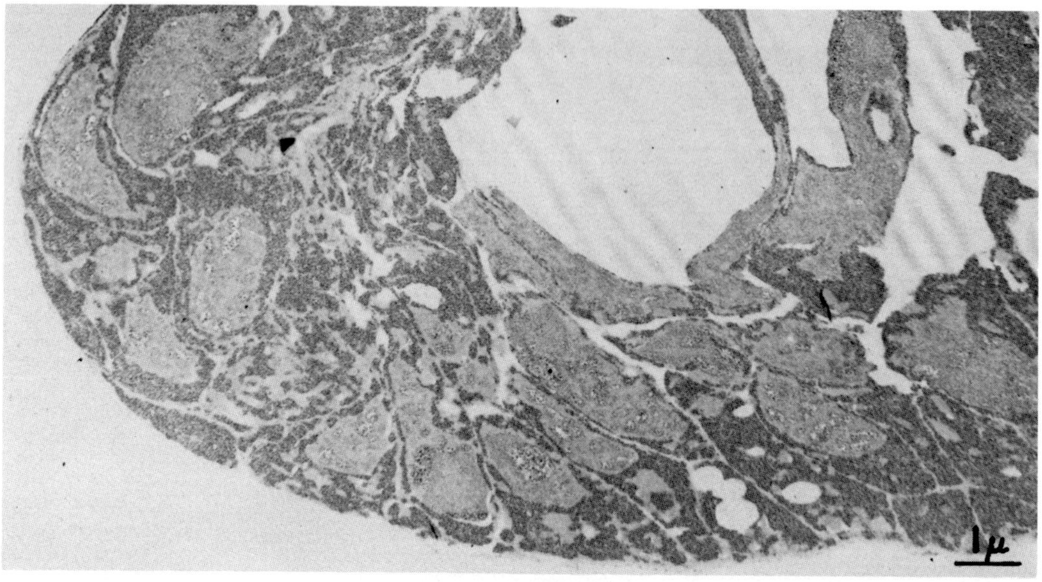

FIG. 6.13. ▲

▼ FIG. 6.14.

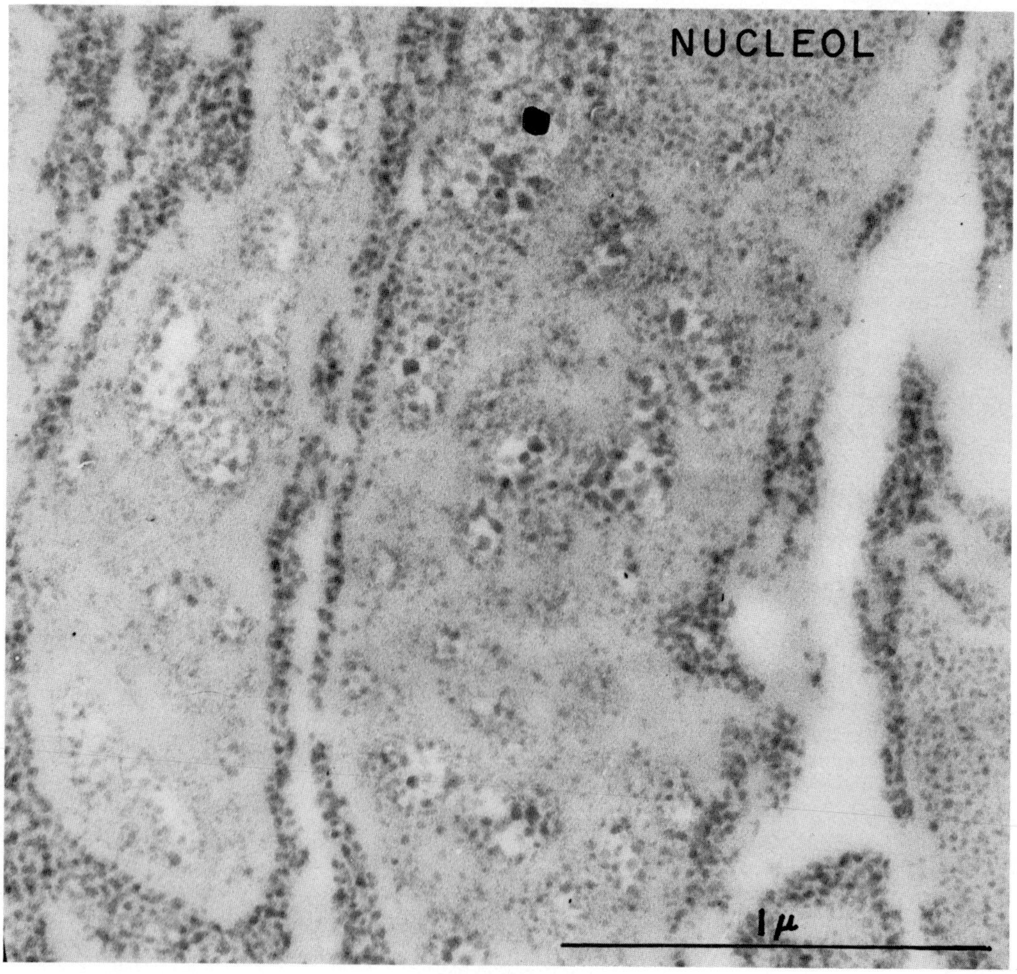

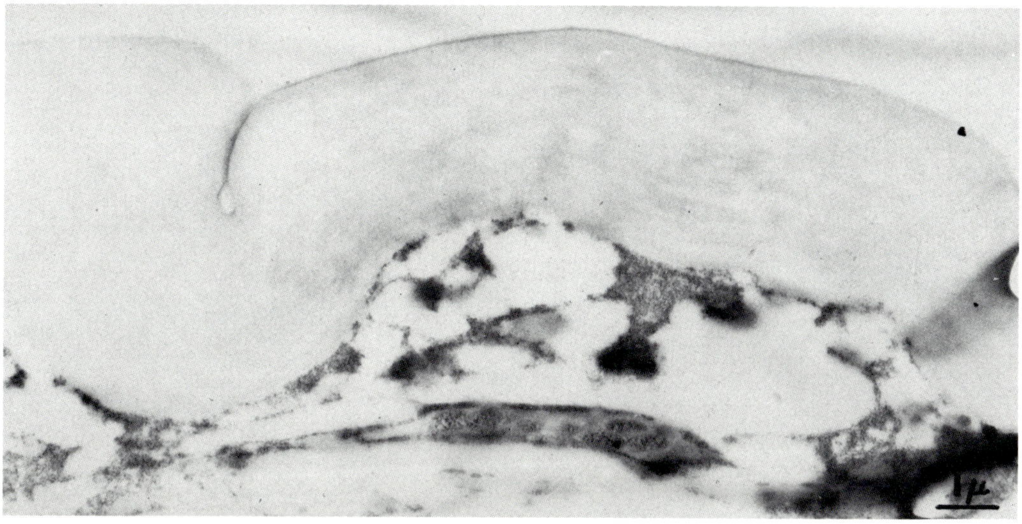

FIG. 6.15.

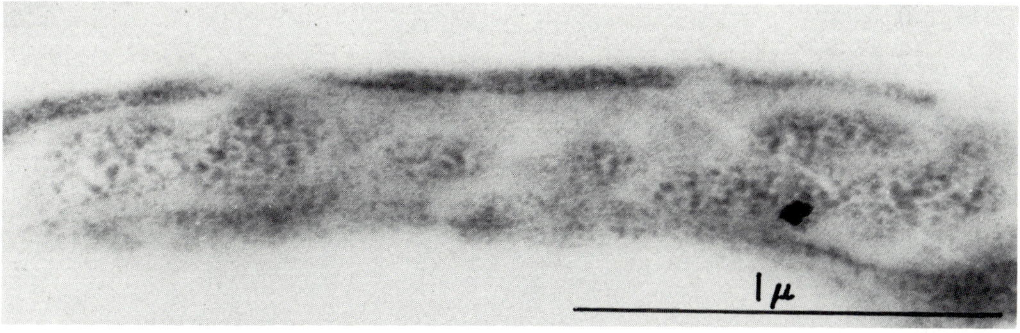

FIG. 6.16.

FIGS. 6.15–6.19. Wild-type, prepupal molt. Figures 6.15 and 6.16, subcuticular cell, to show orientation of the cell and of the DNA pattern in the outstretched cell, located in the wing. Figure 6.17, subcuticular cells to show position of three subjacent cells. The nuclei of two of these are enlarged to show their DNA pattern in Figs. 6.18 and 6.19. Figure 6.15, × 8000; Fig. 6.17, × 14,200; Figs. 6.16, 6.18, and 6.19, × 56,000.

FIGS. 6.13 and 6.14. Wild-type wing disk, early third instar, proximal part. DNA appears in sections in small aggregates which are somewhat outlined by local concentration of RNA granules. Numerous RNA granules occur in the nucleoplasm and, of course, in the nucleolus (NUCLEOL). Figure 6.13 × 8000; Fig. 6.14, × 56,000.

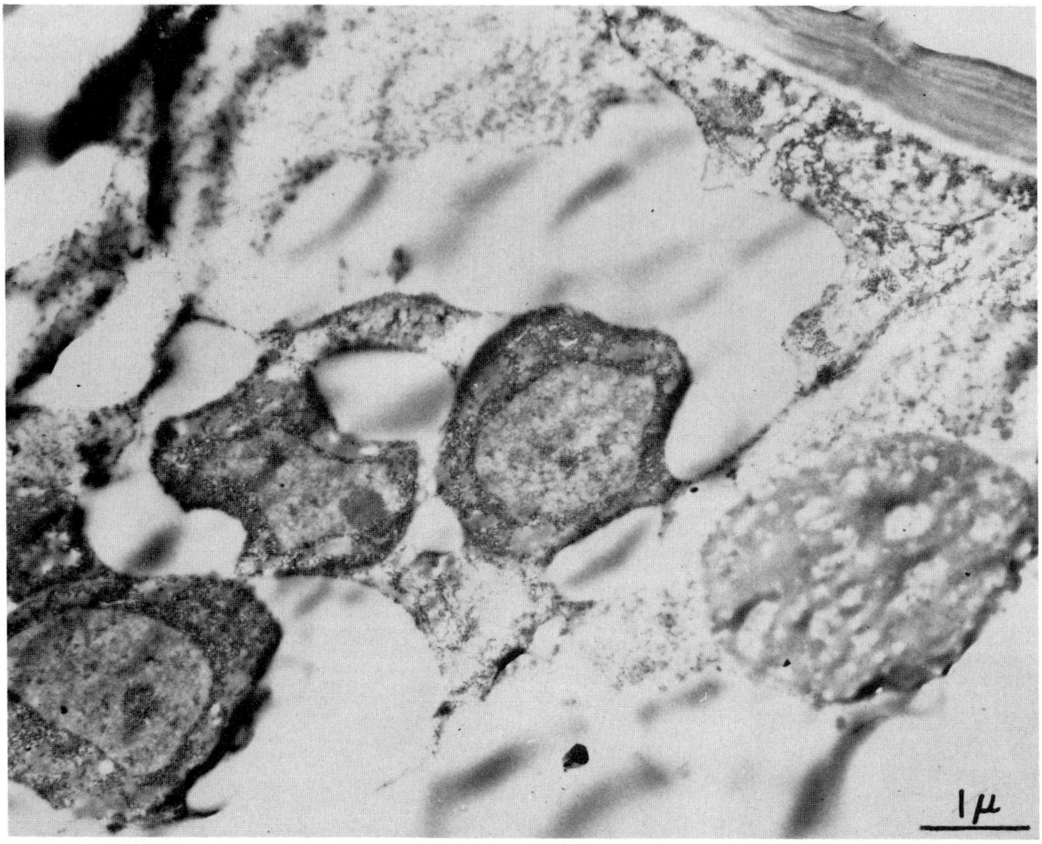

FIG. 6.17. ▲ See p. 175 for complete legend. ▼ FIG. 6.18.

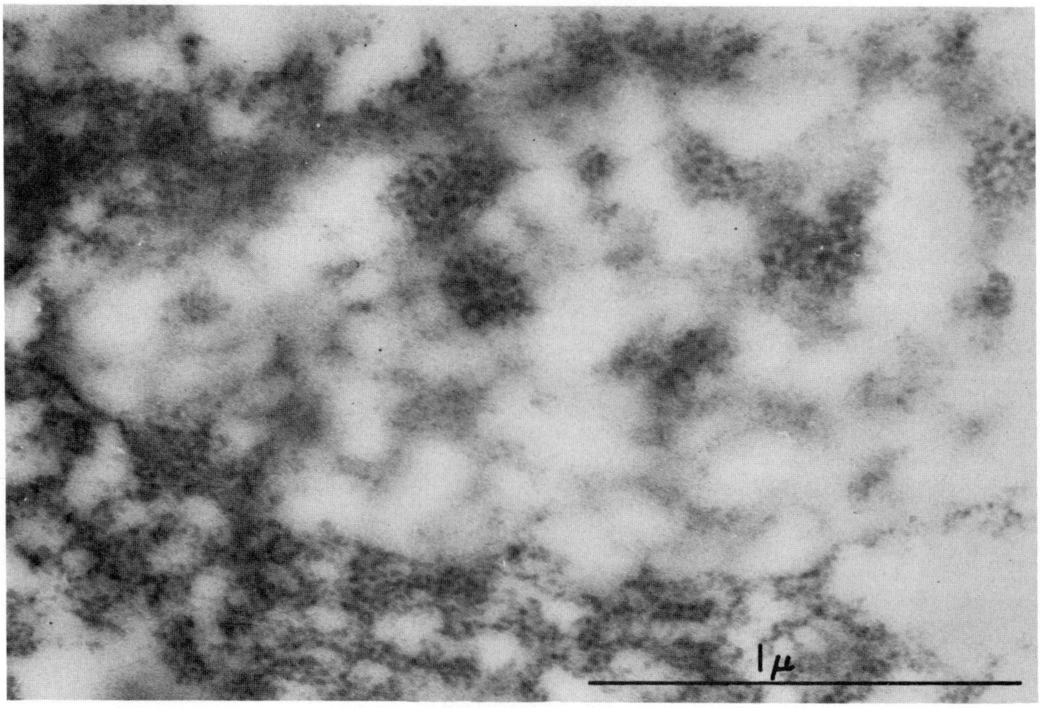

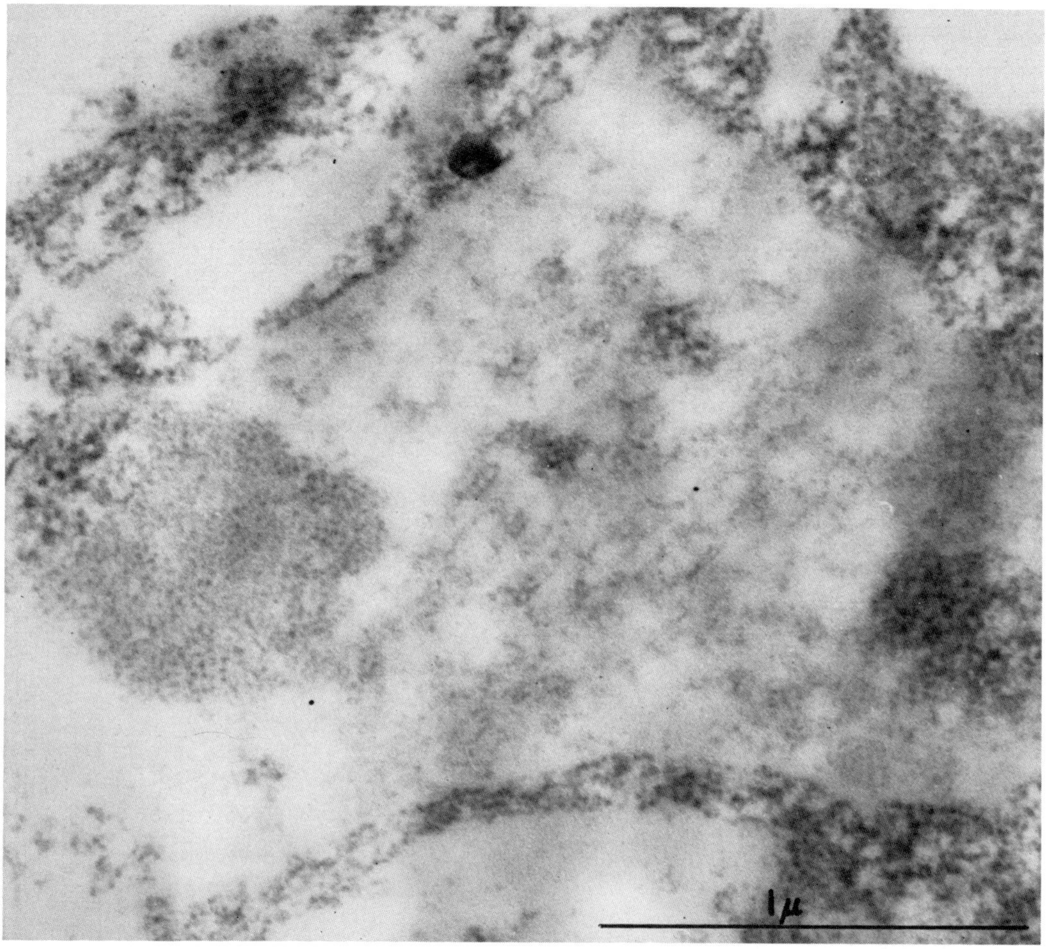

FIG. 6.19. See p. 175 for complete legend.

EARLY THIRD INSTAR WING DISK, PROXIMAL PART (FIGS. 6.13 AND 6.14)

The DNA molecular pattern resembles somewhat that of the corresponding part of the third instar anlage. The DNA appears in sections in small aggregates. These are not as sharply outlined as in the later stage, as the RNA granules are more generally distributed throughout the nucleoplasm.

NUCLEI OF CELLS AT ABOUT THE TIME OF THE PREPUPAL MOLT (FIGS. 6.15–6.19)

The DNA pattern of a subcuticular cell is shown in Figs. 6.15 and 6.16. Its pattern differs from that of other subcuticular nuclei in the thorax (Fig. 6.17), and the pattern in these cells differs in turn from that in somewhat deeper cells (Figs. 6.17–6.19).

Nuclei of Cells in Midpupal Stage When the Wing is Fully Formed (Figs. 6.20–6.27)

This stage corresponds approximately to the 170-hour pupa at 25°C(1). The DNA molecular pattern of the subcuticular nuclei of the wing are of several varieties, some of which are shown in Figs. 6.20–6.23. These differ again from the DNA molecular pattern of the nuclei of subcuticular cells of the thorax (Figs. 6.24 and 6.25), as well as of muscle (Figs. 6.26 and 6.27).

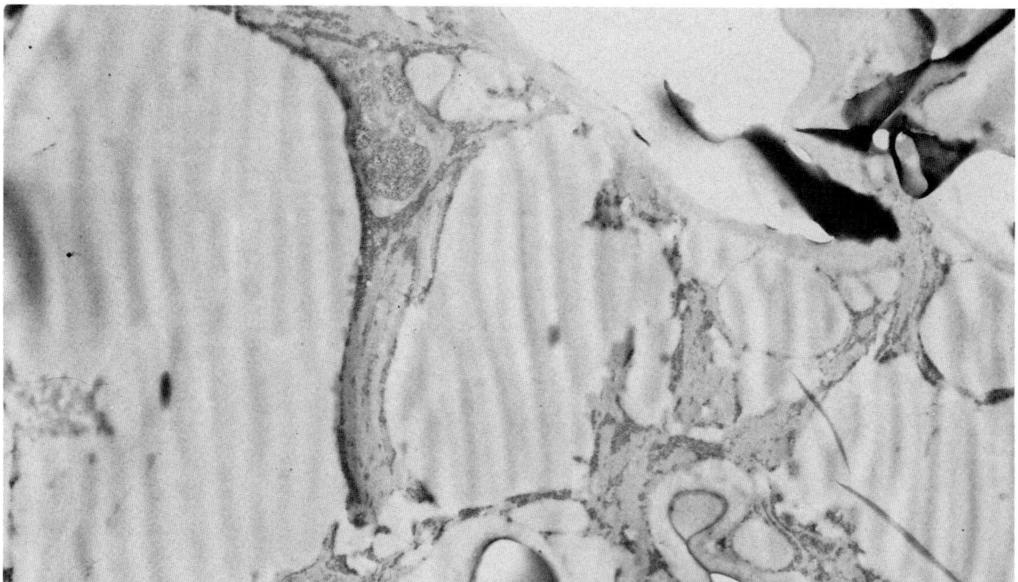

Fig. 6.20.

Figs. 6.20–6.23. Wild-type, midpupa. Figures 6.20 and 6.22 are orientation electron micrographs to show different types of subcuticular cells in the wing. The DNA pattern of the nucleus of one cell in each one of these is shown in Figs. 6.21 and 6.22. Figure 6.20, × 8000; Fig. 6.22, × 14,000; Figs. 6.21 and 6.23, × 56,000.

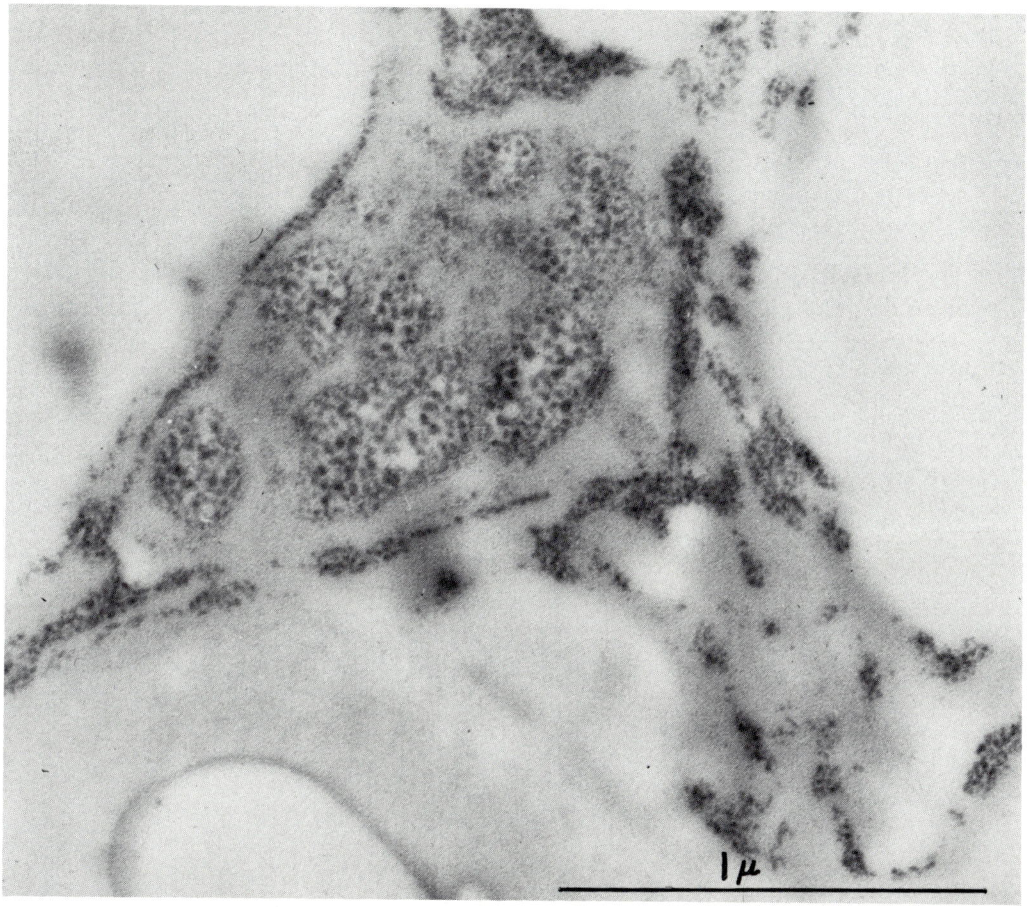

FIG. 6.21.

NUCLEI OF CELLS OF HALTERE DISK, DISTAL PART (FIGS. 6.28–6.30)

The distribution of DNA differs from that in the distal part of the wing disk. The chromatin clumps are mostly widely distributed and interconnected. Numerous RNA granules are in the dense phase of the nucleoplasm.

NUCLEI OF CELLS OF A MUTANT WING DISK (HALTERE-MIMIC/CURLY WING), DISTAL PART (FIGS. 6.31 AND 6.32)

The DNA molecular pattern of these nuclei differs from that in the corresponding part of the wing disk or haltere disk of wild-type larvae. The second-order DNA coils are aggregated in small chromatin clumps which are notably smaller than in comparable nuclei of wild-type wing or haltere anlagen, although at least one clump is frequently larger than the others. The shape of the nuclei is very irregular. Nucleoli are not as condensed; RNA granules are very numerous and diffusely distributed in the nucleoplasm.

Nuclei of Cells of a Second Mutant Wing Disk (Haltere-Mimic/Plum Eye), Distal Part (Figs. 6.33 and 6.34)

The DNA molecular pattern differs from that in comparable cells of the wild type in that the DNA aggregates are smaller and more numerous and less inter-

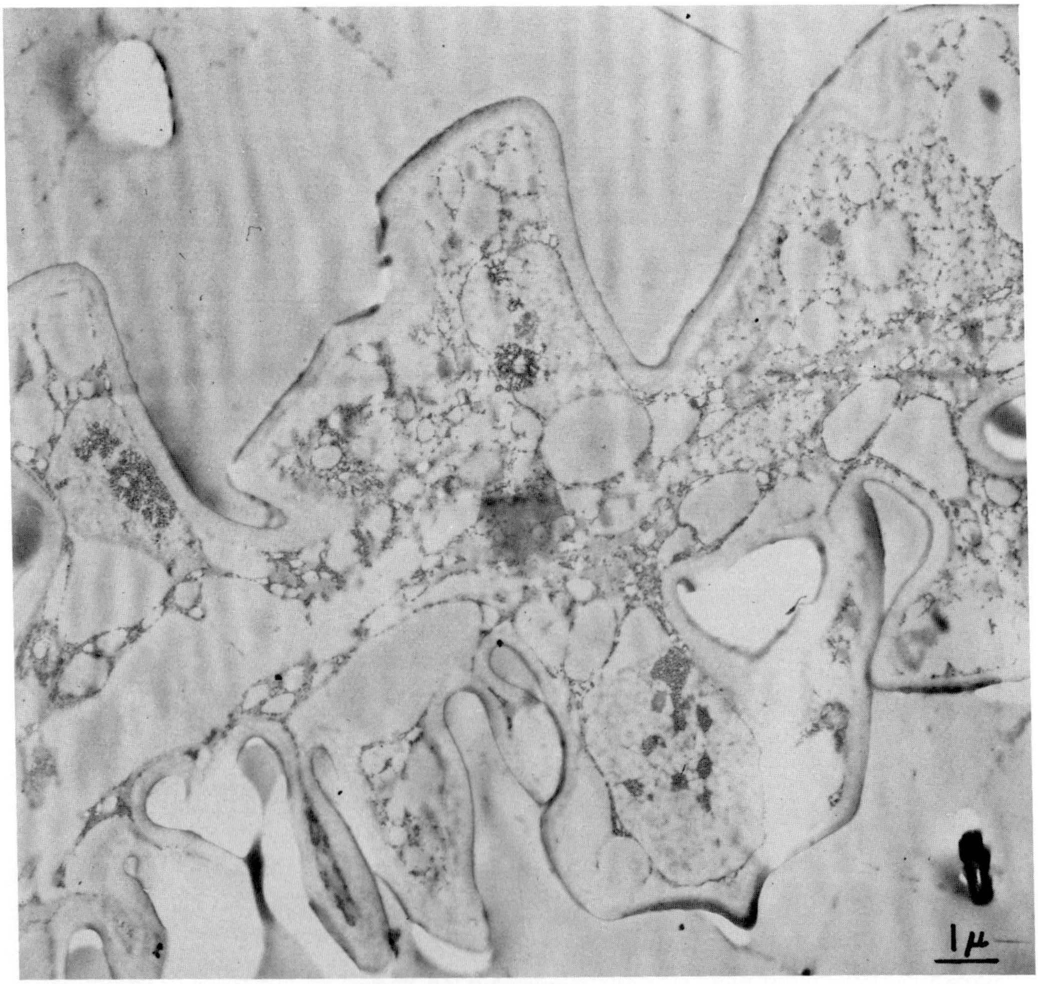

Fig. 6.22. See p. 178 for complete legend.

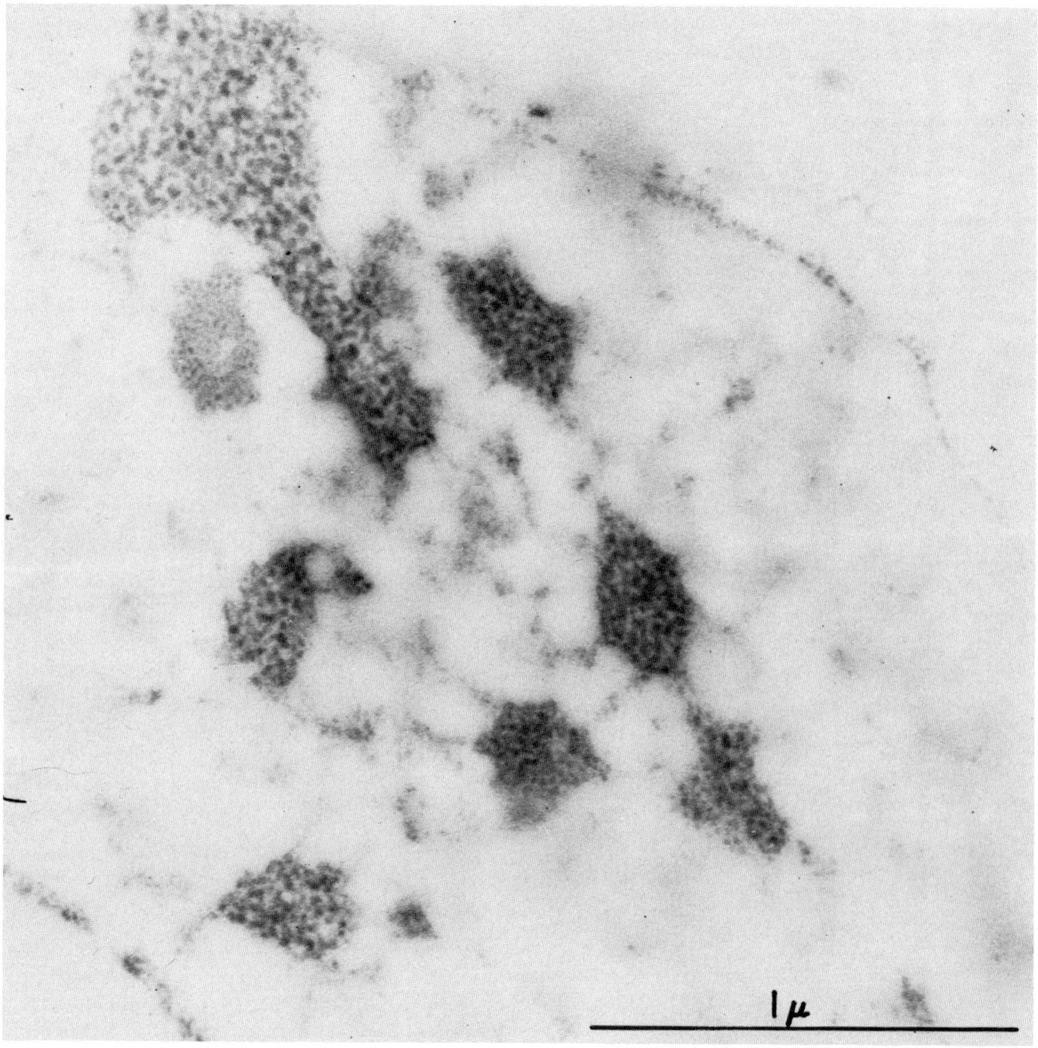

FIG. 6.23. See p. 178 for complete legend.

connected in this mutant than in the wild type. The nucleolus is more open, and shows clearly that the RNA is restricted to the spongelike dense walls of the nucleolus, which appear in sections as branching interconnected cords. Some RNA granules are also present in the nucleoplasmic walls.

NUCLEI OF CELLS OF A THIRD MUTANT WING DISK (CURLY WING), DISTAL PART (FIGS. 6.35–6.37)

The aggregates of second-order DNA coils are even more finely divided than in

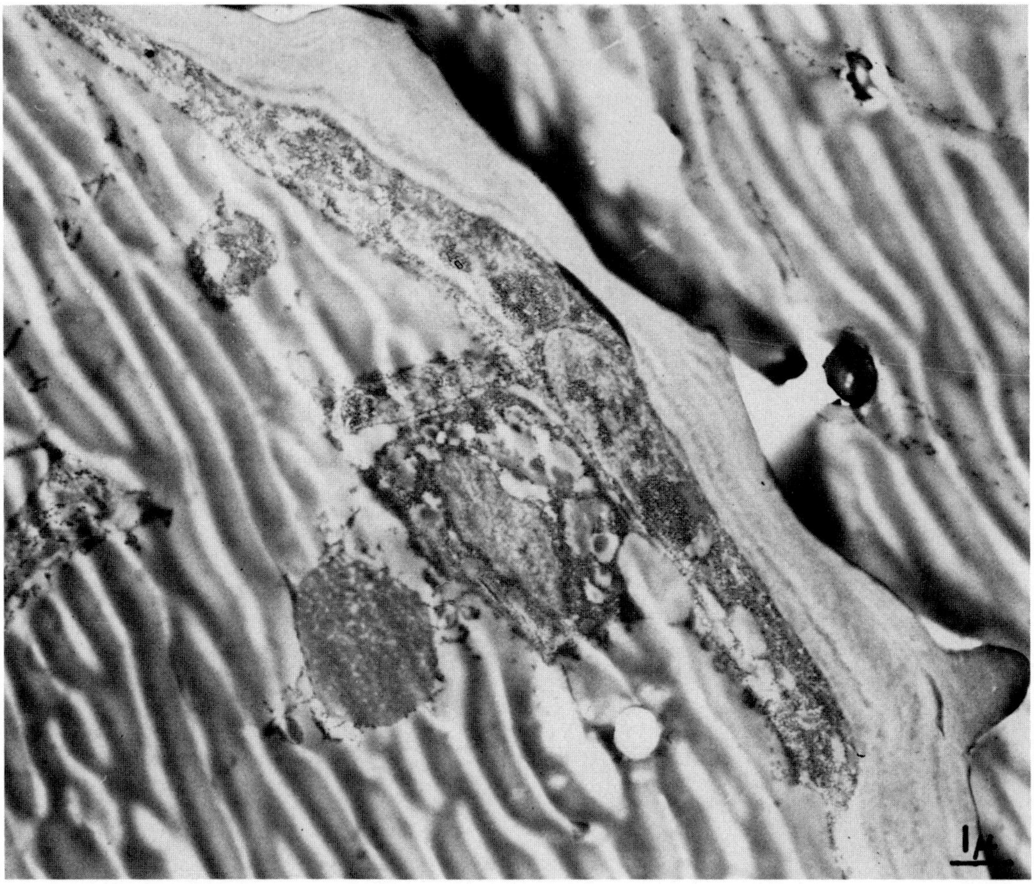

FIG. 6.24.

FIGS. 6.24 and 6.25 (OPPOSITE). Wild-type, midpupa. Figure 6.24 is an orientation electron micrograph to show an outstretched subcuticular cell, with a second attached to it, from the thorax. The first cell is enlarged in Fig. 6.25 to display its different DNA pattern. Figure 6.24, × 8000; Fig. 6.25, × 44,800.

the corresponding parts of the *Hm–Pm* or *Hm/Cy* mutants, to the point where most DNA coils occur singly or in doublets. The nucleolus and nucleoplasm, on the other hand, do not notably differ.

Discussion

In the wild-type wing disk there is a difference between nuclei of the proximal and distal parts, both at early and late third instar. In the distal part, there is a transition from a loosely packed state of the DNA of chromatin at early third in-

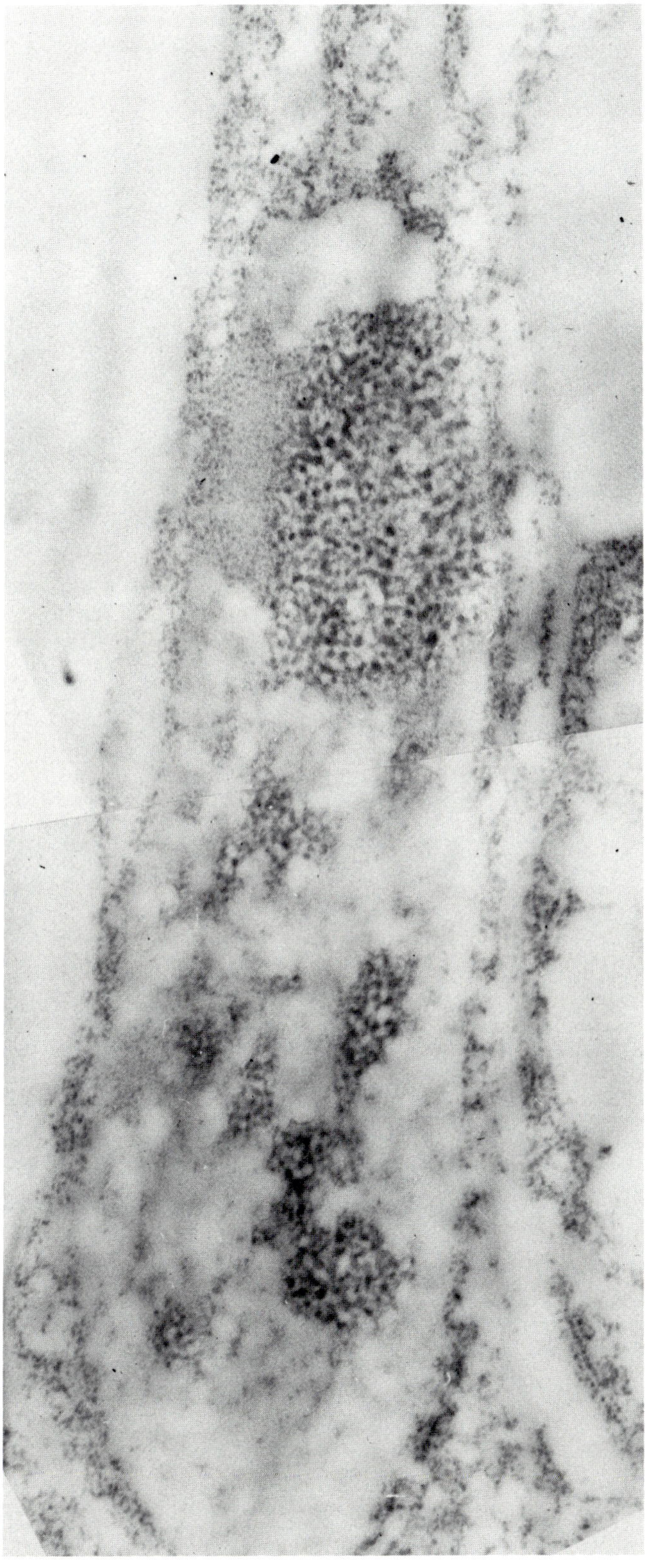

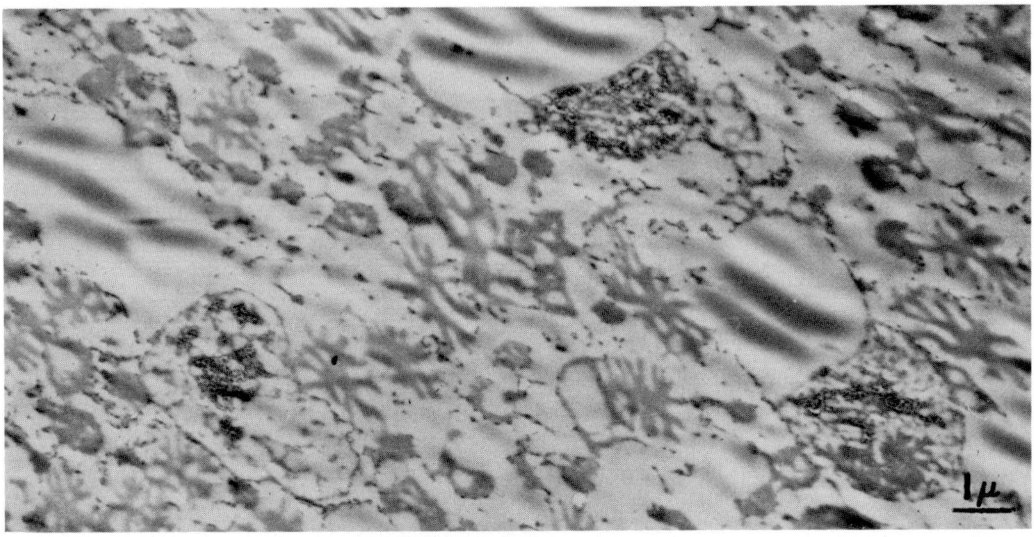

FIG. 6.26. ▲ ▼ FIG. 6.27.

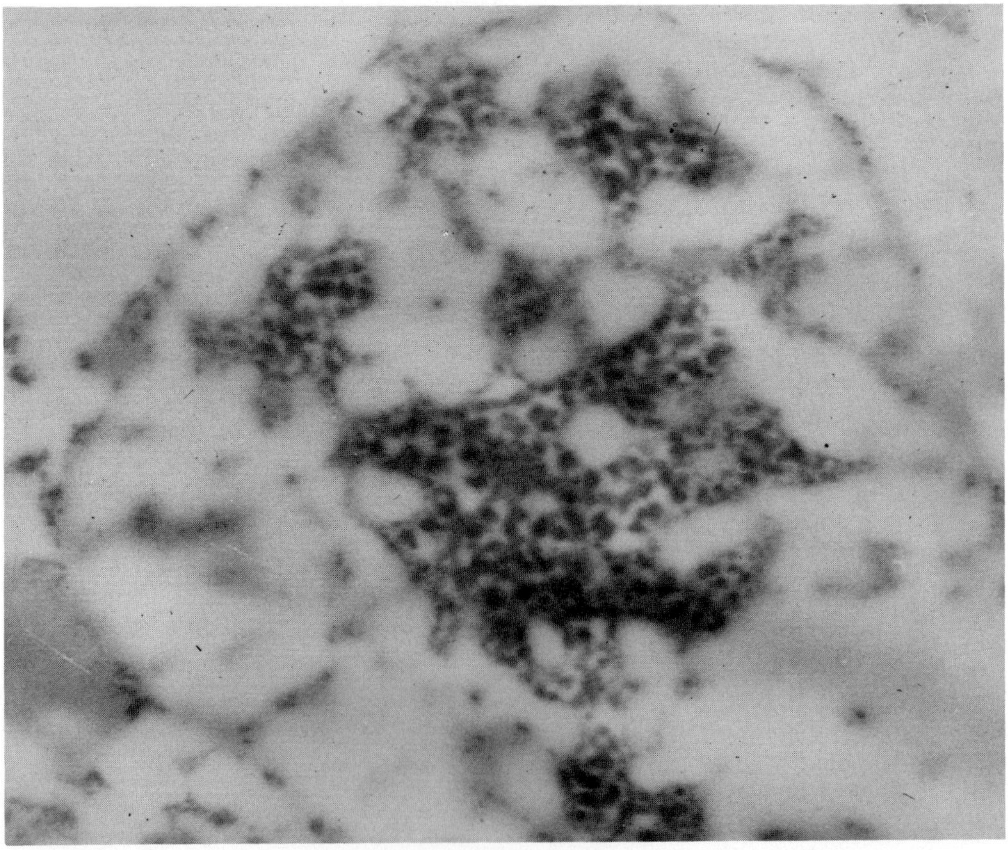

FIGS. 6.26 and 6.27. Wild-type, midpupa. Figure 6.26 is an orientation electron micrograph of transverse section of thoracic muscle to show three nuclei; one of them is shown at higher magnification in Fig. 6.27. Figure 6.26, × 8000; Fig. 6.27, × 56,000.

star to a mostly denser state at the end of the instar. There are more RNA granules in the nucleoplasm at the earlier stage. In the proximal part of the wing disk, the only noticeable change during the third instar is the accumulation of RNA granules around the chromatin masses. At puparium formation and during pupation, striking changes in intranuclear morphology take place. Each differentiated cell type has a characteristic nuclear type.

At late third instar, the molecular pattern of DNA in nuclei of the distal part of the haltere disk differs from that in the distal part of the wing disk.

The distal part of the wing disk in the Haltere-mimic mutant was studied in two genetic combinations. In one *(Hm/Cy)*, the Curly mutant, which also affects wing morphology, was present. In the other *(Hm/Pm)*, no other wing mutant was present. The two types of Haltere-mimic fly have a fairly similar wing nuclear pattern in

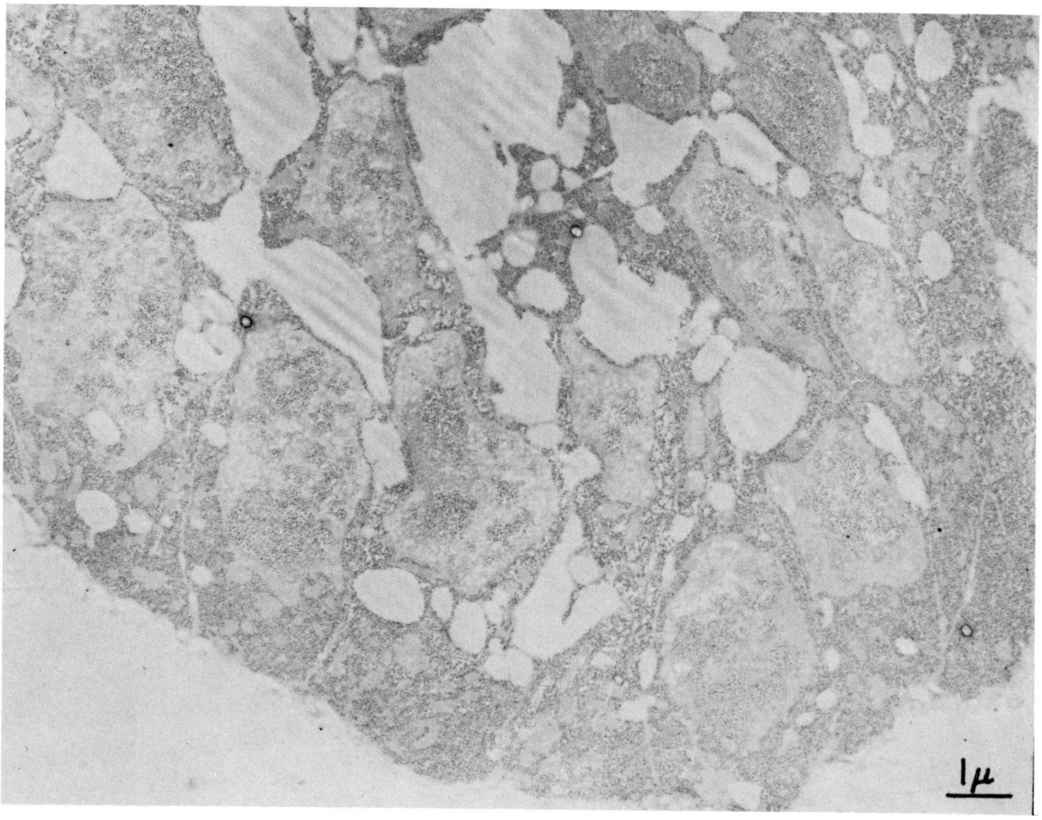

FIG. 6.28.

FIGS. 6.28–6.30. Wild-type haltere disk, late third instar, distal part. As compared with comparable nuclei of the wing disk (in Figs. 6.2–6.6), the chromatin masses are more widely distributed. RNA particles are present throughout the nucleoplasm and are tightly packed in the nucleolus. Figure 6.28, × 8000; Figs. 6.29 and 6.30, × 56,000.

the distal part of the wing disk. This pattern differs from those found in corresponding parts of the wild-type wing and the wild-type haltere at the same stage. The chromatin aggregates are smaller and more numerous in flies with the ***Hm*** mutant and the nucleoli are more diffuse. In flies which have the *Cy* mutant alone (*Cy* 1+), the pattern of DNA coils is even finer and the nucleoli are similarly diffuse. It may be concluded that both the *Hm* and *Cy* mutants cause changes in the DNA molecular pattern. It is unfortunate that chromosomal rearrangements ac-

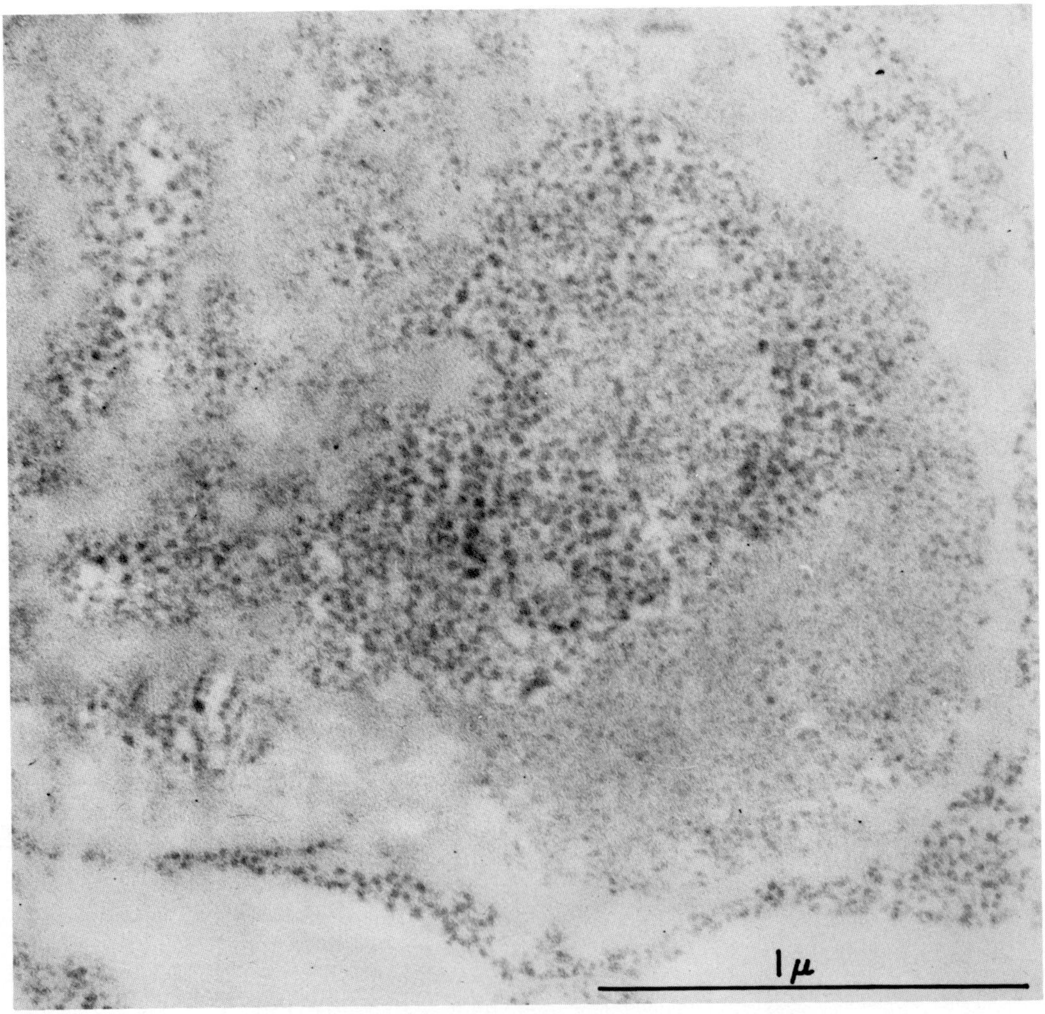

FIG. 6.29. See p. 185 for complete legend.

companying these mutations *(6)* make is impossible to obtain isogenic stocks differing from the wild type only in respect to a single wing mutant at a time.

The elegant experiments of Hadorn and his co-workers *(3, 4)* have shown that the fate of the undifferentiated imaginal disks is already determined in the embryo. The process of determination is distinct from that of differentiation and is normally separated from it by a considerable period of time. Not only is the whole disk determined at the early stage to become a specific structure such as wing or leg, but different parts of the disk are determined by late third instar as prospective parts of the adult structure *(3, 4)*.

Since differentiation takes place under the control of genes, it is to be expected that the very first changes would occur in the nucleus. The differentiation of cells in the wing disk, for example, requires the repression of genes involved in cell division and the derepression of those which control the development of cells each having a specialized function. In our studies, changes in nuclei were indeed found

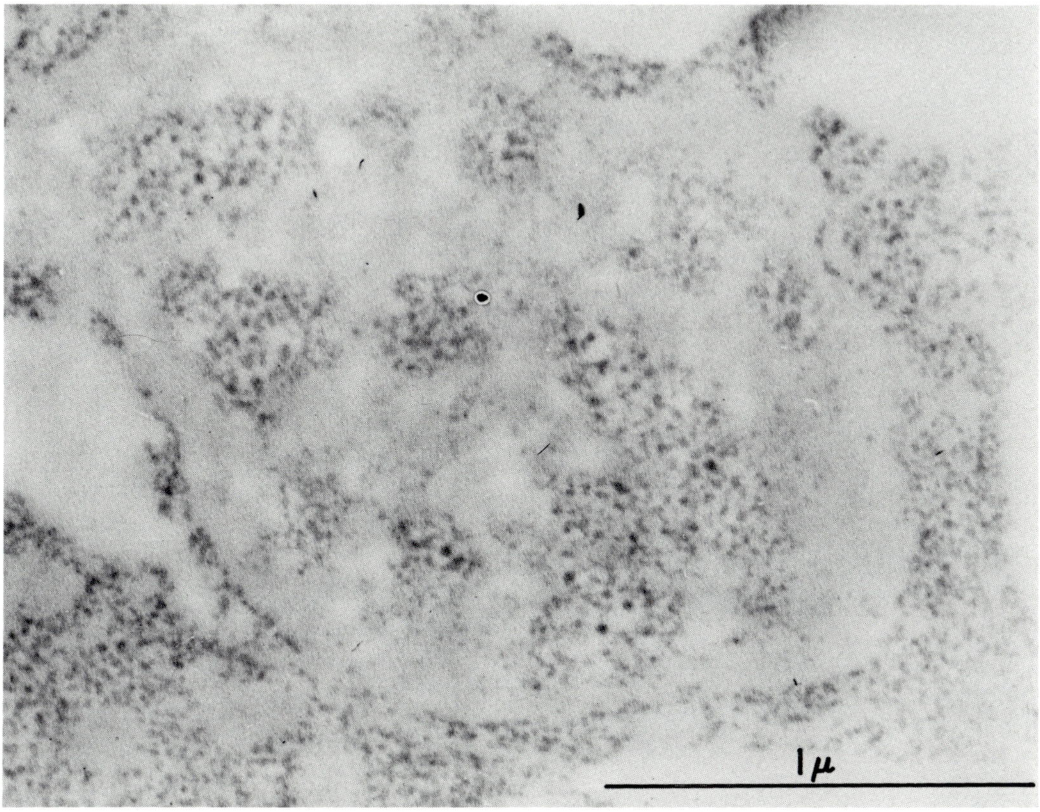

FIG. 6.30. See p. 185 for complete legend.

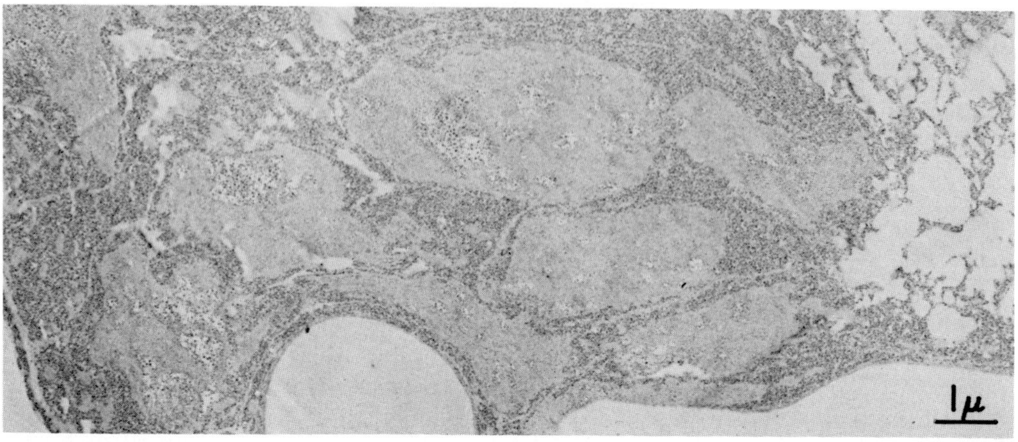

FIG. 6.31. ▲

▼ FIG. 6.32.

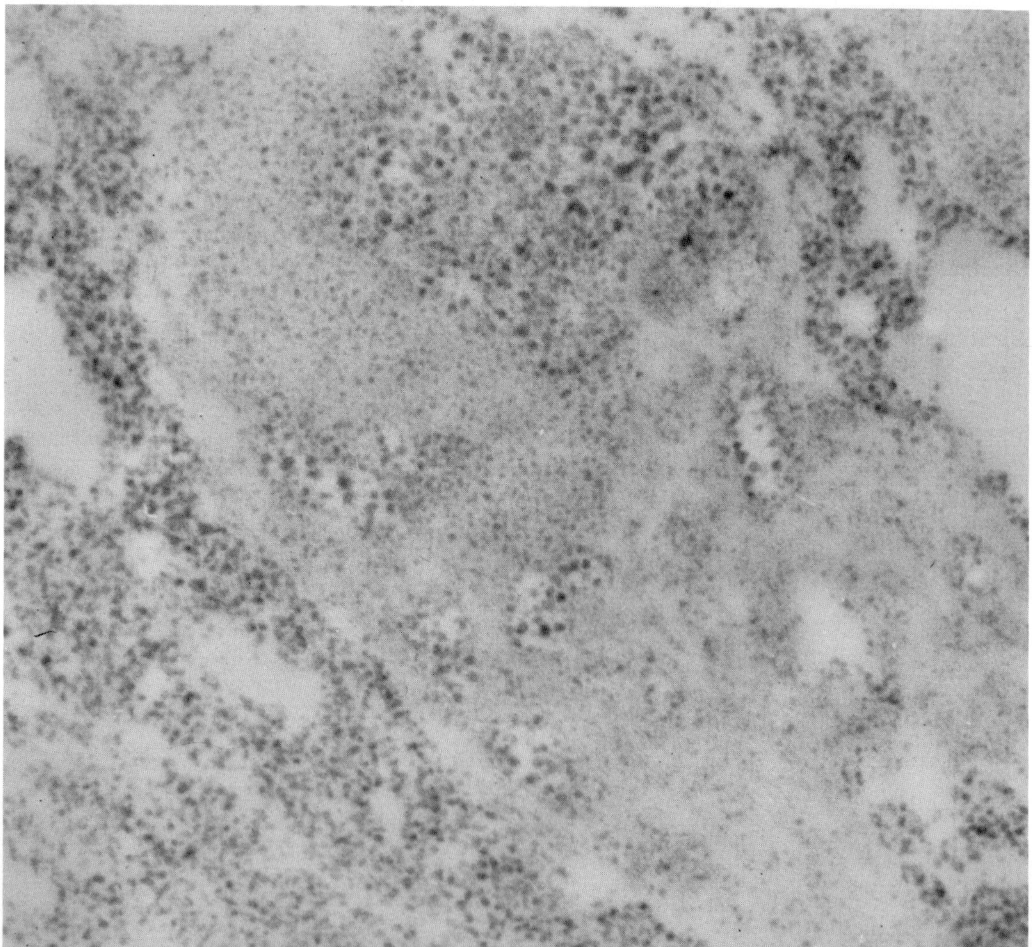

at the very beginning of the period of differentiation of the disks, at or shortly after puparium formation.

Differences between nuclei in different parts of the wing disk were found at an earlier stage, however, much before any cytoplasmic differentiation was detectable. These differences were maintained, with slight changes through the third instar, although no cytoplasmic differentiation took place during this time. We suggest that the earlier differences between nuclei reflect a presetting of the genetic program for a particular set of cell types. This presetting consists of a relatively large-scale reorganization of the genome so that DNA of the genes necessary for the future development of this set of cell types can be quickly activated, while the rest of the genome is segregated and more or less permanently inaccessible. The latter part of the genome is represented in the nucleus by the DNA that is folded and packed into heterochromatic aggregates, while the potentially active part of the genome stretches out as free loops of second-order DNA helix extending in the dense phase of the nucleoplasm. This relatively extended state of the DNA permits genes to be active in transcription of RNA but does not necessarily imply that all genes in the loops are active at one time. Some genes, or groups of genes, may be kept inactive by repressor molecules as in the Jacob–Monod model *(5)* until the appropriate inducers become available. According to this view, each cell type would have, as it were, a coarse and fine adjustment for control. The coarse adjustment would consist in a prearrangement of the chromosomal DNA so as to expose the proper groups of genes and would correspond to the process of determination. The fine adjustment, effective during differentiation, would consist in the sequential activation of genes, or groups of genes, by means of a regulatory system of the Jacob–Monod type.

If our hypothesis is correct, we should find that various differences in cell activity are accompanied or preceded by corresponding differences in the molecular organization of DNA in the nucleus. Results consistent with the hypothesis are provided by studies of three types of variation in cell activity.

First, differentiated cell types have different, specific activities. It is postulated that the nucleus of each such cell type should show a different characteristic pattern of DNA molecular coils. This is what we have found in the nuclei of the wing disk during the pupal period. Further evidence is presented and discussed in Chapter 8 (this volume).

Second, a single cell type can be studied in different states of activity. Although the difference between a quiescent cell and one stimulated to activity might involve only a Jacob–Monod type of derepression, it is postulated that there may also be a conspicious difference in the molecular organization of the DNA. This topic is the subject of Chapter 10 (this volume).

FIGS. 6.31 and 6.32. Wing disk, late third instar, distal part, mutant, Haltere-mimic/Curly wing. The distribution of chromatin differs from that in the comparable part of wild-type larvae. The DNA is in mostly sharply circumscribed, smaller aggregates which are interconnected. Numerous RNA granules are in the nucleoplasm. Figure 6.31, × 8000; Fig. 6.32, × 56,000.

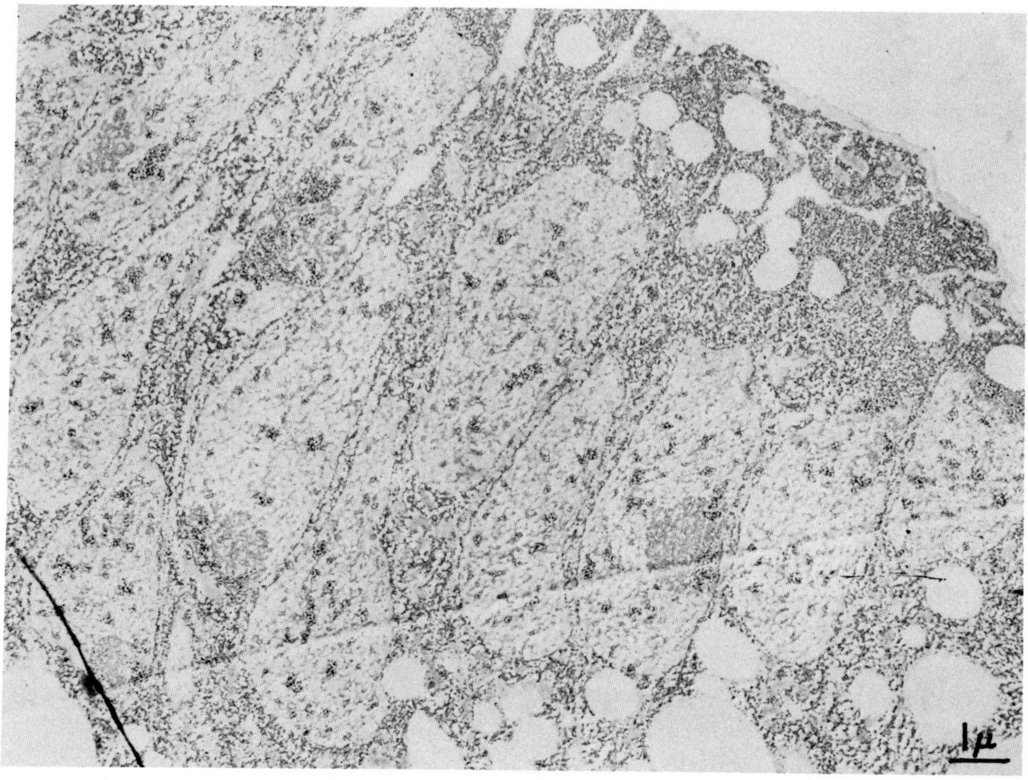

Fig. 6.33. ▲

▼ Fig. 6.34.

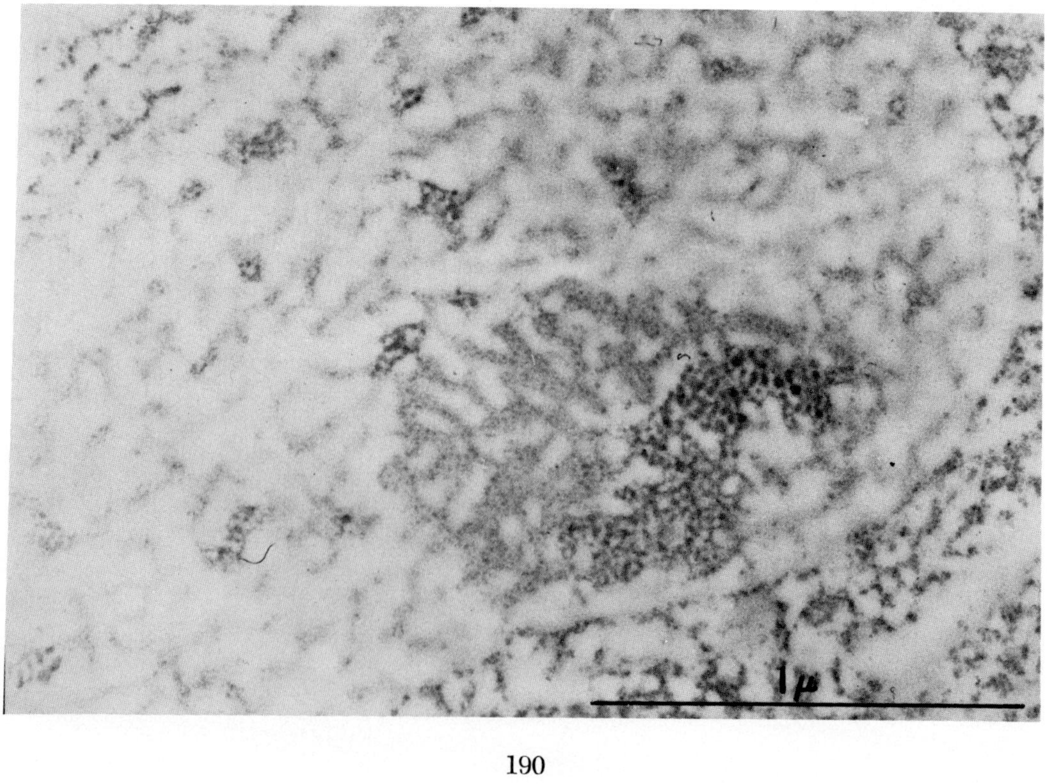

Third, during the process of cell maturation, when cell division slows down and comes to a halt and growth ceases, a large number of genes become inactive. It is postulated that there is a major reorganization of the DNA resulting in a net increase in the amount of aggregated chromatin. The cells of the distal part of the wing disk show nuclear changes of the type expected. The chromatin aggregates are already more compact at the end of the third instar, as compared with the beginning of the instar, and they increase in extent as differentiation proceeds after puparium formation. Further examples are discussed in Chapter 9 (this volume).

Summary

The molecular pattern of DNA was studied in imaginal disks of third instar larvae of *Drosophila* by a method described in Chapter 3 (this volume). The object was to learn whether the DNA molecular pattern is different in cells which are determined but not differentiated. It was found in the wing disk that the DNA molecular pattern in cells destined to become wing differs from that in the cells destined to become thorax. This difference was noted even in early third instar larvae, and became prominent and more richly varied during cell differentiation in the pupal stage. The DNA molecular pattern of the nuclei of cells of the distal part of the wing disk is different from that of cells of the corresponding region of the haltere disk, and the DNA molecular pattern of both differed from that of cells of corresponding regions of certain wing mutants.

From these results we developed a hypothesis on the nature of determination and differentiation at the level of the genome and its morphological expression in terms of DNA molecular pattern.

The hypothesis is that the process of determination involves a presetting of a genetic program for a particular set of cell types. This presetting consists in a relatively large-scale reorganization of the nuclear DNA so that the DNA of the required genes can be quickly activated while the nonessential genes are made inaccessible. The potentially active genome is the DNA that extends out into the nucleoplasm from the larger chromatin masses as first- or second-order DNA helices, while the inactive part of the genome is DNA second-order coils, which are packed into larger heterochromatic aggregates. The relatively extended DNA helix of the second-order coils permits genes to be active in transcription of RNA, but not all genes in such coils need be active at one time. Some genes or groups of

---

FIGS. 6.33 and 6.34. Wing disk, late third instar, distal part, mutant (Haltere-mimic/Plum eye). The distribution of chromatin differs from that of comparable cells of the wild-type and from the *Hm/Cy* larvae. The DNA aggregates are smaller, more widely dispersed, and more numerous in the *Hm/Pm* mutant. The nucleolus is organized in a spongelike manner, with the RNA granules in the denser walls (which appear in section as branching, interconnected cords). The spongelike structure of the nucleoplasm is evident. Figure 6.33, × 8000; Fig. 6.34, × 56,000.

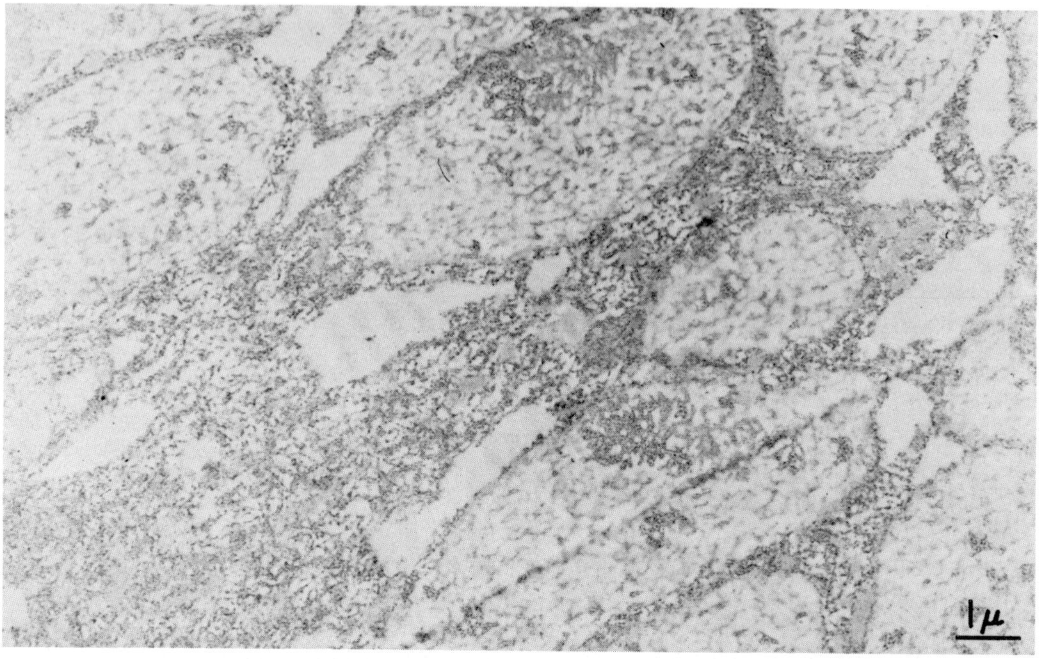

Fig. 6.35.

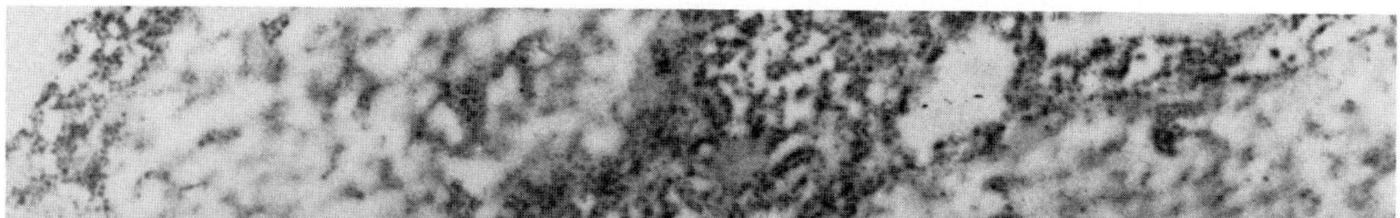

Fig. 6.36.

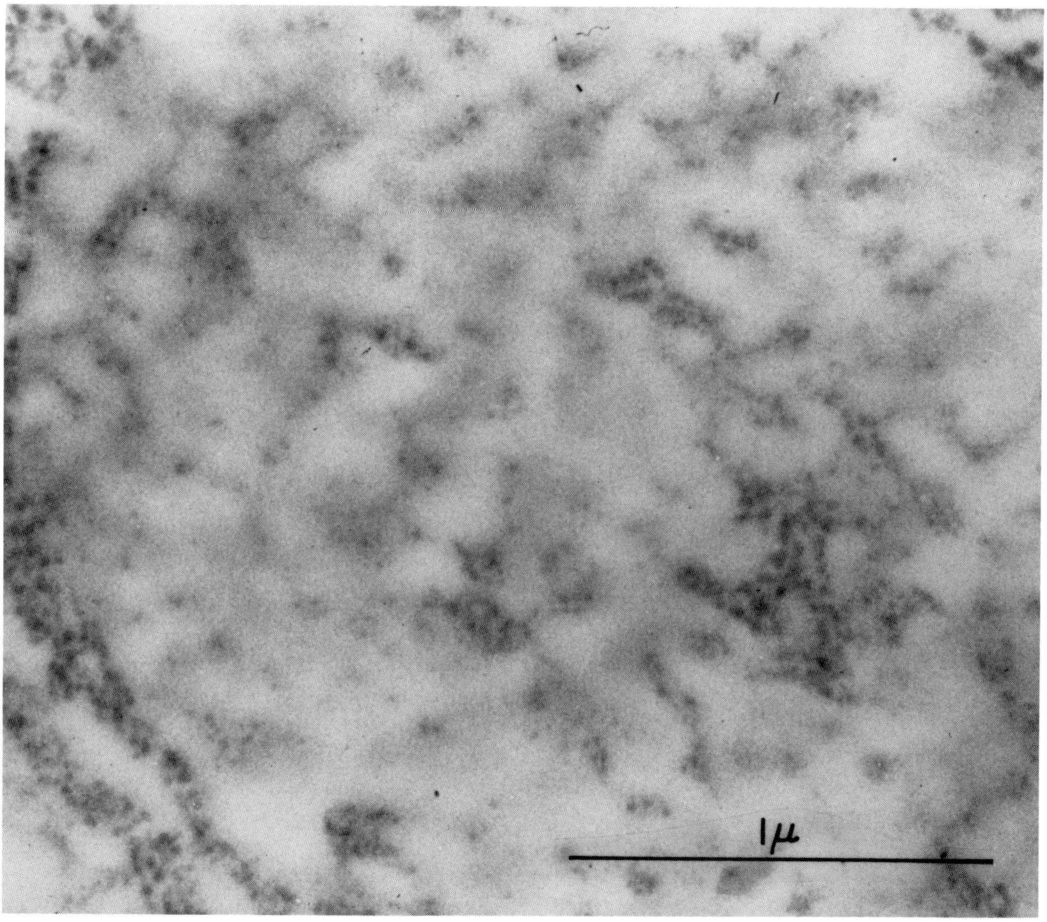

FIG. 6.37.

FIGS. 6.35–6.37. Wing disk, late third instar, distal part, mutant (Curly wing). The DNA aggregates are more finely divided than in comparable cells of the anlagen of the wild-type or the other mutant varieties. Most DNA coils are dispersed as single or double coils which course in the walls of the nucleoplasm. The nucleolus is wide meshed, as in Figs. 6.33 and 6.34. Figure 6.35, × 11,200; Fig. 6.36, × 22,600; Fig. 6.37, × 56,000.

genes on extended second-order DNA coils may be kept inactive by repressor molecules until the appropriate inducers appear. Thus during determination the reorganized or rearranged pattern of second-order DNA coils constitutes the permissive phase. During differentiation, a finer adjustment comes into play, consisting in the sequential activation of genes or groups of genes by means of a regulatory system of the Jacob–Monod type or some similar system involving groups of genes.

References

1. Bodenstein, D. (1950). The postembryonic development of *Drosophila*. *In* "Biology of *Drosophila*" (M. Demerec, ed.), pp. 275–367. Wiley, New York.
2. Fristrom, D. (1969). Cellular degeneration in the production of some mutant phenotypes in *Drosophila melanogaster*. *Mol. Gen. Genet.* **103**, 363–379.
3. Gehring, W. (1968). The stability of the determined state in cultures of imaginal disks in *Drosophila*. *In* "Results and Problems in Cell Differentiation" (H. Ursprung, ed.), Vol. I, pp. 136–154. Springer-Verlag, New York.
4. Hadorn, E., and Buck, D. (1962). Ueber Entwicklungsleistungen Transplantierter Teilstücke von Flügel-Imaginalscheiben von *Drosophila melanogaster*. *Rev. Suisse Zool.* **69**, 302–310.
5. Jacob, F., and Monod, J. (1961). On the regulation of gene activity. *Cold Spring Harbor Symp. Quant. Biol.* **26**, 193–211.
6. Lindsley, D. L., and Grell, E. H. (1968). Genetic variations of *Drosophila melanogaster*. *Carnegie Inst. Wash. Publ.* No. 627.
7. Poodry, C. A., and Schneiderman, H. A. (1970). The ultrastructure of the developing leg of *Drosophila melanogaster*. *Wilhelm Roux Arch. Entwicklungsmech. Organismen* **166**, 1–44.
8. Ursprung, H. (1972). The fine structure of imaginal disks. *In* "The Biology of Imaginal Disks" (H. Ursprung and R. Nöthiger, eds.), pp. 93–107. Springer-Verlag, New York.
9. Ursprung, H., and Schabtach, E. (1968). The fine structure of the male *Drosophila* genital disk during late larval and early pupal development. *Wilhelm Roux Arch. Entwicklungsmech. Organismen* **160**, 242–254.
10. Waddington, C. H., and Perry, M. M. (1960). The ultrastructure of the developing eye of *Drosophila*. *Proc. Roy. Soc. London*, **153**, 155–178.

7

Submicroscopic Distribution of Nucleic Acids in the Barr Body of Chondrocytes in the Proximal Tibial Epiphysis of Female Rats

Isidore Gersh

Literature on the heteropycnotic sex (or X) chromosome has been reviewed fully by Barr *(1)*, who rediscovered this peculiar nuclear body or chromosome, and is responsible for much of the research on it. Pertinent literature has also been reviewed by Brown *(3)* and by Lyon *(7)*. Briefly summarized, the Barr body occurs in nuclei of somatic cells of many invertebrates and vertebrates. In many species of mammals, it may be observed as a single or double body adherent to the nucleolus or the nuclear membrane. It is one of a pair of sex chromosomes which is heterochromatic, at least in part, and is regarded as largely inert genetically. It contains deoxyribonucleic acid (DNA) and ribonucleic acid (RNA). The DNA replicates more slowly than that of other chromosomes. When studied with the electron microscope, its morphology does not differ notably from other chromatin bodies in the same nuclei.

In the course of a cytochemical study of nucleic acids in chondrocytes of the proximal tibial epiphyseal plates of female rats, I noticed that the X chromosome (or Barr body) was prominent when studied with the light microscope and could be readily identified with the electron microscope (see Vol. II, Chapter 5). I thought it might be useful to extend my earlier observations by studying the distribution of nucleic acids in the Barr body of chondrocytes during the maturation and degeneration of the cartilage cells, and thereby broaden the understanding of heterochromatic regions in general.

Material and Methods

Briefly stated, small pieces of the epiphyseal plate were cut about 0.1 mm thick in such a way as to maintain a recognizable orientation of the sequence of maturation stages of the chondrocytes. The specimens were frozen ultrarapidly to avoid the formation of ice crystals, and dried in a vacuum at a low temperature. Without breaking the vacuum, the specimens were treated (postfixed) with a cross-

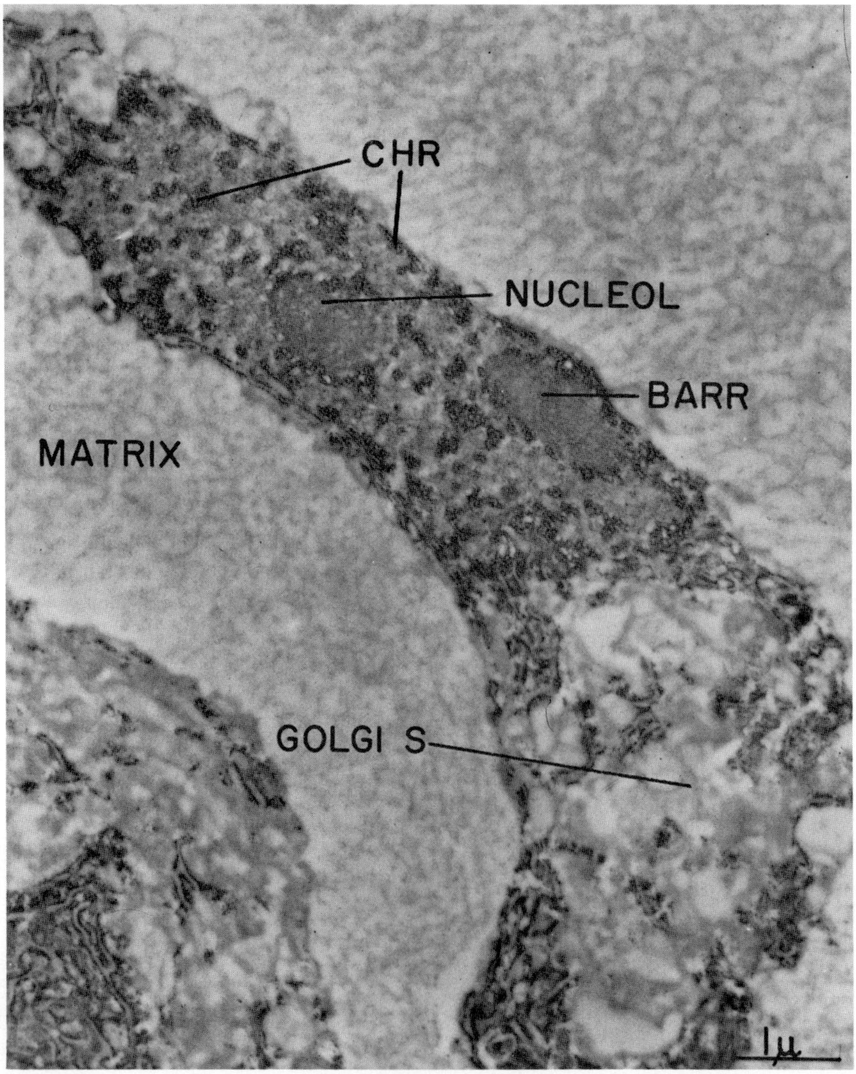

FIG. 7.1.

linking reagent in the vapor state [difluorodinitobenzene (FFDNB)]. These steps were taken to reduce solution, extraction, or molecular displacements, especially during the time when specimens were hardened in alcohol and stained with aqueous gallocyanin–chromalum. Both DNA as well as soluble and insoluble RNAs are selectively and nearly specifically stained by this procedure. The nucleic acids are distinguished one from the other by their morphology and location in the cell, and by their response to ribonuclease (RNase) and deoxyribonuclease (DNase) prior to staining. Details of the methods are given earlier in this volume (Chapters 2 and 3). Ultrathin sections were cut with diamond knives, floated on

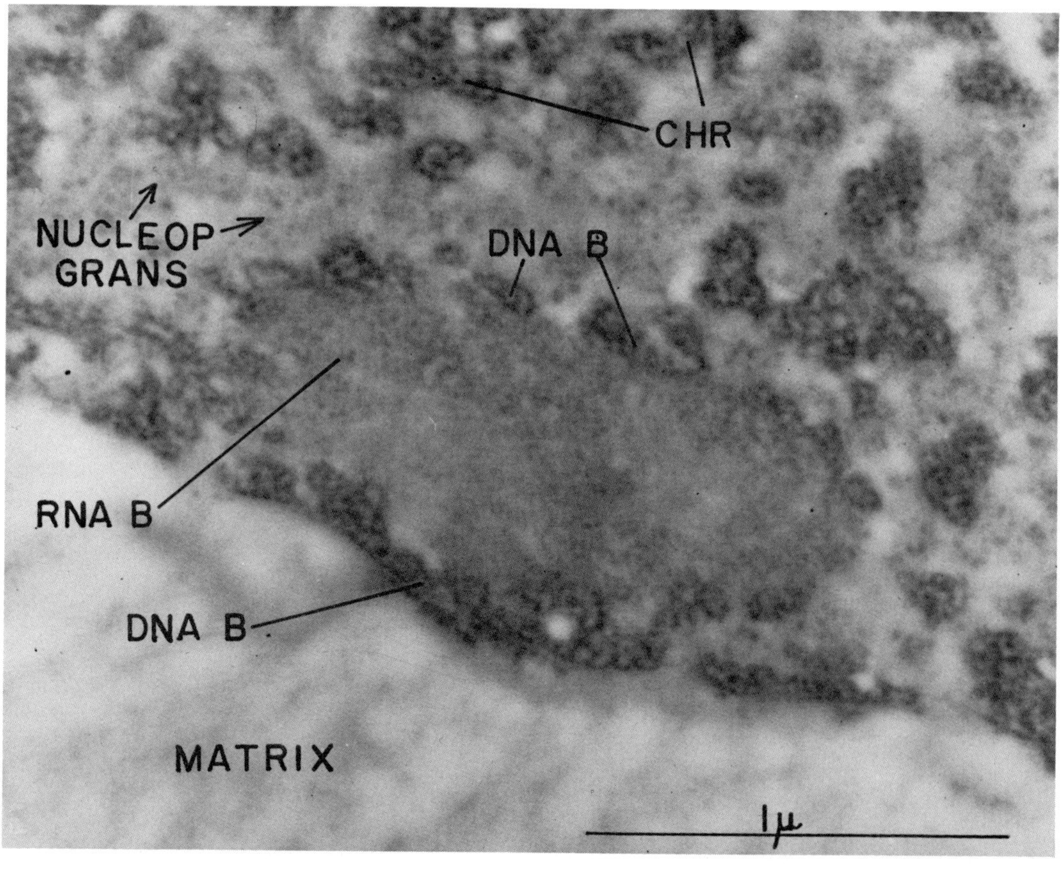

FIG. 7.2.

FIGS. 7.1 and 7.2. Postproliferative cell. The Barr body (BARR) does not differ significantly from the nucleolus (NUCLEOL). Both consist of a central area rich in RNA granules (RNA B) and are surrounded by chromatin (CHR) consisting of secondary coils of DNA (DNA B). The nucleoplasm contains numerous RNA granules (NUCLEOP GRANS). GOLGI S, Golgi site; MATRIX, extracellular matrix. Figure 7.1, × 14,200; Fig. 7.2, × 56,000.

water, mounted on grids, and examined with an Hitachi HU-11A electron microscope operating at an accelerating voltage of 50 kv. Photographic plates were enlarged two times in printing.

Observations

Chondrocytes of the epiphyseal plate were classified according to their secretory activity (Vol. II, Chapter 5).

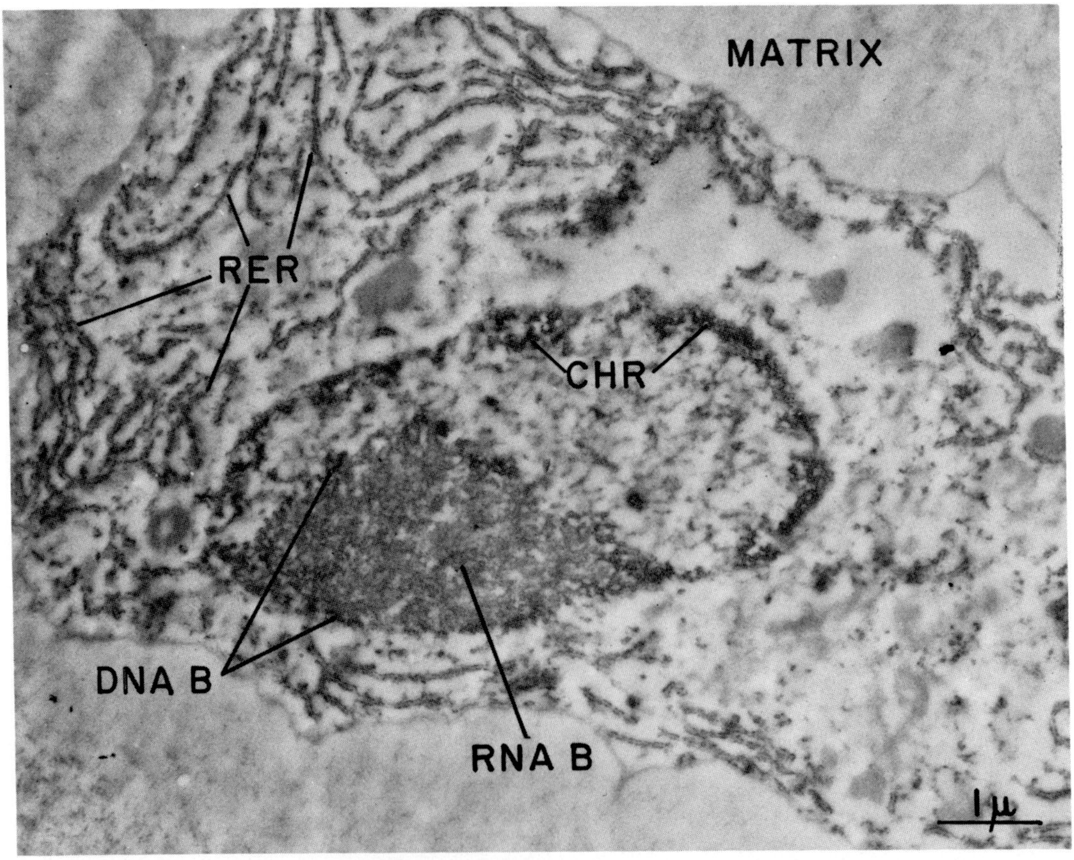

Fig. 7.3. Postproliferative stage. The cell is slightly more mature and the nucleus sectioned nearly transversely. The chromatin (CHR) is mostly more finely distributed than in Fig. 7.1, but the structure of the Barr body is essentially the same as that illustrated in the first two figures. The intracisternal intervals of the RER are narrow. Abbreviations as given in Figs. 7.1 and 7.2. × 14,200.

Postproliferative Stage

This is the stage when chondrocytes are being separated after cell division and aligned to form columns of cells. The Barr body is similar to the nucleolus, except by virtue of the fact that the former is attached to the nuclear surface. Both comprise a dense accumulation of RNA granules partially enclosed by islands of DNA [as they appear in thin sections (Figs. 7.1–7.3)]. The DNA in both sites, as indeed in the rest of the chromatin, is organized in second-order coils, which in sections appear as lines in the form of circles or ovals as noted in an earlier report (this volume, Chapter 3). The coils have a repeating period and diameter of about 400 A. In the earliest stages (Figs. 7.1 and 7.2), the homogeneous background is quite marked and the nucleoplasm is packed with RNA granules. In some cells (Fig. 7.3), spaces are visible in the central portion of the X chromosome which indicate its spongelike nature, which resembles that of nucleoli (Vol. II, Chapter 5).

Preparatory Stage

This is the stage when chondrocytes are lined up in columns and then begin to hypertrophy. The aggregates of second-order coils of DNA are smaller than in the earlier stage, and the nucleoplasm contains fewer RNA granules. In addition, the RNA granules in the X chromosomes are more discrete. Some second-order coils of DNA are visible in the central parts of the X chromosome (Figs. 7.4–7.6). The same general description applies to somewhat later stages (Figs. 7.7–7.12) when the RNA granules in the nucleoplasm are even more reduced in number, and the nucleolus is absent. Spaces in the X chromosome become more apparent, and their continuity with those of the nucleoplasm is more obvious.

Sometimes, two, and rarely three parts of a single Barr body may be present in a single ultrathin section through a nucleus. In some very thick sections observed with the light microscope, the Barr body was seen girdling the nucleus on the inside; but in other nuclei in the same sections, there were one or more bodies in the nucleus which are sometimes seen to be connected by delicate strands.

Very Early Lytic Stage

In this stage the chondrocytes are markedly hypertrophic. Most of the Barr body consists of RNA granules, and granules of DNA in and around the central cluster of RNA granules (Figs. 7.13 and 7.14). At this time, there are few RNA granules in the remainder of the nucleoplasm and no nucleolus is recognizable. The basic spongelike pattern of the X chromosome and its continuity with the enlarged nucleoplasmic spaces is quite obvious.

Lytic Stage

In this stage the hypertropic cells have large lytic spaces. Most of the Barr body consists of RNA granules, with only a few DNA figures persisting (Figs.

7.15 and 7.16). The Barr body persists in later lytic stages, when most of the cytoplasm consists of great spaces and the nucleus is nearly empty.

Identification of Nucleic Acids by Use of RNase and DNase

After the action of RNase, only a few granules in the X chromosome remain barely detectable, most of them not being distinguishable because they have the same low contrast as the background material. RNA granules in the nucleoplasm and in the ribosomes of the rough endoplasmic reticulum (RER) are also faintly

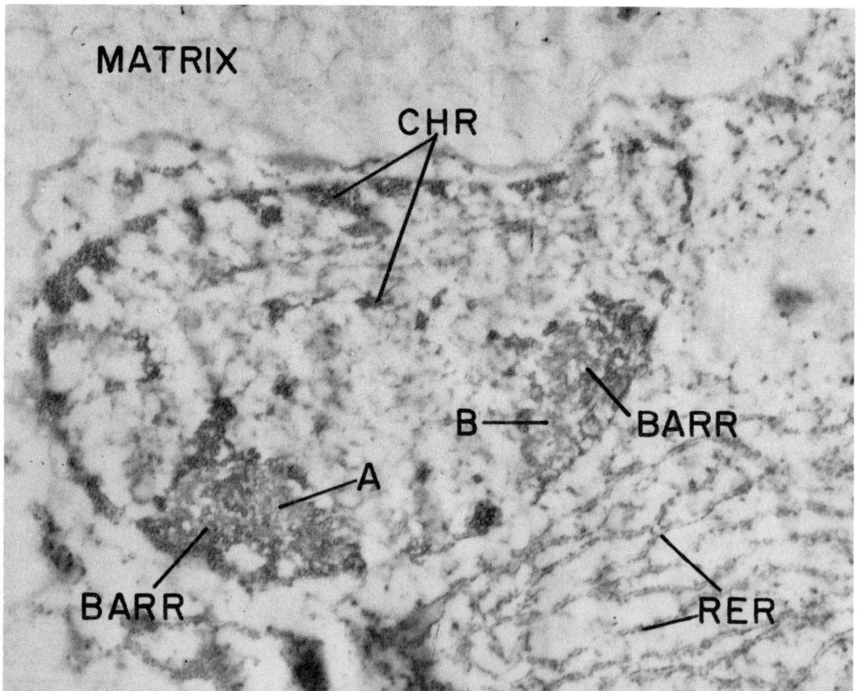

Fig. 7.4.

Figs, 7.4–7.6. Early preparatory stage when cells have just begun to appear in columns. The nucleoplasmic RNA granules are not very numerous, while the RNA granules in the two parts of the Barr body (A and B) are quite numerous. Barr body A is shown at higher magnification in Fig. 7.5, and B in Fig. 7.6. The only new feature is the appearance of clear spaces within the Barr bodies (BARR S). NUC SURF, surface of nucleus. Figure 7.4, × 14,200; Figs. 7.5 and 7.6 × 56,000.

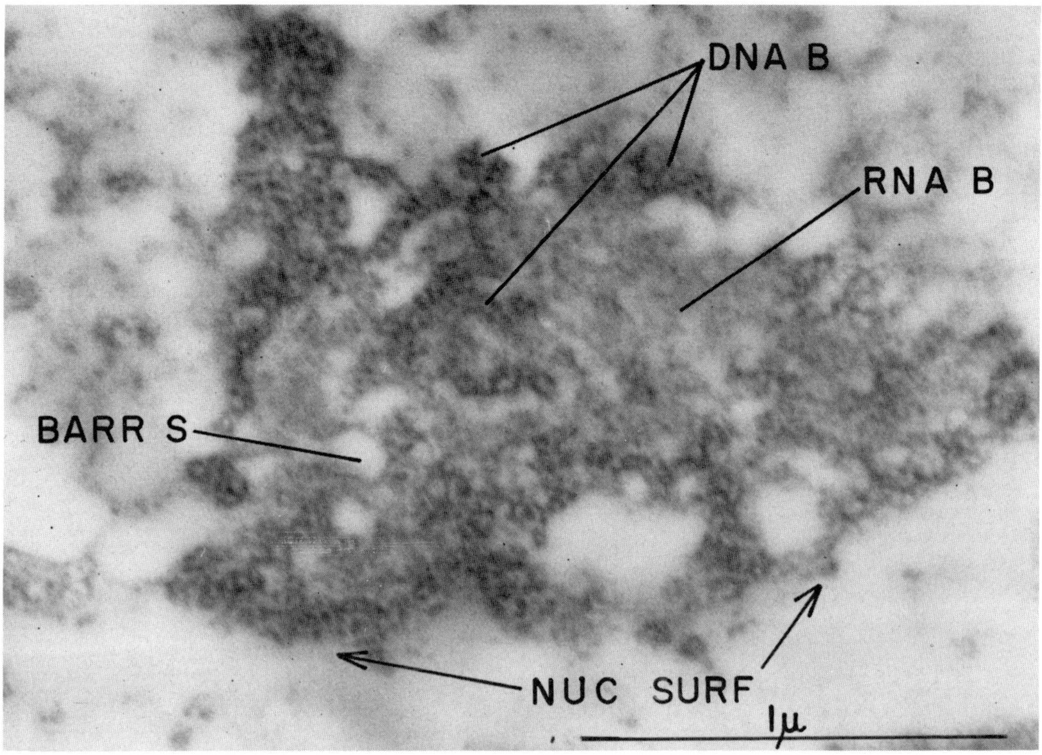

FIG. 7.5. ▲

▼ FIG. 7.6.

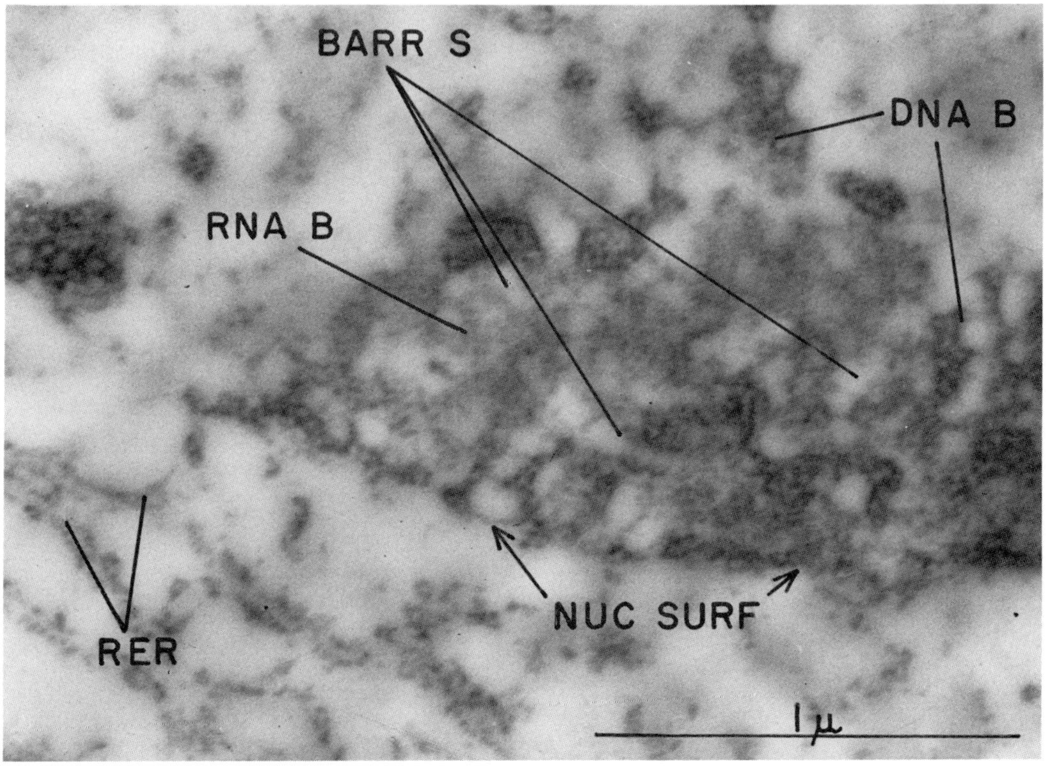

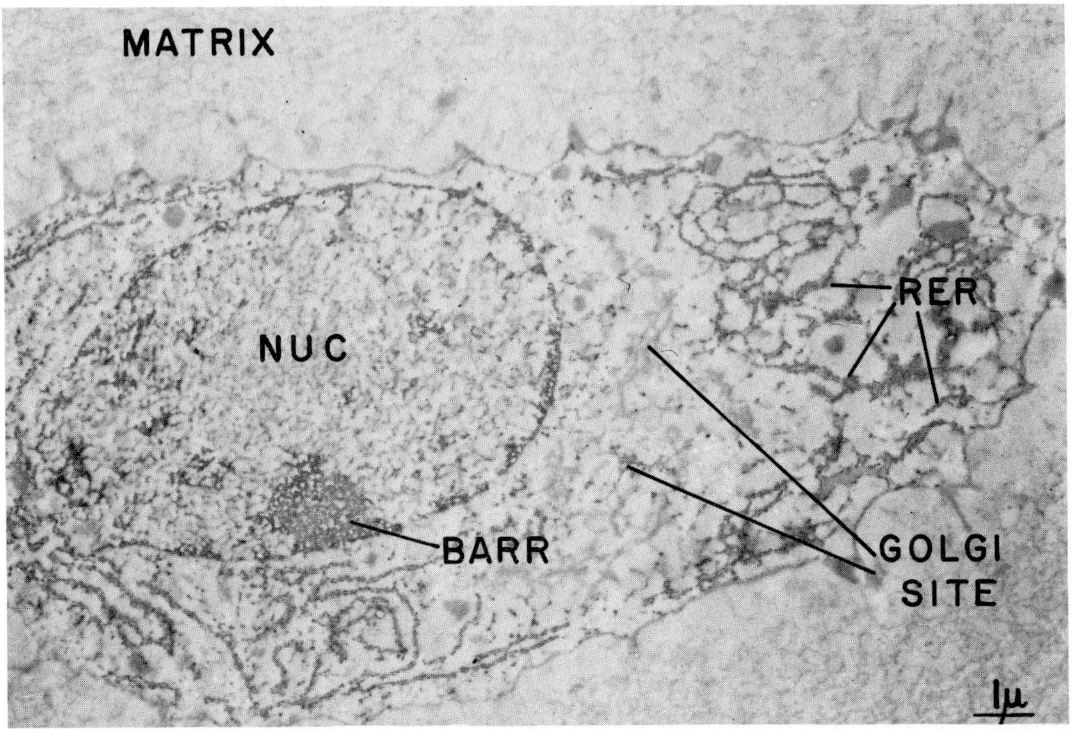
MATRIX
NUC
RER
BARR
GOLGI
SITE
1μ

FIG. 7.7. ▲

▼ FIG. 7.8.

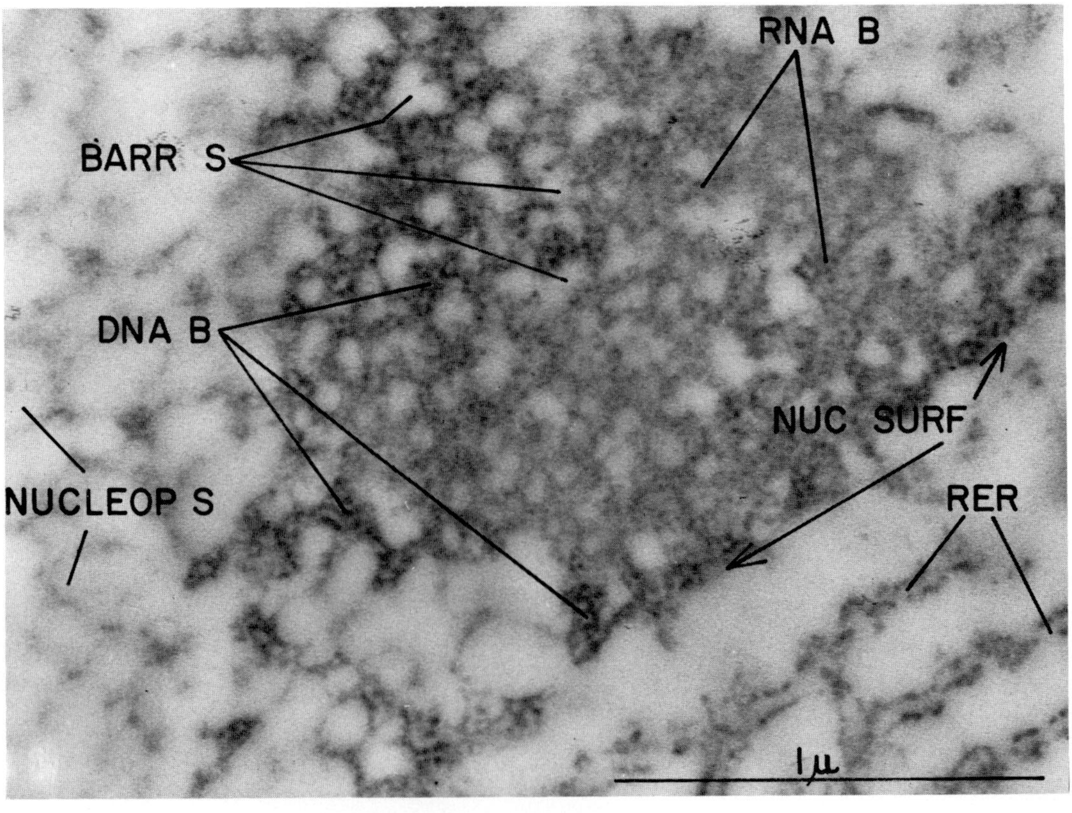
RNA B
BARR S
DNA B
NUC SURF
NUCLEOP S
RER
1μ

discernible. The second-order coils of DNA on the periphery and within the X chromosome are not affected by RNase. This enzymic effect of RNase may be seen in early and late stages of maturation of chondrocytes (Figs. 7.17–7.23). On the other hand, the action of DNase results in loss of DNA, while RNA persists in the central part of tht X chromosome, the nucleolus, and the ribosomes of the RER (Figs. 7.24–7.27).

Discussion

The basic organization of the Barr body and of the nucleolus (including its perinucleolar chromatin) is essentially similar. In each structure, there is a central accumulation of RNA-containing granules enclosed and penetrated by second-order coils of DNA. In early stages of maturation of chondrocytes, the DNA coils are aggregated to form clumps, many of which are large enough to be visible with the light microscope. In later stages of maturation, the DNA coils become separated, and later still, they break down into granules and segments. The major difference between the X chromosome and nucleolus is that the former far outlast the latter, so that residual traces in lytic chondrocytes may be recognized in nuclei which have lost all internal structure except for a few nuclear strands.

Certain inferences may be made from these observations. The first inference is that DNA of the Barr body has the genetic property of making RNA, just as has DNA of the nucleolar-forming chromosomes. This possibility is very clear for early stages of maturation. Whether this continues through later stages of maturation is not clear. In still later stages, when the DNA is reduced to granules and broken segments, it is probable that RNA is not being synthesized in this site. In the absence of kinetic studies on the release of RNA by the X chromosome, it is not possible to assert that new RNA is being continually synthesized and released by the X chromosome, or that no new RNA is being synthesized and that the old simply persists as if mummified because it cannot be hydrolyzed by autodigestive ribonuclease. The first alternative implies continued genetic activity of the RNA-forming genes of the X chromosome. In support of this possibility is the fact that the X chromosome continues to exercise some specific and important genetic activity in certain cells in mice. (See Barr for review of this material.) The activity of the X chromosome in embryos is well documented by Lyons *(7)*. It is highly probable that the presence of RNA in late stages is due to its persistence, without replacement.

FIGS. 7.7 and 7.8. Cell somewhat later in preparatory stage, as indicated by the wider intercisternal spaces of the RER and the finer dispersion of the chromatin of the nucleus (NUC). The spongelike nature of the DNA portion (DNA B) and the RNA portion (RNA B) of the Barr body and its continuity with the nucleoplasm can be inferred by observing the continuity of the spaces in the Barr body (BARR S) and in the nucleoplasm (NUCLEOP S). Other abbreviations as given in previous legends. Figure 7.7, × 8000; Fig. 7.8, × 56,000.

A second inference from the observations reported above is that DNA of the Barr body is resistant to the action of DNase released during cell lysis, since it is destroyed very much more slowly than the DNA in the rest of the nucleus. The persistence of the RNA granules as the major bulk of the X chromosome may be only apparent (new RNA granules being synthesized to replace those released into the nucleoplasm) or it may be real, reflecting a reduction in RNA turnover and/or hydrolysis by cellular RNase. It is thus possible that both DNA and RNA granules are resistant to their hydrolases. It is necessary here to recall that the X

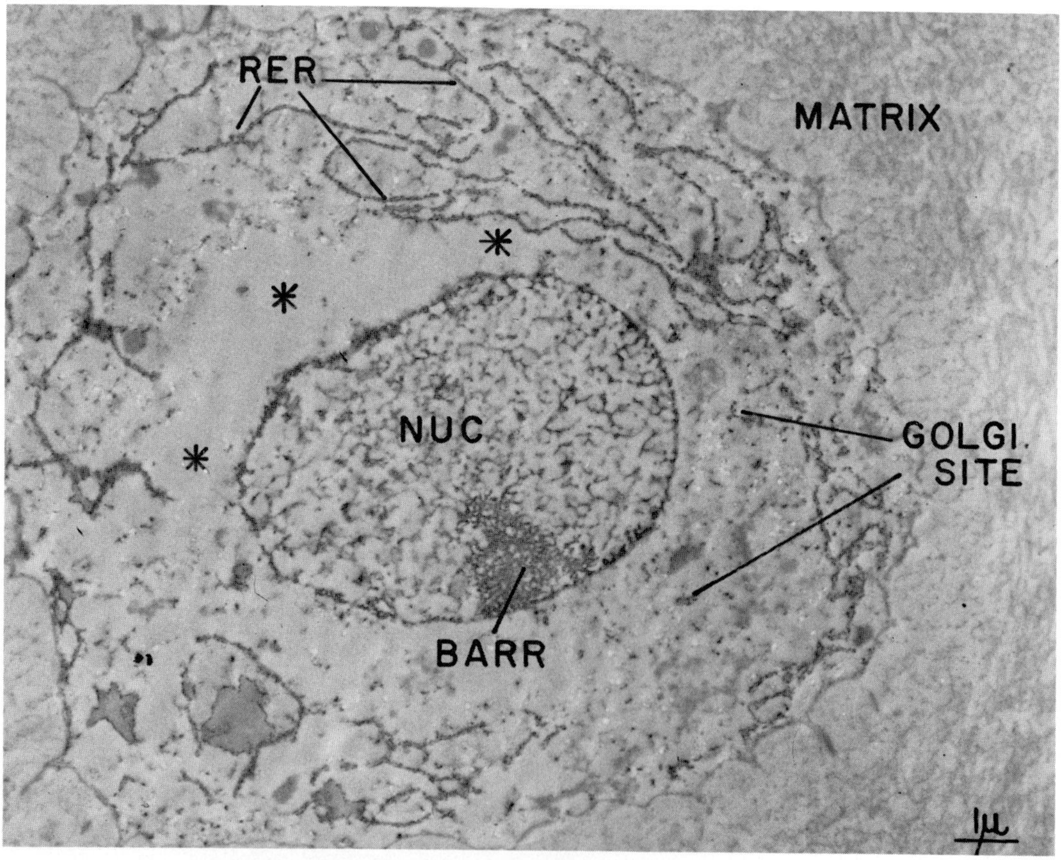

FIG. 7.9.

FIGS. 7.9 and 7.10. Hypertrophied cell later in preparatory stage, as indicated by wider dispersion of RER and finer dispersion of chromatin of the nucleus (NUC). The second-order coils of DNA of the Barr body (DNA B) are now largely in single coils (COIL). The RNA granules of the Barr body (RNA B) are more condensed in the walls which enclose spaces (BARR S), which together reflect the spongelike nature of the Barr body. The DNA coils are also confined to the denser component of the spongelike structure. Asterisks mark artificial tear. Figure 7.9, × 8000; Fig. 7.10, × 56,000.

chromosome is characterized by late replication of DNA as well as persistence as a separate chromosome usually resolvable with the light microscope. The first property could be attributed to an insensitivity or resistance to DNA polymerase or simply to temporary inactivation or lack of this enzyme. This lack of response to these polymerase and hydrolase enzymes (or their absence or inactivation) would seem to constitute a basic property of heterochromatic regions which may have important genetic consequences.

These major inferences to be drawn from the observations described in this report may be interrelated, stemming from a basic factor which may be common to both. The basic factor may be some peculiar property of proteins and/or ribonucleic acid which is associated with the DNA and with the RNA of the granules, that confers on them their special resistance to or inhibitory action of, or synthesis of certain enzymes. The RNA referred to in the preceding sentence may be part of the heterogeneous nuclear RNA (HnRNA) [see reviews by Darnell *(4)* and

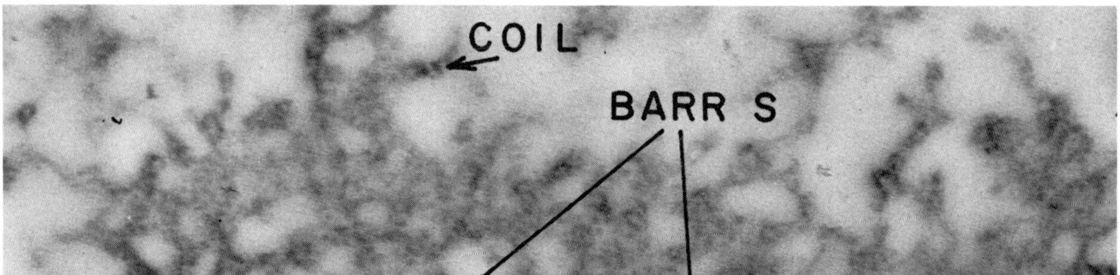

Fig. 7.10.

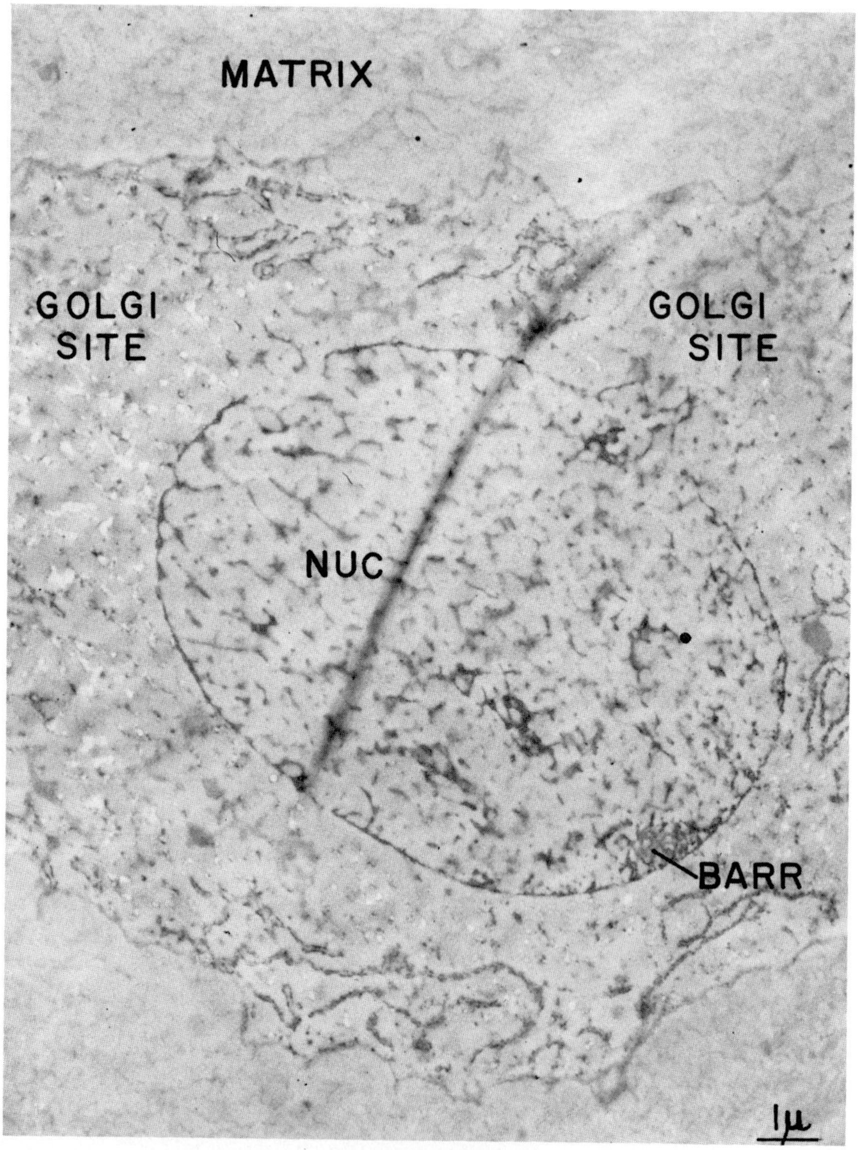

FIG. 7.11.

FIGS. 7.11 and 7.12. Cell later in preparatory stage, hypertrophied. The structure of the Barr body (BARR) is essentially the same as in the preceding cell. All abbreviations as given earlier. Figure 7.11, × 8000; Fig. 7.12, × 56,000.

Georgiev *(5)* or more specifically chromosomal RNA (cRNA) as reported by Bekhor, Bonner and Dahmus *(2)*]. In very general terms, such interfering substances would account for the relative genetic inertness of the sex chromosome. This suggestion is in line with that proposed by Lyon *(7)*.

The X chromosome is significant in considerations of the relations of chromosomes and chromatin to the nuclear body. Evidence is summarized in Chapter 14 (this volume) to support the thesis that chromosomes have a certain relationship to each other and to the nuclear membrane during all stages of the mitotic cycle, the orderly relationship being required for the exact division of DNA to take place reproducibly during successive cell divisions. It is also essential for the proper and unique coordination of gene activity that characterizes the specific cell type. This is consistent with the observation that the chromatin pattern of somatic

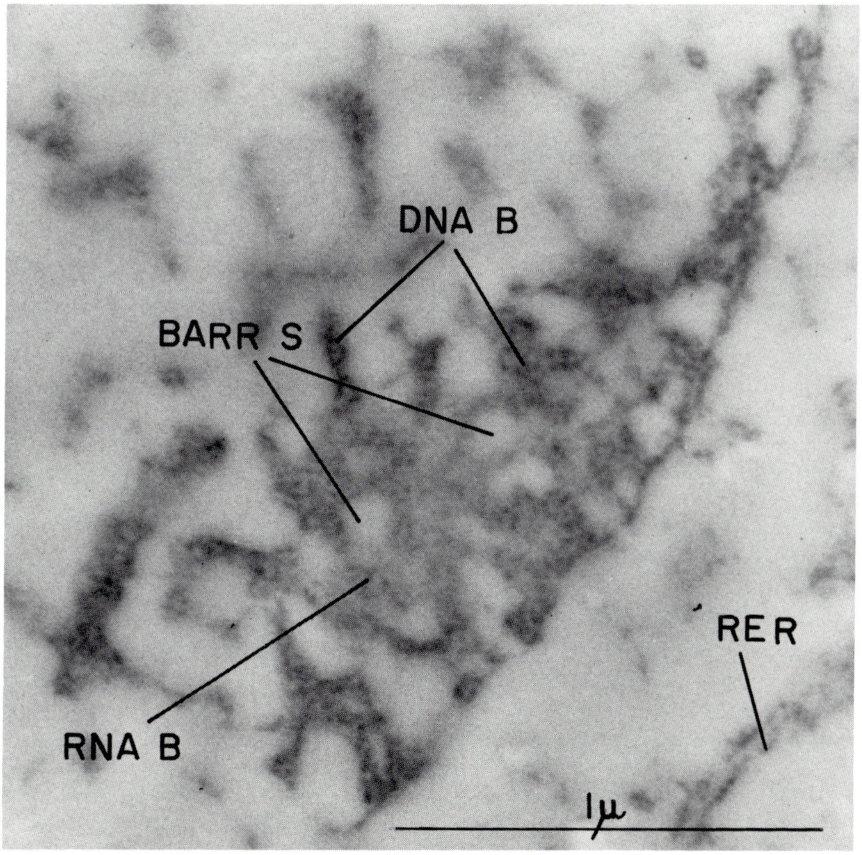

Fig. 7.12.

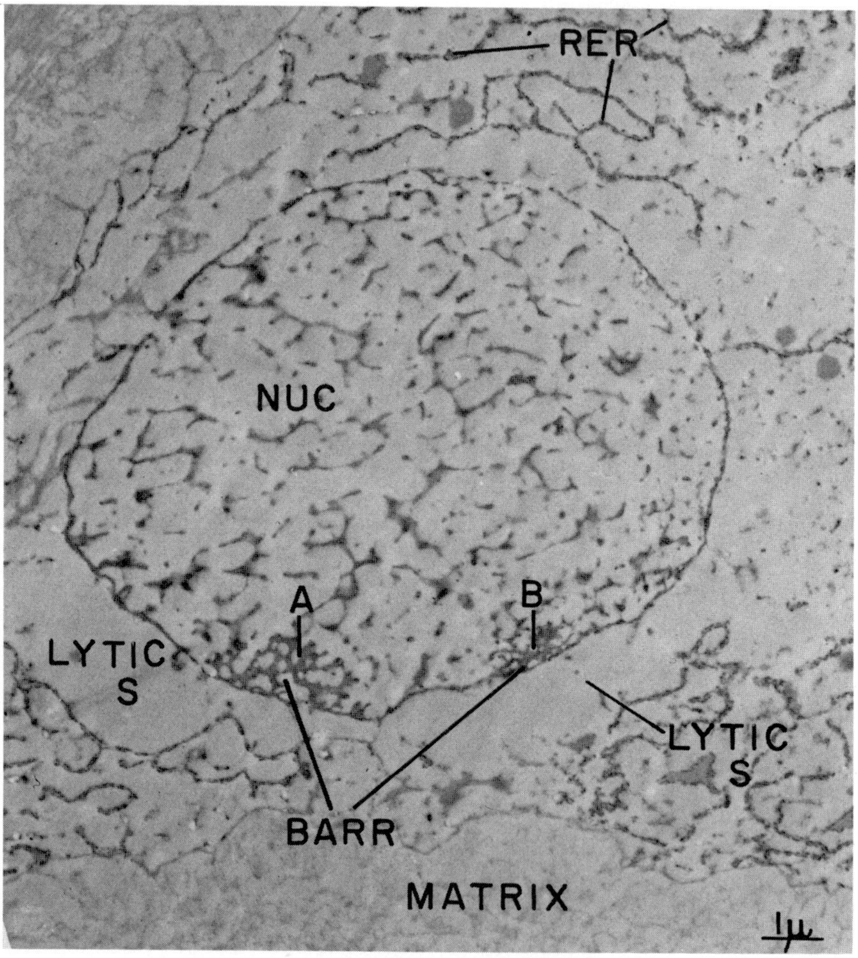

FIG. 7.13.

FIGS. 7.13 and 7.14. Cell in very early lytic stage. One of the two parts of the Barr body (A) is magnified in Fig. 7.14. The DNA is now largely in the form of granules (DNA B) which are larger and denser than the RNA granules (RNA B). The spaces in the spongelike structure of the Barr bodies (BARR S) are enlarged, but not nearly as much as the nucleoplasmic spaces (NUCLEOP S). Other abbreviations as indicated earlier. Figure 7.13, × 8000; Fig. 7.14, × 56,000.

cell nuclei is characteristically constant in unstimulated cells. This implies that each chromatin mass has a definite relation to the others, and that their relative positions reflect the relative positions of the corresponding chromosomes giving rise to them. The nuclear membrane must play a significant role in maintaining this relative constancy of pattern. From this point of view, the Barr body has a special, heuristic importance, for the X chromosome is the only restricted recognizable chromosome which spreads out as chromatin and retains its individuality throughout the cell cycle *(6)*. The continued cell cycling of somatic cells of a particular type surely demands a functional pattern of relationship of the chromosomes based on a spatial pattern of relationship. The Barr body is the only chromatin mass restricted to a single chromosome which can possibly be separated and used as a spatial marker for study of the factors which are responsible for the attachment of chromosomes to the nuclear membrane. Some of the factors possibly involved are discussed in Chapters 13 and 14 (this volume).

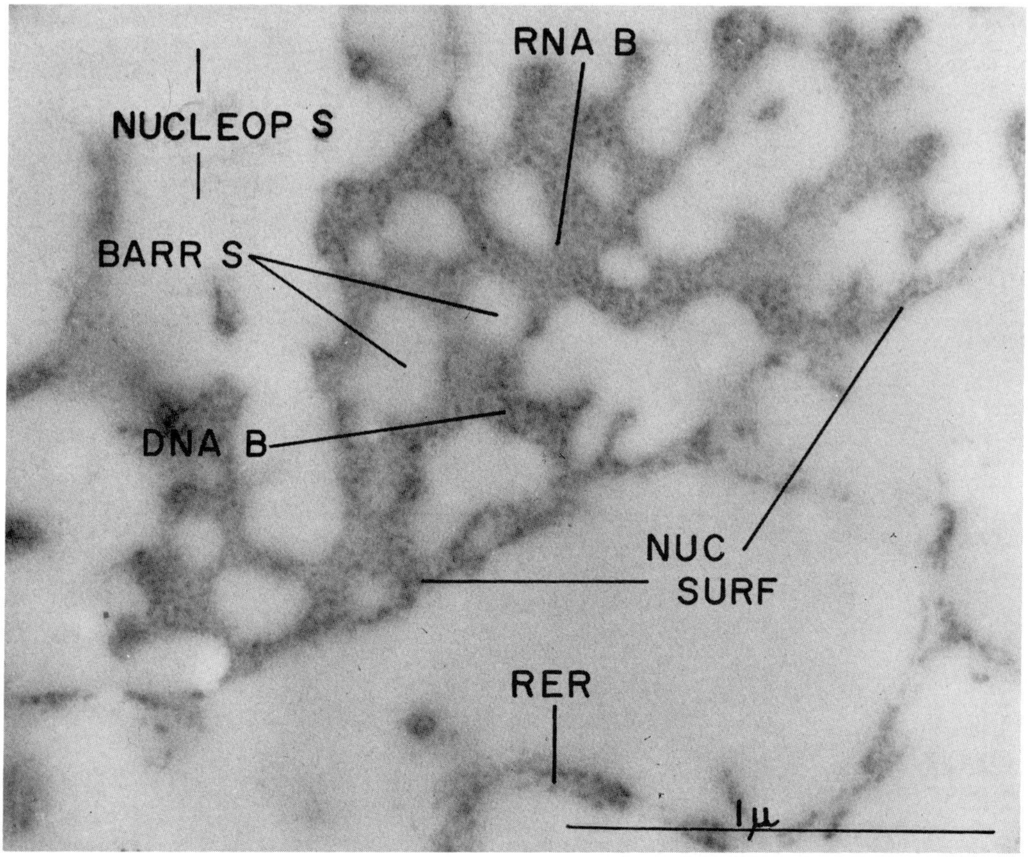

FIG. 7.14.

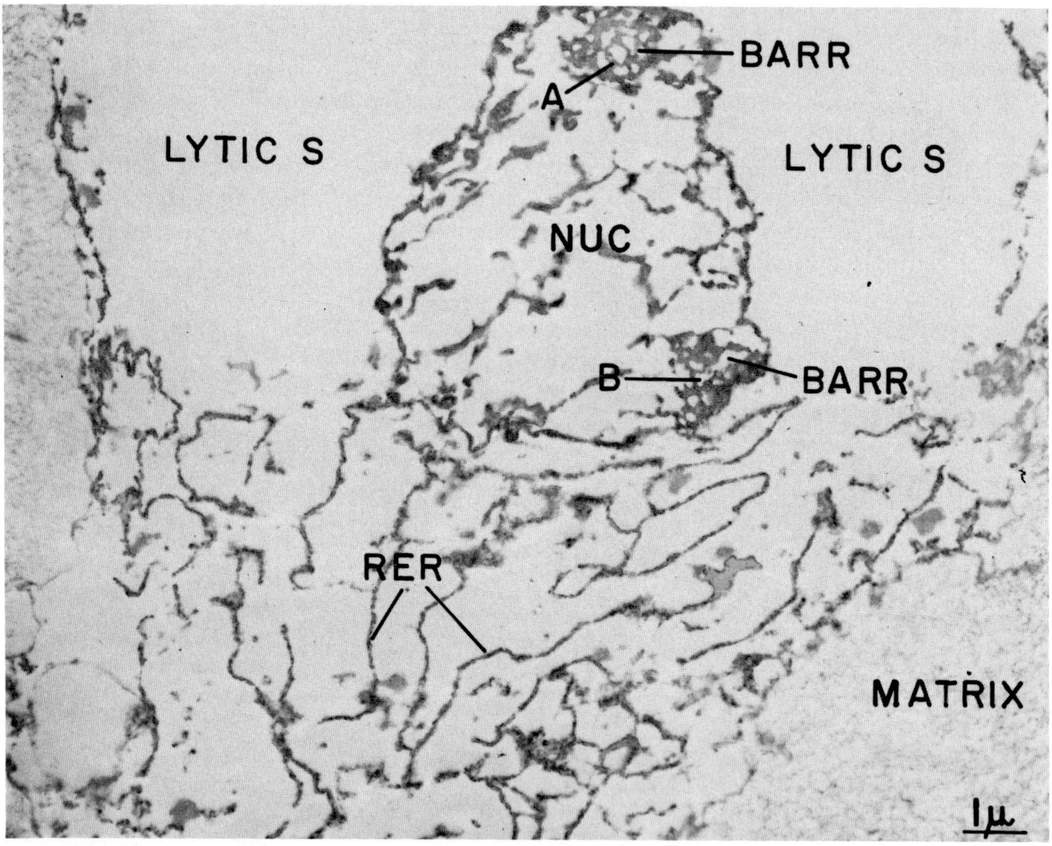

FIGS. 7.15 and 7.16. Cells in lytic stage (that in Fig. 7.16 being later than in Fig. 7.15). The two portions of the Barr body in Fig. 7.15, and the single portion in Fig. 7.16 persist

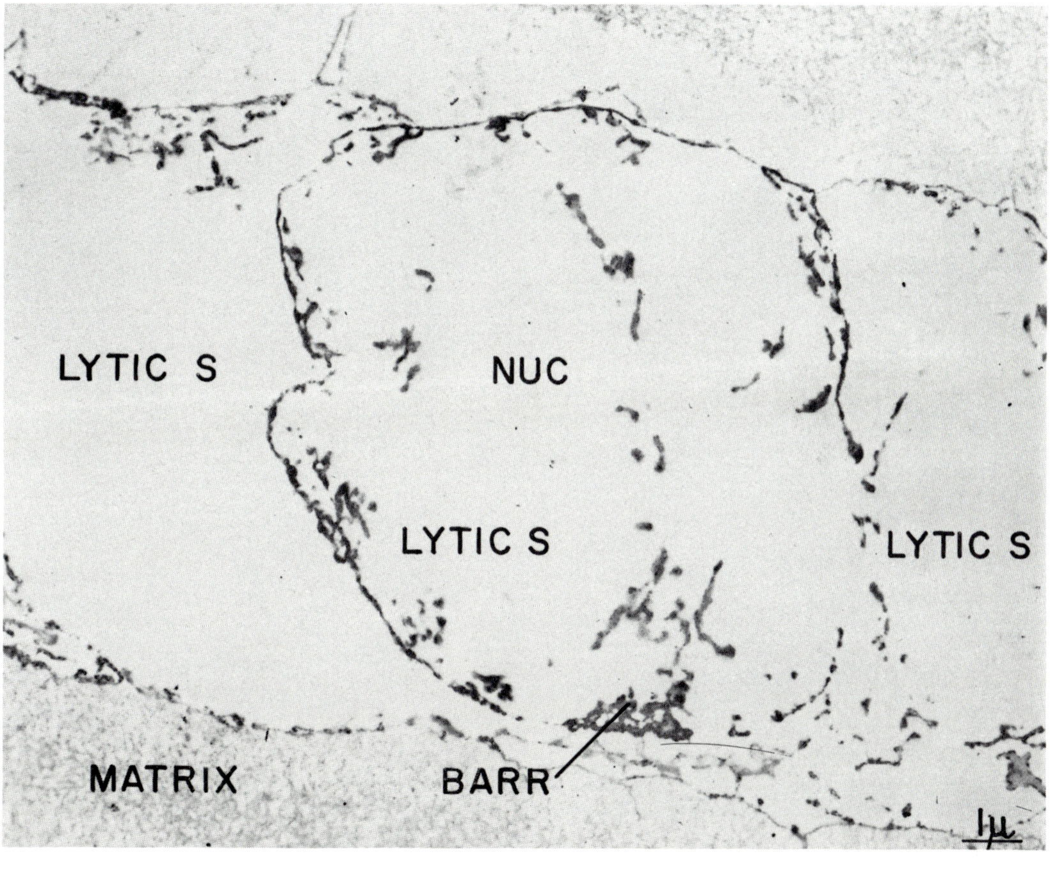

despite the enormous lytic spaces (LYTIC S) in the cytoplasm and nucleus (NUC). The RER seems to be greatly reduced in the more advance stage. Other abbreviations as noted earlier. Fig. 7.15, × 8000; Fig. 7.16, × 56,000.

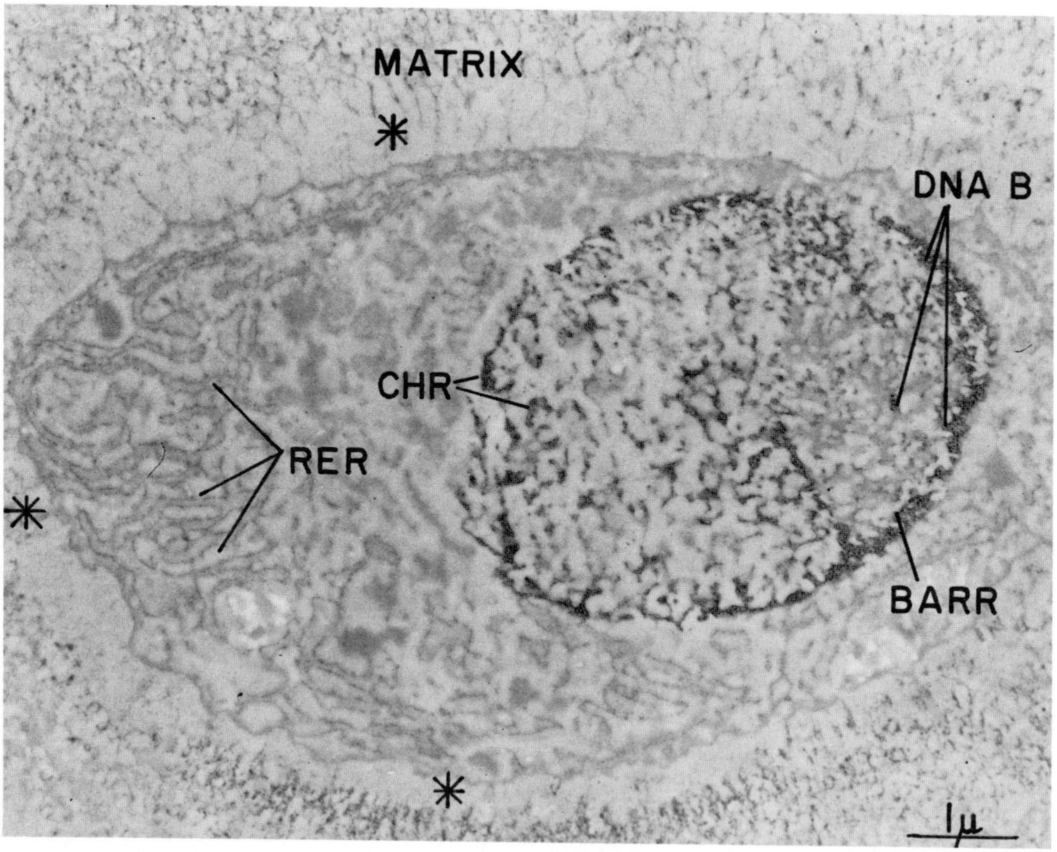

FIGS. 7.17 and 7.18. The effect of ribonuclease (staining and postfixation as in text) on nucleic acids of a postproliferative cell corresponding with those in Figs. 7.1–7.3. RNA in the RER, the nucleoplasm, and the Barr body is barely detectable. The second-order coils of DNA

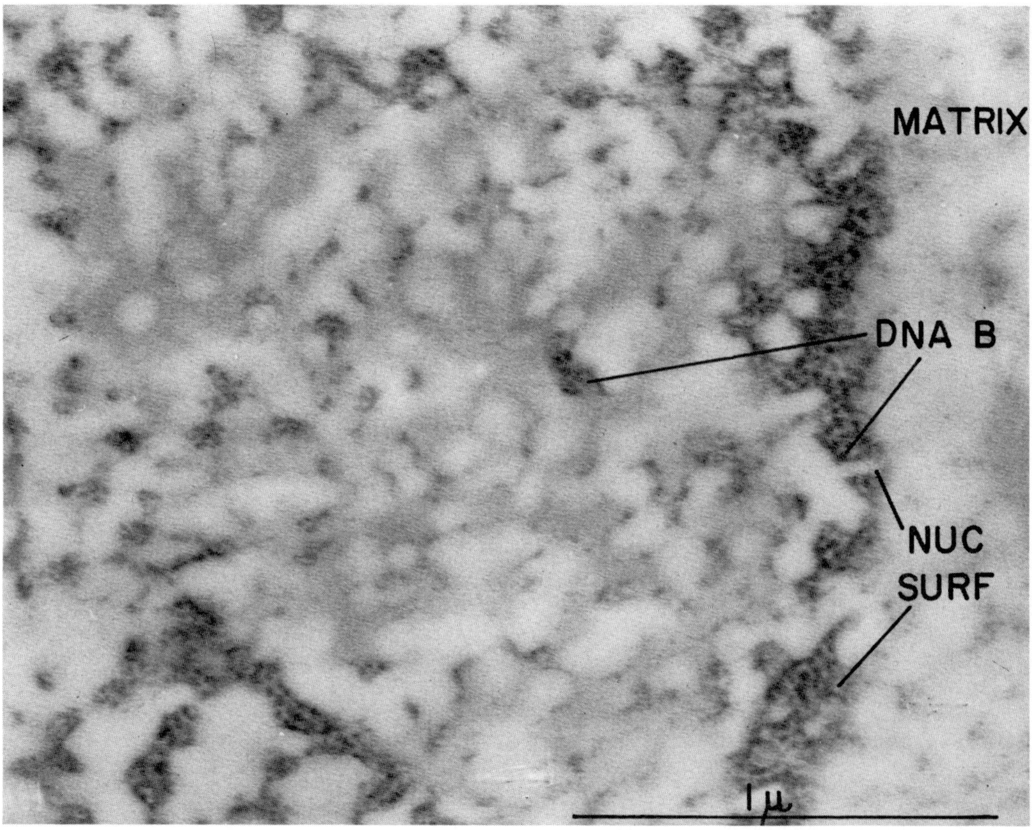

in the chromatin (CHR) and the outer and inner parts of the Barr body (DNA B) remain densely stained. Asterisks mark shrinkage space. Abbreviations as given previously. Figure 7.17, × 14,200; Fig. 7.18, × 56,000.

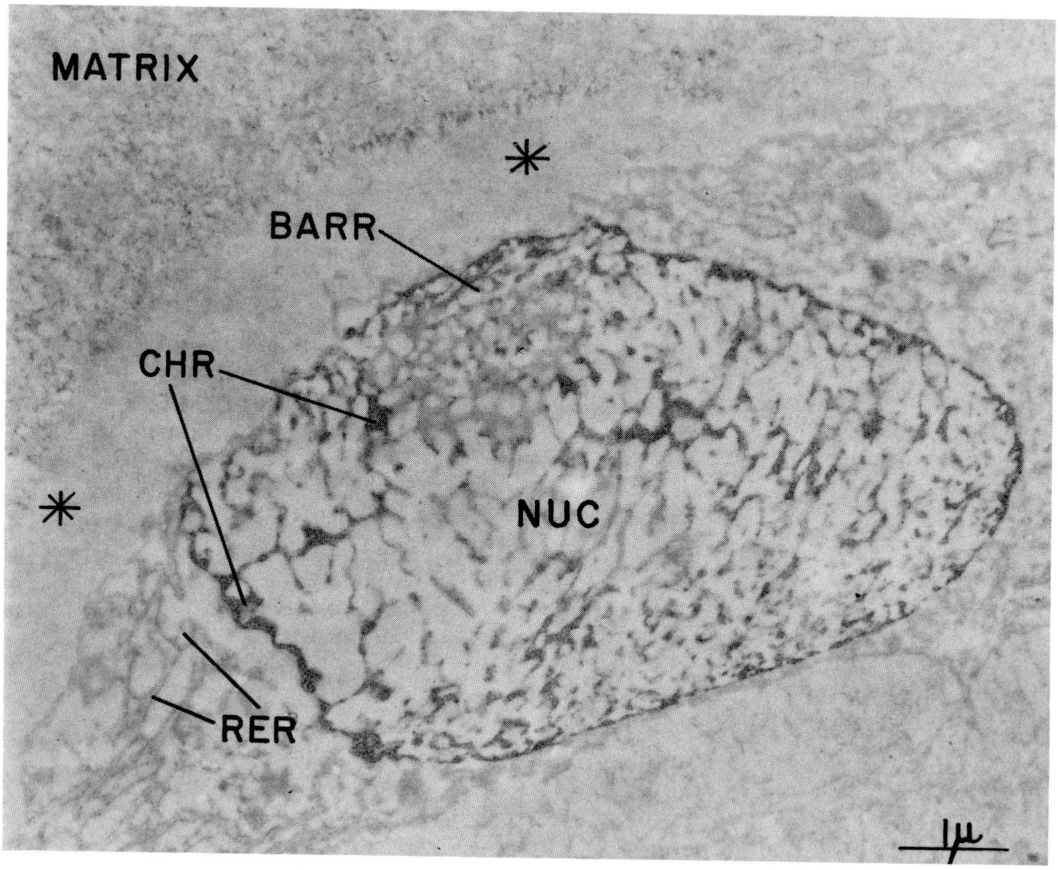

FIGS. 7.19 and 7.20. The effects of ribonuclease (staining and postfixation as in text) on nucleic acids of a cell in an early preparatory stage, corresponding with that in Figs. 7.4–7.6. The second-order coils of DNA in the Barr body (DNA B) and in the chromatin (CHR) of

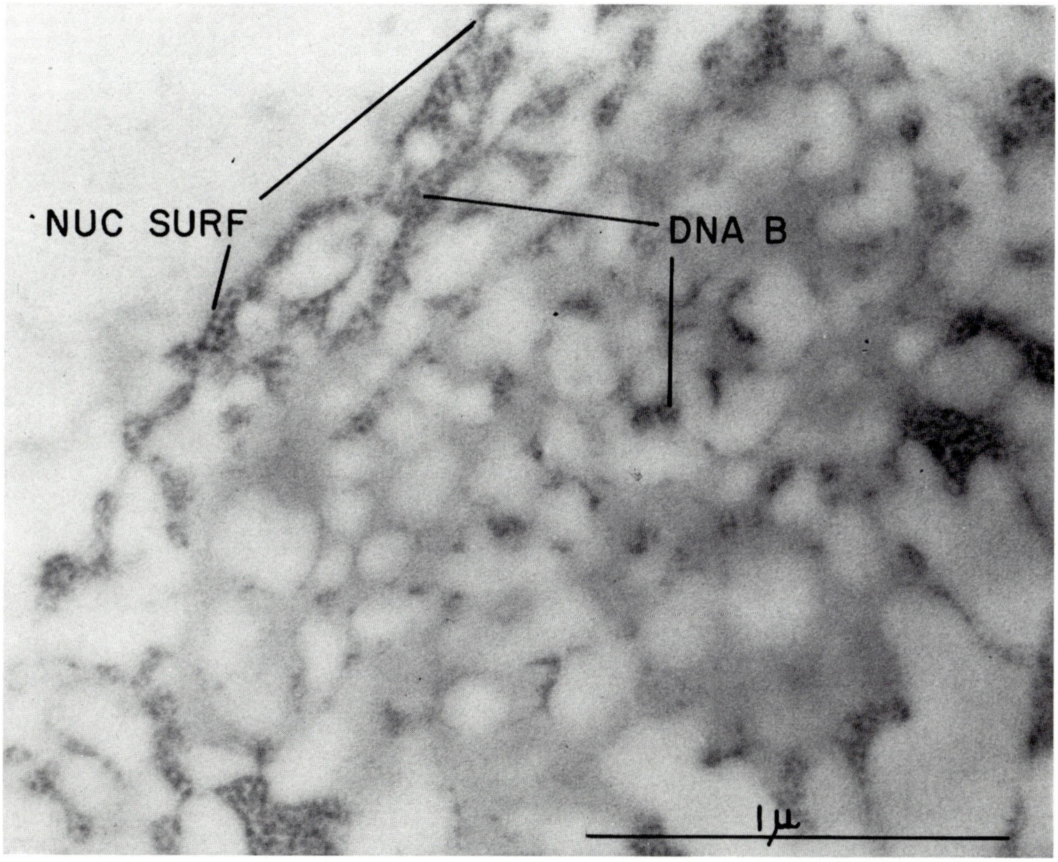

the nucleus persist, while the RNA in the RER, the nucleus, and the Barr bodies is greatly reduced in density, to the vanishing point in some parts. Asterisks mark shrinkage space. All other abbreviations as noted previously. Figure 7.19, × 14,200; Fig. 7.20, × 56,000.

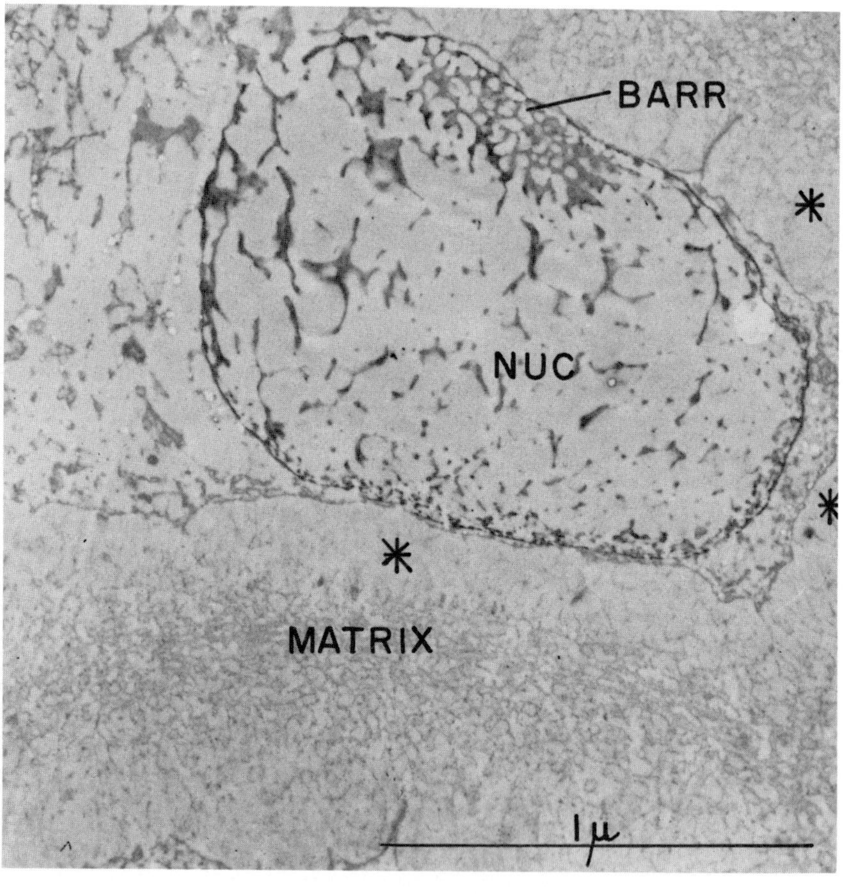

FIG. 7.21.

FIGS. 7.21 and 7.22. The effects of ribonuclease (staining and postfixation as noted in text) on nucleic acids of a cell in a prelytic stage corresponding with one a little younger than that in Figs. 7.13 and 7.14. Some RNA granules in the Barr body are perhaps barely distinguished, while the DNA persists largely as granules (DNA B). Asterisks mark shrinkage space; abbreviations as noted previously. Figure 7.21, × 8000; Fig. 7.22, × 56,000.

FIG. 7.23. Effects of ribonuclease on a cell in a prelytic or early lytic stage. Staining and postfixation as explained in text. The DNA largely enclosing the Barr body (BARR) is more prominent than usual because of the removal of RNA from the granules in the Barr body. The RNA has also been largely removed from the RNA granules of the RER. Other abbreviations as noted. × 8000.

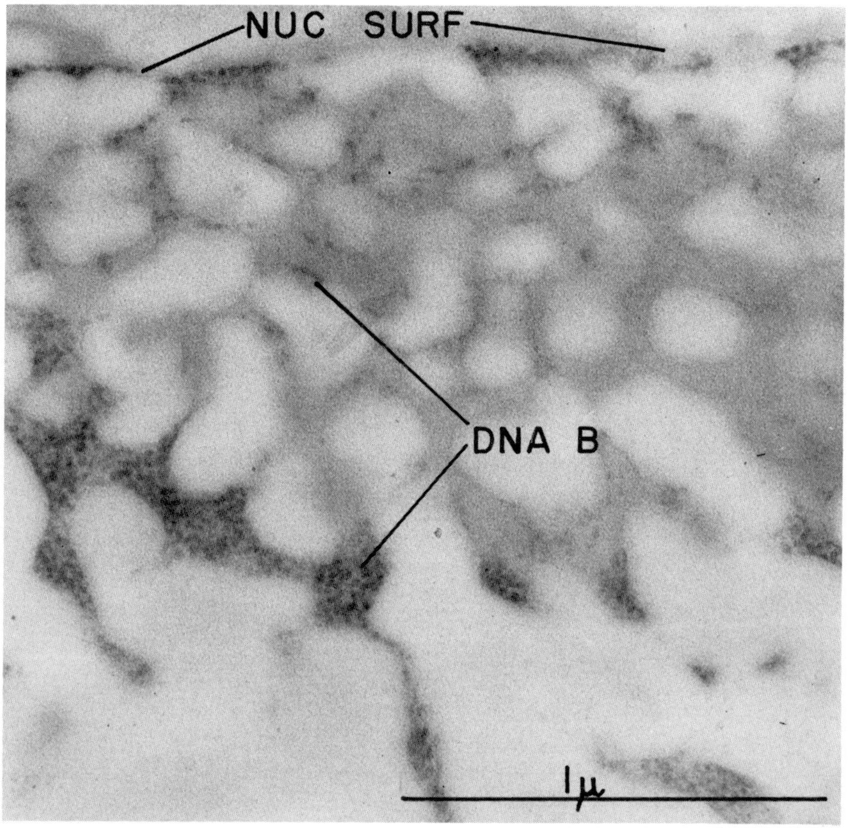

FIG. 7.22. ▲

▼ FIG. 7.23.

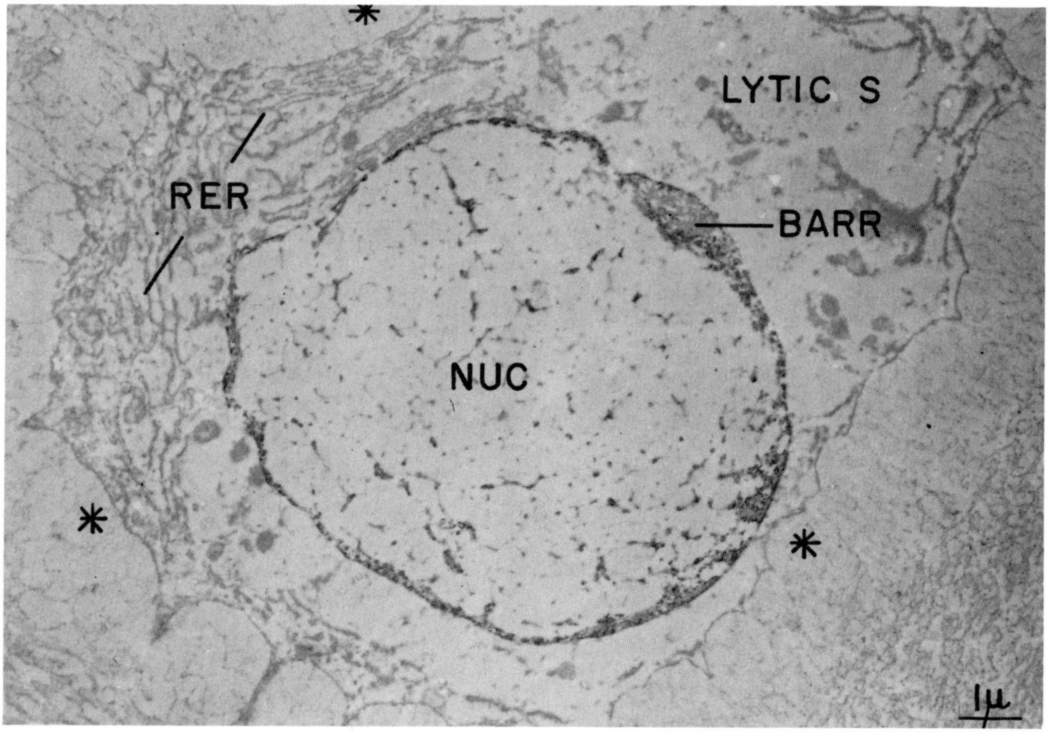

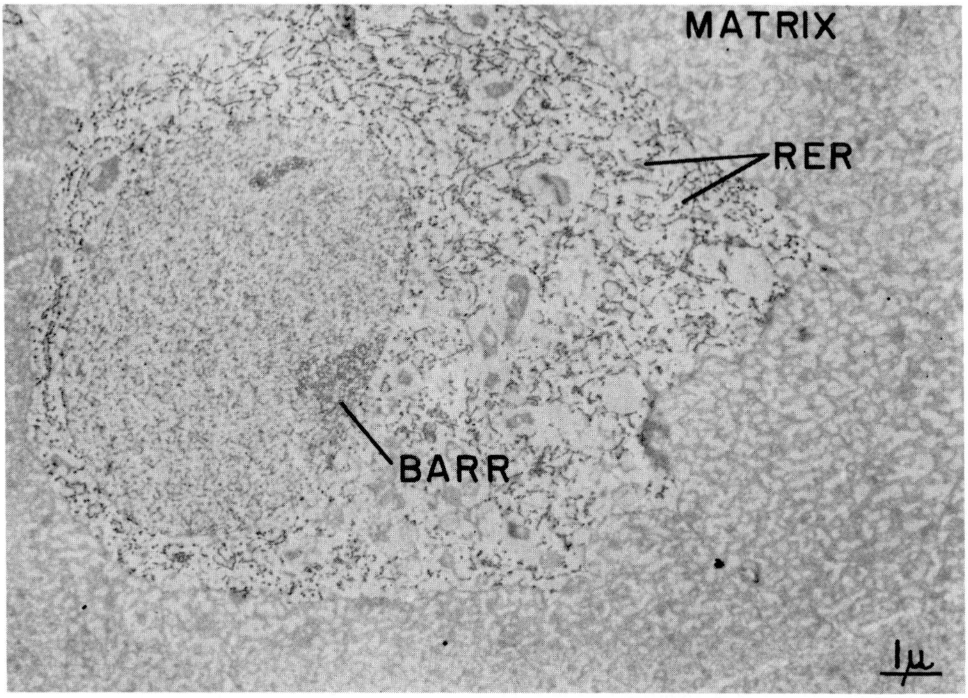

FIG. 7.24. Effects of deoxyribonuclease (staining and postfixation as noted in text) on nucleic acids of a cell in a postproliferative or early preparatory stage. Chromatin masses are no longer detected, while the RNA of the Barr body (BARR) and the RER persist in large part. × 8000.

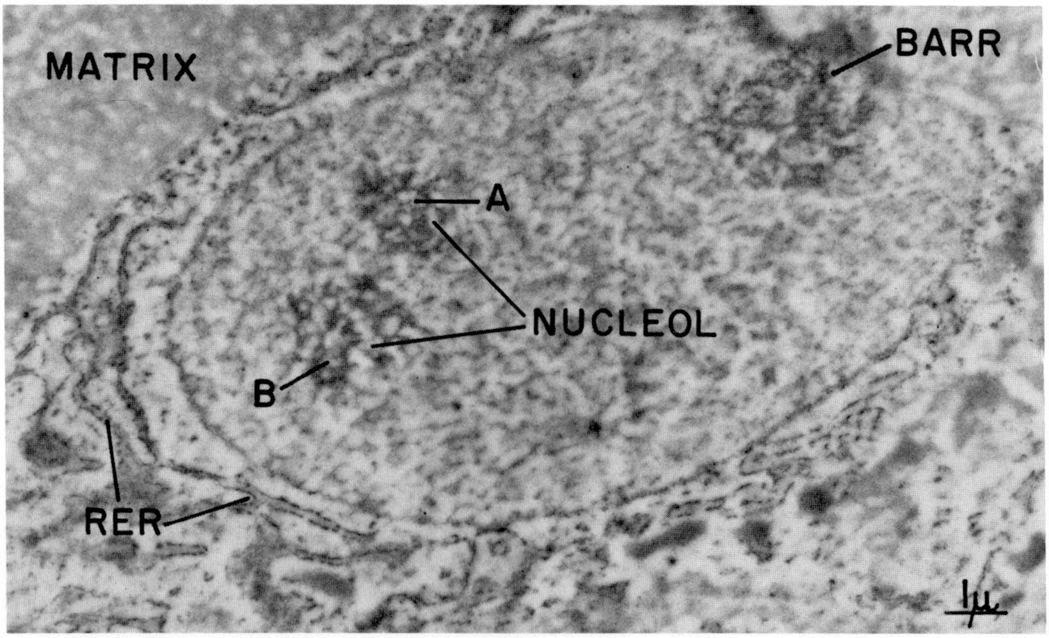

FIG. 7.25.

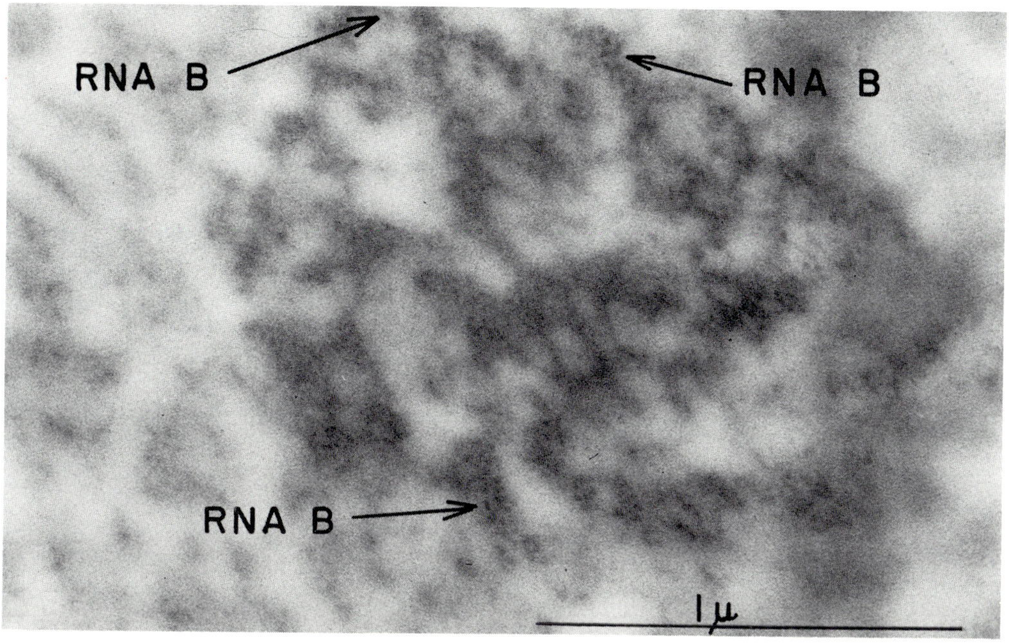

FIG. 7.26. ▲

▼ FIG. 7.27.

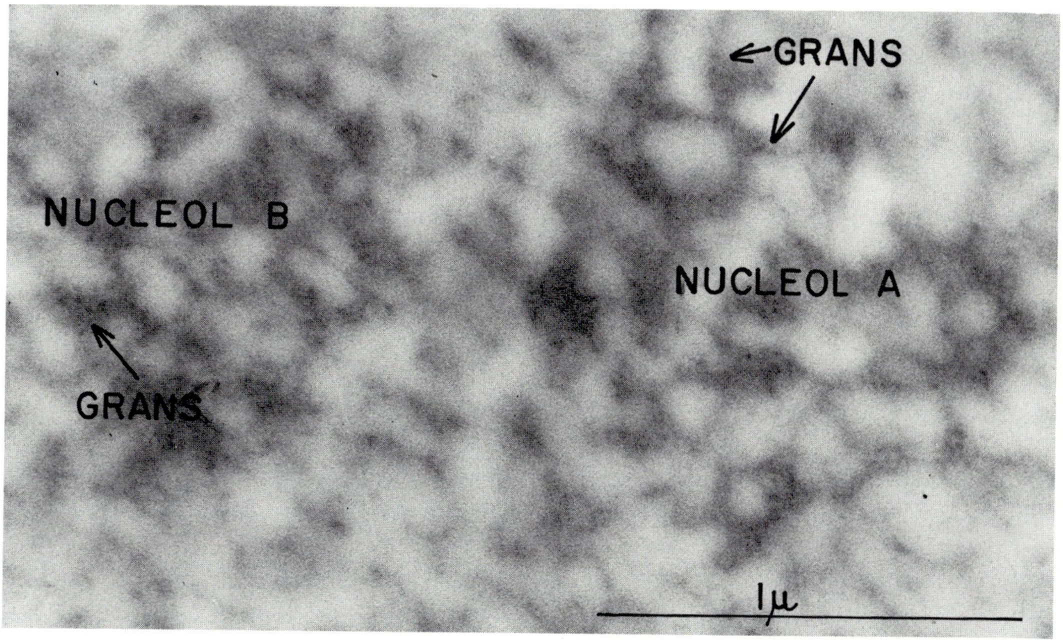

FIGS. 7.25–7.27. Effects of deoxyribonuclease on nucleic acids of a postproliferative cell. Staining and postfixation as noted in text. While the RER, central portion of the Barr bodies, and nucleoli (A and B) persist, the DNA of the chromatin, perinucleolar chromatin, and the X chromosome body are no longer visible. RNA persists in the Barr body (Fig. 7.26) and in the nucleoli (Fig. 7.27) as granules (RNA B and GRANS). Figure 7.25, × 8000; Figs 7.26 and 7.27, × 56,000.

Summary

In chondrocytes of the proximal tibial epiphyseal plate of female mice, the X chromosome represented by the Barr body is distinguishable from nucleoli in the postproliferative stage only by position, the Barr body being attached to the nuclear surface. The RNA granules and DNA coils of both structures are indistinguishable. While the nucleolus soon disappears as the cell matures, the X chromosome persists even in late stages of cell lysis. In these stages, the DNA coils deteriorate and persist as granules. Possible mechanisms for these observed changes are suggested.

References

1. Barr, M. L. (1966). The significance of the sex chromatin. *Int. Rev. Cytol.* **19**, 35–95.
2. Bekhor, I., Bonner, J., and Dahmus, G. K. (1969). Hybridization of chromosomal RNA to native DNA. *Proc. Nat. Acad. Sci. U. S.* **62**, 271–277.
3. Brown, S. W. (1966). Heterochromatin. *Science* **151**, 417–425.
4. Darnell, J. E. Jr. (1968). Ribonucleic acids from animal cells. *Bacteriol. Rev.* **32**, 262–290.
5. Georgiev, G. P. (1968). The regulation of the biosynthesis and transport of messenger RNA in animal cells. *In* "Regulatory Mechanisms for Protein Synthesis in Mammalian Cells" (A. San Pietro, M. R. Lamborg, and F. T. Kenney, eds.), pp. 24–48. Academic Press, New York.
6. Gillot, S. (1968). Hétérogénéités fonctionnelles dans l' ADN de noyaux géants. *Exp. Cell. Res.* **50**, 388–402.
7. Lyon, M. F. (1968). Chromosomal and subchromosomal inactivation. *Annu. Rev. Genet.* **2**, 31–52.

8

Preliminary Study of the Specificity of the DNA Molecular Patterns of Differentiated Cells of the Young Adult Mouse

Isidore Gersh

DNA is recognized in electron micrographs as filaments magnified to about 45 A by the use of the dye gallocyanin–chromalum. These filaments represent the DNA double helix of first-order helix. They are organized as second-order helices with a diameter and periodicity of generally about 400 A. The helices may be separated from each other, or may be aggregated as submicroscopic or microscopic chromatin masses (this volume, Chapter 3; Chapter 5, Vol. II). (The term "microscopic" is used in describing structures which can be seen with the light microscope; when structures cannot be resolved with the light microscope, they are referred to as submicroscopic.) A hypothesis was proposed that the single loops of single second-order helices, or very small aggregates could be considered the potentially active (or euchromatic) DNA, while larger aggregates of second-order helices could be considered the inactive or less active (or heterochromatic) DNA (Chapter 4, this volume). This is related to the probability that interchange between DNA and the nucleoplasm (and hence access to the cytoplasm) would be greater in the former than the latter. This hypothesis was strengthened by demonstrating that the pattern of aggregation of second-order DNA helices was different in the proximal and distal parts of imaginal wing disks of *Drosophila* when they were determined, but not yet differentiated (Chapter 6, this volume). Further, considering only the distal part of the imaginal disk of third instar larvae, which is destined to form the wing of the mature fly, the pattern of aggregation of second-order DNA of the wild type differs from that of several wing mutants as also from that of the distal part of the haltere imaginal disk. These findings constitute, in part, the basis for a second hypothesis, i.e., the rearrangement of the second-order helices of DNA may expose some previously heterochromatic loops as euchromatic regions or may convert some previously euchromatic loops to the heterochromatic state, thus consituting a mechanism for turning on the activity of some (formerly heterochromatic) DNA coils while reducing or turning off the activity of other (formerly euchromatic) DNA coils. This hypothesis interposes another factor between the possibly pertinent repressor–derepressor system and the control of genetic activity in nucleated cells, i.e., while the latter type of system *(8)* was originally conceived as controlling individual genes or groups of genes, these must first be made accessible by rearrangement and reaggregation of the second-order DNA coils. The latter make possible coarse adjustment while the former control the fine adjustment of gene action.

There are two major corollaries of the second hypothesis: (1) In the nuclei of differentiated cells, which perform different functions, the DNA molecular patterns should be recognizably different from each other, but persistent and constant in each differentiated cell type. (2) In any specific cell type, the second-order DNA should be more separated (euchromatic) when the cell is more active than when it is less active, in which state the DNA should be more aggregated (heterochromatic). Evidence in support of these corollaries will be presented in this and the following two chapters, respectively.

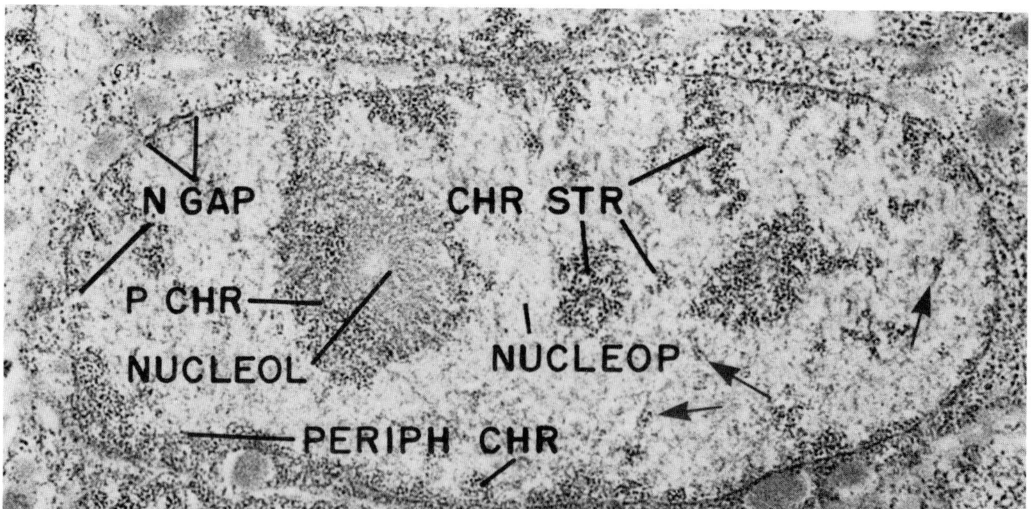

FIG. 8.1. Columnar cell of intestinal gland of duodenum. The peripheral chromatin (PERIPH CHR) varies considerably in thickness and gives rise to many chromatin strands (CHR STR) which course in the nucleoplasm (NUCLEOP). Some of them, marked by arrows, are fine. The perinucleolar chromatin (P CHR) is continuous with the peripheral chromatin and encloses the nucleolus (NUCLEOL). The nuclear gaps (N GAP) are small and numerous. Numerous ribosomes are in the cytoplasm. Nuclei of equivalent cells in the colon resemble these very closely.

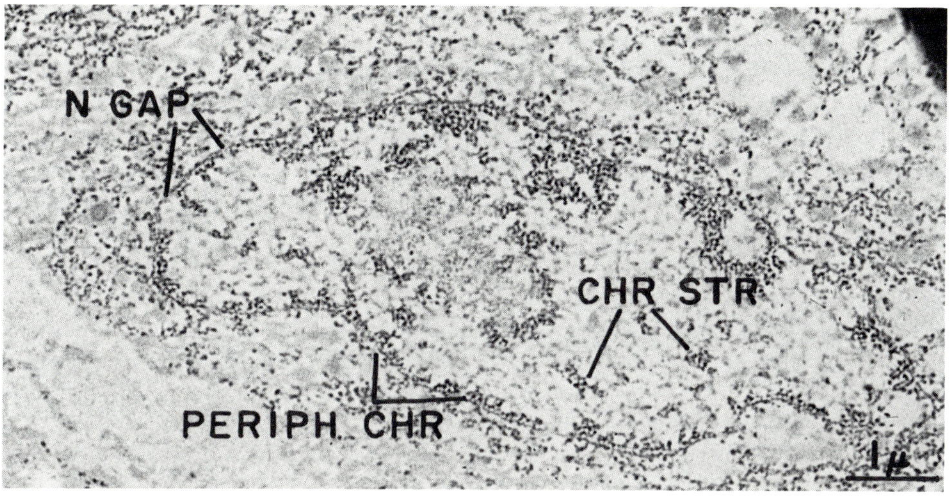

Fig. 8.2. Chief cell of fundic gland of stomach. The peripheral chromatin (PERIPH CHR) is more uniformly distributed and gives rise to fewer chromatin strands (CHR STR) than in Fig. 8.1. There are small and large nuclear gaps (N GAP). Numerous ribosomes are present in the cytoplasm.

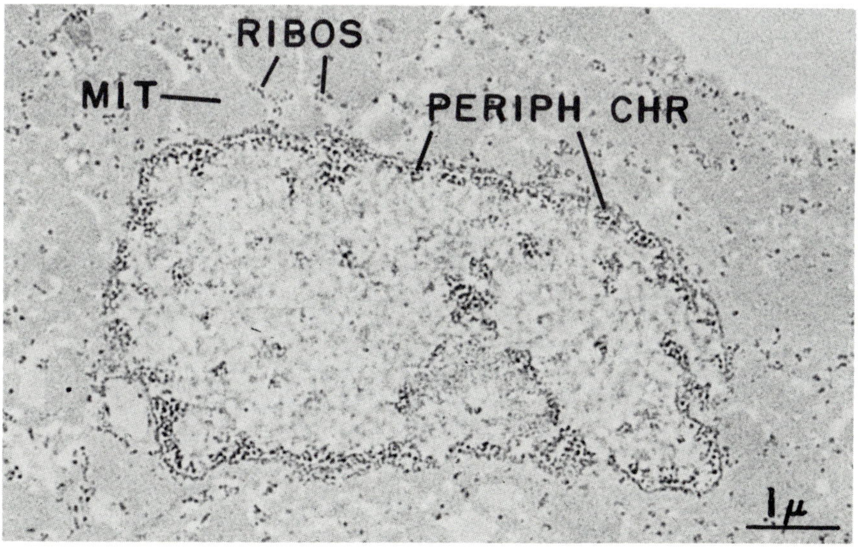

Fig. 8.3. Parietal cell of fundic gland of stomach. Peripheral chromatin (PERIPH CHR) is more uniform than in zymogenic cells. There are relatively few nuclear gaps and these are of small dimensions. Ribosomes (RIBOS) are scarce, and are scattered chiefly between mitochondria (MIT).

Method and Materials

It is of prime importance in a study of this sort that the method of fixation of DNA be sufficiently sensitive and the method of staining of DNA sufficiently selective or specific to reflect accurately enough the pattern of distribution of DNA in the nucleus of the living cell. The evidence presented in Chapter 3 (this volume) and Chapter 5 (Vol. II) support the belief that the methods employed probably satisfy the requirements. Specimens are frozen ultrarapidly in an attempt to avoid significant dislocations caused by the formation of ice crystals, and dried in a vacuum at a low temperature. Proteins in the specimens are cross-linked by introducing a vapor reagent, mercury hexafluoroacetylacetonate (Hg hfac). These hardened specimens are subsequently stained with gallocyanin–chromalum under conditions known to be highly selective and virtually specific for nucleic acids. Closer identification as DNA or RNA was made by morphological criteria as well as the use of RNase and DNase before staining (Chapters 3 and 4, this volume; Chapter 5, Vol. II). Specimens were embedded in water-soluble Durcupan.

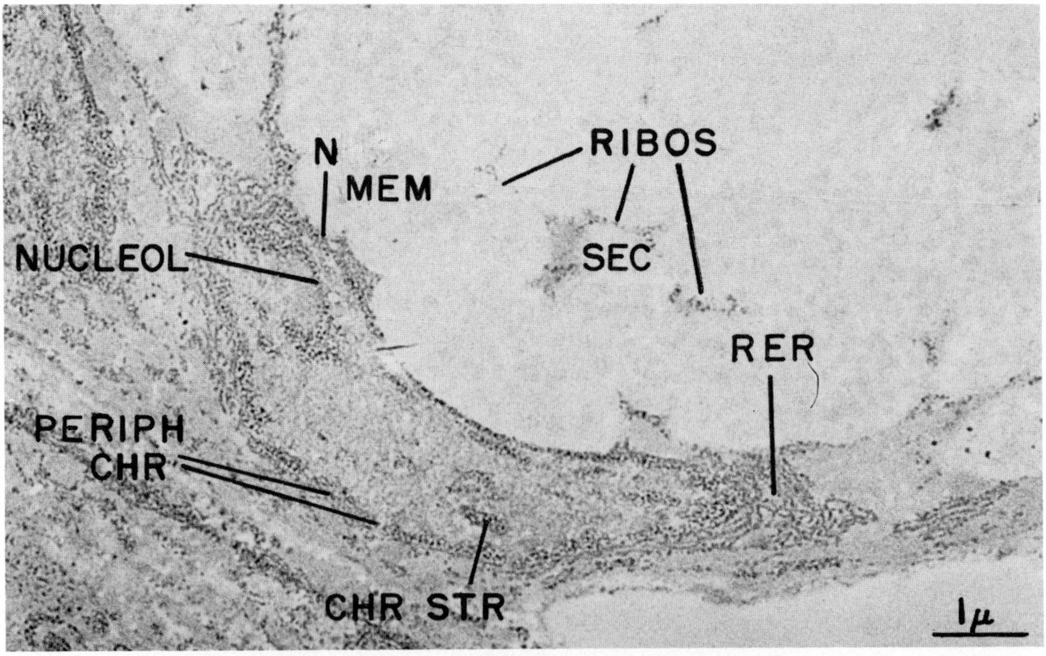

FIG. 8.4. Mature goblet cells of intestinal gland of colon. The peripheral chromatin (PERIPH CHR) covers nearly the whole internal surface of the nucleus. The chromatin strands (CHR STR) are few in number and coarse. Nucleoli (NUCLEOL) are close to the nuclear membrane (N MEM) of the thinned nucleus. There are few gaps in the membrane. A rough endoplasmic reticulum (RER) is present at the base of the cell near the nucleus. Ribosomes (RIBOS) also occur between the unstained mucigen granules (SEC), but they are very scarce.

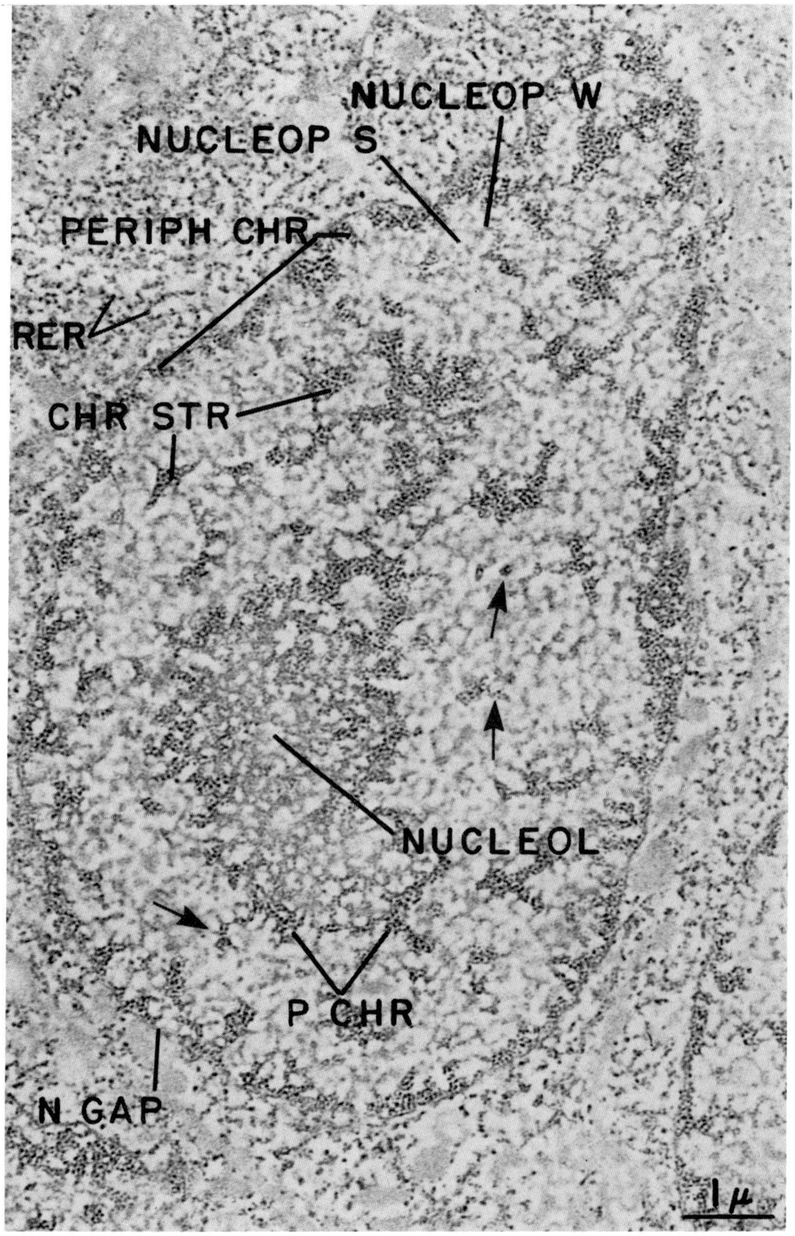

FIG. 8.5. Parotid gland cells. The peripheral chromatin (PERIPH CHR) is of variable thickness, and is joined with many chromatin strands (CHR STR). Some of these are also connected with the perinucleolar chromatin (P CHR) which nearly encloses the nucleolus (NUCLEOL). Some very thin strands are marked with arrows. The pseudovacuolar system of the nucleoplasm is particularly clear. It consists of denser walls (NUCLEOP W) and less dense spaces (NUCLEOP S). The chromatin strands course in the nucleoplasmic walls, which becomes thicker when they accommodate the thicker strands. The nuclear gaps (N GAP) are small and scarce. RER in the cytoplasm is prominent.

In order to make possibly valid comparisons of the DNA molecular patterns in different cell types, it was deemed necessary to accumulate fifty electron micrographs of nuclei of each cell type at approximately their maximum diameter. The only practical way to collect such large numbers was to take electron micrographs at a low magnification and enlarge selected nuclei, the diameter being the only criterion in the selection. It was assumed that all other factors would be kept con-

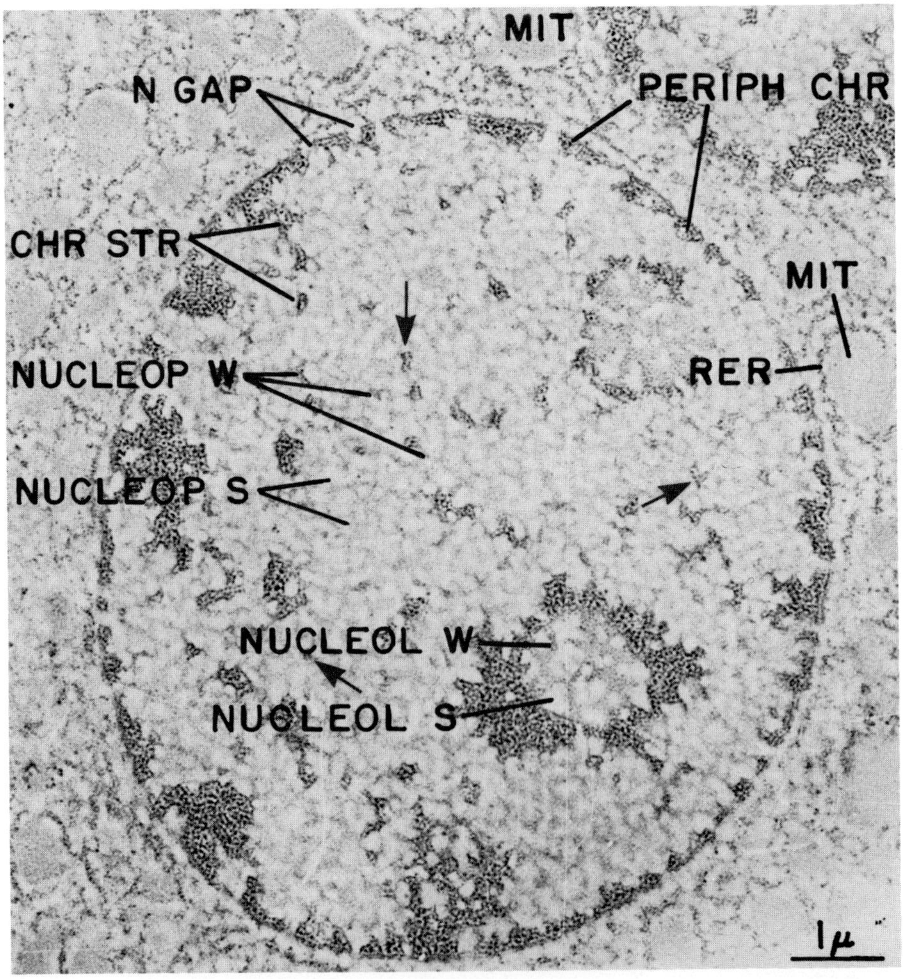

Fig. 8.6. Hepatic cell. The peripheral chromatin (PERIPH CHR) is in the form of small aggregates some of which are connected to chromatin strands (CHR STR). These occur separately in the walls of the nucleoplasm (NUCLEOP W), which may contain one or two second-order DNA coils (arrows), separated by nucleoplasmic spaces (NUCLEOP S). The nuclear gaps (N GAP) are numerous and rather large. The nucleolar walls (NUCLEOL W) and spaces (NUCLEOL S) are very clear. RER is especially to be seen around mitochondria, but is present elsewhere.

stant as much as possible, i.e., the age, state of hydration and of activity of the animals, the process of freezing, drying, postfixation, staining, embedding, and sectioning. It is not practical to enlarge the electron micrographs sufficiently to show single DNA molecules, as was done for some cells in Chapters 3, 4, and 7 (this volume) and Chapter 5 (Vol. II). The enlargement is, however, adequate to distinquish DNA regions sufficiently for tracing and the extraction of quantitative data from such tracings. The end result of this process is a set of electron micrographs, through approximately their maximum diameter, showing the distribution of DNA in all conceivable planes through the nuclei. Precautions were taken to avoid duplication of nuclei. It was (and is) hoped that eventually ways would be found to reconstruct from these electron micrographs the actual distribution of

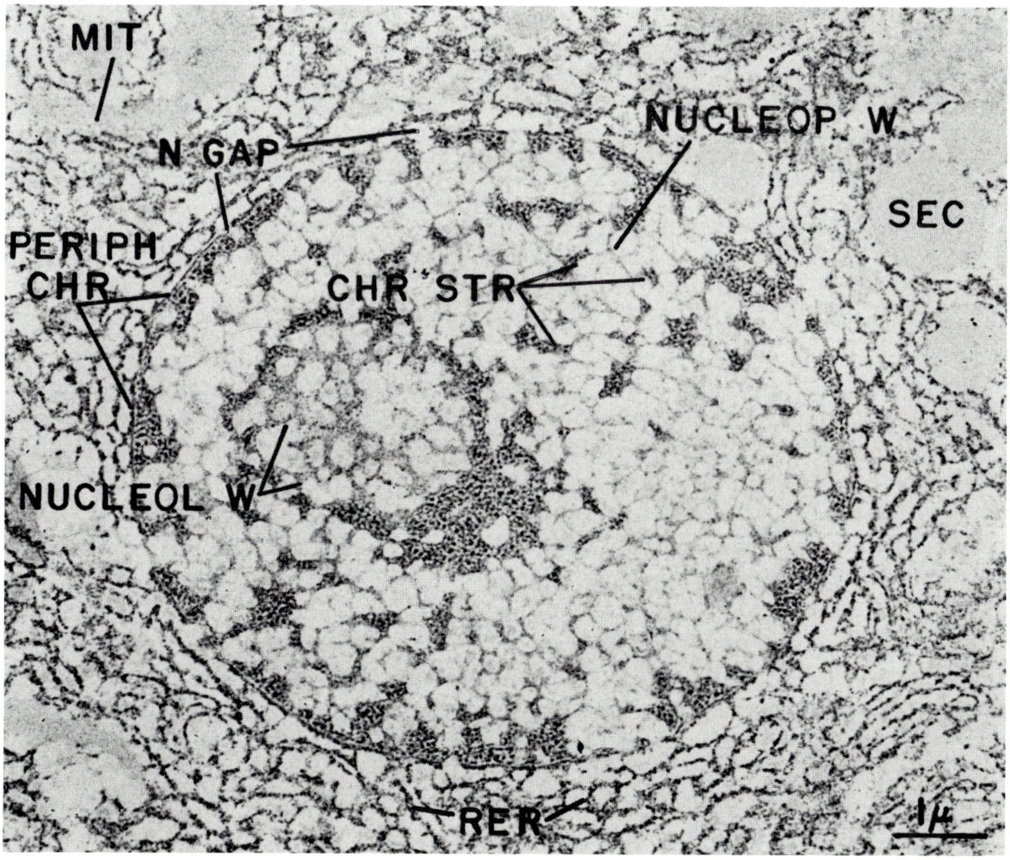

FIG. 8.7. Exocrine cell of the pancreas. The peripheral chromatin (PERIPH CHR) is coarser than in the liver, and the aggregates are more widely separated by the nuclear gaps (N GAP) than in the hepatic cell. The chromatin strands (CHR STR) are disposed in the walls of the nucleoplasm (NUCLEOP W). Nucleoplasmic walls are continuous with the nucleolar walls (NUCLEOL W). The RER is extensive. Mitochondria (MIT) and secretory granules (SEC) are not stained.

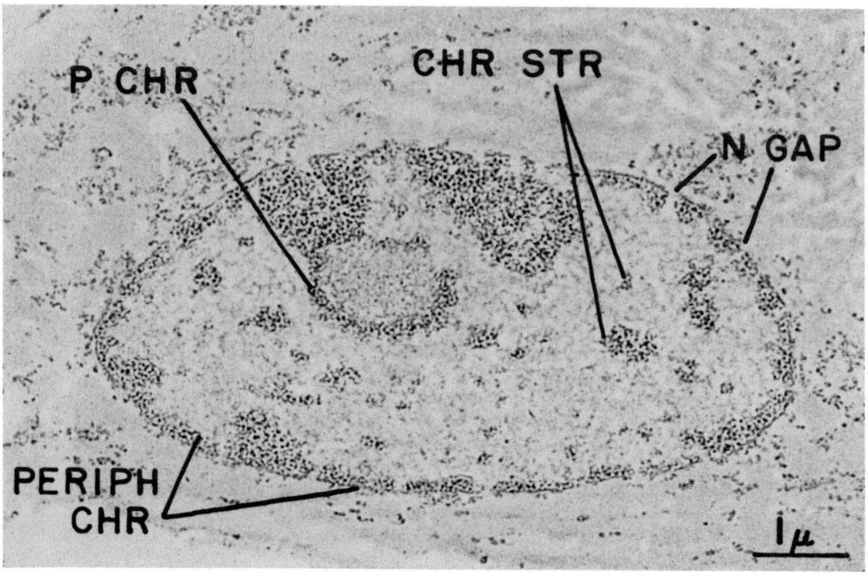

FIG. 8.8. Proximal convolution of the kidney. There are few narrow gaps (N GAP) in the relatively uniform layer of the peripheral chromatin (PERIPH CHR). The chromatin strands (CHR STR) vary greatly in size. The thick connection of perinucleolar (P CHR) and peripheral chromatin is common.

DNA in each type of nucleus and to quantify some of the observations which could be significant in gene action. Pending the development of a computer program adequate for the analysis of the DNA patterns of nuclei, the electron micrographs will be used qualitatively, as in this chapter.

Several ultrathin sections were cut, and selected cells in one section were photographed on Gevaert Scientia plates at a magnification of × 2000 with a Siemens EMU-1 electron microscope operating at an accelerating voltage of 60 kv. The nuclei were enlarged × 6 in printing. All electron micrographs are printed at × 12,000, except for the spinal ganglion cell which is printed at × 8000. This enlargement was just adequate to obtain the quantitative information as enumerated below. Following a series of ultrathin sections, a 4-μ section was cut and mounted on a slide for comparative study with the light microscope. Thick sections (usually 15–30) were then cut so that a new set of nuclei at their maximum diameter appeared, and another group of ultrathin sections was cut for study with the electron microscope, only one of which was photographed. This process was repeated until about fifty electron micrographs were accumulated of different nuclei sectioned at or near their maximum diameter. These will be the basis for the eventual three-dimensional reconstruction of the distribution of the second-order DNA coils in the nucleus by a computer technique yet to be worked out.

Trials have been made of the possibility of collecting significant quantitative information from the electron micrographs. It seems feasible to quantitate certain data which may have a bearing on the genetic activity of somatic cells. The following nuclear details are measurable from electron micrographs:

1. Area of nucleus and nucleoli.
2. Total area of chromatin, i.e., the area of the nucleus occupied by second-order DNA coils. This includes the peripheral chromatin, which is attached to the nuclear membrane, the perinucleolar chromatin, and the "unattached" or intermediate chromatin (or chromatin strands) which intervenes between these.
3. Length of the perimeter of all chromatin surfaces exposed to nucleoplasm. This is an indicator of the degree of dispersion of DNA, and of the exposure of DNA, to transfer of substances between it and the nucleoplasm. Surfaces thus exposed could conceivably be genetically active.

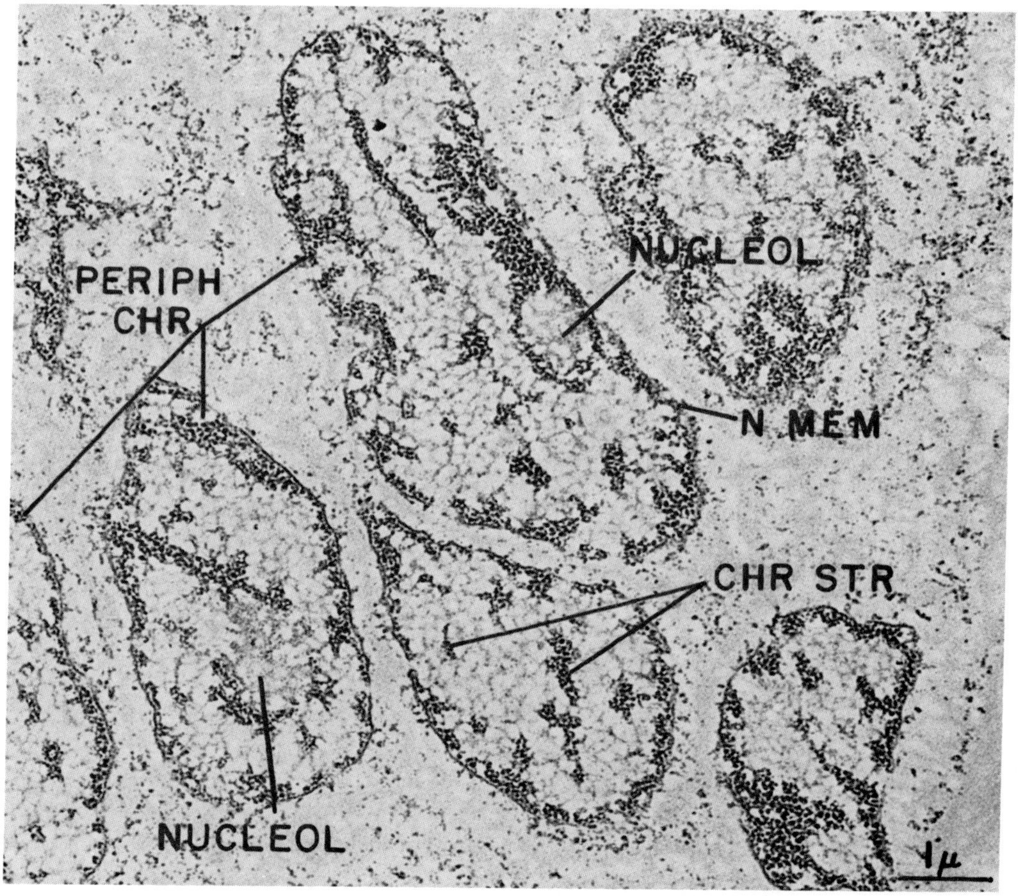

FIG. 8.9. Epithelium of the ventral abdominal skin near the pubis. The epithelium is one or two cells thick, and the nuclei are rather similar. They are somewhat irregular, and the thickness of the peripheral chromatin (PERIPH CHR) varies markedly. Sometimes the nucleolus (NUCLEOL) is close to the nuclear membrane (NUC MEM). The chromatin strands (CHR STR) are also variable. Nuclear gaps are irregularly scattered.

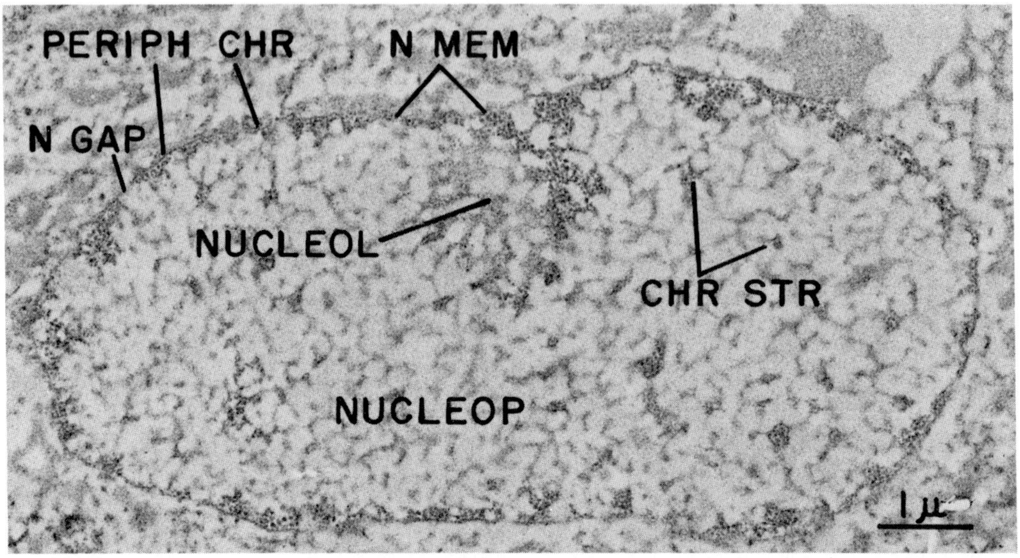

FIG. 8.10. Epithelium of the urinary bladder. The peripheral chromatin layer (PERIPH CHR) is very thin and interrupted by many nuclear gaps (N GAP). The chromatin strands (CHR STR) course through the walls of the nucleoplasm (NUCLEOP) as mainly very fine strands. The nucleolus (NUCLEOL) is attached to the nuclear membrane (N MEM).

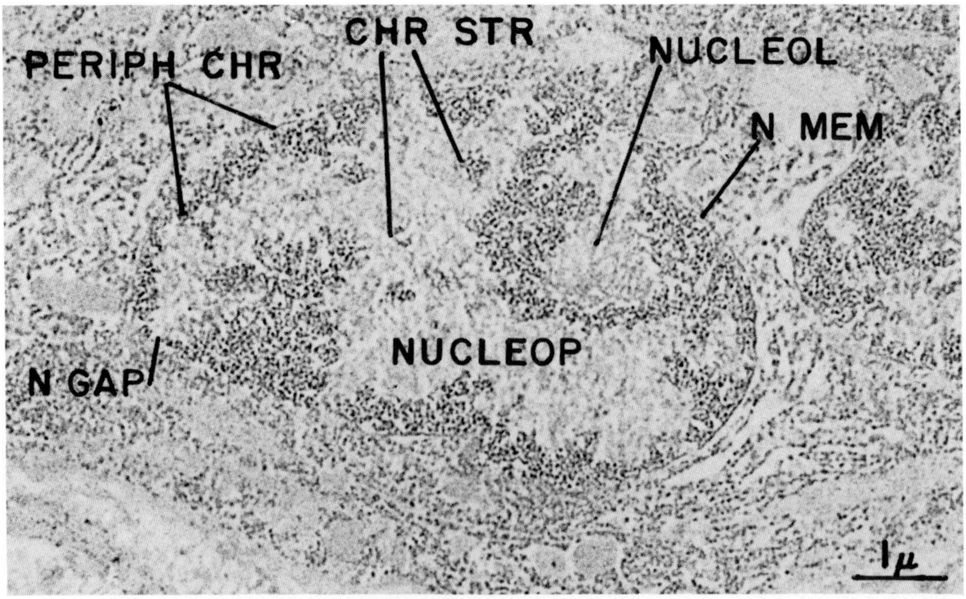

FIG. 8.11. Epithelium of the seminal vesicle. The peripheral chromatin (PERIPH CHR) is made up of large, thick aggregates of second-order coils with few nuclear gaps (N GAP) of variable size. Some coarse and some very fine chromatin strands (CHR STR) extend through the nucleoplasm (NUCLEOP). The nucleolus (NUCLEOL) is commonly attached by a broad base of DNA coils to the nuclear membrane (N MEM).

4. Number and length of loops of isolated second-order DNA coils and length of exposed surfaces of minimally aggregated DNA coils. This is also a measure of potentally genetically active sites.

5. Number and length of gaps in the nuclear membrane, in relation to the total length of the nuclear membrane in sections. A gap in the nuclear membrane is defined as a portion of the nuclear membrane free of attached DNA aggregates. The relation to nuclear "pores" is undetermined.

These measurements are essentially indications of accessibility of DNA for the transfer of metabolites and final products of the activity of DNA between it and the nucleoplasm and then to the cytoplasm. The assumption is that when DNA is actively making RNA, it will do so more efficientlyy when the DNA coils are separated from each other and exposed more fully to the nucleoplasm, and will do so less efficiently or not at all when the DNA coils are tightly packed in aggregations recognizable in the light microscope as chromatin masses. The potential activity of DNA is considered to be reflected in terms of the degree of packing of the second-order DNA coils. Other possible factors, such as repressors, derepressors,

TABLE 8.1
LIST OF MOUSE CELL TYPES WHOSE NUCLEAR DNA MOLECULAR PATTERN WAS STUDIED WITH LIGHT AND ELECTRON MICROSCOPES[a]

Epithelium	Epithelium (continued)
Duodenum	Vagina*
Columnar cells	Uterus
Stomach	Thyroid
Chief cells	Adrenal medulla
Parietal cells	
Colon	Nerve tissue
Columnar cells	Spinal ganglion
Goblet cells	Nerve cells
Parotid	Glial cells*
Serous gland cells	
Liver	Muscle tissue
Hepatic cells	Skeletal muscle
Pancreas	Cardiac muscle
Zymogen cells	Smooth muscle cells of urinary bladder, colon, and uterus
Kidney	
Cells of proximal convolution	
Skin	Connective tissues
Ventral abdominal wall	Polymorphonuclear neutrophils of the circulating blood and bone marrow*
Urinary bladder	
Seminal vesicle	Fibroblast of dermis, subcutaneous fascia
Dorsal prostate	Lymphocytes of circulating blood, lymph node, and thymus
Ventral prostate	
Epididymis	Red blood cell series of the bone marrow*
Head	Fat cells*
Tail	

[a] Cell types marked with an asterisk are not included in this chapter. Except when stated, the mice were 3 months old, and deprived of food and water for 18 hours prior to their death.

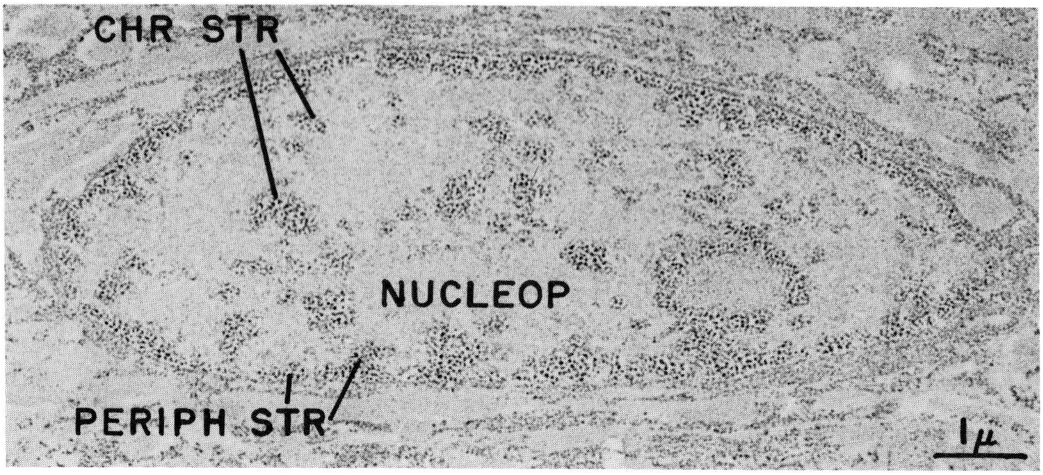

FIG. 8.12. Epithelium of the dorsal prostate. The peripheral chromatin (PERIPH STR) appears irregular in thickness, probably because of the numerous thick strands (CHR STR) which originate from it and lie in the nucleoplasm (NUCLEOP).

operons, and controlling elements are not considered at this time. The cell nuclei studied are presented in Table 8.1.

The following notes pertain to the collection and preparation of the specimens. Epithelia of the duodenum, stomach, and colon were prepared by making a longi-

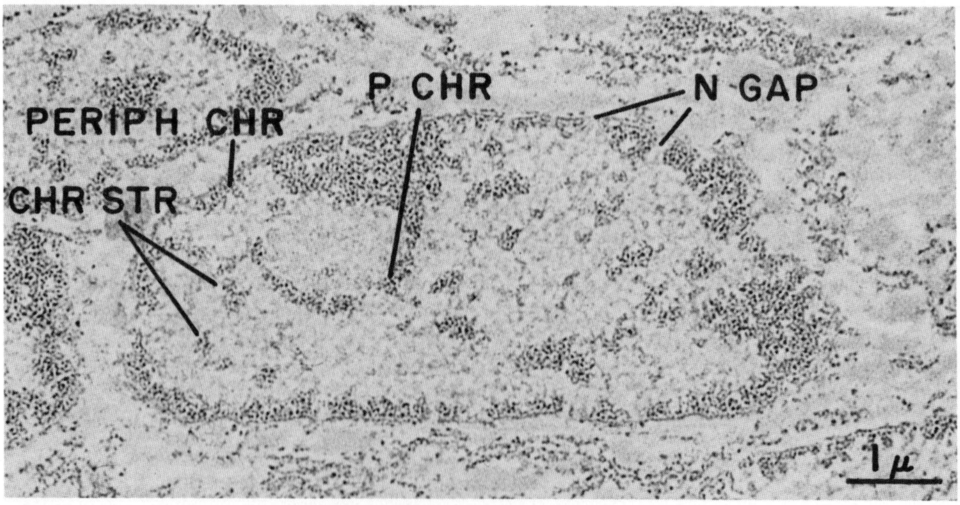

FIG. 8.13. Epithelium of the ventral prostate. The peripheral chromatin (PERIPH CHR) is more uniform than in Fig. 8.12, and the chromatin strands (CHR STR) are fewer and thinner. Also, nuclear gaps (N GAP) are larger and more numerous. The perinucleolar chromatin (P CHR) is very closely attached to the peripheral chromatin but is more interrupted over the nucleolus.

tudinal incision through the viscera. When the intestinal surface was everted, a small piece was cut with a fine scissor, and this was subsequently sliced with a fine razor blade in a moist box. The proximal convoluted tubules were obtained from a slice of the kidney separating the cortex corticis. This was further sliced in a moist box. The skin was removed, placed on its epithelial surface in the moist box, and sliced thin. The vagina was removed, stretched flat in a moist box, and cut into small squares with a fine razor blade. The uterine horns were removed, slit lengthwise, and flattened. Small squares were cut from this, transferred to foil, stretched, and frozen. The thyroid tissue was exposed with the animal in a moist box, and both lobes were removed. The isthmus was freed from its attachments, transferred to foil, and frozen. The adrenal medulla was separated by quickly slicing away most of the cortex. The medulla was subsequently sliced. The spinal ganglia were exposed in a moist box, and sliced further prior to freezing. Skeletal muscle was removed from the quadriceps femoris, sliced further, and frozen. Cardiac muscle was from the left ventricle. A thin slice was placed on its flat side and sliced further. Smooth muscle was prepared separately only from the colon, from which a longitudinal strip was removed, and further sliced for freez-

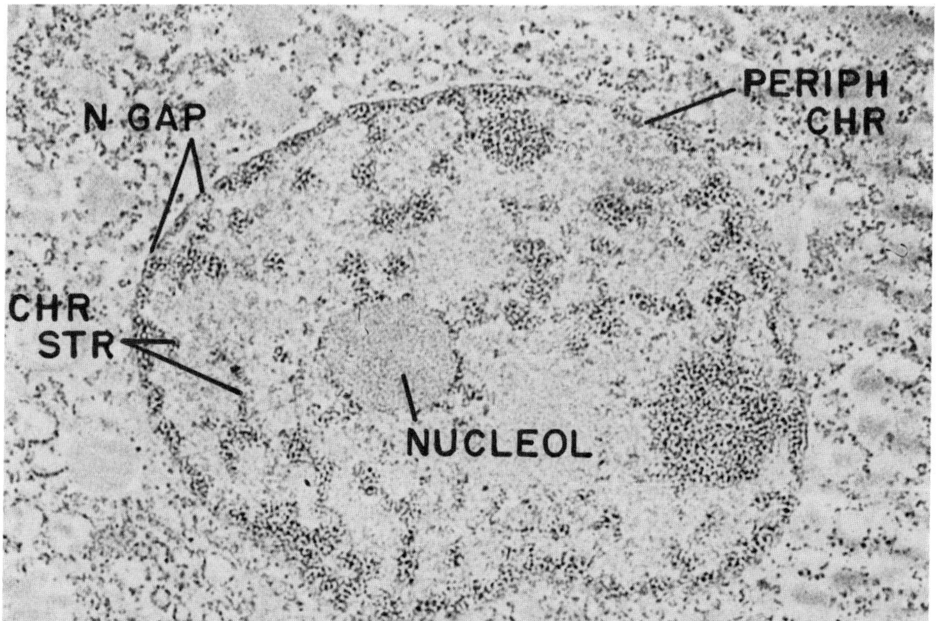

FIG. 8.14. Epithelium of the head of the epididymis. The peripheral chromatin (PERIPH CHR) is more or less uniformly thin, and seems to vary only as the thick chromatin strands (CHR STR) attach to it at frequent points. There are few very small nuclear gaps (N GAP). The nucleolus (NUCLEOL) is attached to the strands.

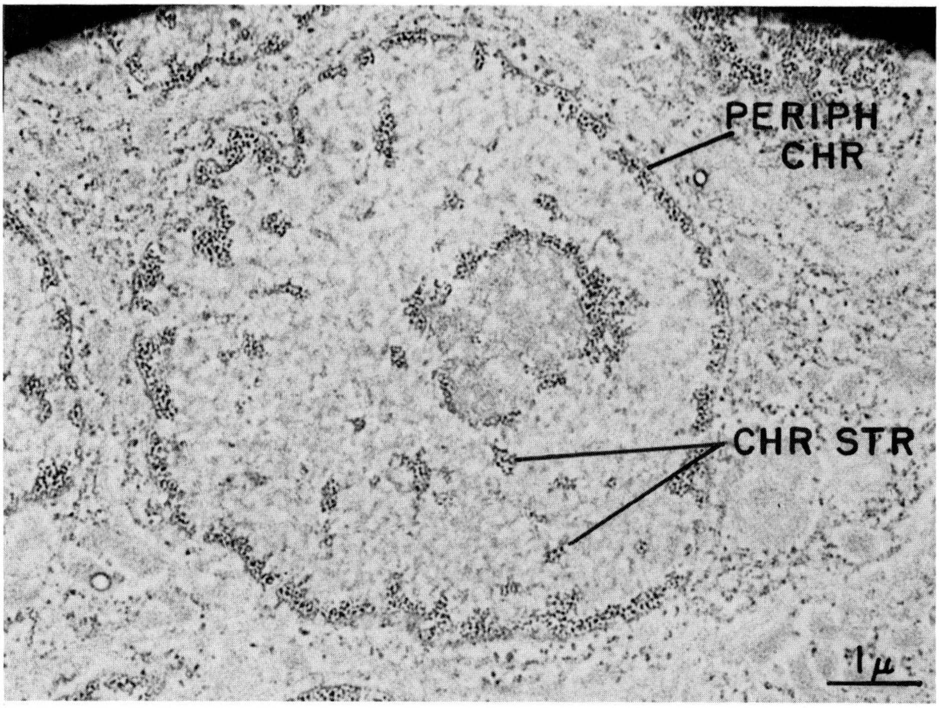

Fig. 8.15. Epithelium of the tail of the epididymis. The peripheral chromatin (PERIPH CHR) forms a very thin layer; parts of it are separated by more numerous, wider gaps than in Fig. 8.14. There are fewer strands (CHR STR), which are relatively uniform.

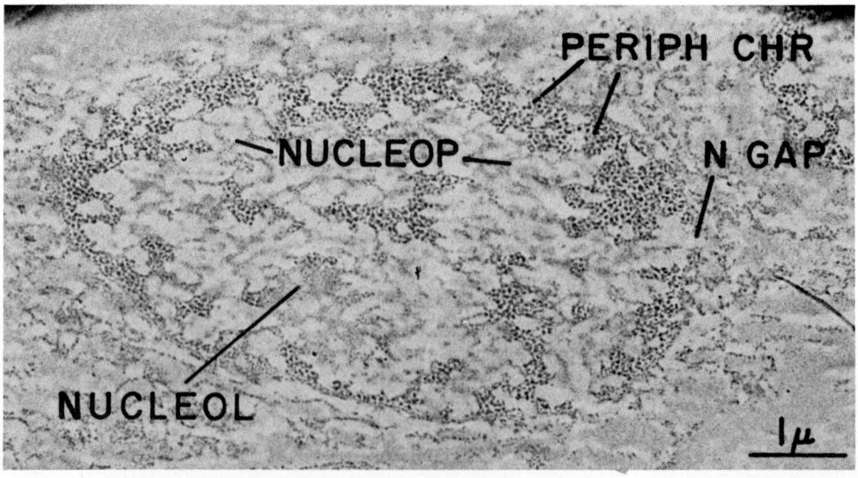

Fig. 8.16. Epithelium of thyroid follicle. The second-order DNA coils of the peripheral chromatin (PERIPH CHR) are not delimited, but extend as a wide-meshed net in the nucleoplasm (NUCLEOP), with few nuclear gaps (N GAP). Perinucleolar chromatin is very restricted around the nucleolus (NUCLEOL).

ing. Polymorphonuclear neutrophils and lymphocytes of the circulating blood were prepared from the buffy coat by the method described by Shelley *(11)*. Lymphocytes were obtained also from the popliteal lymph node by excising the gland, slicing off pieces of cortex, and slicing these further. Bone marrow was obtained from the tibia, which was removed and sliced with a razor blade. Some bone marrow was from normal female mice. Others were injected with turpentine (0.5 ml subcutaneously, on the dorsum, once daily for 2 days, with bone marrow removed 3 days after the second injection), or with phenylhydrazine (3 mg phenylhydrazine hydrochloride, subcutaneously, once daily for 4 days, with bone marrow removed the next day). Blood smears of turpentine-treated mice contained numerous large polymorphonuclear leukocytes with pale nuclei and basophilic

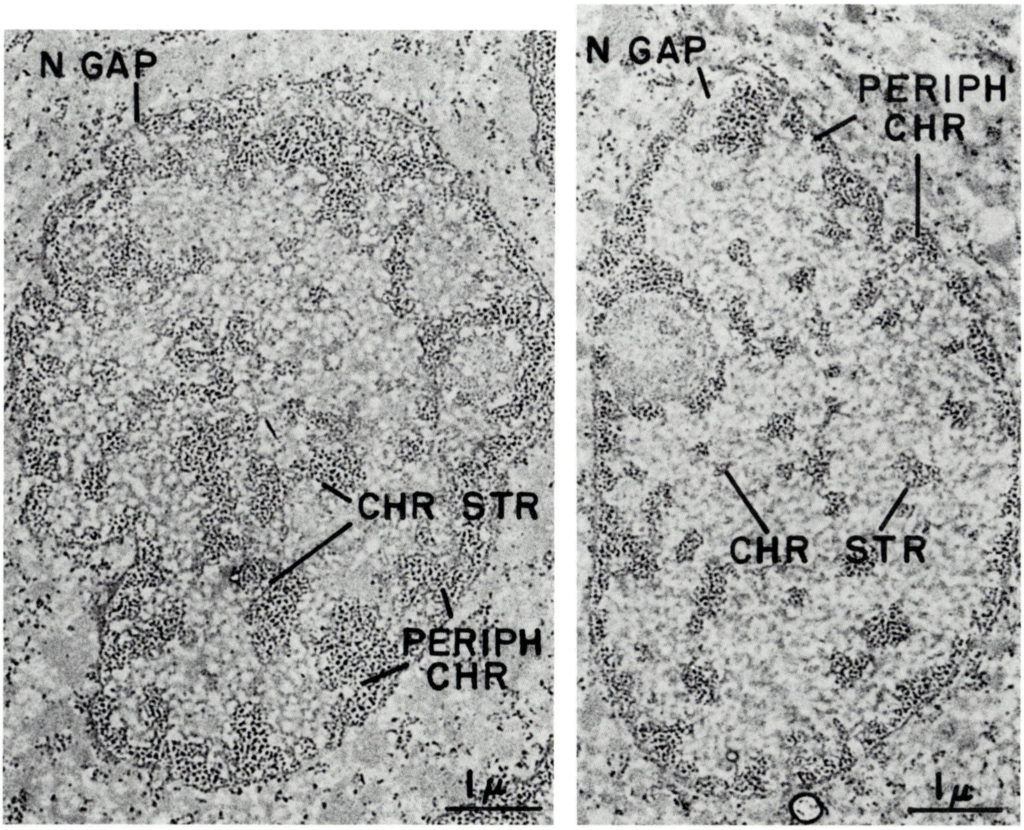

FIGS. 8.17 (LEFT) AND 8.18 (RIGHT). Gland cells of the adrenal medulla. There are at least two kinds of nuclei which may correspond to those described in the literature *(2, 7)* or may be in different secretory stages. In one (Fig. 8.17), the peripheral chromatin (PERIPH CHR) is irregularly thick, with some narrow nuclear gaps (N GAP), and the chromatin strands (CHR STR) are coarse and thick. In the other (Fig. 8.18) the peripheral chromatin is thinner and more uniform, and the chromatin strands are also finer.

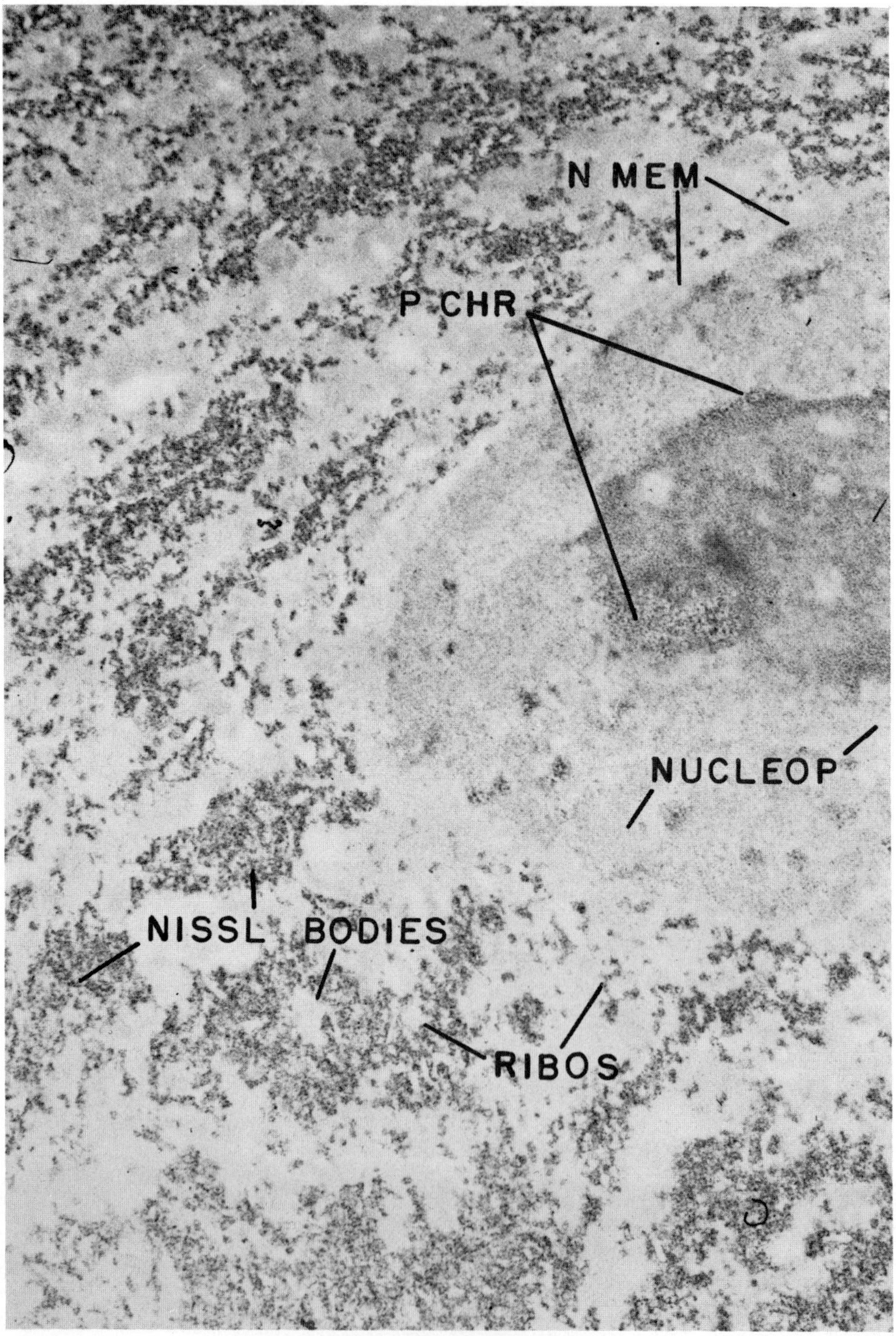

FIG. 8.19. Spinal ganglion nerve cell. Most of the nuclear membrane (N MEM) is free of attached chromatin. The peripheral chromatin (PERIPH CHR) consists of a few widely separated small clusters of secondary DNA coils. Chromatin clumps are attached to the periphery

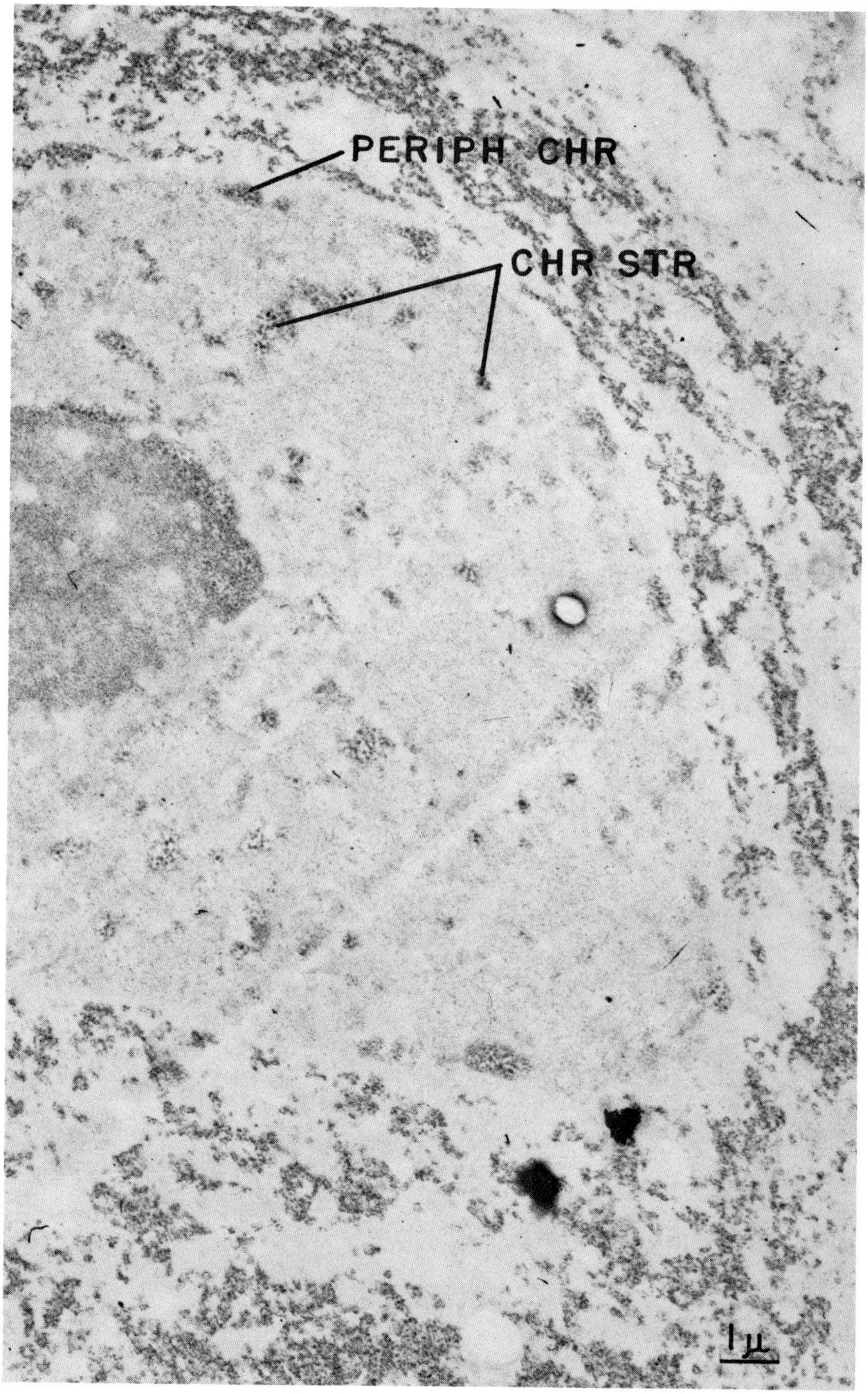

of the enlarged nucleolus (P CHROM). Fine chromatin strands (CHR STR) are cut in various planes in the nucleoplasm (NUCLEOP) and are widely separated. Ribosomes (RIBOS) are clustered in groups which are sections through Nissl bodies. × 8000.

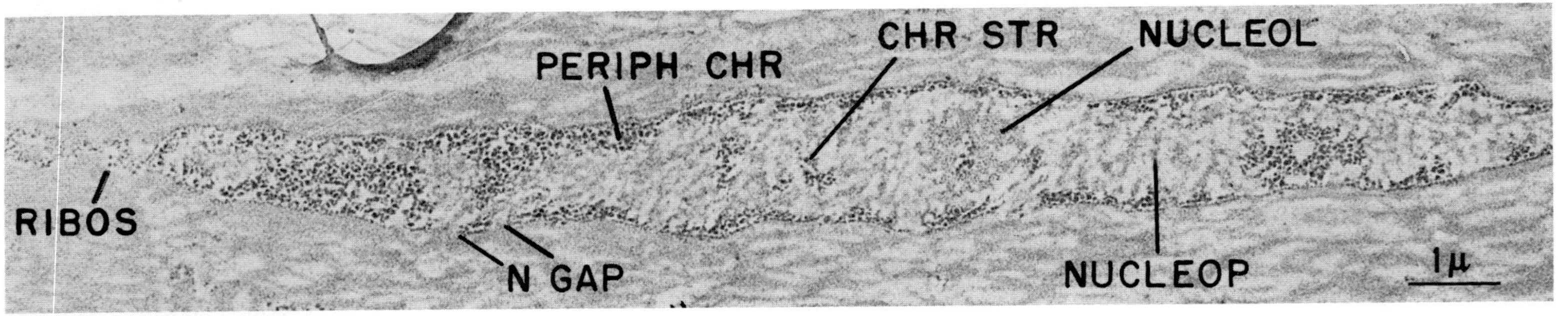

Fig. 8.20. Skeletal muscle. The peripheral chromatin (PERIPH CHR) is a thin layer of second-order coils irregularly interrupted by large and small nuclear gaps (N GAP). There are occasional thickenings, which project into the nucleoplasm (NUCLEOP), and some strands (CHR STR). The nucleolus (NUCLEOL) contains some DNA strands. Ribosomes (RIBOS) are concentrated at the nuclear poles.

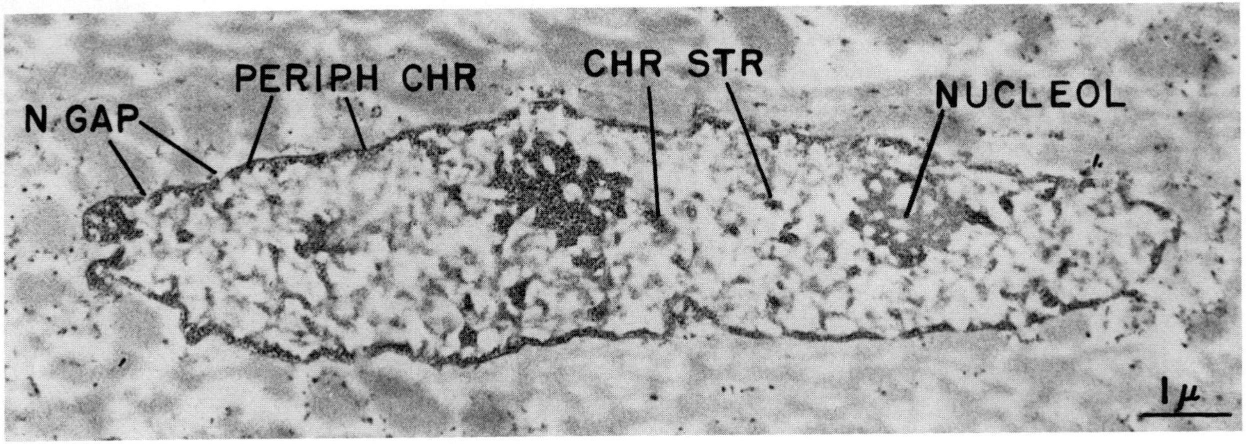

Fig. 8.21. Cardiac muscle. The layer of peripheral chromatin (PERIPH CHR) is thin here also, as in Fig. 8.20, and is frequently interrupted by small and large nuclear gaps (N GAP). There are numerous fine chromatin strands (CHR STR) as compared with skeletal muscle. The nucleolus (NUCLEOL) is partly enclosed by small aggregates of second-order DNA coils.

cytoplasm and very few, usually small, granules. Blood smears of phenylhydrazine-treated mice had a very high proportion of large, basophilic, nonnucleated erythrocytes. Organs and tissues not mentioned in this paragraph, but listed in Table 8.1 were prepared from specimens removed from the animal to a moist box where they were finely sliced with a new, thin, double-edged razor blade. All mice were bled before specimens were removed.

General Observations

The presentation of several thousand electron micrographs for descriptive comparison is extremely difficult. Out of many electron micrographs of sections of any single cell type, I have selected one which seemed to be typical of that cell type. When comparing such typical electron micrographs, it is readily apparent that the DNA molecular pattern of some nuclei is quite different from that of other nuclei, as, for example, hepatic cells, spinal ganglion nerve cells, and skeletal muscle. At first glance, however, the DNA molecular patterns of some nuclei seem to be similar, as, for example, columnar epithelium of the duodenum and chief cells of the fundic glands. The DNA molecular patterns are, however, readily identified by comparison of specific items: the thickness and uniformity of the peripheral chromatin, the number and thickness of the chromatin strands, the number, length, and regularity of the nuclear gaps, etc. Even the DNA molecular patterns of nuclei of epithelial cells as closely related as dorsal and ventral prostate, and the head and tail of the epididymis can, be distinguished from each other, and from the DNA molecular patterns of all other cell types. The impression is that the aggregation pattern of the second-order DNA helices of the nucleus of each cell type differs, sometimes subtly, from that of nuclei of all other cell types. The final proof awaits a three-dimensional reconstruction of the DNA molecular pattern with the quantitative data to be derived, hopefully, from electron micrographs already in hand.

As stated previously, the DNA double helix appears in electron micrographs after suitable fixation and staining as filaments about 45 A in diameter. This is the first-order helix. It is arranged as a second-order helix with, generally, a diamater and period of about 400 A. In electron micrographs the gyres of this helix are recognized as mainly circular or oval figures, whose specific morphology depends on the relation of the plane of the second-order coil to the plane of the section. Third- and fourth-order coils will be described in connection with chromosomes (Chapter 13, this volume). In somatic nondividing cells, the DNA occurs only as second-order helices, the sole exceptions noted thus far being in the nucleus of nerve cells at the height of the chromatolytic reaction (this volume, Chapter 12) and in interphase of the syncytial cells of the developing egg of *Drosophila* (this volume, Chapter 13). The relation of the second-order helices in the chromatin of somatic nuclei to chromasomes is unknown, except for the X chromosome or Barr body (Chapter 7, this volume). In other words, it has not been possible to

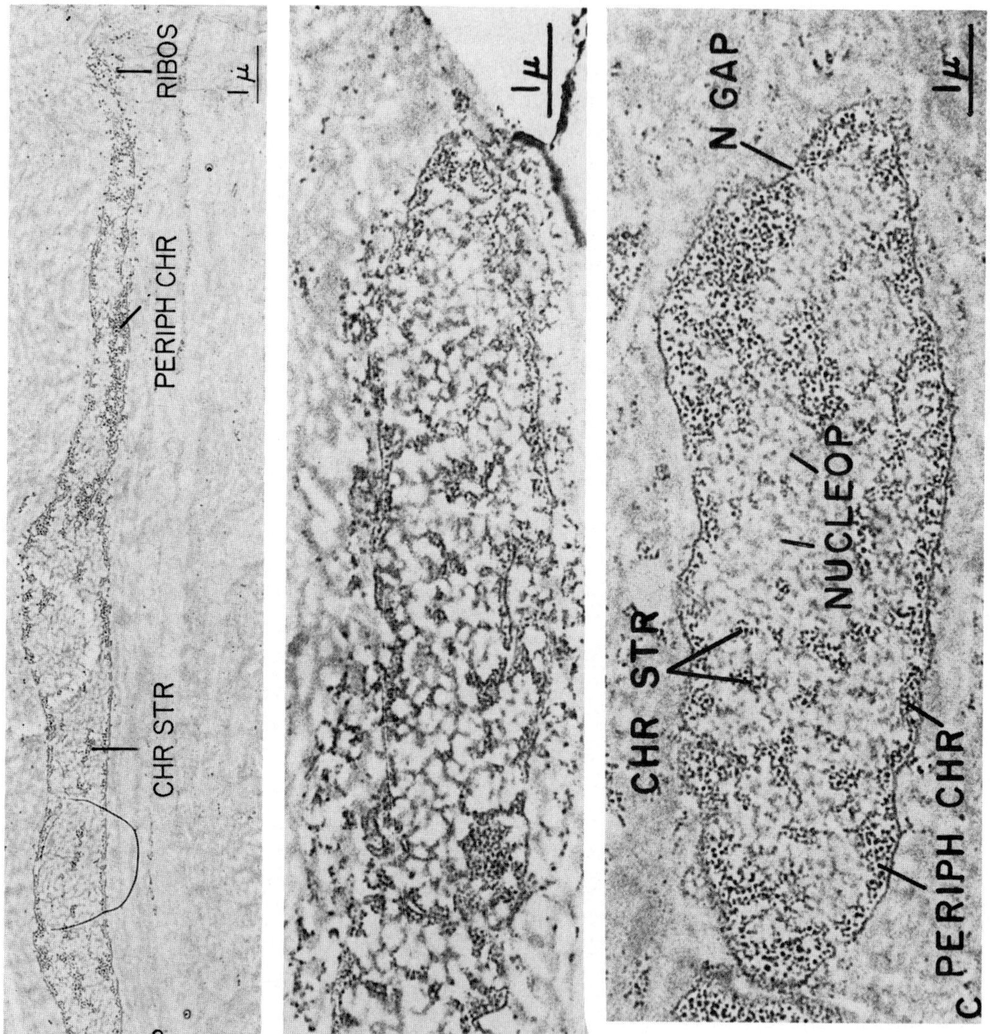
A
RIBOS
N GAP
CHR STR
PERIPH CHR
RIBOS
1μ
B
N MEM
1μ
C
PERIPH CHR
CHR STR
NUCLEOP
N GAP
1μ

identify any group of second-order coils with any specific chromosome, with the single exception just noted.

Most of the chromatin (which is to say, most of the second-order DNA coils) is attached to the nuclear membrane and is referred to as peripheral chromatin. Some coils are attached to the nucleolus, when this is present, and a variable but usually small amount of chromatin is dispersed in the nucleoplasm between these two sites. In the peripheral chromatin, the second-order DNA coils may be aggregated as submicroscopic or microscopic clumps, which are separated by gaps whose dimensions vary in different cell types, or the aggregated DNA coils may line the nuclear membrane continuously, with no notable gaps. In many cell types, a DNA clump at the periphery extends into the nucleoplasm and encloses one or several nucleoli; such clumps are said to contain the genes responsible for the formation of the nucleolus. Or there may be several such "nucleolar organizing clumps" of DNA, each with one or more nucleoli. In addition, in most cell types, small (mostly submicroscopic) aggregates arise from the peripheral chromatin and course in the nucleoplasm to reattach to the perinucleolar chromatin clumps or to another part of the peripheral chromatin. Sometimes these aggregates comprise only two second-order DNA helices. It is also possible that these "unattached" helical DNA coils arise in the peripheral chromatin and form one or several loops in the nucleoplasm to return close to the point of origin in the peripheral chromatin.

In some chromatin clumps, there appear to be blank areas or "holes" free of DNA coils. These have been interpreted as sections through surface irregularities of the clumps, and as comprising nucleoplasmic components.

In all instances, the nucleoplasm consists of a spongelike structure with a denser component which may contain RNA-rich granules and the second-order DNA coils. In all nuclei in which they occur, nucleoli contain granules rich in RNA. The nuclear membrane is never selectively stained for nucleic acids by the method used. Its presence is indicated in the material presented in this report by the occurrence of DNA coils on the inner surface, the occurrence of cytoplasmic structures on the outer surface (such as ribosomes, RER, or mitochondria), or by differences in the pattern of pseudovacuoles on both sides. The nuclear membrane is similarly difficult to demarcate when the cells are stained for protein ex-

---

FIGS. 8.22(A–C). Smooth muscle cell of: (A) Colon. The second-order DNA coils are one or two coils thick in the peripheral chromatin (PERIPH CHR) with very numerous, fine nuclear gaps (N GAP). Chromatin strands (CHR STR) are nearly all submicroscopic. Ribosomes (RIBOS) are accumulated near the nuclear poles. Because the nucleus is too long to fit on the printed page in the usual way, the original figure was reduced as compared with other related nuclei. (B) Urinary bladder. The distribution of second-order DNA coils is similar to that shown in Fig. 8.22A. Whatever differences may appear can be attributed to the more irregular outline of the nuclear membrane (N MEM) brought about by greater contraction of the muscle fiber prior to freezing. (C) Uterus. The peripheral layer of chromatin (PERIPH CHR) is thicker and the nuclear gaps (N GAP) are very scarce and fine. Numerous coarser chromatin strands (CHR STR) run in the nucleoplasm (NUCLEOP).

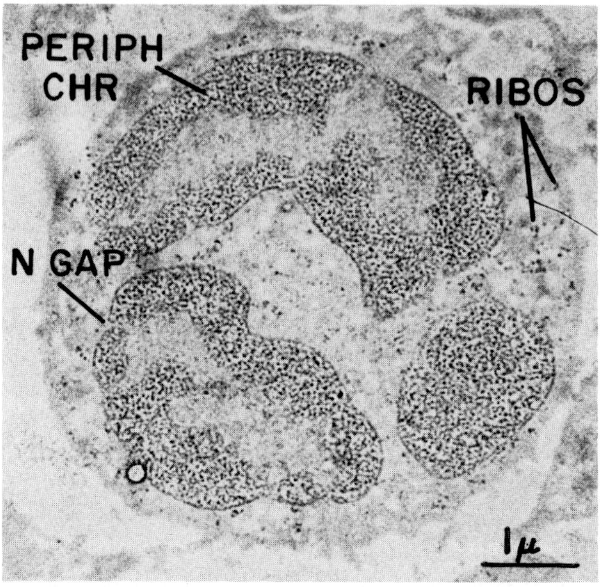

Fig. 8.23. Polymorphonuclear neutrophil. The DNA is restricted to a thick peripheral layer of chromatin (PERIPH CHR), with very few (usually small) gaps in the nuclear membrane (N GAP). There are few ribosomes (RIBOS) between the neutrophilic granules.

cept for salivary gland nuclei of *Drosophila* (Chapter 5, this volume). The nuclear membrane has been clearly demarcated in freeze-dried specimens only when its lipid components are suitably postfixed and stained or after the use of certain vapor protein reagents. (Chapter 1, Vol. II).

This general description applies only to the nucleus of the mature, unstimulated cell. During and after stimulation and during development, the pattern of organization of second-order DNA coils may be markedly different, as shown in the next two chapters.

Specific descriptions of individual cell nuclei are found in the figure legends.

Discussion

In general, nuclear structure conforms to a single pattern. The nucleoplasm is spongelike in structure, with continuous branching walls separated by less dense spaces. The walls contain RNA granules and some less readily resolvable RNA components (sRNA), and the second-order DNA coils of rather uniform dimensions—a repeat unit and diameter of about 400 A. These are most commonly aggregated in local regions where the nucleoplasmic walls become thicker and eventually obliterate the nucleoplasmic spaces. Some of these aggregated DNA coils are always found as a layer on the periphery of the nucleus, attached to the nuclear membrane. The thickness of this layer of peripheral chromatin varies in different cell types. The peripheral chromatin is nearly always separated into smaller

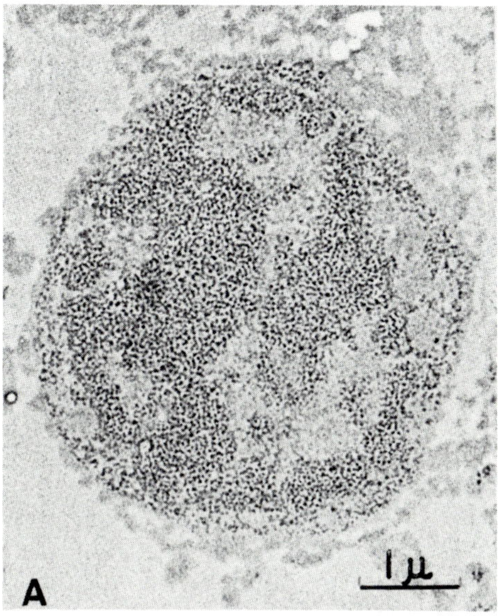

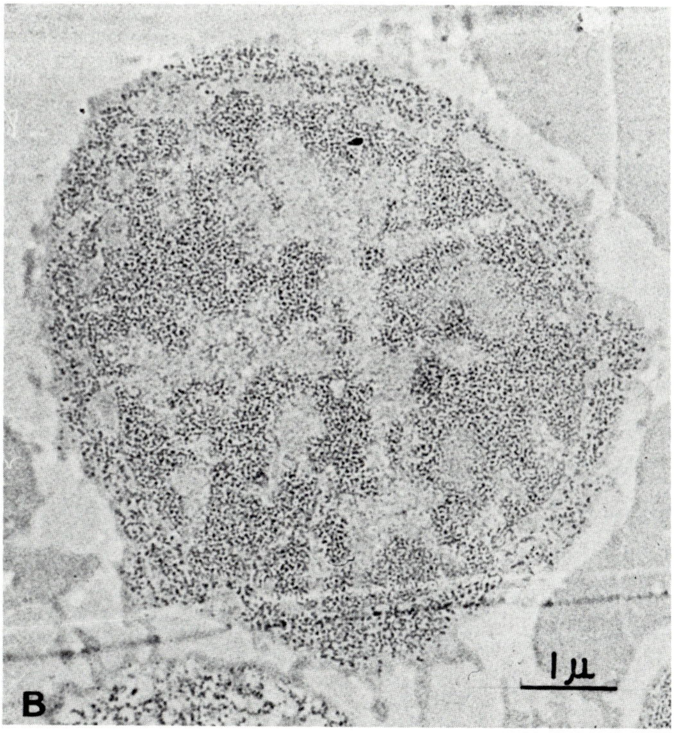

FIGS. 8.24(A–F). Lymphocytes in circulating blood (A and B). The chromatin consists of an irregularly thick layer of second-order DNA coils, which bulges centrally into the nucleus, leaving little space for the nucleoplasm. There are few nuclear gaps. (C and D) Lymph node. The chromatin clumps are not as thick as in (A and B); the surface of the chromatin aggregates is more irregular where they are exposed to the nucleoplasm, accordingly, they enclose numerous spaces or "holes," which really contain nucleoplasm. Nuclear gaps are more numerous than in (A and B). (E and F) Thymus. The chromatin appears to be intermediate between the cells of the circulating blood (A and B) and of the lymph node (C and D).

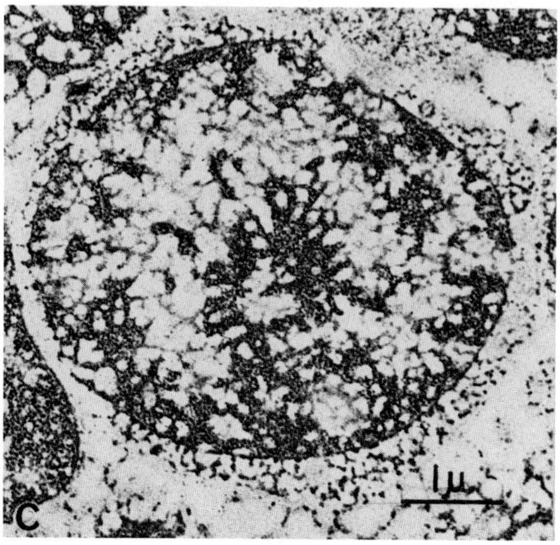

FIG. 8.24C.

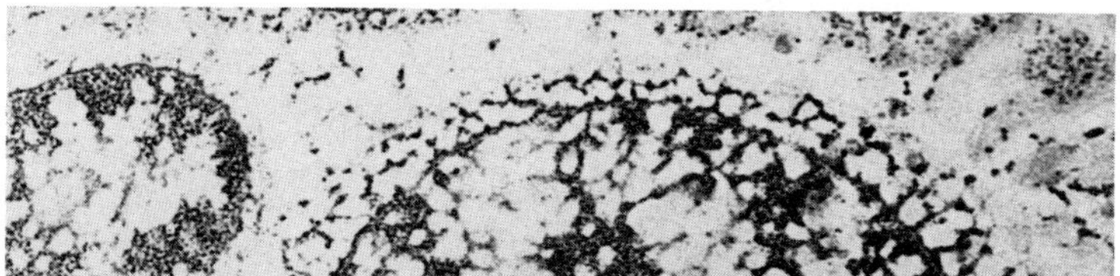

FIG. 8.24D.

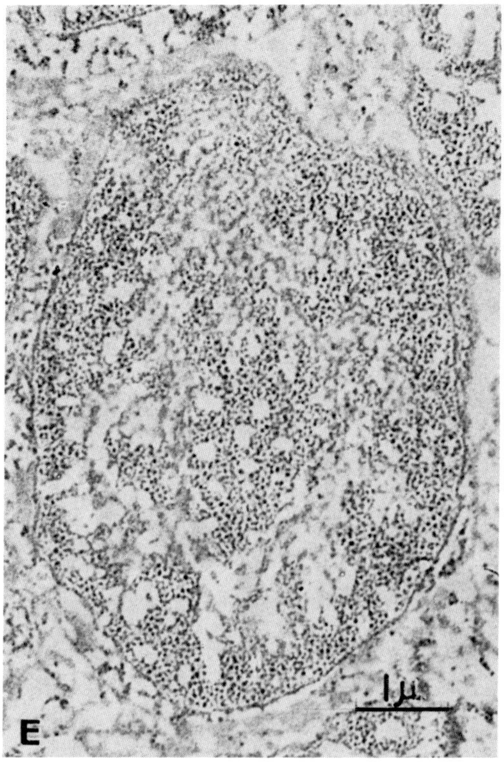

Fig. 8.24E.

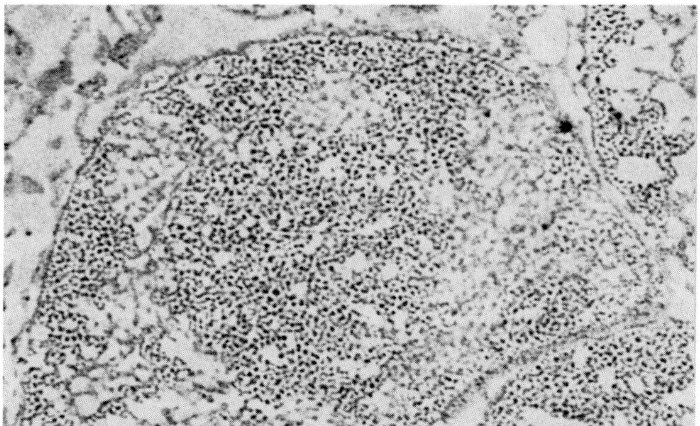

Fig. 8.24F.

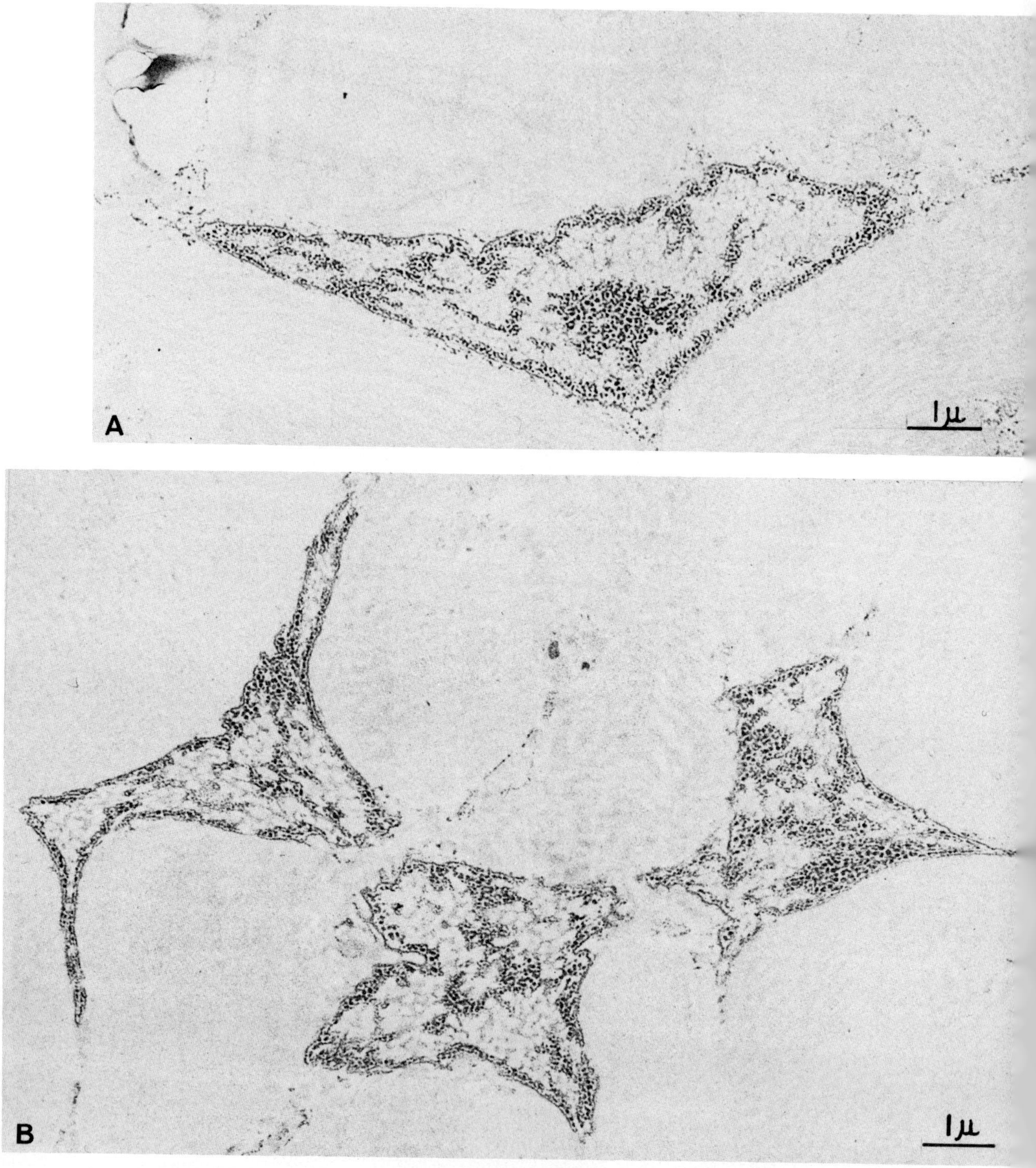

FIGS. 8.25(A and B). Fibroblasts. Apart from differences in shape, the distribution of second-order DNA coils in the nuclei is similar in fibroblasts of (A) dermis (B) and the subcutaneous loose connective tissue.

aggregates by gaps free of DNA, where transfer of metabolites and small species of nucleic acids may take place between nucleoplasm and cytoplasm. Extending centrally from this layer are variable numbers of chromatin strands of varying diameters. Both characteristics seem, however, to be specific for each mature cell type in the unstimulated state. Extending also from the peripheral chromatin are aggregates of DNA coils which enclose, to varying degrees, one or more nucleoli. The perinucleolar chromatin is thus continuous with the peripheral chromatin generally through a broad base and stalk. The structure of the nucleolus is also spongelike, and the denser walls are continuous with those of the perinucleolar chromatin and surrounding nucleoplasm, as are the less dense spaces. This general pattern is the same when stains are used which are specific for nucleic acids or for proteins. The spaces are not empty—they contain protein which may be stained after certain procedures (Chapters 2 and 5, this volume).

The tightly packed aggregates of second-order DNA coils are considered to be heterochromatin. The tight packing might have the effect of severely restricting the transport of metabolites, especially rRNA and mRNA, as well as limiting the movement (and activity) of larger molecules, such as enzymes *(10)*. To the extent that this takes place, the genes of such regions would be inactive, and the genes may be said to be turned off. On the surface of these aggregates, as well as in the thinner chromatin strands and perinucleolar chromatin, where exchange with nucleoplasmic constituents is or could be freer or less restricted, the second-order DNA coils are considered to be potentially euchromatic. This definition is primarily morphological, and is limited to the extent that transfer may be affected by the nature of the protein associated with the DNA coils, as well as by the association of RNAs different from mRNA or rRNA. Such protein and RNA, postulated as components of a repressor system *(5)*, may override the obvious morphological aspects.

The thesis tested in this chapter is a corollary of the hypothesis that the pattern or distribution of second-order DNA coils is a recordable indicator of the activity of genes. This report shows that these DNA patterns are indeed recognizably different in different cell types, sometimes to a surprising degree. Even the DNA patterns of epithelial cells, as similar as those of the dorsal and ventral prostate, differ in subtle ways, as do those of the epithelial cells of the head and tail of the epididymis.

There are some startling differences in the DNA molecular patterns of cells which are essentially similar, based on morphological criteria, as smooth muscle of the uterus in contrast with other smooth muscle fibers studied. Perhaps the DNA pattern of smooth muscle cells of uterus should be different from other smooth muscle fibers despite their morphological similarity because the former reacts almost selectively to various hormonal stimuli (see Chapter 10, this volume).

As expected, the observations of thin sections with the light microscope are in agreement with the electron micrographs of thinner sections of the same cell type. For example, in those cell types where the peripheral chromatin layer or chromatin strands are thick in electron micrographs, they can be recognized as clumps in photomicrographs, and where the chromatin is very fine, it is either invisible or appears as a thin line. However, the nuclei of different cell types with thin layers of peripheral chromatin cannot be distinguished one from the other in photomicrographs, when they are readily separable in electron micrographs. The same applies also to nuclei of different cell types which have large aggregates of DNA. The electron microscope, with its greater resolving power, is essential for the investigation of the DNA molecular patterns of different cell types. The studies of Strangeways and Canti *(12)*, of Martens *(9)*, and of others deal with structures one or two orders of magnitude too large for the problem presented and analyzed in this chapter. Questions concerning whether the nucleus is homogeneous, granular, or reticulated at the level of resolution of the light microscope disappear when considered at the level of organization of the DNA molecule and reappear as problems concerning patterns of organization of second-order coils of DNA molecules.

What has impressed me most in this study is the layer of peripheral chromatin adherent to the nuclear membrane. This is the location of most of the nuclear DNA and, in most instances, by any definition, of most heterochromatin. In agreement with the observations in this chapter, changes in the DNA pattern in different cell types probably require that some special patches of heterochromatic DNA be transformed to the euchromatic state and vice versa. One might conclude that here, associated with the nuclear membrane, must lie the basis for such transformations as well as for the constancy of each cell type. I wish to propose that underlying the transformations are the proteins (and perhaps also some forms of RNA) associated with DNA, on the one hand, and the nuclear membrane, on the other. How strong is this bond can be surmised from the work of Brenner *(6)*, who studied the "chromatic nuclear membrane" (for which may be read the peripheral layer of chromatin or of second-order DNA molecules) after ultracentrifugation of hepatic cells. While nearly all the chromatin was displaced centrifugally, a thin layer resisted prolonged centrifugation at 150,000 g and remained attached to the centripetal portion of the nuclear membrane, connected to the main mass of displaced chromatin by a few strands. This strong attachment of chromatin had been noted earlier by Beams, and was confirmed more recently *(3, 4)*. This powerful attachment of the DNA to the nuclear membrane probably takes place through membrane attachment proteins, which may be part of the stained protein adjacent to the nuclear membrane which is illustrated in Chapter 2 (this volume).

A topic which had been studied for many years and then dropped, is the theory of chromosome individuality *(1)*. If the activity of genes is the basis of persistent, specific cellular activity, and if the genes are contained in chromosomes, then chromosomes must mantain their individuality continuously through cell cycles of growth, differentiation, activity, and mitosis (or meiosis), even when the chromosomes disappear as visible units into the optical homogeneity or reticular network of vesicular nuclei. Wilson *(13)* reviewed the evidence and summarized the situation as follows: ". . . we can hardly avoid the conclusion that even vegetative nuclei of the ordinary type must consist of definite areas or regions, each the product of a single chromosome and each the fundamental basis of a future corresponding chromosome." It is very difficult to imagine how a chromosome could be transformed to a patch of DNA on the surface of the nuclear membrane unless the chromosome is considered to be divisible into separable portions of DNA which occupy specific areas of the nuclear membrane. I wish to propose that these separable portions of DNA are the second-order DNA coils, and that they are held in their appropriate relations primarily by the attachment protein referred to in the preceding paragraph. This attachment to the membrane provides the basis for the aggregation of second-order DNA coils during interphase, ensuring the orderly disposition of chromosome regions at this stage. The second-order DNA coils are aggregated perhaps by virtue of the activity of a second-protein, the coil-aggregating protein.

One can imagine that a change in form, charge, or size of one or more such protein molecules could release some or all DNA coils. Depending on the circumstances this could result in altered cell activity, cell maturation, cell death, cell transformation (metaplasia), or mitosis. These topics will be discussed in the next chapters.

Summary

DNA molecular patterns were studied in a large number of different cell types. The first-order DNA helix is organized as a second-order helix with a diameter and period of about 400 A. The patterns in which these are organized appeared different in all cell types studied in one or more of the following aspects: (1) total area occupied by second-order DNA coils; (2) size and number of chromatin masses; (3) the relative proportions of second-order coils attached to the nuclear membrane or to the nucleolus, or existing as chromosome strands lying in the nucleoplasm not clearly attached to either (in sections); (4) the degree of dispersion of second-order DNA coils; and (5) the number and dimensions of gaps at the nuclear membrane free of DNA (in sections). These criteria are essentially functional, reflecting the readiness with which transfer may take place of metabolites and final products of DNA activity.

Most of the DNA in all cells was attached to the nuclear membrane. This was related to the relatively constant pattern of a disposition of chromosomes during the mitotic cycle which permits individuality of chromosomal material even in chromatin of the interphase nucleus.

These findings are significant in consideration of cellular differentiation.

References

1. Baltzer, F. (1967). Theodor Boveri. University of California Press, Berkeley, California.
2. Bänder, A. (1950). Über zwei verschiedene chromaffine Zelltypen im Nebennierenmark und ihre Beziehung zum Adrenalin und Arterenolgehalt. *Anat. Anz. Supple.* **97**, 172–176.
3. Beams, H. W., and Kessel, R. G. (1968). Properties of the Ehrlich ascites tumor cell as

determined by electron microscopy, ultracentrifugation, and hydrostatic pressure. *Cancer Res.* **28**, 1944–1951.

4. Beams, H. W., and Mueller, S. (1970). Effects of ultracentrifugation on the interphase nucleus of somatic cells with special references to the nuclear envelope–chromatin relationship. *Z. Zellforsch. Mikrosk. Anat.* **108**, 297–308.
5. Bonner, J., and P. Ts'o, (eds.). (1964). "The Nucleohistones." Holden-Day, San Francisco, California.
6. Brenner, S. (1953). The chromatic nuclear membrane. *Exp. Cell Res.* **5**, 257–260.
7. Hillarp, N-Å., and Hökfelt, B. (1953). Evidence of adrenaline and noradrenaline in separate adrenal medullary cells. *Acta Physiol. Scand.* **30**, 55–68.
8. Jacob, F., and Monod, J. (1961). On the regulation of gene activity. *Cold Spring Harbor Symp. Quant. Biol.*, **26**, 193–211.
9. Martens, P. (1928). Recherches expérimentales sur la cinèse dans la cellule vivante. *Cellule* **38**, 69–174.
10. Mirsky, A. E., and Silverman, B. (1972). Blocking by histones of accessibility to DNA in chromatin. *Proc. Nat. Acad. Sci. U. S.* **69**, 2115–2119.
11. Shelley, W. B. (1963). Rapid technic for obtaining leukocytes from small samples of blood. *Amer. J. Clin. Pathol.* **39**, 433–435.
12. Strangeways, T. S. P., and Canti, R. G. (1927). The living cell *in vitro* as shown by dark ground illumination and the changes induced in such cells by fixing reagents. *Ouart. J. Microsc. Sci.* **71**, 1–14.
13. Wilson, E. B. (1947). "The Cell in Development and Heredity," *3rd* ed., p. 895. MacMillan, New York.

9

Changing DNA Molecular Patterns during Cell Maturation in the Mouse

Isidore Gersh

In the preceding chapter, the DNA molecular pattern of many cell types was described. The patterns were found to be qualitatively different in the different cell types when they are unstimulated. The lifetime of mature cells is limited, and nearly all cell types are being replaced by maturing or developing cells *(4, 10)*. The state of maturation is easier to identify in some cell types, and these are the ones discussed in this chapter. One would expect that as cells mature, the nature of the DNA molecular pattern should change as the synthesis of some proteins is turned off or reduced. One would expect the euchromatin to reach its lowest and the heterochromatin to reach its highest peak as the cells are fully matured. One would also expect a progressive decrease in RNA formation. Morphologically, these changes would appear as a progressively increased aggregation of second-order DNA coils, as the fine chromatin strands and patches of peripheral chromatin decrease in number, and as a decrease in nucleolar and nucleoplasmic RNA. The ribosomal concentration in mature cells would also be expected to be reduced. These statements are made despite the findings showing that the DNA content of certain cells is not always constant or equally accessible to stains *(1, 8, 14, 15)*. These deviations are not large, and would probably not be detectable qualitatively or quantitatively in the preparations examined. When the loss of DNA is large as in karyolysis (see this volume, Chapter 11; Vol. II, Chapter 5), it is, of course, readily detectable.

Method and Material

Four cell types were selected for special study: goblet cells of the duodenum and colon, vaginal epithelium, polymorphonuclear neutrophils, and developing erythrocytes. The earliest stages were difficult to identify with certainty, except in the vaginal epithelium. In the other cell types, the earliest stages selected were marked by the first appearance of their specific product of cell synthesis—mucigen granules for goblet cells, neutrophilic granules for polymorphonuclear leukocytes, and, for erythrocytes, diffusely distributed hemoglobin whose intrinsic mass frequently resulted in a pale background density to the cytoplasm. The observation of adjacent ¼-μ sections stained only with gallocyanin-chromalum, or with Wright's stain (for blood cells and bone marrow) was helpful, especially in the identification of early stages of leukocytes and erythrocytes.

The plan was the same as in Chapter 8 (this volume). Specimens were freeze-dried, treated *in vacuo* with a cross-linking reagent in vapor form, and stained for nucleic acids with gallocyanin-chromalum. Specimens were then embedded in water-soluble Durcupan. Technical procedures were kept as nearly constant as possible so that electron micrographs would be comparable. Electron micrographs were made with a Siemens EMU–1 operating at an accelerating voltage of 60 kv at a magnification of × 2000 and enlarged × 6 in printing.

Observations

Goblet Cells

In the earliest stage identified (Fig. 9.1), the nucleus is large, with a thin layer of peripheral chromatin with clear nuclear gaps. The nucleoplasmic walls contain numerous chromatin strands and RNA granules. Ribosomes are arranged mainly in parallel lamellae, as in the pancreas. In a slightly more advanced stage (Fig. 9.2), the peripheral chromatin is thicker, as are also the chromatin strands. Ribosomes and the RER are restricted mainly to the perinuclear region. In a later stage (Fig. 9.3), the nucleus appears disk-shaped—the chromatin strands are uniformly thick and the RNA granules in the nucleoplasm are somewhat fewer. In the nearly mature stages (Figs. 9.4 and 9.5), the second-order DNA coils of the peripheral chromatin and of the chromatin strands are tightly packed. RNA granules in the nucleoplasmic walls are scarce or nearly absent. The RER is restricted to the perinuclear region, but there is less of it than in earlier stages. Nucleoli may be present in all stages of cell maturation. From the earliest appearance of mucigen granules and in all subsequent stages, the cytoplasm between the mucigen granules contains very few ribosomes.

Figs. 9.1–9.5. Electron micrographs of progressive stages of maturation of goblet cells of duodenum and colon.

Fig. 9.1. Peripheral chromatin (PERIPH CHROM) is thin and interrupted by numerous nuclear gaps (N GAP). The chromatin strands (CHR STR) are attached to some of the peripheral chromatin clumps. The chromatin strands lie in nucleoplasmic walls, which also contain RNA granules. Ribosomes lie on the walls of the rough endoplasmic reticulum (RER). The Golgi site (GOLGI SITE) is free of ribosomes. Mucigen granules (SEC) are separated by thin walls of cytoplasm containing few ribosomes (RIBOS).

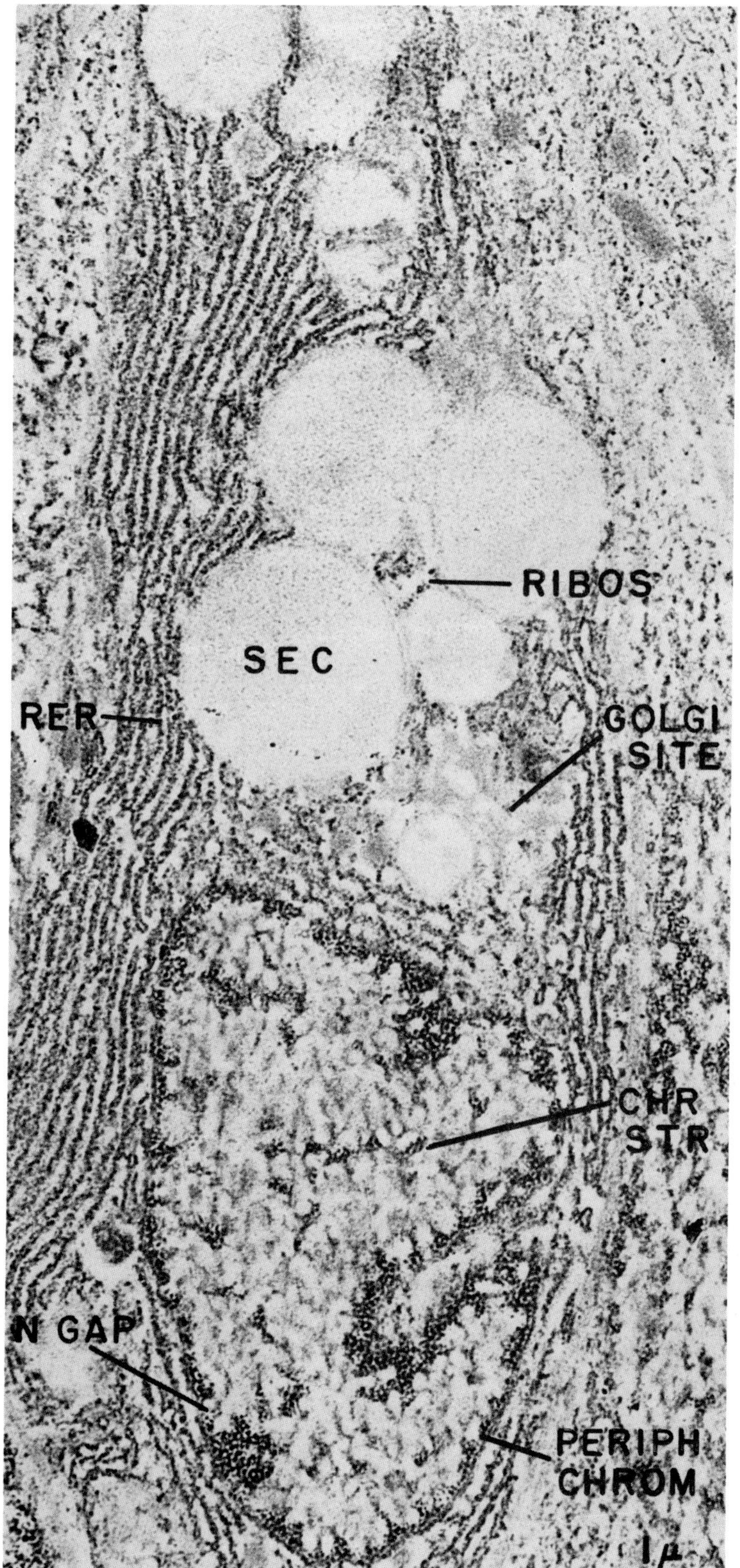
RIBOS
SEC
RER
GOLGI
SITE
CHR
STR
N GAP
PERIPH
CHROM
1μ

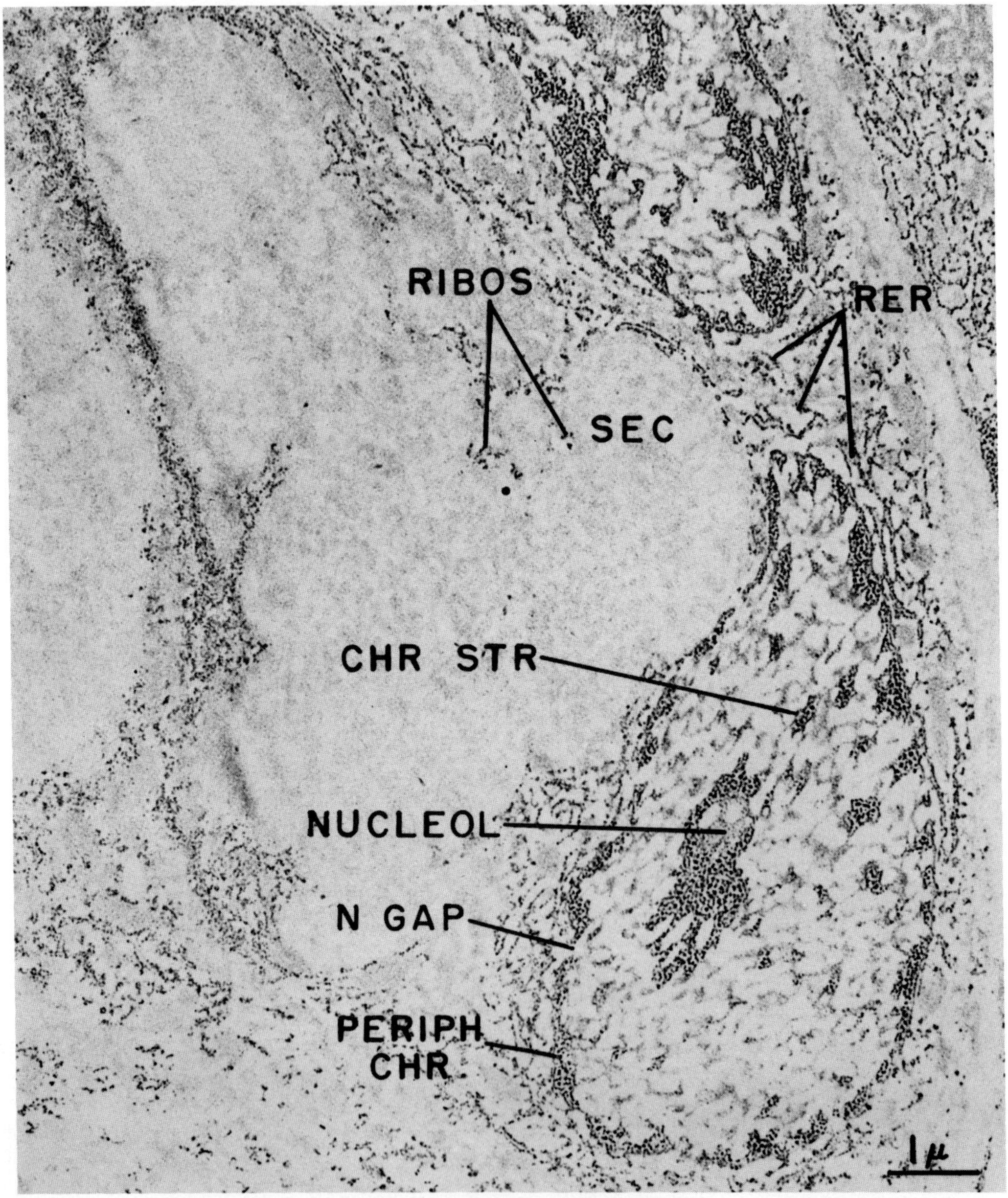

FIG. 9.2. The peripheral chromatin (PERIPH CHR) is thicker, as are most chromatin strands (CHR STR). A small part of a nucleolus (NUCLEOL) is present. RNA granules are present in the nucleolus. The RER and ribosomes (RIBOS) are more restricted. Mucigen granules (SEC) are more numerous. N GAP, nuclear gap.

Vaginal Epithelium

In the basal cells (Fig. 9.6), the layer of peripheral chromatin is irregular in thickness, varying from one or two second-order DNA coils to many layers. This is probably because the patches of peripheral chromatin give rise to numerous chromatin strands, whose diameters vary correspondingly. Nuclear gaps are numerous. In the intermediate layer of cells (Fig. 9.7), the second-order DNA coils are similarly disposed, but the nucleoli, even when multiple, seem to be enormously enlarged. The superficial cells (Figs. 9.8–9.12) are progressively more outstretched, and the DNA coils tend to be more condensed as thickened peripheral chromatin and thickened chromatin strands. In some nuclei, the periphery is quite thick, with reduced nucleoplasm, which is almost free of RNA granules and of nucleoli. The nuclear shape may be distorted. When the nucleus is maximally condensed, cytoplasmic ribosomes are nearly absent.

Polymorphonuclear Neutrophils

In young cells (Figs. 9.13 and 9.14), the peripheral chromatin layer is thin and interrupted by numerous small and large nuclear gaps. From the small patches of

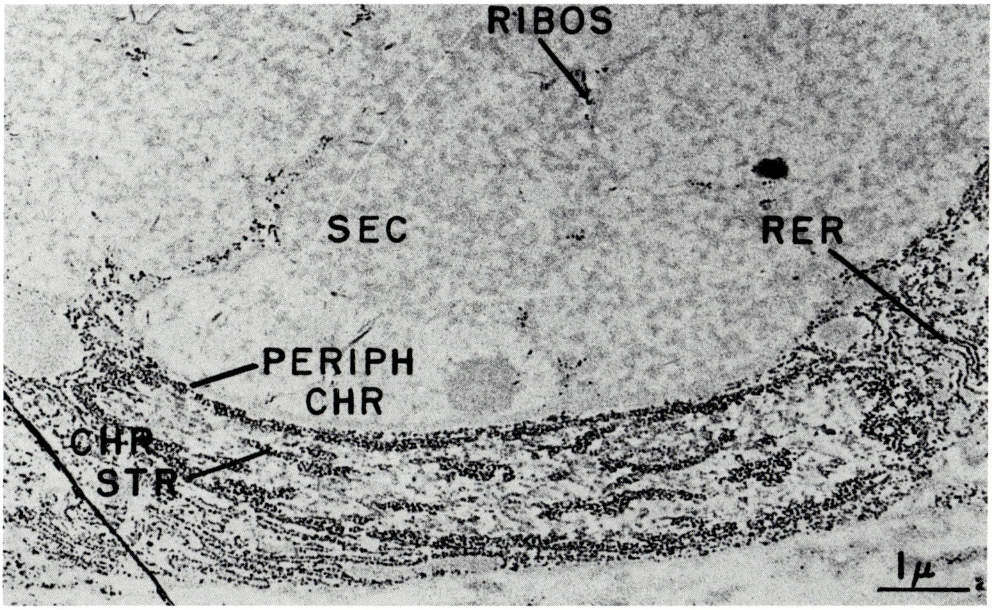

Fig. 9.3. Chromatin strands (CHR STR) are more nearly uniformly thick. Peripheral chromatin (PERIPH CHR) is also more uniformly thick than in earlier stages. RNA granules are present in the walls of the nucleoplasm. The RER is more restricted at the base of the cell. Mucigen granules (SEC) are more numerous and are separated by thin layers of cytoplasm containing few ribosomes (RIBOS).

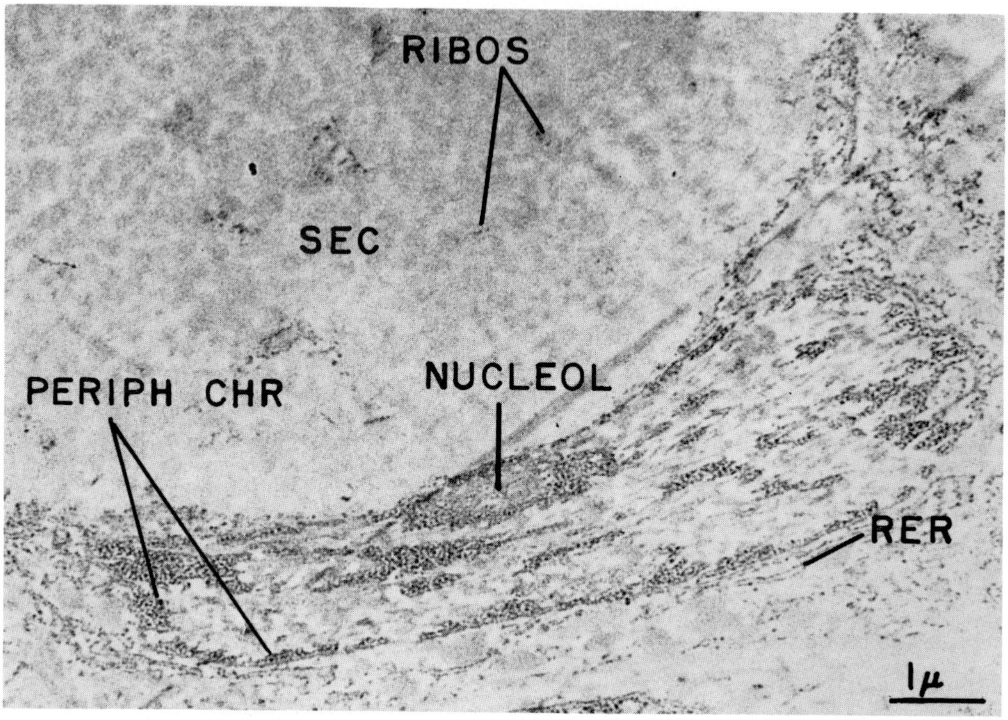

FIG. 9.4. ▲ ▼ FIG. 9.5.

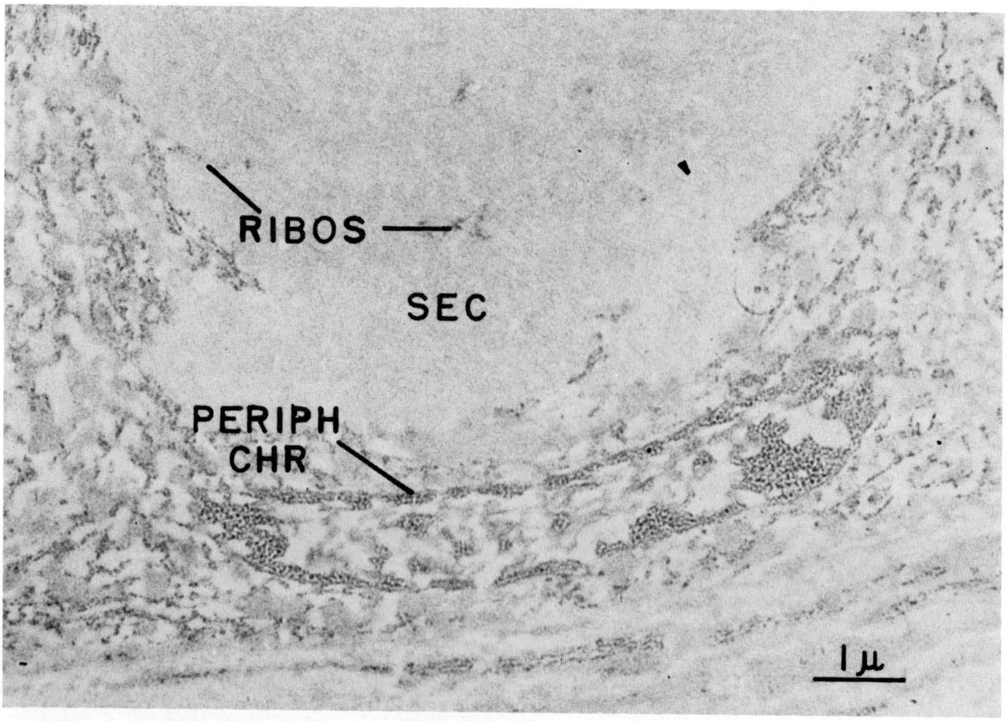

FIGS. 9.4 and 9.5. Nearly mature stages. Peripheral chromatin (PERIPH CHR) is progressively more tightly packed. RNA granules are scarce (Fig. 9.4) or nearly absent (Fig. 9.5) from the nucleus. A nucleolus (NUCLEOL) is present in Fig. 9.4. Ribosomes (RIBOS) are again largely restricted to the base of the cell, with some visible in the cytoplasmic layers which separate the mucigen granules (SEC).

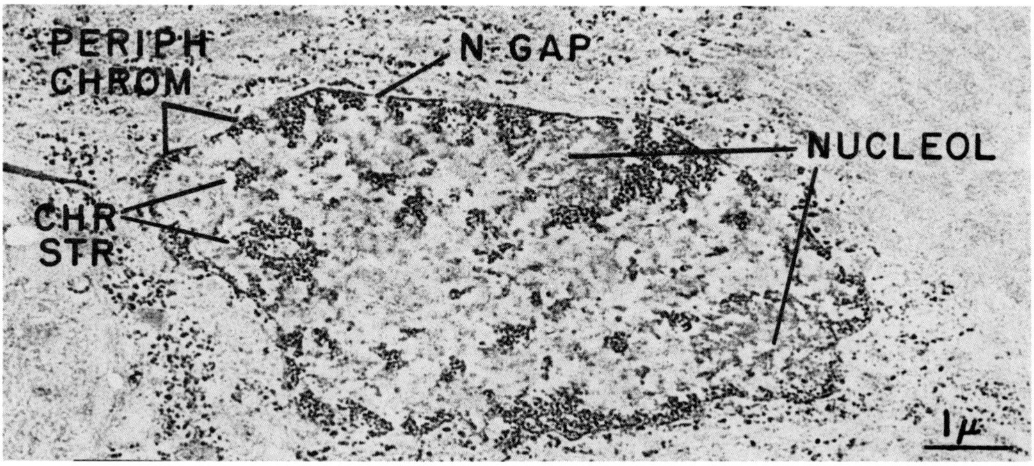

FIGS. 9.6–9.12. Electron micrograph of progressive stages of development of cells of vaginal epithelium.

FIG. 9.6. Basal cell. The layer of peripheral chromatin (PERIPH CHROM) is irregular in thickness, and separated into numerous patches by many nuclear gaps (N GAP). Chromatin strands (CHR STR) vary greatly in thickness. Nucleoli (NUCLEOL) are close to the nuclear membrane.

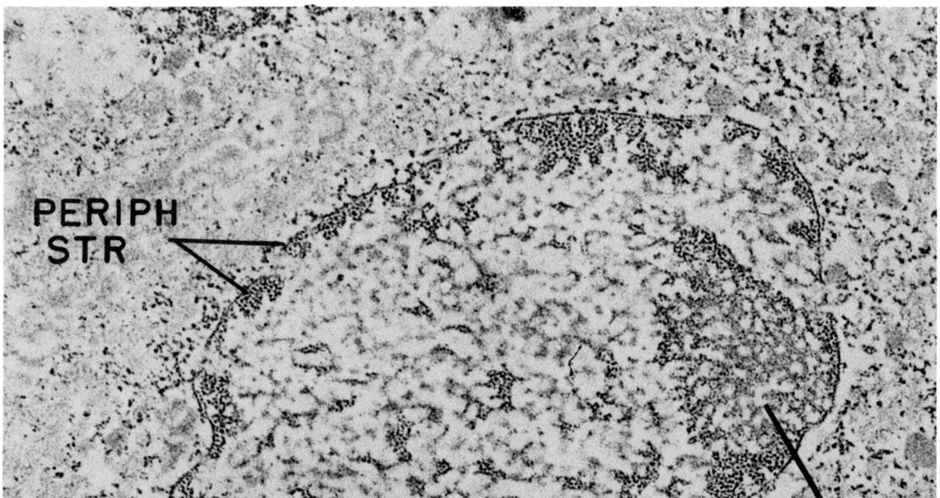

FIG. 9.7. Cell of intermediate layer. The peripheral chromatin (PERIPH STR) is more uniformly thick as compared with Fig. 9.6, and the nucleolus (NUCLEOL) seems enlarged.

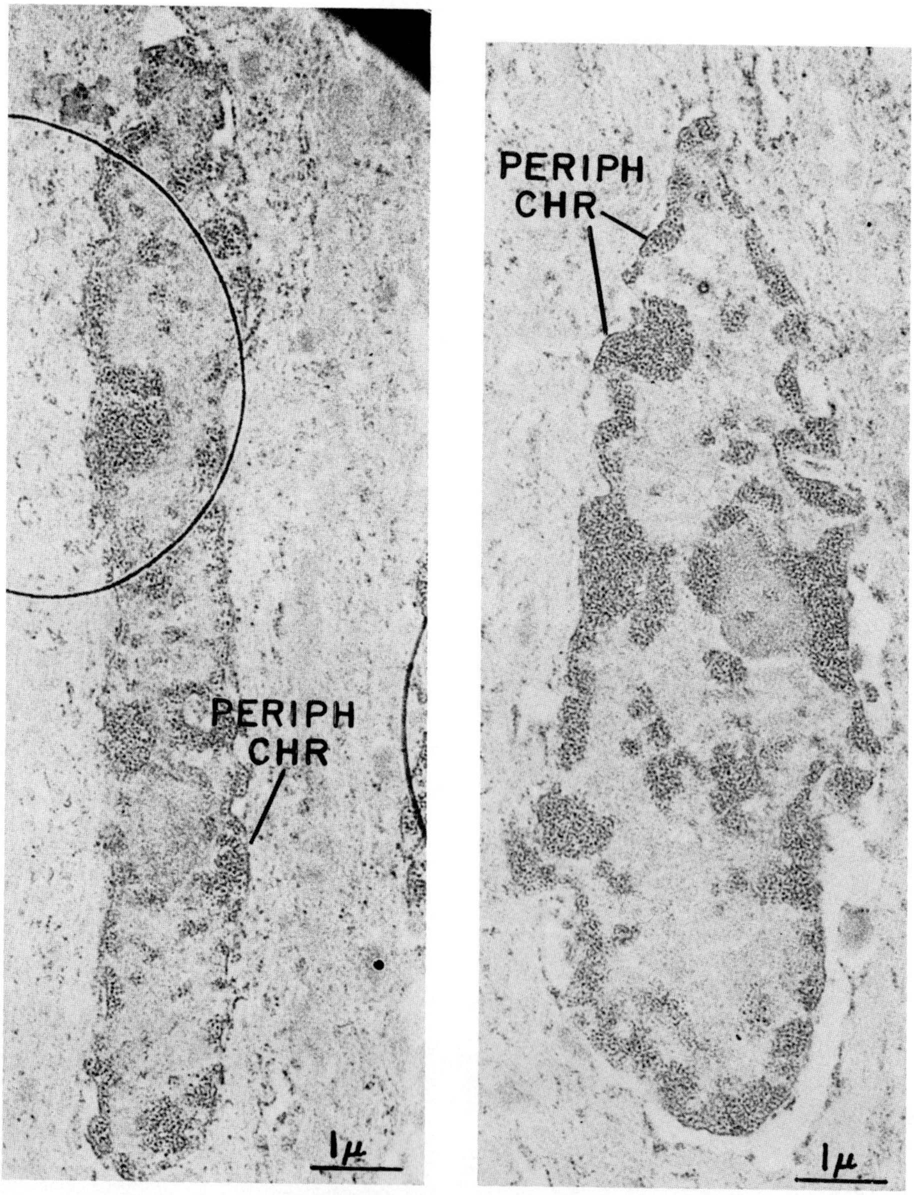

FIGS. 9.8 and 9.9. Elongated cells near surface. Peripheral chromatin (PERIPH CHR) is thick.

second-order DNA coils arise numerous fine chromatin strands which course in the denser nucleoplasmic walls where there are many RNA granules. Chromatin strands are also attached to the perinucleolar chromatin which encloses the nucleoli. Ribosomes are very numerous. In a somewhat later stage, still marked by the presence of nucleoli (Fig. 9.15), the peripheral layer of DNA is denser, thicker, and more irregular, and the nuclear gaps are reduced in number. The chromatin strands are also fewer and thicker. The reduced number of ribosomes corresponds to the increased number of cytoplasmic granules. In a still later stage (Fig. 9.16), when nuclear lobulations appear, the peripheral chromatin layer begins to assume the form characteristic of the adult, fully mature cell—the second-order DNA coils are densely packed at the nuclear membrane. Nuclear strands are scarce and thick, and nucleoli are absent. There are few ribosomes between the

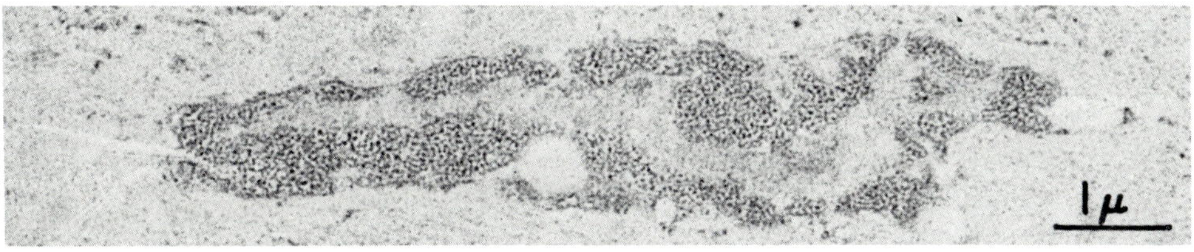

Fig. 9.10.

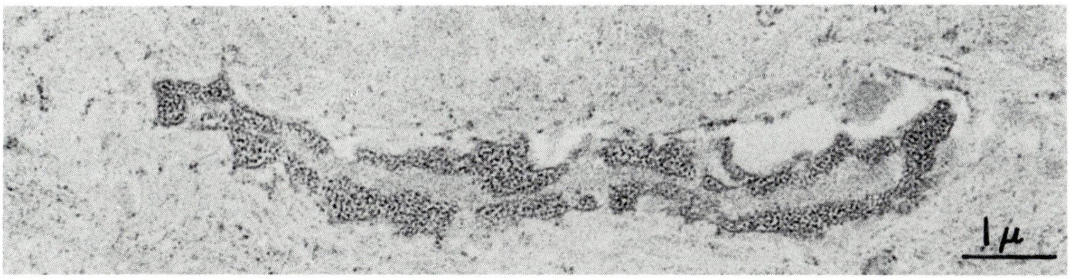

Fig. 9.11.

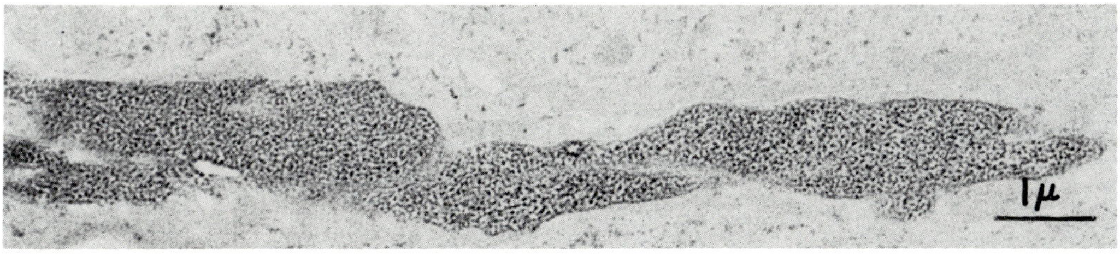

Fig. 9.12.

Figs. 9.10–9.12. Superficial cells. The chromatin is condensed and confined largely to the periphery of the nucleus. The DNA seems to occupy the whole of the nucleus with virtually no nucleoplasm visible. It is possible that this may be exaggerated in Fig. 9.12 because the plane of sectioning may be through the peripheral part of the nucleus.

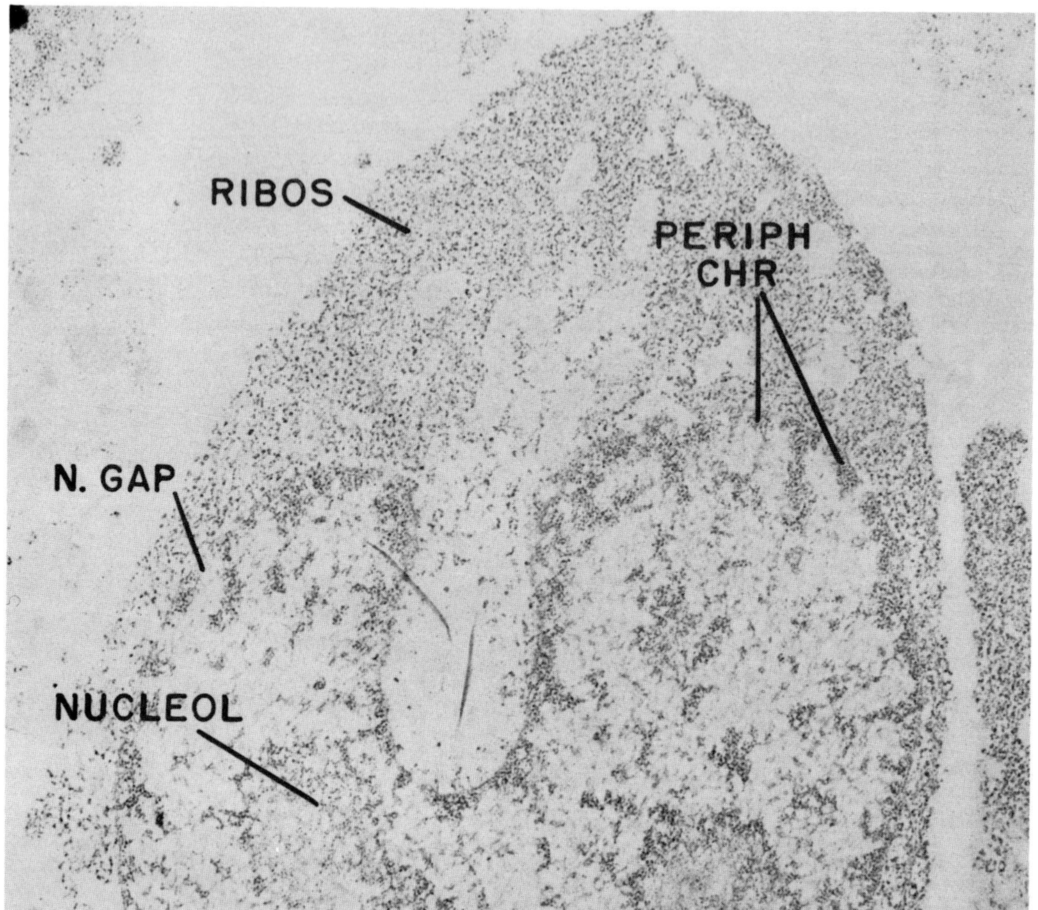

FIG. 9.13.

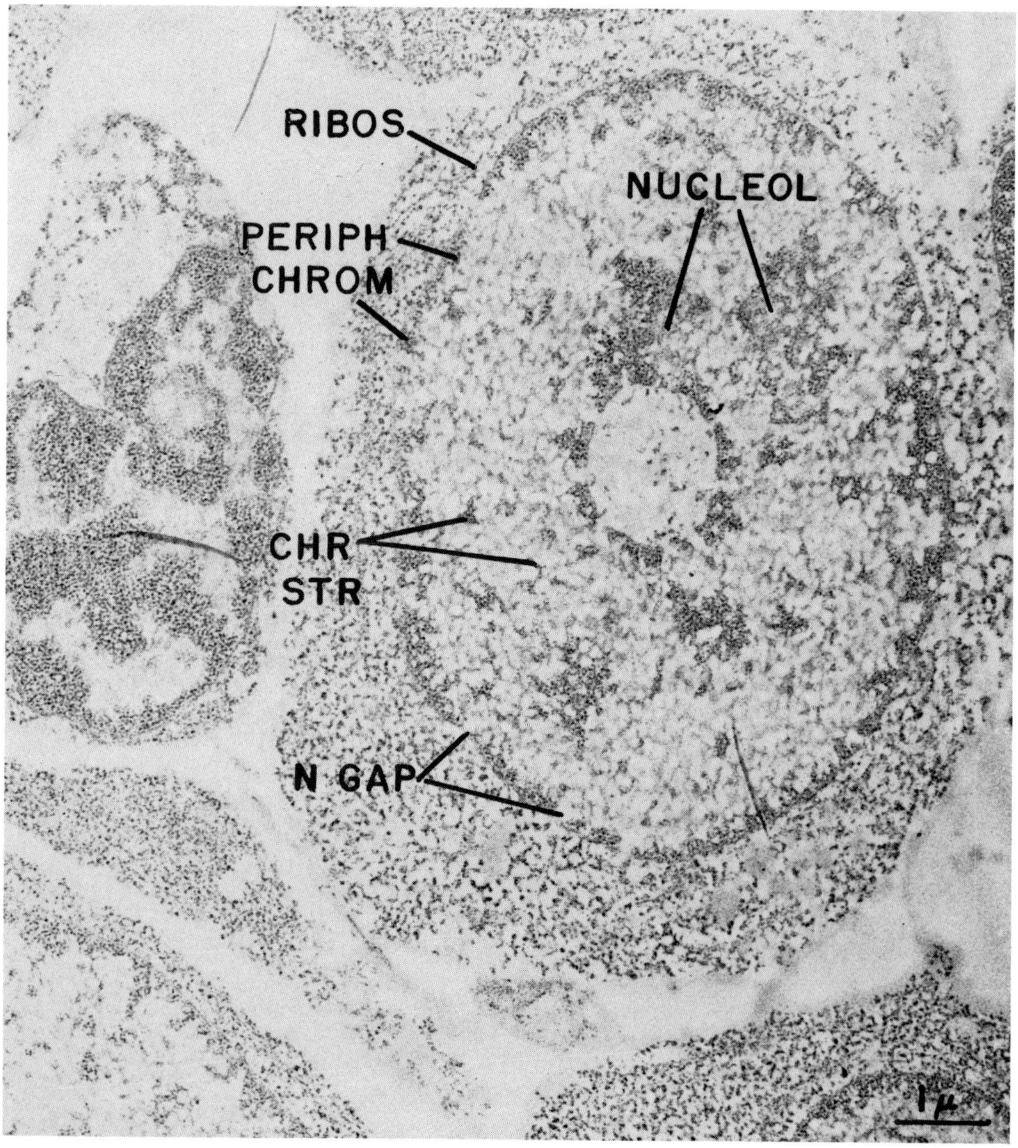

Fig. 9.14.

Figs. 9.13–9.18. Electron micrographs of progressive stages of development of polymorphonuclear neutrophils.

Figs. 9.13 and 9.14. Early stage with the nuclei sectioned in different planes. Peripheral chromatin layer (PERIPH CHROM CHR) is thin and interrupted by many nuclear gaps (N GAP). Fine chromatin strands (CHR STR) run in nucleoplasmic walls which also contain many RNA granules. Nucleoli (NUCLEOL) are large. Numerous ribosomes (RIBOS) are present in the cytoplasm.

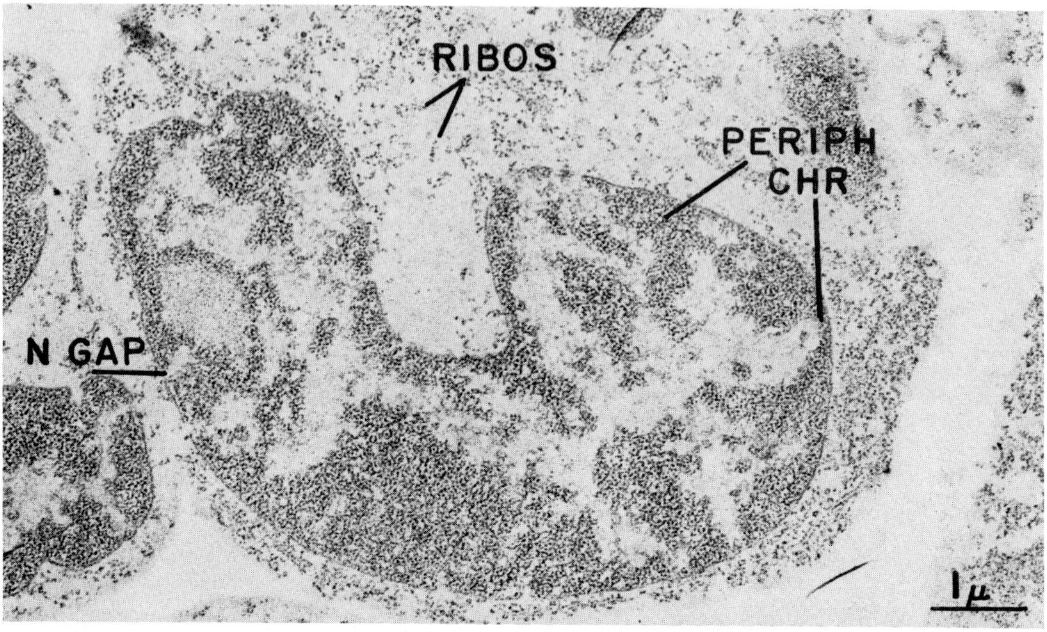

FIG. 9.15. Later stage. Peripheral chromatin layer (PERIPH CHR) is thicker, and there are fewer nuclear gaps (N GAP). Chromatin strands are thicker. Ribosomes (RIBOS) are reduced in number where granules are more numerous.

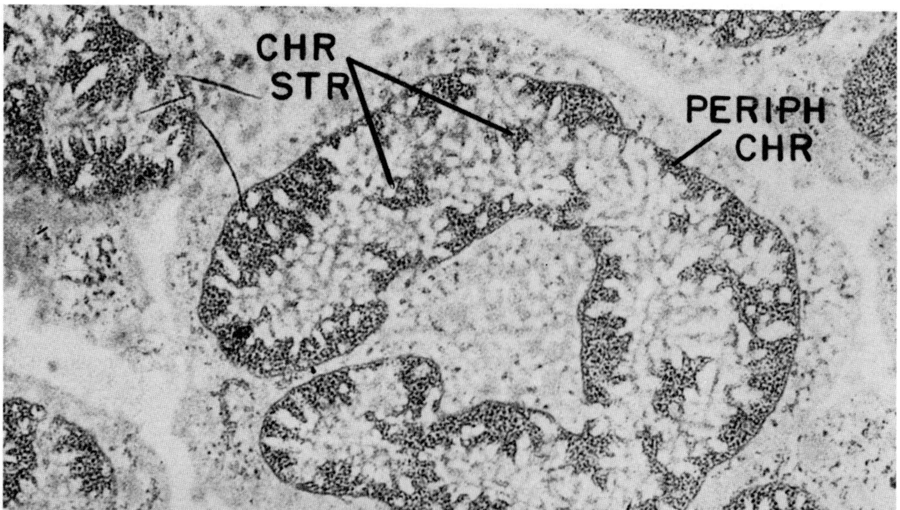

FIG. 9.16. A still later stage. The beginnings of lobulations are present. Chromatin assumes the appearance characteristic of the mature state with dense peripheral chromatin (PERIPH CHR) and few coarse central chromatin strands (CHR STR). Nucleoli are absent or very small.

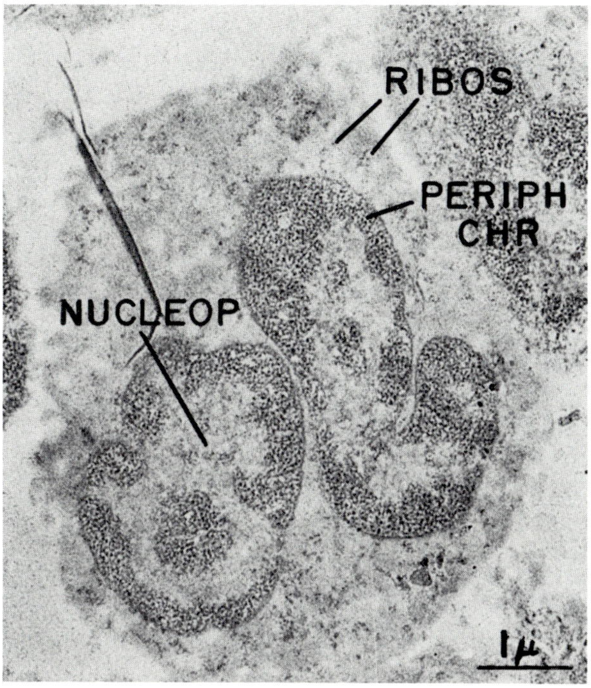

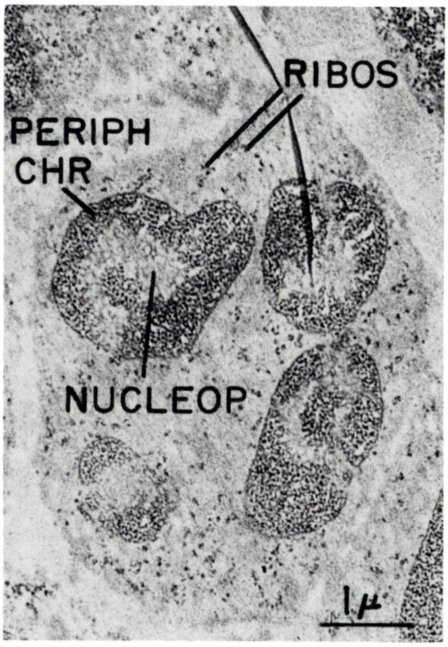

FIGS. 9.17 (LEFT) and 9.18 (RIGHT). Mature cells. Peripheral chromatin (PERIPH CHR) is dense, and the nucleoplasm (NUCLEOP) is reduced in amount. Ribosomes (RIBOS) are confined to the cytoplasmic spaces between granules.

neutrophilic granules. When the morphology of the fully mature circulating leukocyte is reached (Figs. 9.17 and 9.18), the peripheral chromatin is very thick, as are also the chromatin strands, if they occur. There are few RNA granules in the centrally disposed nucleoplasm.

DEVELOPING ERYTHROCYTES

In the earliest cells of this series (Figs. 9.19 and 9.20), the layer of peripheral chromatin is variable, and is joined with great chromatin strands to form a rather characteristic pattern which becomes accentuated in subsequent stages. The cytoplasm contains numerous ribosomes. The second-order DNA coils in later stages (Figs. 9.21 and 9.22) become condensed into large lumps of peripheral and central chromatin joined by equally dense strands. Nuclear gaps, which are large and relatively numerous, are restricted in a later stage (Fig. 9.23). The nuclei resemble those which seem about to be extruded from the cell. In these, as in extruded nuclei (Fig. 9.24), the second-order DNA coils are tightly packed, with the nucleoplasm confined to only a small fraction of the nucleus. RNA granules are not identifiable in sections of extruded nuclei.

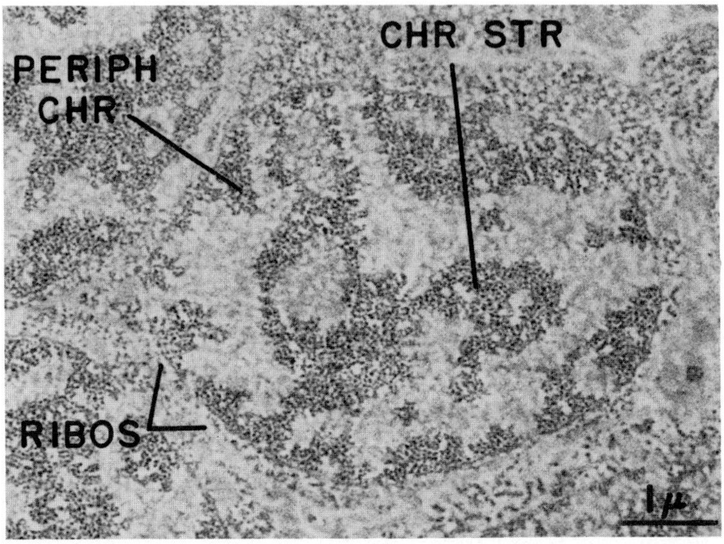

FIG. 9.19.

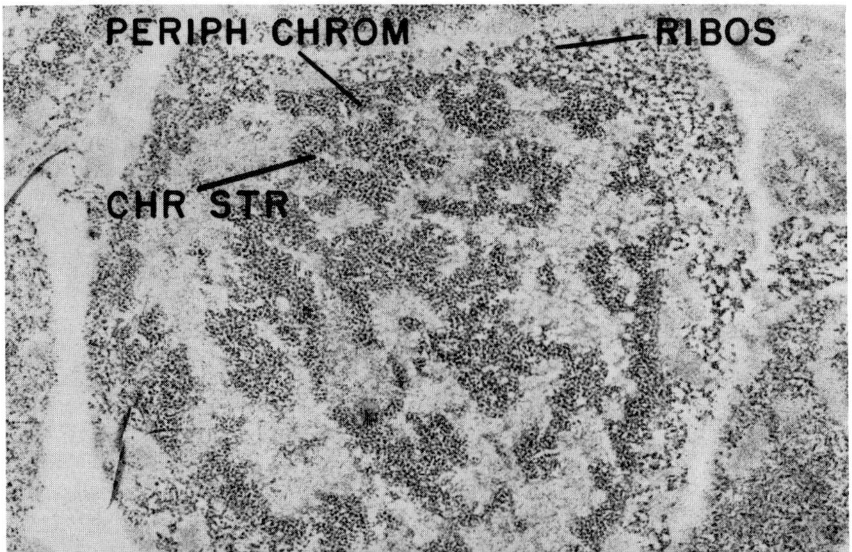

FIG. 9.20.

FIGS. 9.19–9.24. Electron micrographs of developmental stages of erythrocytes.

FIGS. 9.19 and 9.20. Early stages. The peripheral chromatin (PERIPH CHR) layer is thick and irregular. Attached to it are coarsely branching chromatin strands (CHR STR). Numerous ribosomes (RIBOS) are present in the cytoplasm.

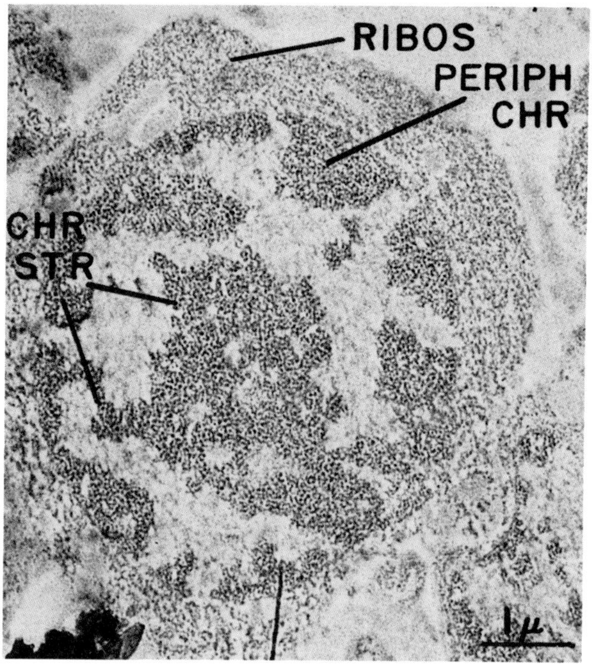

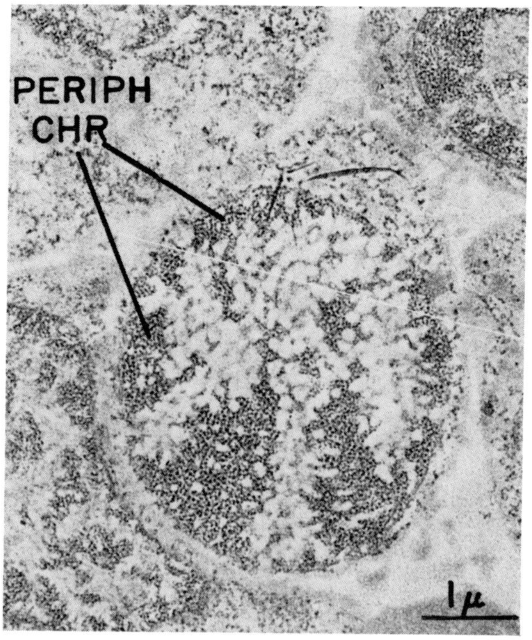

FIGS. 9.21 (LEFT) and 9.22 (RIGHT). Intermediate stages. Peripheral chromatin (PERIPH CHR) and centrally clumped chromatin are dense and joined by equally dense chromatin strands (CHR STR). RIBOS, ribosomes.

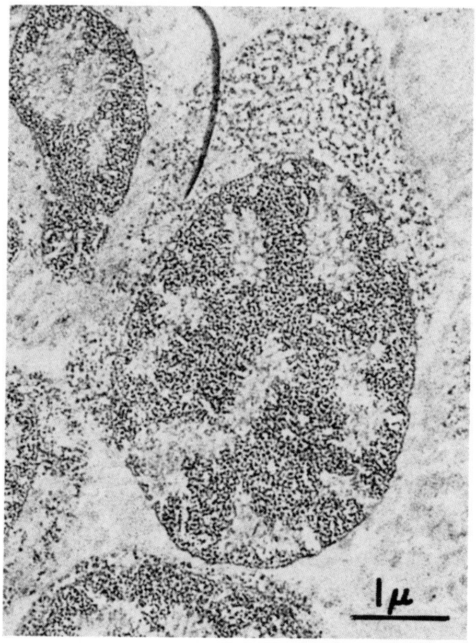

FIG. 9.23. Nucleus nearly extruded. The chromatin is condensed and the nucleoplasm is reduced in amount.

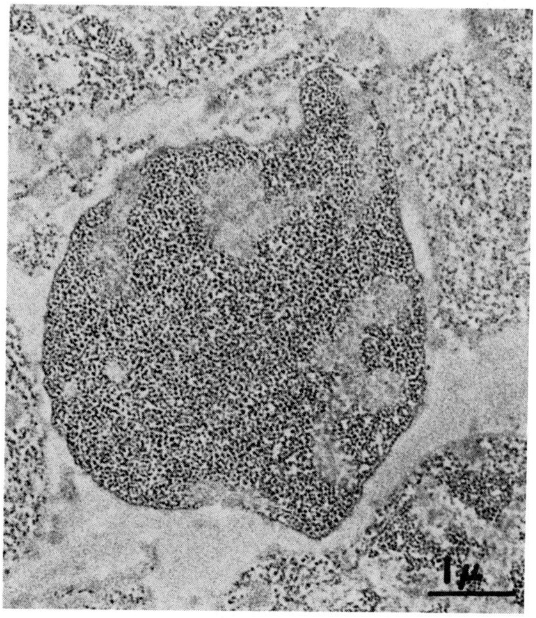

FIG. 9.24. Extruded nucleus. The chromatin is highly condensed and the nucleoplasm further reduced. RNA granules are not visible in the nucleoplasm.

Discussion

In the early stages of cell maturation, one would expect the DNA to be programmed for the production of many kinds of proteins at a relatively high rate. By contrast, in mature cells, the DNA is called on to produce very much the same kinds of proteins, but sufficient for replacement and maintenance only. All proteins are produced but only a few molecules and species of molecules may be generated at any one time, so that the overall rate of synthesis would be expected to be lower. Under such conditions, one would expect that the changes in DNA molecular pattern are related primarily to the rate of synthesis of proteins and to the numbers and kinds of molecules to be produced at the particular moment of fixation.

The following changes were observed to take place in the DNA molecular pattern of maturing goblet cells and polymorphonuclear neutrophils: (1) Condensation of and increased thickness of the layer of second-order DNA coils in the peripheral chromatin and in the chromatin strands. This change implies that more second-order DNA coils are closely packed as heterochromatin, and that the activity of this closely packed DNA may be reduced because of the restricted possibilities of exchange with the nucleoplasmic environment. This change also implies that there is a reduction in the fine strands of second-order DNA coils which have ready access to the nucleoplasmic environment. (2) Reduction of RNA granules in nucleoplasmic walls and walls of nucleoli. (3) Reduction in cytoplasmic ribosomes. Our findings are consistent with morphological observations of maturing polymorphonuclear leukocytes *(2)* and of goblet cells *(13)*.

In the superficial cells of the vagina, the end of DNA activity associated with pyknosis is accompanied by a marked reduction in the number of ribosomes. Presumably, in such cells, keratin formation nears the end. This does not necessarily mean that the direction of DNA activity is irreversible. It happens that, under conditions not well investigated, in certain animals (such as the mouse and guinea pig) the superficial cells with their pyknotic-like nuclei, which had produced keratin, are known to be transformed to cells producing mucigen. The evidence for this change is presented later in Chapter 11 (this volume).

In the maturing erythrocyte, the residual activity of the DNA terminates before nuclear extrusion, although the synthesis of hemoglobin in the remaining portion may continue because it is mediated by the cytoplasmic RNAs. The densely packed second-order DNA coils and the paucity of RNA granules of the pyknotic extruded nucleus seem to be irreversible in mammals. It is as if the DNA had been programmed for extinction of its activity. Thus, extinction of cellular activity may take place by conversion of euchromatin to heterochromatin, as in irreversible pyknosis of erythrocytic nuclei in mammals, or by the fragmentation and eventual lysis of second-order DNA coils as in karyolysis in certain chondrocytes (see Chapter 5, Vol II). Another example of programming for self-extinction is the stratified squamous epithelium, where the pyknotic nuclei of the superficial cells may disappear slowly or very suddenly (see Chapter 11, this volume).

The progressive crowding of the DNA molecules during maturation of the erythrocytes is consistent with the morphological changes *(7)*. In nucleated erythrocytes of mature vertebrates, the chromatin is dense, and the DNA is ineffective

(5, 9). The DNA, as described in cells fixed with aqueous solutions, consists of masses of granules (6). However, the DNA can be reactivated under certain complicated and particular conditions, as judged by the increased uptake of tritiated uridine. In such nuclei, the chromatin is progressively dispersed and even nucleoli reappear in the last stages of nuclear enlargement (3, 12). If these findings may be reinterpreted at the molecular level in terms of the observations reported in this chapter, I would predict that the densely packed second-order DNA coils of the mature cell are separated when the nucleus is activated and that the peripheral layer of chromatin and the chromatin strands eventually become thin and well dispersed, perhaps even as separated helices, with the appearance of nuclear gaps, nucleoplasmic RNA granules, and nucleoli. The sequence of events is predicted to be the same as those described for maturing erythrocytes, but in the reverse order. The same sequence of events is predicted also following transplantation of blastula nuclei and of adult brain nuclei into maturing oocytes of *Xenopus laevis* (11).

In the preceding chapter, I suggested that the second-order DNA helices of the peripheral chromatin are bound to the nuclear membrane by one or several attachment proteins. The observations of this chapter on changes in the DNA patterns of maturing cells extend this suggestion so that during maturation more second-order DNA helices are bound to the nuclear membrane and inactivated by more, or more effective, attachment-protein molecules. The second-order helices are also held packed together and are presumably attached to one another by coil-aggregating protein molecules.

Summary

The DNA molecular pattern was studied in the following cells during maturation: goblet cells, stratified squamous epithelium, polymorphonuclear neutrophils, and erythrocytes. The second-order DNA molecular patterns change in a characteristic fashion during maturation: increasingly dense packing of the second-order DNA coils at the periphery of the nucleus, with a corresponding restriction in the number and size of the nuclear gaps between the peripheral masses of chromatin and decrease in the amount of nucleoplasm and of the number of RNA granules in both nucleoplasm and nucleolus. These morphological criteria are related to a progressive decrease in synthesis and transfer of RNAs and the associated restriction in protein synthesis. The increasingly dense packing of the second-order DNA coils is attributed to an increase in the number or effectiveness of attachment and coil-aggregating proteins which bind the second-order DNA coils to the nuclear membrane and possibly to each other. These findings are of interest in consideration of control mechanisms for gene activity in somatic cells.

References

1. Adrian, E. K., Jr. (1971). The metabolic stability of nuclear DNA. *In* "Cellular and Molecular Renewal in the Mammalian Body" (I. L. Cameron and J. D. Thrasher, eds.), pp. 25–44, Academic Press, New York.

2. Bainton, D. F., and Farquhar, M. G. (1966). Origin of granules in polymorphonuclear leukocytes. *J. Cell Biol.* **28**, 277–301.
3. Bolund, L. N., Ringertz, N. R., and Harris, H. (1969). Changes in the cytochemical properties of erythrocyte nuclei reactivated by cell fusion. *J. Cell Sci.* **4**, 71–87.
4. Cameron, I. L. (1971). Cell proliferation and renewal in the mammalian body. *In* "Cellular and Molecular Renewal in the Mammalian Body" (I. L. Cameron and J. D. Thrasher, eds.), pp. 45–85. Academic Press, New York.
5. Cameron, I. L., and Prescott, D. M. (1963). RNA and protein metabolism in the maturation of the nucleated chicken erythrocyte. *Exp. Cell Res.* **30**, 609–612.
6. Davies, H. G. (1968). Electron microscope observations on the organization of heterochromatin in certain cells. *J. Cell Sci.* **3**, 129–150.
7. Fawcett, D. W. (1966). "The Cell." Saunders, Philadelphia, Pennsylvania.
8. Garcia, A. M. (1970). Stoichiometry of dye binding versus degree of chromatin coiling. *In* "Introduction to Quantitative Cytochemisty II" (G. L. Wied and G. F. Bahr, eds.), pp. 153–170. Academic Press, New York.
9. Grasso, J. A., Woodard, J. W., and Swift, H. (1963). Cytochemical studies of nucleic acids and proteins in erythrocytic development. *Proc. Nat. Acad. Sci. U. S.* **50**, 134–140.
10. Gross, R. J. (1964). "Adaptive Growth." Academic Press, New York.
11. Gurdon, J. B. (1968). Changes in somatic cell nuclei inserted into growing and maturing amphibian oocytes. *J. Embryol. Exp. Morphol.* **20**, 401–414.
12. Harris, H. (1967). The reactivation of the red cell nucleus. *J. Cell Sci.* **2**, 23–32.
13. Moe, H. (1968). The goblet cells, Paneth cells, and basal granular cells of the epithelium of the intestine. *Int. Rev. Gen. Exp. Zool.* **3**, 241–287.
14. Pelc, S. R. (1968). Turnover of DNA and function. *Nature (London)* **219**, 162–163.
15. Roels, H. (1966). "Metabolic" DNA: a cytochemical study. *Int. Rev. Cytol.* **19**, 1–34.

10

Alteration of DNA Molecular Patterns with Activation of Certain Cells in the Mouse

Isidore Gersh

One would expect that when protein synthesis is activated to a high degree, the changes in the nuclear DNA molecular pattern of the stimulated cells would be the reverse of those observed during cell maturation. One would expect to find a large increase in the potentially euchromatic regions (as manifested by the presence of a larger number of more finely aggregated DNA helices having free access to the nucleoplasmic environment) and a corresponding reduction in the size and number of heterochromatic regions (characterized by their tightly aggregated second order DNA helices and their restricted possibilities for exchange with the nucleoplasm and cytoplasm). These expectations were tested by study of the distribution of DNA in nuclei of five cell types appropriately stimulated: (1) exocrine cells of the pancreas stimulated with pilocarpine; (2) cells of the adrenal medulla stimulated with reserpine; (3) cells of the seminal vesicle stimulated with testosterone; and (4) and (5) cells of the uterus stimulated with estradiol.

Methods

Group 1

Female mice were injected intraperitoneally with 0.3 ml of a solution of 0.2% pilocarpine nitrate. They were bled at the following time intervals (in hours) after the injection and small pieces of the pancreas of each animal were frozen ultrarapidly: ½, 1, 2, 2½, 3, 4, and 5. Pilocarpine is, of course, a time-honored parasympatheticomimetic used in many studies to accelerate glandular secretion.

Group 2

Mice were injected intraperitoneally with 0.2 mg of reserpine as Serpasil (Ciba) and repeated 24 hours later. Reserpine depletes the stores of catecholamines of the adrenal medulla, and its action is long-lasting *(7)*. The mice were somewhat depressed, and were bled 48 hours after the first injection, when small pieces of the adrenal medulla were frozen ultrarapidly.

Group 3

Mice were castrated when 7 weeks old. Four weeks later, some mice were injected intramuscularly with 10 mg testosterone. They were bled 2 hours later, and small pieces of seminal vesicle of these and untreated mice were frozen ultrarapidly. The effects of testosterone on epithelial cells of the seminal vesicle have been exhaustively reviewed, and submicroscopic studies have been supplemented by some related biochemical findings *(5, 6, 21)*.

Group 4

Ovariectomized mice about 4 weeks old were injected intramuscularly and subcutaneously with 50 mg of 17β-estradiol in corn oil. They were bled 3½ hours later, and small pieces of the uterus of these mice and of untreated mice were frozen ultrarapidly. The action of estrogen on the epithelium and smooth muscle of the uterus has been reviewed in numerous articles, with particular attention to cytoplasmic changes in RNA and other biochemical accompaniments *(4, 10, 13, 15–17, 22, 25, 32, 33)*. The outer longitudinal layer of smooth muscle of a mouse in very late pregnancy was also prepared and compared with the same layer of the uterus of a 4-month-old virgin mouse of the same weight.

All specimens were prepared as described in Chapter 3 (this volume) for nucleic acids. Specimens were frozen ultrarapidly, dried at a low temperature, treated *in vacuo* with vapors of mercury (II) hexafluoroacetylacetonate, and stained with gallocyanin–chromalum. Efforts were made throughout to repeat procedures as nearly exactly as possible so that electron micrographs could be comparable. Electron micrographs were made at a magnification of × 2000 with a Siemens EMU–1 at an accelerating voltage of 60 kV and enlarged × 6 in printing. The second-order DNA helices are not clear at this magnification, and the reader should refer to Chapters 3 and 5 (this volume) and Chapter 5 (Vol. II), where they are shown at higher magnifications. As explained in Chapter 8 (this volume), the particular magnification was selected as a practical maneuver to make it possible to present this material in published form. Publication at greater enlargement would be prohibitive, and, in any case, would not add to the analysis of the pattern of second-order DNA helices.

All electron micrographs of activated cells were compared with those of the same cell types of untreated mice of the same age presented in this chapter or in Chapter 8 (this volume).

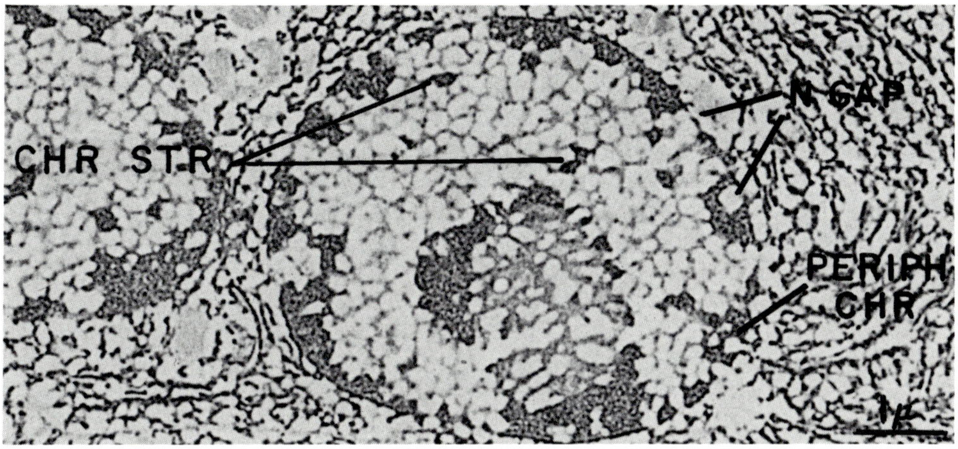

FIGS. 10.1–10.3. Electron micrographs of sections of exocrine pancreatic cells of mice injected with pilocarpine. (Compare with electron micrograph of normal cell in Chapter 8, Fig. 8.7, this volume.)

FIG. 10.1. One-half hour after injection. Nuclear gaps (N GAP) markedly enlarged. Chromatin strands (CHR STR) mostly very fine. Peripheral chromatin clumps (PERIPH CHR) are small.

Observations

STIMULATED EXOCRINE CELLS OF THE PANCREAS

At ½ hour after administration of pilocarpine, the aggregates of DNA coils are small, and the nuclear gaps are numerous and large (Fig. 10.1). The chromosomal strands are nearly all very fine. At 2½ hours the peripheral chromosomal aggregates and many chromosomal strands are coarser (Fig. 10.2). At 4 hours (Fig. 10.3), the DNA pattern is coarser and indistinguishable from that of normal cells.

STIMULATED CELLS OF THE ADRENAL MEDULLA

Nearly all nuclei are very pale (Fig. 10.4), as compared with the dark and pale nuclei of normal cells (this volume, Chapter 8, Figs. 8.17 and 8.18). The layer of peripheral chromatin is very thin, as compared with the dark and light nuclei of normal medulla. Many nuclear gaps interrupt this layer. Many chromatin strands are finer than in normal cells.

CELLS OF THE SEMINAL VESICLE OF THE CASTRATE MOUSE WITH AND WITHOUT STIMULATION WITH TESTOSTERONE

The second-order DNA molecular pattern of the epithelial cell nucleus after castration (Fig. 10.5) is coarser than that of the normal cell (this volume, Chap-

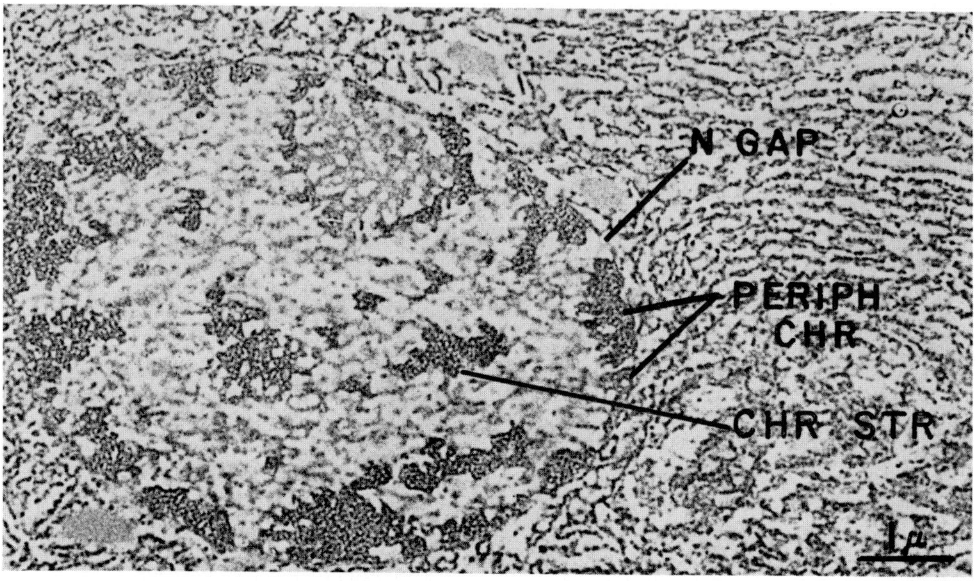

FIG. 10.2. Two and one-half hours after injection. Nuclear gaps (N GAP) still enlarged. Numerous very fine chromatin strands (CHR STR). PERIPH CHR, peripheral chromatin.

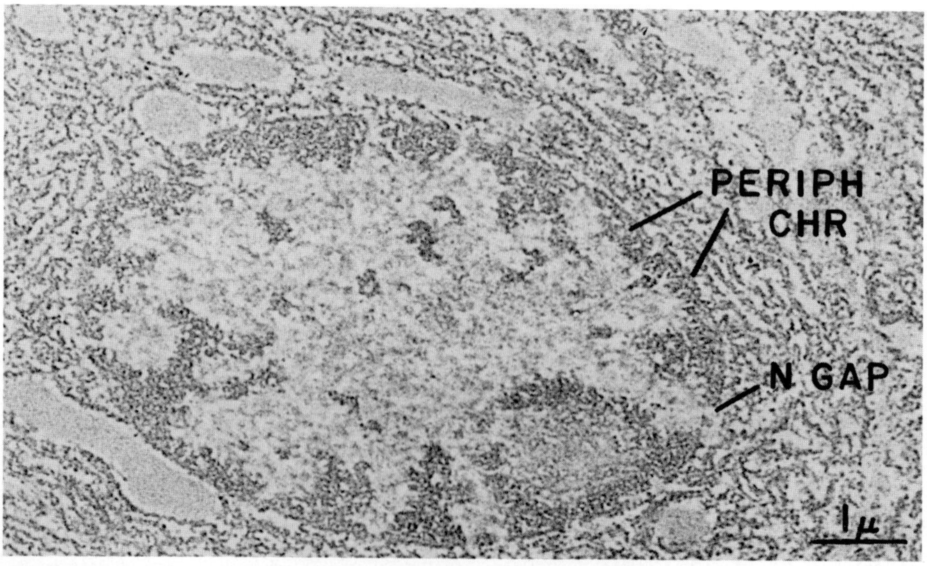

FIG. 10.3. Four hours after injection. Nuclear chromatin in the normal range, with the peripheral chromatin (PERIPH CHR) possibly thicker than normal.

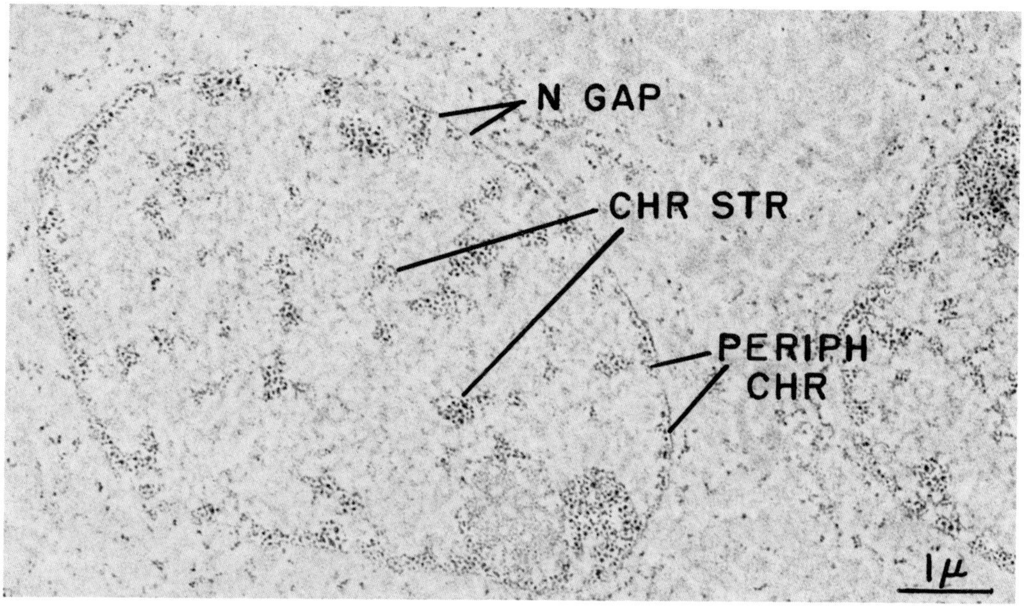

Fig. 10.4. Pale nucleus of adrenal medulla after injection of reserpine. The layer of peripheral chromatin (PERIPH CHR) is very thin and very much interrupted by numerous small and large nuclear gaps (N GAP). Chromatin strands (CHR STR) are thinner and more numerous than in normal cells. (The latter are seen earlier in this volume, Chapter 8, Figs. 8.17, and 8.18.)

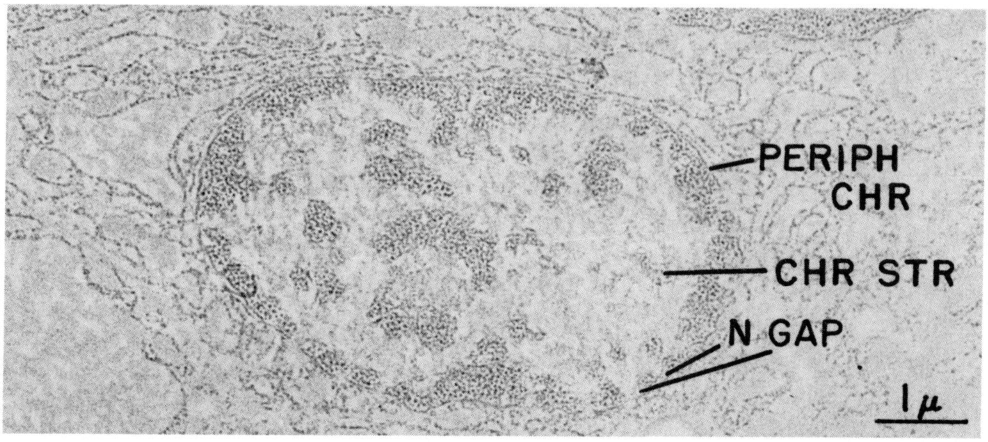

Fig. 10.5. Nucleus of seminal vesicle epithelium of castrate mouse. The peripheral chromatin (PERIPH CHR) and chromatin strands (CHR STR) are coarser than in the cells of the normal animal shown in Chapter 8, Fig. 8.11 (this volume). The nuclear gaps (N GAP) are reduced in number and size, and there are fewer RNA granules in the nucleoplasm.

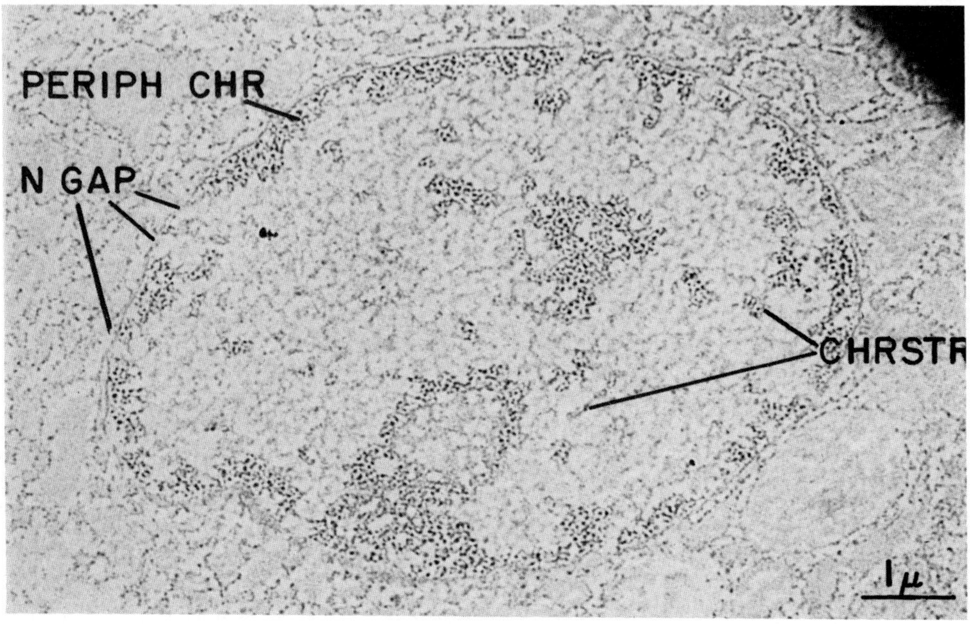

FIG. 10.6. Injection of testosterone into a castrate mouse results in a dramatic change in the DNA pattern of the epithelial cells of the seminal vesicle. The layer of peripheral chromatin (PERIPH CHR) is thin, with numerous nuclear gaps (N GAP). There are many chromatin strands (CHR STR) and, of these, many are thin.

ter 8, Fig. 8.11)—the chromatin strands being thicker and the nuclear gaps fewer and smaller. Also, there are fewer RNA granules in the nucleoplasmic walls of the cells after castration. After stimulation with testosterone (Fig. 10.6), the peripheral chromatin layer is thinner and the chromatin strands are more numerous and finer. RNA granules in the nucleoplasmic walls are more numerous than in the normal. Nuclear gaps are very numerous.

Smooth Muscle Cells of Uterus

Control Uterus of Mice 4 Weeks Old, Unstimulated (Fig. 10.7). The nuclei of smooth muscle cells contain a dense peripheral layer of chromatin, with few narrow gaps. Chromatin strands are small in number and rather thick.

Stimulated Uterus of Mice 4 Weeks Old, Treated with Estradiol (Fig. 10.8). The peripheral layer of chromatin is irregularly thin and interrupted by numerous gaps. The chromatin strands are mainly fine, some being reduced to single second-order coils of DNA. Nucleoli are common, as compared with control cells, where the nucleoli are scarce and small.

Fully Developed, Nonpregnant Uterus (Fig. 10.9). The nucleus is intermediate in appearance between that of the immature and of the immature stimulated uterus.

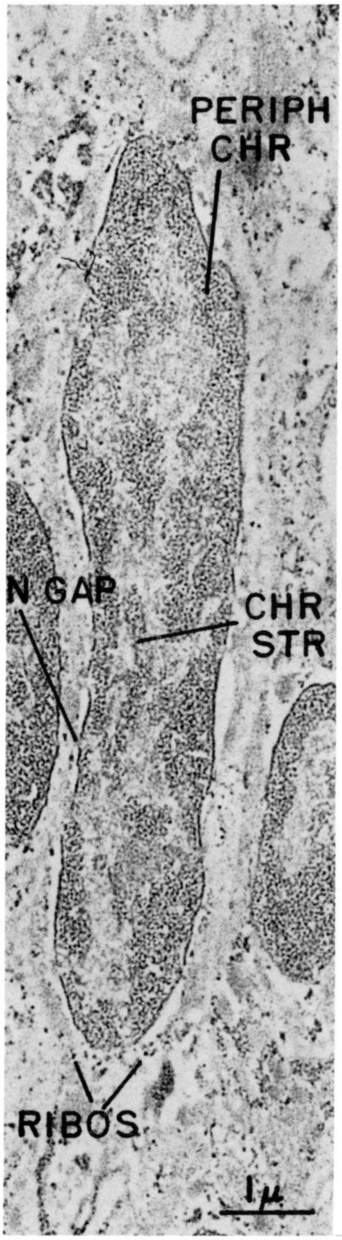

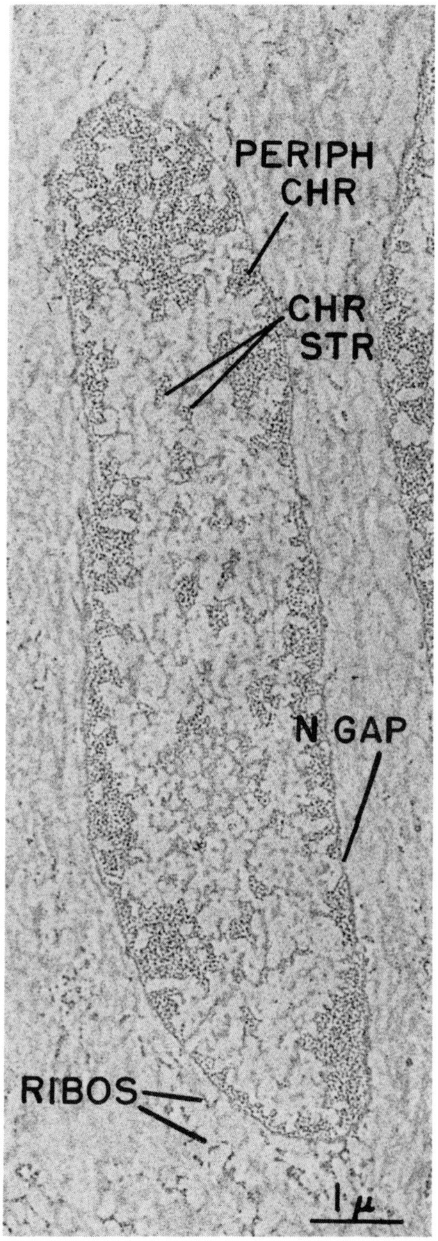

FIG. 10.7 (LEFT). Nucleus of smooth muscle of unstimulated uterus of mouse 4 weeks old. Peripheral chromatin (PERIPH CHR) is thick, with few fine nuclear gaps (N GAP). Chromatin strands (CHR STR) are few in number and coarse. Ribosomes (RIBOS) are scarce.

FIG. 10.8 (RIGHT). Nucleus of smooth muscle of uterus of mouse 4 weeks old stimulated with estrogen. Peripheral chromatin (PERIPH CHR) is irregular in thickness, with fine interruptions. Chromatin strands (CHR STR) are fine. Nucleolus is prominent. Ribosomes (RIBOS) are scarce.

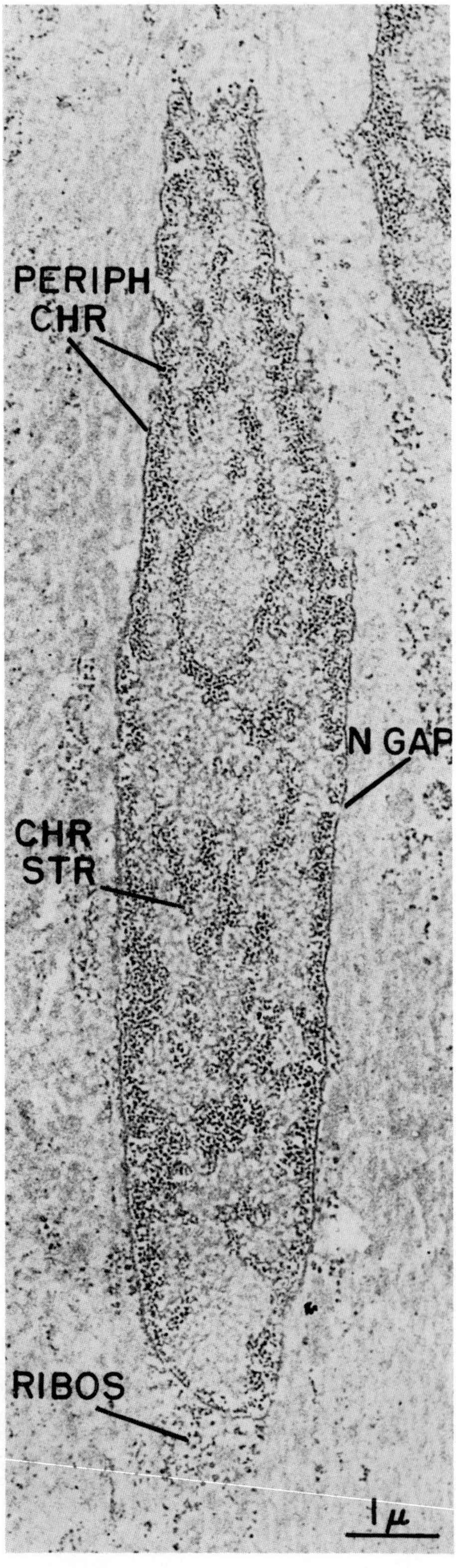

FIG. 10.9. Nucleus of smooth muscle of mature uterus, control for Fig. 10.10. The layer of peripheral chromatin (PERIPH CHR) is thinner than that of the unstimulated less mature uterus, as are the chromatin strands (CHR STR). In addition, the nucleolus is prominent, and ribosomes (RIBOS) are more numerous.

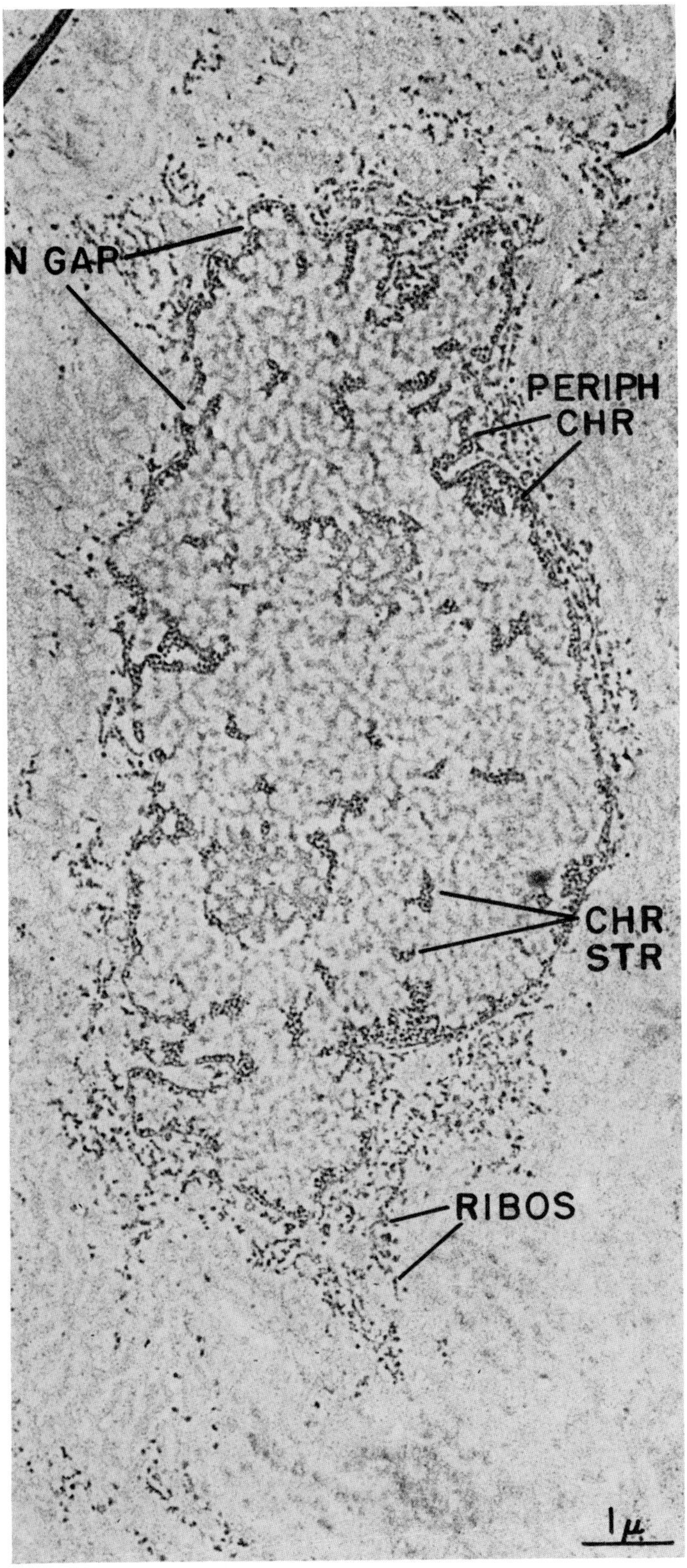

FIG. 10.10. Nucleus of smooth muscle of pregnant mouse near term. The layer of peripheral chromatin (PERIPH CHR) is thin, with numerous fine interruptions (GAP). The chromatin strands (CHR STR) are very fine, consisting of one or two second-order DNA coils dispersed in the nucleoplasm. Ribosomes (RIBOS) are very numerous and arranged largely as RER.

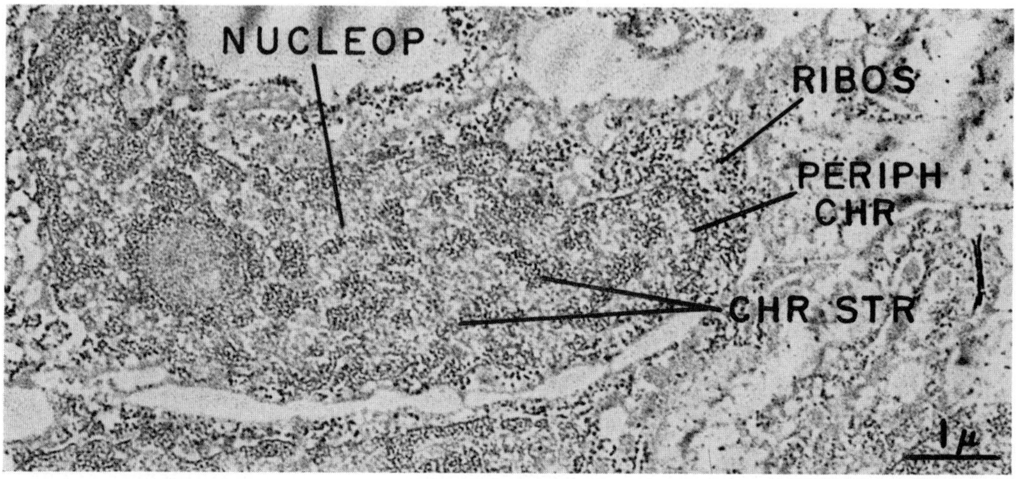

FIG. 10.11. Nucleus of surface epithelium of uterus of untreated, control mouse, 4 weeks old. Peripheral chromatin (PERIPH CHR) is irregular in thickness, with small interruptions. Chromatin strands (CHR STR) are relatively uniform and coarse. NUCLEOP, nucleoplasm; RIBOS, ribosomes.

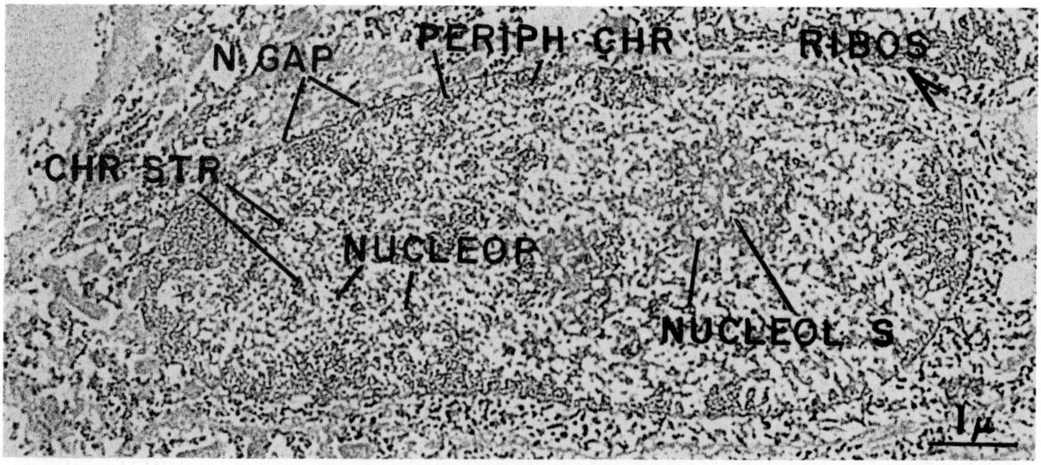

FIG. 10.12. Nucleus of surface epithelium of uterus of mouse 4 weeks old, injected with estradiol. Peripheral chromatin (PERIPH CHR) is very irregular in thickness, with numerous fine interruptions. Chromatin strands (CHR STR) are very fine. Nucleoli are very prominent. The spongelike structure of the nucleus is very prominent, as compared with the control, because of the enlarged nucleoplasmic spaces (NUCLEOP) and nucleolar spaces (NUCLEOL S). Ribosomes (RIBOS) are much more numerous than in control cells. × 16,000.

Ribosomes are less numerous than in the cytoplasm of smooth muscle cells of the hypertrophied uterus of late pregnancy.

Hypertrophied Uterus of Late Pregnancy (Fig. 10.10). The peripheral layer of chromatin is extremely thin, consisting of second-order DNA coils one or two coils thick, with frequent nuclear gaps. The chromatin strands are uniformly fine as they are distributed in the nucleoplasmic walls. Nucleoli are large and may be multiple. RNA granules in the nucleoplasmic walls are more numerous than in the uterine muscle nuclei of other types of animals. Ribosomes arranged in ordered patterns are most prominent in these cells.

Epithelial Cells of Uterus

Control Uterus of Mice 4 Weeks Old, Unstimulated (Fig. 10.11). The peripheral chromatin is variable in thickness with numerous, chiefly coarse, chromatin strands.

Stimulated Uterus of Mice 4 Weeks Old, Treated with Estradiol (Fig. 10.12). The nucleus is paler, and the peripheral chromatin is interrupted by numerous fine gaps. The chromatin strands are mostly very fine. The spongelike structure of the nucleus is prominent, as compared with nuclei of unstimulated cells. Ribosomes in the stimulated cells are much more numerous than in unstimulated cells.

Discussion

In all five kinds of stimulated cells, the DNA patterns are clearly affected in similar ways. The large aggregates of second-order DNA helices are reduced to larger numbers of lesser aggregates, at the same time that the chromatin strands become more numerous and finer. The nuclear gaps are also increased in number and size. Frequently, the RNA granules in the nucleoplasm are also increased. Similar changes were noted also in cartilage cells at the prelytic level of the epiphyseal plate which are actively secreting components of cartilage matrix. The changes appear in lytic cells to an extreme degree, as if in a caricature (see Chapter 5, Vol. II).

These observations can be interpreted to mean that when cells are strongly stimulated to synthesize proteins at a high rate, second-order DNA helices are separated from large aggregations, and are in a better position to generate and release mRNA to the nucleoplasm, and, via the nuclear gaps, to the cytoplasm. These changes are the opposite of those observed in maturing cells (this volume, Chapter 9), when the rate of protein synthesis is lower. One may also suggest here (i.e., in stimulated cells) that a basic mechanism involved is a change in a coil-aggregating and attachment-protein molecules, which attach DNA helices to each other and to the nuclear membrane. These may be altered or removed, and in this way permit modification of the second-order DNA helical pattern in the ways described in this and the preceding chapter. It is not known whether this postulated change in protein associated with DNA is a primary effect of the stimulating agent, or whether the effect is achieved after one or more intermediate steps. What seems clear is that, on the one side, there are rearrangements of second-order DNA coils with accompanying changes in distribution and number of RNA particles in the nucleus and cytoplasm, and, on the other side, that there are fluctuations in a large number of enzymic and energetic phenomena, accompanied by the flow of many small molecules. The exact relationships between the stimulus and these two sets of happenings remain mysterious despite the numerous reviews on this subject *(8, 9, 11, 14, 16–20, 25–29)*.

Often, increased cellular activity is accompanied by enlargement of the nucleus. This was observed most strikingly in the results recorded in this chapter in the smooth muscle cells of the uterus. Nuclear swelling has been found to be accompanied by an accumulation of protein in the nucleus *(2, 24)*. It probably occurred also in the nuclei of the other activated cells studied, and would have been more readily detected in sections 8 μ or more thick. Localized swelling of parts of cells not enclosed in membranes also occur, as in chromosomal puffs (Chapter 4, this volume) and in the cytoplasm of chromatolytic ganglion cells (Chapter 12, this volume). There is obviously more water in such swollen parts, but the effect is probably not one of simple dilution. In the nuclei of stimulated uterine cells, there is more total protein *(1)*. In at least one puff measured cytochemically *(23)*, and one other studied cytochemically *(24)*, there was also more total protein. In chromatolytic regions of cytoplasm, the same was also found to be true *(3, 12)*. In the work reported in this book, the puffs were found also to contain more insoluble RNA and tRNA (Chapter 4, this volume). In addition, in the nuclei of stimulated cells, as in the puffs, the secondary DNA helices are separated and in a better position to produce and release more mRNA. Thus, in all instances, the morphological and cytochemical evidence points to an improved accessibility of the machinery for the synthesis of new insoluble RNAs and new protein. Whether the new protein is made in the nucleus or in the cytoplasm is irrelevant for the conclusion that gene-activated processes are probably involved. The new protein, depending on its nature and its charge, could attract and hold cations. Together they would all tend to attract and hold water. In no instance, is it necessary to invoke alterations in membrane permeability or the activity of "pumps." This is especially so because the same (or essentially similar) changes take place in sites not enclosed by membranes (cytoplasm of chromatolytic nerve cells and puffs of *Drosophila* salivary gland chromosomes) as in nuclei of activated cells.

Summary

Changes in the organization of second-order DNA coils were observed when the activity of certain cells was stimulated: pancreatic acinar cells, cells of the adrenal medulla, epithelial cells of the seminal vesicle, and smooth muscle and epithelial cells of the uterus. The changes were in the same direction in all stimulated cells: (1) separation of second-order DNA coils to form increased numbers of finer aggregates or of separate coils; (2) increase in the number and size of nuclear gaps between residual masses of peripheral chromatin; and (3) increased numbers of RNA granules in the nucleoplasm and, sometimes, nucleoli. These morphological changes reflect conditions more favorable to the synthesis of RNAs and their transfer from chromatin to nucleoplasm and cytoplasm. The general "loosening" of second-order DNA coils to form less compact patterns is attributed to a decrease in the number of coil-aggregating and attachment proteins or to a loss in their effectiveness. These findings are the obverse of those associated with reduced cellular activity, and are considered in relation to control mechanisms for gene activity in somatic cells.

References

1. Alfert, A., and Bern, H. A. (1951). Hormonal influence on nuclear synthesis. I. Estrogen and uterine gland nuclei, *Proc. Nat. Acad. Sci. U. S.* **37**, 202–205.
2. Auer, G. A., Zetterberg, A., and Foley, G. E. (1970). The relationship of DNA synthesis to protein accumulation in the cell nucleus. *J. Cell Physiol.* **76**, 357–363.
3. Brattgård, S.-O., Edström, J.-E., and Hydén, H. (1957). The chemical changes in regenerating neurons. *J. Neurochem.* **1**, 316–325.
4. Carsten, M. E. (1968). Regulation of myometrial composition, growth, and activity. *In* "Biology of Gestation." (N. S. Assali, ed.), Vol. 1, pp. 355–425. Academic Press, New York.
5. Deane, H. W. (1963). Electron microscopic observations on the mouse seminal vesicle. *Natl. Cancer Inst. Monogr.* **12**, 63–83.
6. Deane, H. W., and Porter, K. R. (1960). A comparative study of cytoplasmic basophilia and the population density of ribosomes in the secretory cells of mouse seminal vesicle. *Z. Zellforsch. Mikrosk Anat.* **52**, 697–711.
7. Goodman, L. S., and Gilman, A. (eds.) (1965). "The Pharmacological Basis of Therapeutics." 3rd ed., p. 569. MacMillan, New York.
8. Gorski, J., and Notides, A. (1969). Estrogen control of uterine growth: Synthesis of specific uterine proteins. *In* "Biochemistry of Cell Division" (R. Baserga, ed.), pp. 57–76. Thomas, Springfield, Illinois.
9. Hamilton, T. H. (1968). Control by estrogen of genetic transcription and translation. *Science* **161**, 649–661.
10. Hamilton, T. H., and Teng, C.-S. (1969). Regulation by estrogen of synthesis of chromatin-directed RNA and of non-histone chromatin patterns. *Genetics Suppl.* **61**, 381–390.
11. Hechter, O., and Halkerston, I. D. K. (1964). On the action of mammalian hormones. *In* "The Hormones" (G. Pincus, K. V. Thimann, and E. B. Astwood, eds.), pp. 697–825. Academic Press, New York.
12. Hydén, H. (1943). Protein metabolism in the nerve cell during growth and function. *Acta Physiol. Scand.* **6**, *Suppl. 17,* 1–136.
13. Jensen, E. V., Numata, M., Smith, S., Suzuki, T., Brecher, P. I., and DeSombre, E. R. (1969). Estrogen-receptor interaction in target tissues. *Develop. Biol. Suppl.* **3**, 151–171.
14. Kuo, J. F. and Greengard, P. (1969) . Cyclic nucleotide-dependent protein kinases. IV. Widespread occurrence of adenosine 3′, 5′-monophosphate-dependent protein kinase in various tissues and phyla of the animal kingdom. *Proc. Nat. Acad. Sci. U. S.* **64**, 1349–1355.
15. Laguens, R. (1964). Effect of estrogen upon the fine structure of the uterine smooth muscle cell of the rat. *Ultrastruct. Res.* **10**, 578–584.
16. McKerns, K. W. (1967). Genetic, biochemical, and hormonal mechanisms in the regulation of uterine metabolism. *In* "Cellular Biology of the Uterus" (R. M. Wynn, ed.), pp. 71–113. Appleton-Century-Crofts, New York.
17. Nemetschek-Gansler, H. (1967). Ultrastructure of the myometrium. *In* "Cellular Biology of the Uterus" (R. M. Wynn, ed.), pp. 353–385. Appleton-Century-Crofts, New York.
18. O'Malley, B. W. (1971). Mechanisms of action of steroid hormones. *New Engl. J. Med.* **284**, 370–377.

19. O'Malley, B. W., McGuire, W. L., Kohler, P. O., and Korenman, S. G. (1969). Studies on the mechanism of steroid hormone regulation of synthesis of specific proteins. *Recent Progr. Horm. Res.* **25**, 105–160.
20. O'Malley, B. W., Spelsberg, T. C., Schrader, W. T., Chytil, F., and Steggles, A. W. (1972). Mechanisms of interaction of a hormone-receptor complex with the genome of a eukaryotic cell. *Nature (London)* **235**, 141–144.
21. Price, D., and Williams-Ashman, H. G. (1961). The accessory reproductive glands of mammals. *In* "Sex and Internal Secretions" (W. C. Young, ed.), 3rd ed., Vol. I, pp. 366–468. Williams & Wilkins, Baltimore, Maryland.
22. Ross, R., and Klebanoff, S. J. (1967). Fine structural changes in uterine smooth muscle and fibroblasts in response to estrogen. *J. Cell Biol.* **32**, 155–167.
23. Rudkin, G. T. (1964). The proteins of polytene chromosomes. *In* "The Nucleohistones" (J. Bonner and P. T'so, eds.), pp. 184–192. Holden-Day, San Francisco, California.
24. Schultze, B., Citoler, P., Hempel, K., Citoler, K., and Maurer, W. (1965). Cytoplasmic protein synthesis in cells of various types and its relation to nuclear protein synthesis. *In* "The Use of Radioautography in Investigating Protein Synthesis" (C. P. Leblond and K. B. Warren, eds.), pp. 107–138. Academic Press, New York.
25. Segal, S. J., and Scher, W. (1967). Estrogens, nucleic acids, and protein synthesis in uterine metabolism. *In* "Cellular Biology of the Uterus" (R. M. Wynn, ed.), pp. 114–150. Appleton-Century-Crofts, New York.
26. Singhal, R. L., Vjayvargiya, R., and Ling, G. M. (1970). Cyclic adenosine monophosphate: Andromimetic action on seminal vesicular enzymes. *Science* **168**, 261–263.
27. Smellie, R. M. S. (ed.). (1971). "The Biochemistry of Steroid Hormone Action." Academic Press, New York.
28. Spelsberg, T. C., Steggles, A. W., and O'Malley, B. W. (1971). Changes in chromatin composition and hormone binding during chick oviduct development. *Biochim Biophys. Acta* **254**, 129–134.
29. Sutherland, E. W., Robinson, G. A., and Butcher, R. W. (1968). Some aspects of the biological role of adenosine 3', 5'-monophosphate (cyclic AMP). *Circulation* **37**, 279–306.
30. Teng, C.-S., and Hamilton, T. H. (1969). Role of chromatin in estrogen action in the uterus. II. Hormone-induced synthesis of nonhistone acidic proteins which restore histone-inhibited DNA-dependent RNA synthesis. *Proc. Nat. Acad. Sci. U. S.* **63**, 465–472.
31. Wynn, R. M. (1967). Current problems in uterine cellular biology. *In* "Cellular Biology of the Uterus" (R. M. Wynn, ed.), pp. 475–500. Appleton-Century-Crofts, New York.

11

DNA Molecular Pattern during Metaplasia of Stratified Squamous Epithelium of Mouse Vagina to Mucin-Producing Cells

Isidore Gersh

During maturation of the stratified squamous epithelium of the vagina, changes take place in the DNA molecular pattern of the nuclei. The well dispersed second-order helices of the basal and intermediate layers are condensed in the superficial cell nuclei as a thick layer of peripheral chromatin while the volume of nucleoplasm is greatly reduced (see this volume, Chapter 9). In the mouse, rat, and guinea pig, for reasons which are not well understood, the superficial cells may be transformed to produce mucin *(14, 16, 20, 30)*. That most of the transformed mucous cells arose by metaplasia rather than by cell division followed by differentiation was demonstrated by the use of tritiated thymidine (1). It was thought to be important to learn whether the DNA molecular pattern is altered in such metaplastic cells.

Mucification of the stratified squamous epithelium of the vagina occurs normally in some rodents in prepuberty and pregnancy. The cells which produce mucin are derived from the superficial layers. These cells, instead of being shed or developing into nonnucleated cornified cells produce mucus rather than keratin. A few attempts have been made to induce this metaplastic change experimentally, but they have not been altogether successful *(6, 11, 14, 16, 20)*. A similar metaplasia was produced in skin ectoderm in tissue cultures by the addition of vitamin A to the nutrient medium *(7, 8)*, and in the cheek pouch of hamsters with implanted pellets of vitamin A *(9, 12)*. On further investigation, this was interpreted to mean that brief contact with excess vitamin A had switched the cell mechanism of differentiation into an alternate pathway *(29)*.

Materials and Method

Observations are recorded in this chapter on the DNA molecular pattern of nuclei of the vaginal epithelium of six mice where a sufficiently large area could be prepared free of visible ice crystals. In one 35-day-old mouse, the stratified squamous epithelium was thickest and there were no mucigen cells. In a second immature mouse, the vaginal epithelium contained some poorly developed mucous cells. In a third mouse, mucigen cells were present in the vaginal epithelium as a surface layer one or two cells thick. In a fourth immature mouse, and in a pregnant mouse (embryos 12 mm, crown-rump length), the superficial layer of mucigen cells in the vagina was two or three cells thick. In a second pregnant mouse (embryos 18–19 mm, crown-rump length), the superficial mucous layer of the vaginal epithelium was four or more cells thick. Thus, accidentally rather than through design, a good series of stages of metaplasia was obtained. Small pieces were frozen ultrarapidly, dried at a low temperature, treated *in vacuo* with vapors of mercury (II) hexafluoroacetylacetonate, and stained for nucleic acids with gallocyanin–chromalum. Efforts were made throughout to repeat procedures as nearly exactly as possible, so that electron micrographs could be comparable. Electron micrographs at a magnification of 2000 were made of large areas of a section, with a Siemens EMU-1 at an accelerating voltage of 60 kV and, from these, a photomontage was prepared for orientation purposes. Prints of selected nuclei were enlarged × 6 for detailed study of their DNA molecular pattern. The second-order DNA helices are not clear at this magnification, and the reader should refer to Chapters 3 and 4 (this volume), and Chapter 5 (Vol. II) where they are shown at higher magnification. As explained previously in this volume (Chapter 8) the particular magnification (× 12,000) was selected for practicality.

Further orientation was achieved by study of $\frac{1}{4}$-μ sections mounted on slides and stained with either toluidine blue for nucleic acids or periodic acid leukofuchsin for mucigen. In addition, sections 6-μ thick were prepared from formalin-fixed specimens embedded in paraffin, and stained with hematoxylin and eosin or with periodic acid leukofuchsin.

Observations

Descriptions are presented in the order of increasing number of mucigen granules in the superficial epithelial cells.

1. The superficial cells are nonnucleated, squamous, fully keratinized cells, free of mucigen granules (Figs. 11.1–11.5). The epithelium is about eight (nucleated) cells thick. Basal cell nuclei are oval, with dense peripheral chromatin clumps irregularly disposed on the inner surface of the nuclear wall, and coarse chromatin strands more centrally placed (Fig. 11.1). The nuclei are more outstretched in the intermediate cells (Fig. 11.2). In the next layers, where the nuclei are flattened, the layer of peripheral chromatin is more uniformly distributed,

there are few chromatin strands, and the nucleoplasmic volume is reduced (Fig. 11.3). The most superficial nuclei are irregularly compact and fragmented (Fig. 11.4). Abruptly after this layer the cells are nonnucleated. The ribosomal pattern is as follows: In the basal and intermediate cells ribosomes are numerous. In the more superficial cells with flattened nuclei, there are fewer ribosomes. In the most superficial nucleated cells, the number of ribosomes is further reduced. In the nonnucleated cells, there are no ribosomes (Fig. 11.5).

2. The epithelium is three cells thick, with some superificial cells containing a few mucigen granules (Figs. 11.6–11.8). The basal nuclei are ovoid, with a thin layer of peripheral chromatin interrupted by numerous narrow nuclear gaps (Fig. 11.6). Many fine to coarse chromatin strands arise from the peripheral chromatin aggregates. In nuclei of intermediate cells, the layer of peripheral chromatin is more easily defined because there are fewer (coarse) chromatin strands connected with the peripheral clumps (Fig. 11.7). In the nuclei of superficial cells, the nucleoplasmic volume is reduced as are also the nuclear gaps. The layer of peripheral chromatin and chromatin strands is not notably different from those in intermediate cells (Fig. 11.8). Nucleoli may be present in nuclei of all cell layers. Mucigen granules in the superficial cells vary from zero to four in sections of most cells. Ribosomes are present in all cells, and ring most mucigen granules.

3. The epithelium is about six cells thick, of which the outer one or two cells contain masses of mucigen granules (Figs. 11.9–11.13). Basal cell nuclei may be nearly spherical, and intermediate cell nuclei are oval. The layer of peripheral chromatin in both is thin, with numerous nuclear gaps and fine chromatin strands (Figs. 11.9 and 11.10). Nucleoli are prominent but are scarce or not visible in more superficial cells. Somewhat more superficially, the nuclei are outstretched. In these nuclei, the layer of peripheral chromatin becomes more uniform, mainly because fewer (and coarser) chromatin strands arise from its inner surface (Figs. 11.11 and 11.12). In the first (deeper) mucigen cells, there may be many mucigen granules. Here the DNA molecular pattern changes, with a denser (tighter) packing of second-order DNA coils and a few very coarse chromatin strands (Fig. 11.13). The nucleoplasm is markedly reduced in volume and is relatively free of RNA granules. In the superficial cells, the nucleus is irregular in shape, with condensed peripheral chromatin and few, if any, chromatin strands. Ribosomes are most numerous in the cytoplasm of basal and intermediate cells, and are also present in mucous cells.

4. The epithelium is six or seven cells thick, and of these the outer two or three cells contain large numbers of mucigen granules (Figs. 11.14–11.19). The nuclei of the deepest of these cell layers are ovoid, and may contain prominent nucleoli (Figs. 11.14 and 11.15). The layer of peripheral chromatin of the nuclei is thin and may be interrupted by nuclear gaps. Some RNA granules are present in the nucleoplasm. In more superficial cells, the nucleus is outstretched and there may be few or many mucigen granules, and nucleoli are very uncommon (Figs. 11.16 and 11.17). The layer of peripheral chromatin is compact and the chromatin strands, if present, are coarse. In the most superficial cells, the mucigen granules are most abundant. The nucleus is irregularly shaped, and the layer of

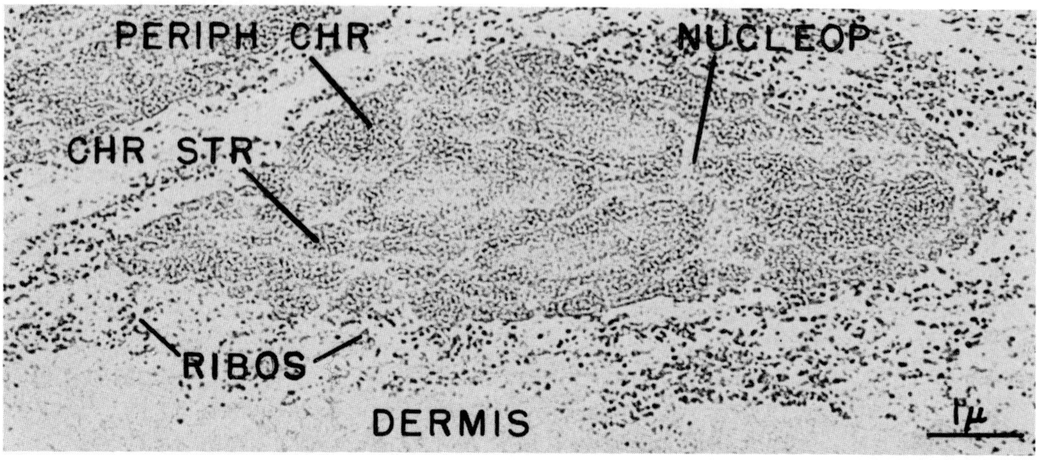

FIGS. 11.1–11.5. Successive nuclear regions of keratinized stratified squamous epithelium of vagina.

FIG. 11.1. Basal cell nucleus, with layer of peripheral chromatin (PERIPH CHR) and chromatin strands (CHR STR). RNA granules are present in the nucleoplasm (NUCLEOP). Ribosomes (RIBOS) are most numerous. Connective tissue is at the base of the cell (DERMIS).

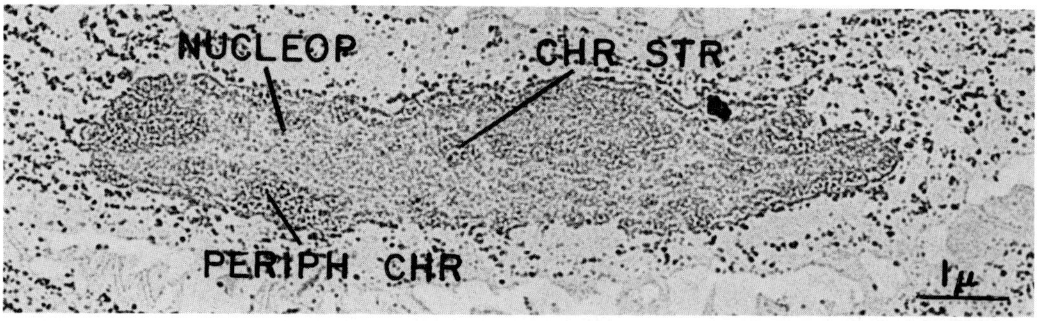

FIG. 11.2. Intermediate cell nucleus, with layer of peripheral chromatin (PERIPH CHR). Chromatin strands (CHR STR) are coarse. Nucleoplasm (NUCLEOP) contains numerous RNA granules.

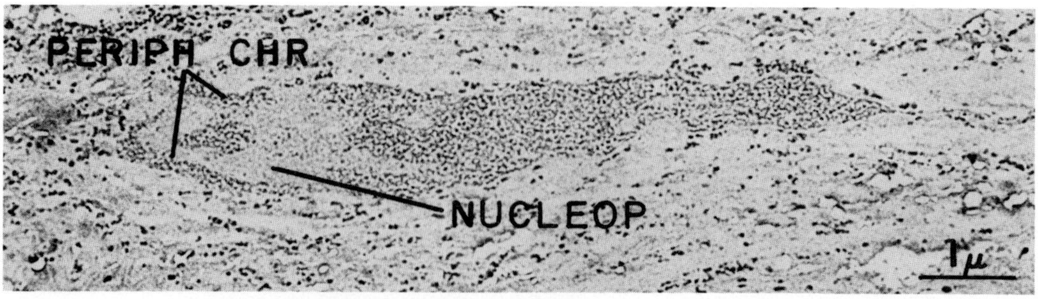

FIG. 11.3. Condensed nucleus of an adjacent cell. Nucleoplasm (NUCLEOP) is greatly reduced, and contains some RNA granules. Chromatin confined largely to a thick surface layer (PERIPH CHR).

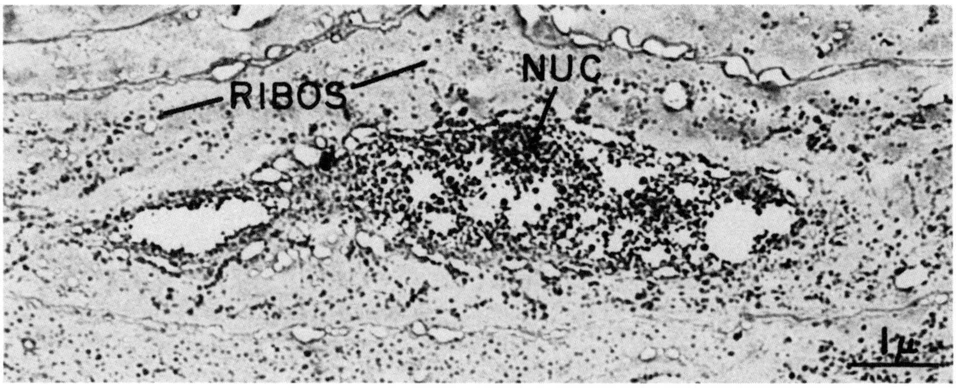

Fig. 11.4. Nuclear region of the most superficial nucleated cell (NUC). The DNA is condensed and granular. Ribosomes (RIBOS) are present in the cytoplasm.

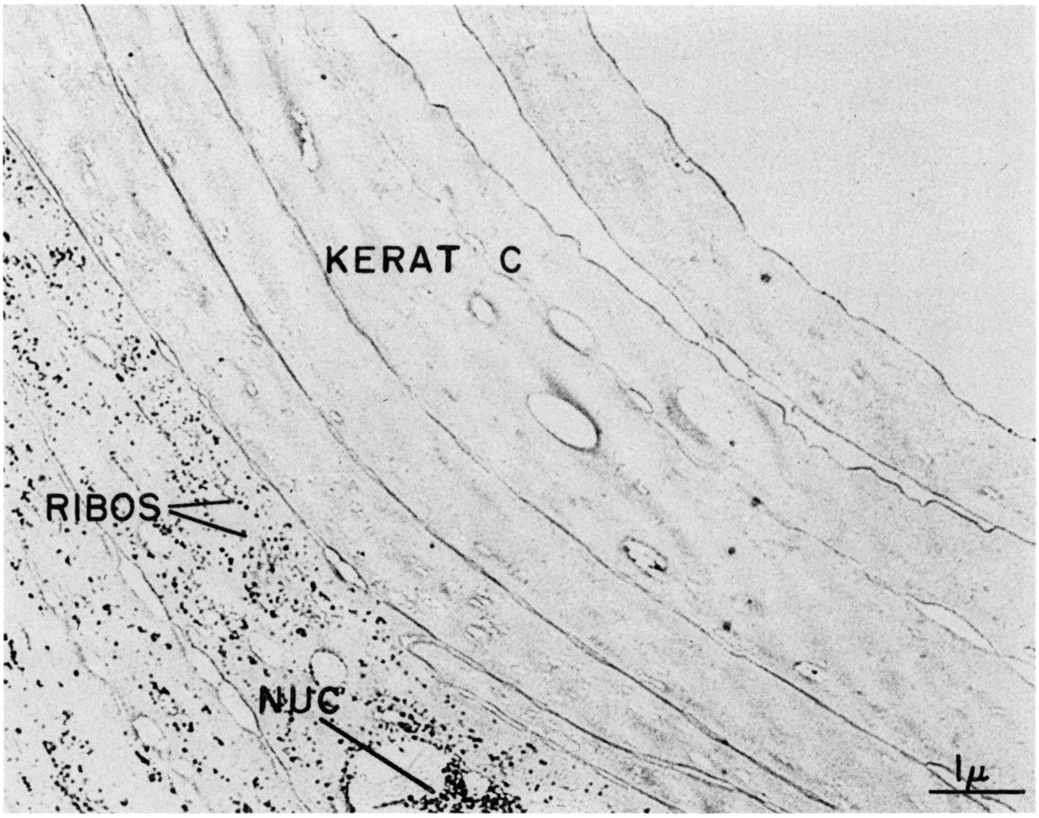

Fig. 11.5. Superficial nonnucleated layers. In the lowest layer is a portion of a cell with a fragmented nucleus (NUC) and cytoplasmic ribosomes (RIBOS). Nuclei and ribosomes are lacking in all the keratinized cells (KERAT C).

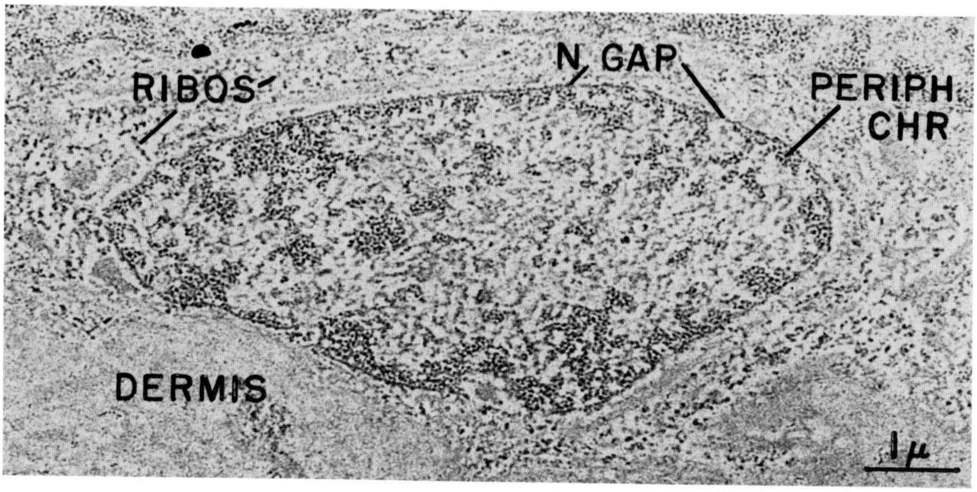

FIGS. 11.6–11.8. Successive nuclear regions of thin vaginal epithelium with some few, small mucigenic cells.

FIG. 11.6. Basal cell nucleus. The layer of peripheral chromatin (PERIPH CHR) is thin and interrupted by numerous nuclear gaps (N GAP). Chromatin strands vary from thick to thin. There seem to be fewer ribosomes (RIBOS) than in more superficial cells. Connective tissue (DERMIS) underlies basal cell.

peripheral chromatin is condensed (Figs. 11.18 and 11.19). There are few RNA granules in the reduced volume of nucleoplasm. Ribosomes are present in the cytoplasm of all cells, and are commonly observed in the cytoplasmic layers between mucigen granules.

5. The epithelium could be divided into two layers—an inner layer with no mucigen granules, about four cells thick, and an outer layer, four or more cells thick, whose cells are enormously distended with mucigen granules (Figs. 11.20–11.24). In the basal cells, the layer of peripheral chromatin is thin and interrupted by numerous nuclear gaps. Chromosomal strands are numerous and fine, and numerous RNA granules are present in the nucleoplasmic walls (Fig. 11.20). Immediately superficial to this is a layer of cells whose nuclei are somewhat larger and oval, with prominent nucleoli (Fig. 11.21). The layers of peripheral chromatin are more compact than in the basal cell nuclei, and the chromatin strands are more uniformly thick. The nucleoplasmic walls contain many RNA granules. In the more immediately superficial cells, the nuclei are oval. The distribution of second-order DNA coils is essentially the same, but nucleoli are scarce (Fig. 11.22). In the most superficial of the deeper half of the epithelium, the nucleus is somewhat irregular, and the layer of peripheral chromatin is more condensed (Fig. 11.23, lower cell). There are fewer chromatin strands and they are more uniformly coarse. The cells of the outer portion usually contain numerous mucigen granules. The nuclei of the deeper cells are irregular in shape, with condensed peripheral chromatin and chromatin strands (Fig. 11.23, upper cells). RNA granules are present in the nucleoplasm. In the more superficial cells, there

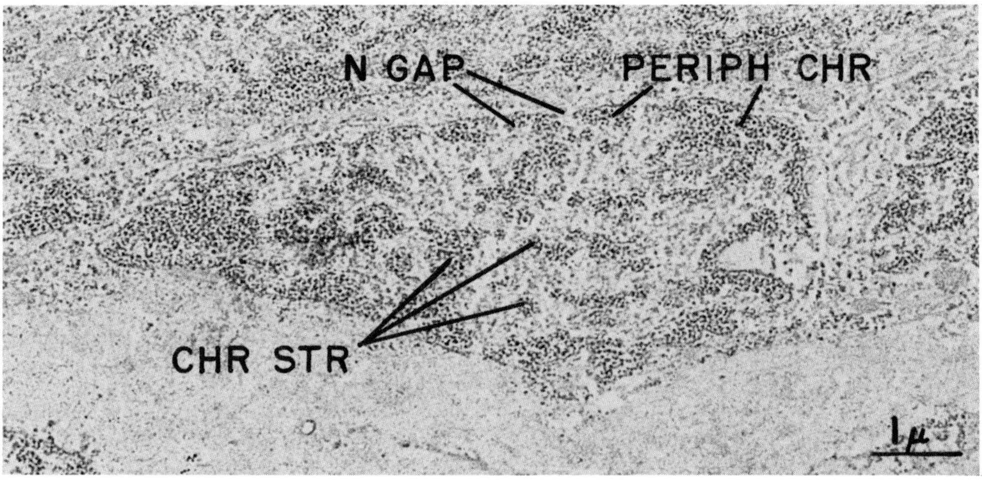

Fig. 11.7. Intermediate cell nucleus. Layer of peripheral chromatin (PERIPH CHR) is irregular in thickness. Chromatin strands (CHR STR) are intermediate in thickness between those in Figs. 11.6 and 11.8. Nuclear gaps (N GAP) are numerous.

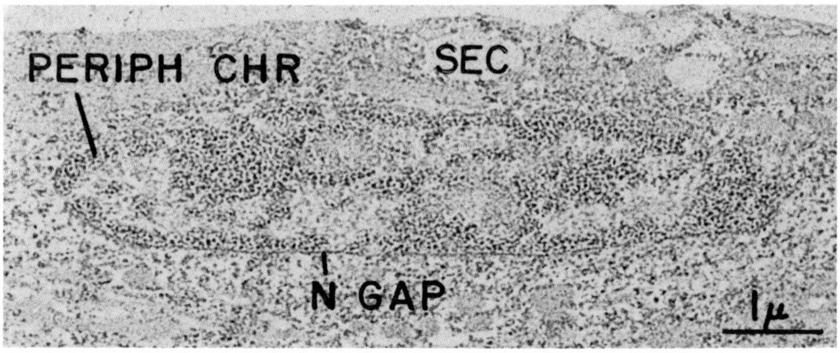

Fig. 11.8. Superficial nucleus of cell containing a few mucigen granules (SEC). Layer of peripheral chromatin (PERIPH CHR) is more uniform than in deeper cells, with fewer, coarser strands (CHR STR). Nuclear gaps (N GAP) are scarce.

are fewer chromatin strands and the layer of peripheral chromatin is well delineated (Fig. 11.24). There are relatively few RNA granules in the nucleoplasm. Ribosomes are present in the cytoplasm of all cells, and are visible in thin layers of cytoplasm between mucigen granules.

Discussion

The observations reported here confirm those described previously (Chapter 9), where changes in the DNA pattern of maturing epithelial cells of the vagina were described. The thin layer of second-order DNA helices of the peripheral

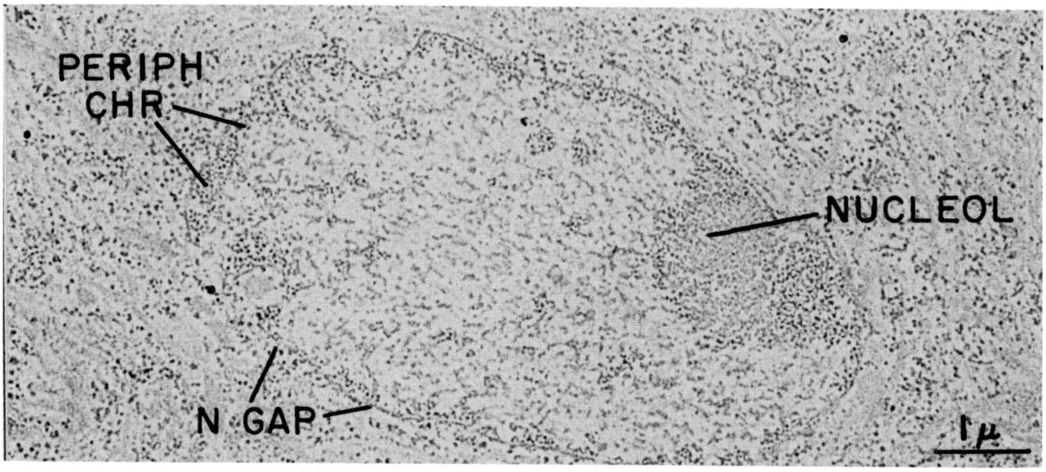

FIG. 11.9. ▲ ▼ FIG. 11.10.

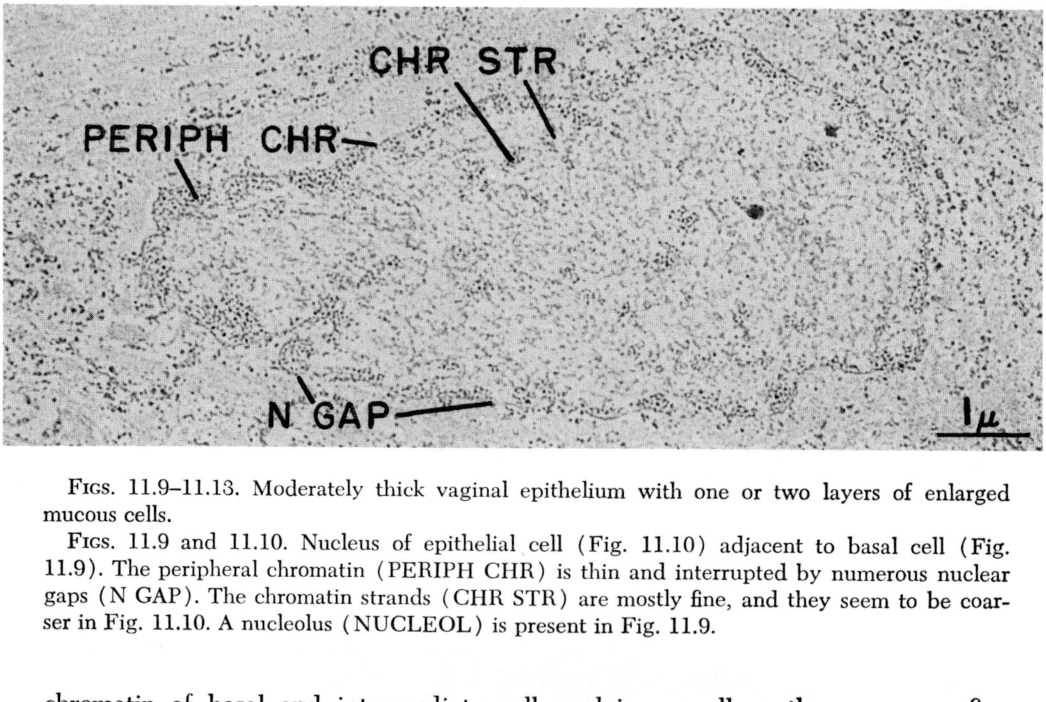

FIGS. 11.9–11.13. Moderately thick vaginal epithelium with one or two layers of enlarged mucous cells.

FIGS. 11.9 and 11.10. Nucleus of epithelial cell (Fig. 11.10) adjacent to basal cell (Fig. 11.9). The peripheral chromatin (PERIPH CHR) is thin and interrupted by numerous nuclear gaps (N GAP). The chromatin strands (CHR STR) are mostly fine, and they seem to be coarser in Fig. 11.10. A nucleolus (NUCLEOL) is present in Fig. 11.9.

chromatin of basal and intermediate cell nuclei, as well as the numerous fine chromatin strands, are condensed in the superficial cell nuclei into a well delimited layer of peripheral chromatin with a reduced number of chromatin strands, which are also coarser. Nucleolar and nucleoplasmic RNA granules are reduced and eventually the nucleoplasmic volume is similarly reduced. At this point, the cells are shed and their natural history as epithelium is ended. If their life is extended, for whatever reasons, the cells may develop in either of two directions:

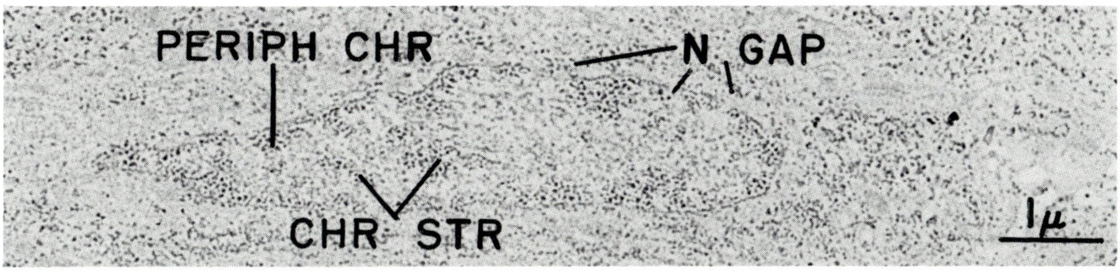

Fig. 11.11. More superficial, outstretched cell. The peripheral chromatin (PERIPH CHR) is very irregular in thickness and interrupted by numerous fine and large nuclear gaps (N GAP). Chromatin strands (CHR STR) are also variable in thickness.

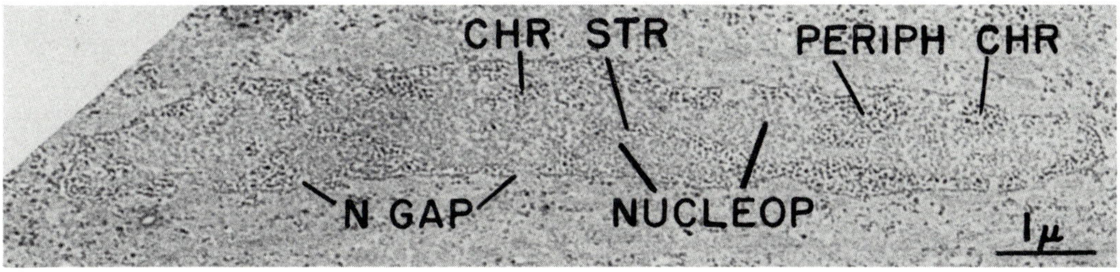

Fig. 11.12. Another nucleus at the same level of the epithelium. Peripheral chromatin (PERIPH CHR) and chromatin strands (CHR STR) are condensed, but there are numerous large nuclear gaps (N GAP). Nucleoplasm (NUCLEOP) is filled with RNA granules.

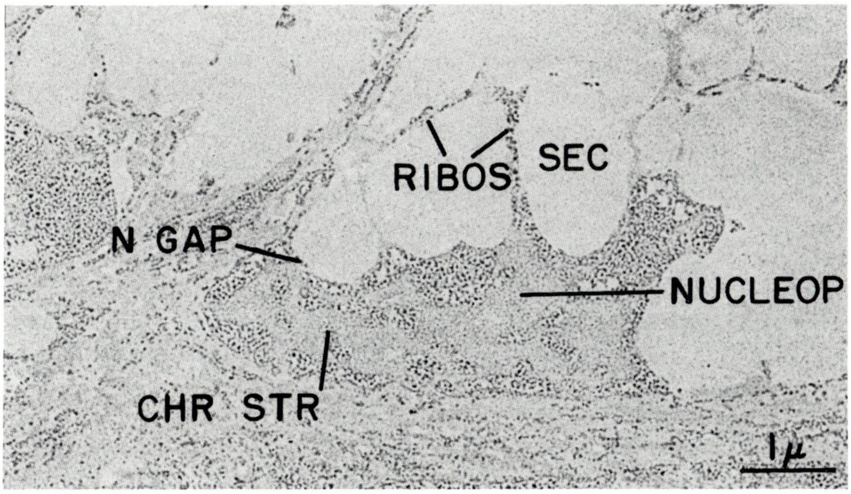

Fig. 11.13. Superficial mucous cell with numerous secretory granules (SEC) many of which are outlined by cytoplasmic layers containing ribosomes (RIBOS). The layer of peripheral chromatin (PERIPH CHR) is condensed and more or less uniformly thick, with a few nuclear gaps (N GAP). There are few, coarse chromatin strands (CHR STR). The compact nucleoplasm (NUCLEOP) contains RNA granules.

1. They may develop further and become highly keratinized and nonnucleated. In the nuclei of such cells, the layer of peripheral chromatin becomes further condensed, nucleoplasmic RNA is reduced, and the nucleoplasm is reduced almost to the vanishing point. Cytoplasmic ribosomes are also less numerous. In the most superficial layer of cells, nuclei become most condensed, then fragmented, and abruptly disappear, presumably by lysis. In the adjacent layer, the cells are free of nuclei and of ribosomes, as are also all the more superficial cells. This final morphological change also takes place very abruptly, in the time required for a cell to be displaced by one layer.

2. They may develop further in another direction, becoming transformed to metaplastic mucigen cells. In these cells, the DNA molecular pattern undergoes little change as the first mucigen granules appear. Only when there are many layers of mucigen cells packed with masses of secretory granules do nuclear changes appear. These include continued condensation of the layer of peripheral second-order DNA helices, a reduction in number of coarse chromatin strands, absence of nucleoli, and reduction in the number of RNA granules in the nucleoplasm. This is progressively reduced in volume, but never, as in the lysed nuclei, to the vanishing point. The changes resemble closely those described in maturing goblet cells (Chapter 9, this volume).

If the cells become keratinized rather than mucified, the eventual lysis of the peripheral nuclei of stratified squamous epithelium may take place in thickened parts of the epidermis. The facts that nuclear lysis is abrupt and that cytoplasmic ribosomes disappear at the same time suggests that lysosomes have erupted into activity. This is in marked contrast with the more leisurely self-destruction of hypertrophic cartilage cells of the epiphyseal plate of cartilage, where progressive stages in the process of reduction of second-order DNA helices to broken fragments and eventually granules could be observed (Chapter 5, Vol. II). Presumably, the DNA program for the synthesis of keratin by the penultimate layer of nucleated cells of the stratified squamous epithelium is terminated when the layer of peripheral chromatin becomes completely condensed and is unable to release mRNA (at the least) into the cytoplasm. Lysis of protoplasmic components during keratin transformation may be programmed, but it seems more probable that it is the consequence of lysosomal activity released because of progressive dehydration of the superficial cells, lack of nutrients, change in pH, or some other factors.

FIGS. 11.14–11.19. Moderately thick vaginal epithelium, of which the outer two or three cell layers are distended mucous cells.

FIGS. 11.14 and 11.15. Nucleus of epithelial cell (Fig. 11.14) adjacent to basal cell (Fig. 11.15). The distribution of nucleic acids in both nuclei is similar. The thin, irregular peripheral chromatin (PERIPH CHR) is interrupted by numerous fine gaps (N GAP). Chromatin strands (CHR STR) are mostly very fine. Nucleoli (NUCLEOL) are present. Nucleoplasm (NUCLEOP) clearly shows the spongelike dense walls and the intervening spaces. Ribosomes (RIBOS) are rather numerous.

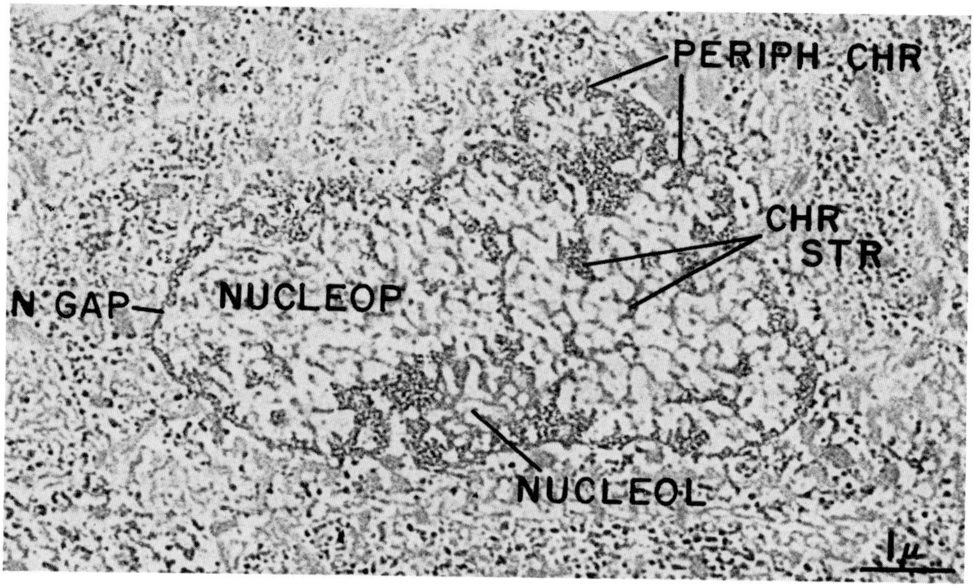

FIG. 11.14. ▲

▼ FIG. 11.15.

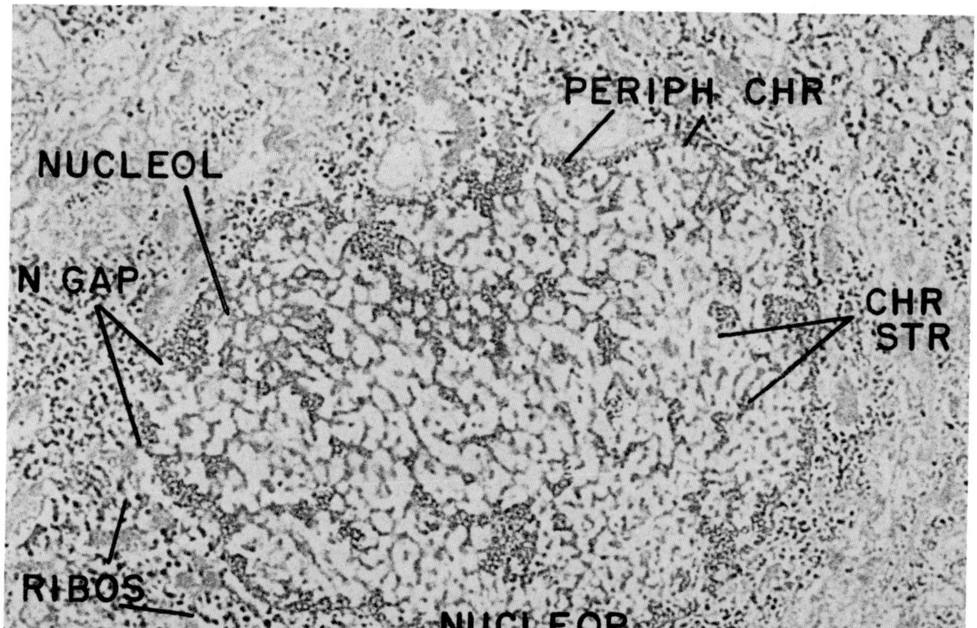

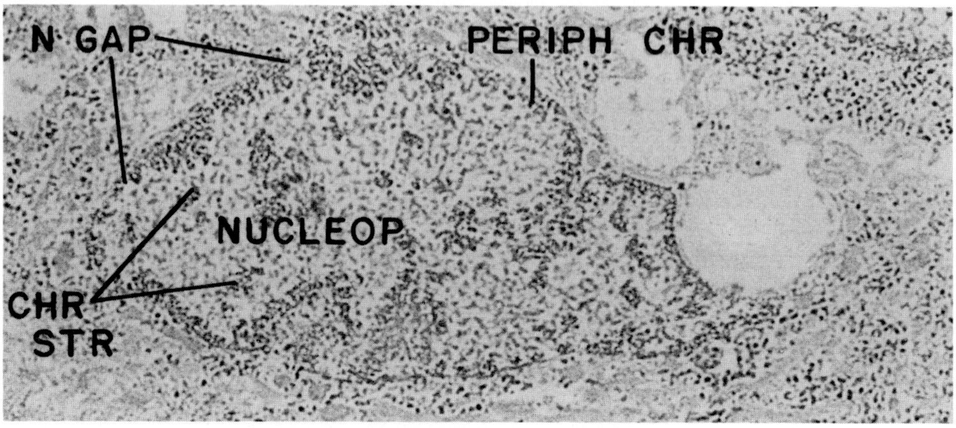

Fig. 11.16. Nucleus of another cell at the same level, with at least two vacuoles in the cytoplasm. Abbreviations as previously noted.

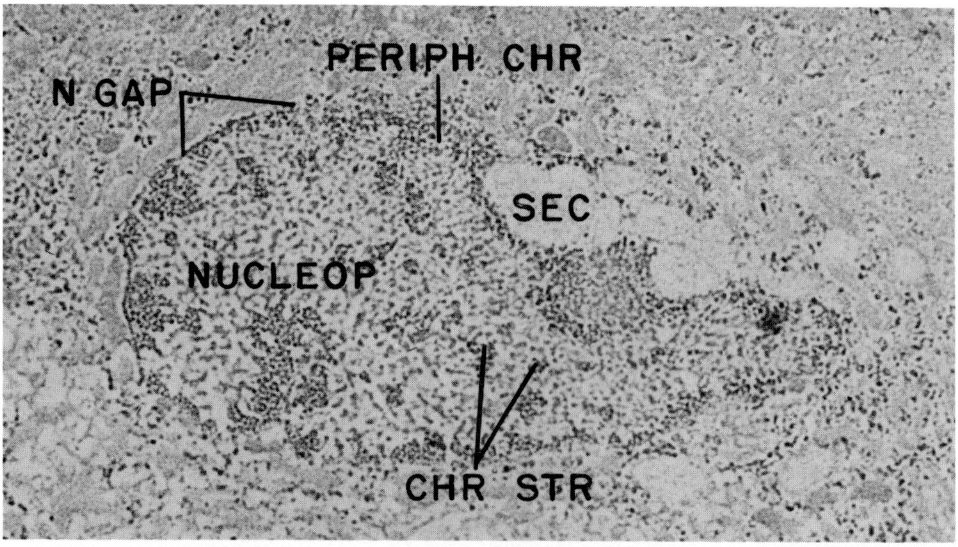

Fig. 11.17. Nucleus of a deeper cell containing few small secretory granules (SEC). Peripheral chromatin (PERIPH CHR) layer is irregular, but, on the whole, thin, with numerous fine nuclear gaps (N GAP). Chromatin strands (CHR STR) are irregular, but more of them are fine. The two phase system of the nucleoplasm (NUCLEOP) appears clearly.

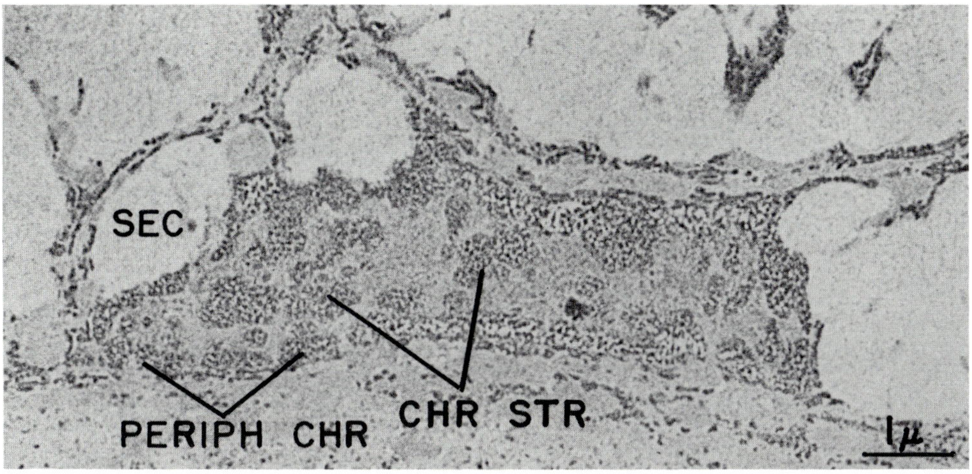

FIG. 11.18. Nucleus of a cell which seems to have more mucigen granules (SEC). Peripheral chromatin (PERIPH CHR) sharply delimited as in Fig. 11.19, but there seems to be more nucleoplasm (NUCLEOP). CHR, chromatin strand.

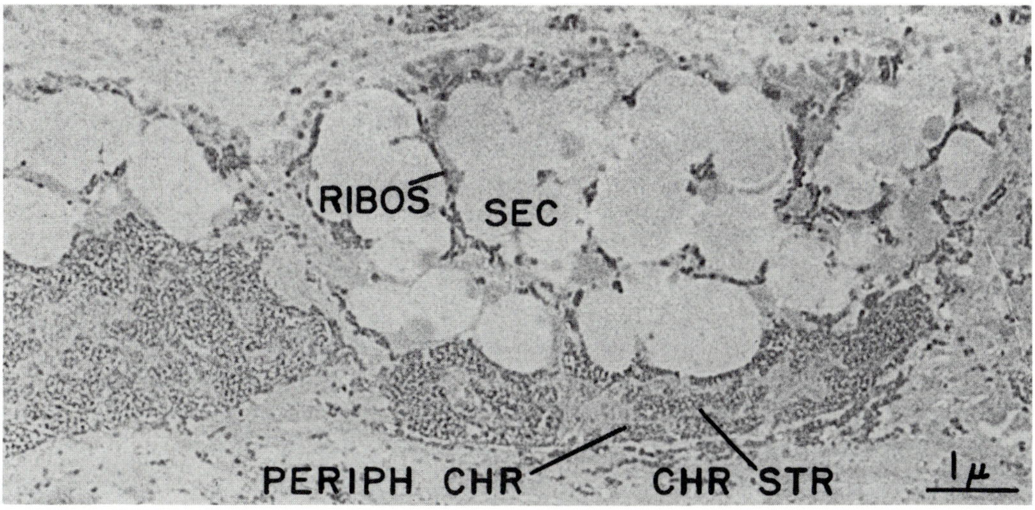

FIG. 11.19. Superficial mucous cell with secretory granules (SEC) separated by cytoplasm containing ribosomes (RIBOS). Peripheral chromatin (PERIPH CHR) and chromatin strands (CHR STR) sharply delimited from nucleoplasm, which contains some RNA granules.

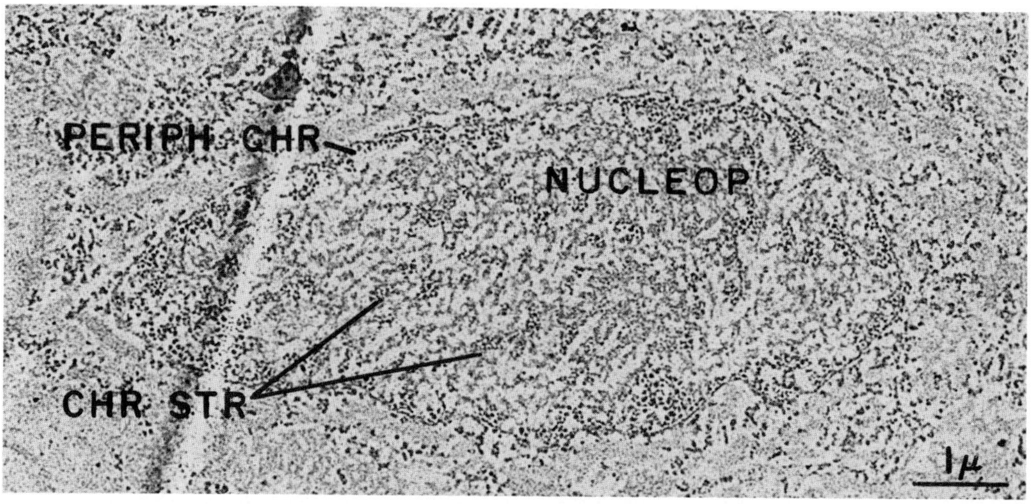

FIGS. 11.20–11.24. The vaginal epithelium is thick and composed of two layers—a deeper layer of four stratified squamous nonmucigenous cells, and a superficial layer consisting of mucous cells four or more cells thick.

FIG. 11.20. Basal cell nucleus with irregularly thin peripheral chromatin (PERIPH CHR) and chromatin strands (CHR STR). Numerous RNA granules are present in the nucleoplasm (NUCLEOP).

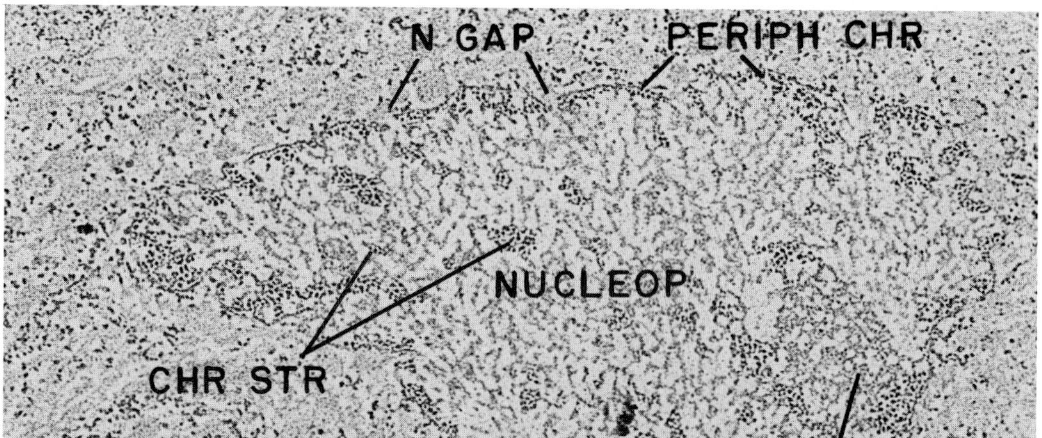

FIG. 11.21. Nucleus of a cell adjacent to basal cell. The peripheral chromatin layer (PERIPH CHR) is thin and interrupted by numerous nuclear gaps (N GAP). Chromatin strands (CHR STR) are somewhat irregular in thickness, but the proportion of fine strands is greater than in Fig. 11.22 and smaller than in Fig. 11.20. The nucleoplasm (NUCLEOP) contains RNA granules. The nucleolus (NUCLEOL) also shows the two phase system.

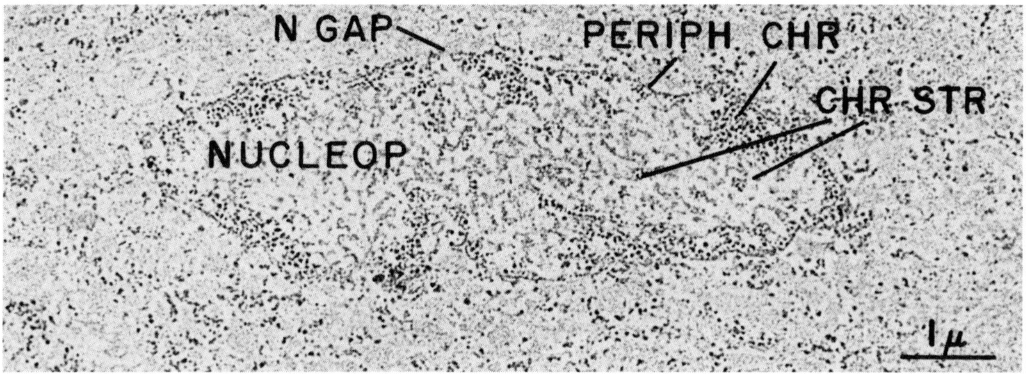

FIG. 11.22. Intermediate cell nucleus. The peripheral chromatin layer (PERIPH CHR) is highly irregular, and the chromatin strands (CHR STR) are mostly very fine. Numerous nuclear gaps (N GAP) interrupt the peripheral chromatin. Numerous RNA granules are present in the nucleoplasm (NUCLEOP).

The development of the process of keratinization of stratified squamous epithelium to the stage when several layers of nonnucleated cornified plates develop has provoked numerous studies which are reviewed in recent books *(3, 7, 9, 13, 15, 18, 32)*. Here I wish to discuss briefly only one pertinent set of observations made in the course of this study, i.e., the startlingly abrupt change from a nucleated to a nonnucleated condition, and the accompanying loss of ribosomes. This is different from the gradual loss of DNA with progressive differentiation, as determined quantitatively, which may be observed morphologically in skin, where the more superficial nuclei may be obviously depleted of chromatin as compared with deeply lying cells and where the intervening stages may show intermediate stages of depletion *(5, 23)*. In the specimen described in this report (p. 284), the change was not gradual, but sudden. It is not only a question of sudden disappearance of nuclei and ribosomes, but of cytoplasmic structures other than keratin *(2, 10)*. The absence of DNA and RNA from cornified cells may be mediated by the suggested action in cells of their respective hydrolytic enzymes *(27, 28)*.

The possibility has been raised that the DNase in epidermis may differ in mechanism of action from endonucleases of other cells *(17)*. If so, the enzymes are presumably released from lysosomes, as has been suggested *(21, 25, 26)*, but a clear demonstration of their presence and discharge at the correct moment is lacking *(9, 22, 31)*. Similar changes in nucleic acids in the hair bulb are also described *(15)*. I assume that these observations on the epidermis of the skin apply in principle to the same processes in the epidermis of the vagina.

The cytochemical observations on nucleic acids in cells of the vaginal epithelium reported here are consistent with morphological descriptions by various authors on their distribution in stratified squamous epithelia from various sources *(2, 4, 15, 18, 24, 32)*.

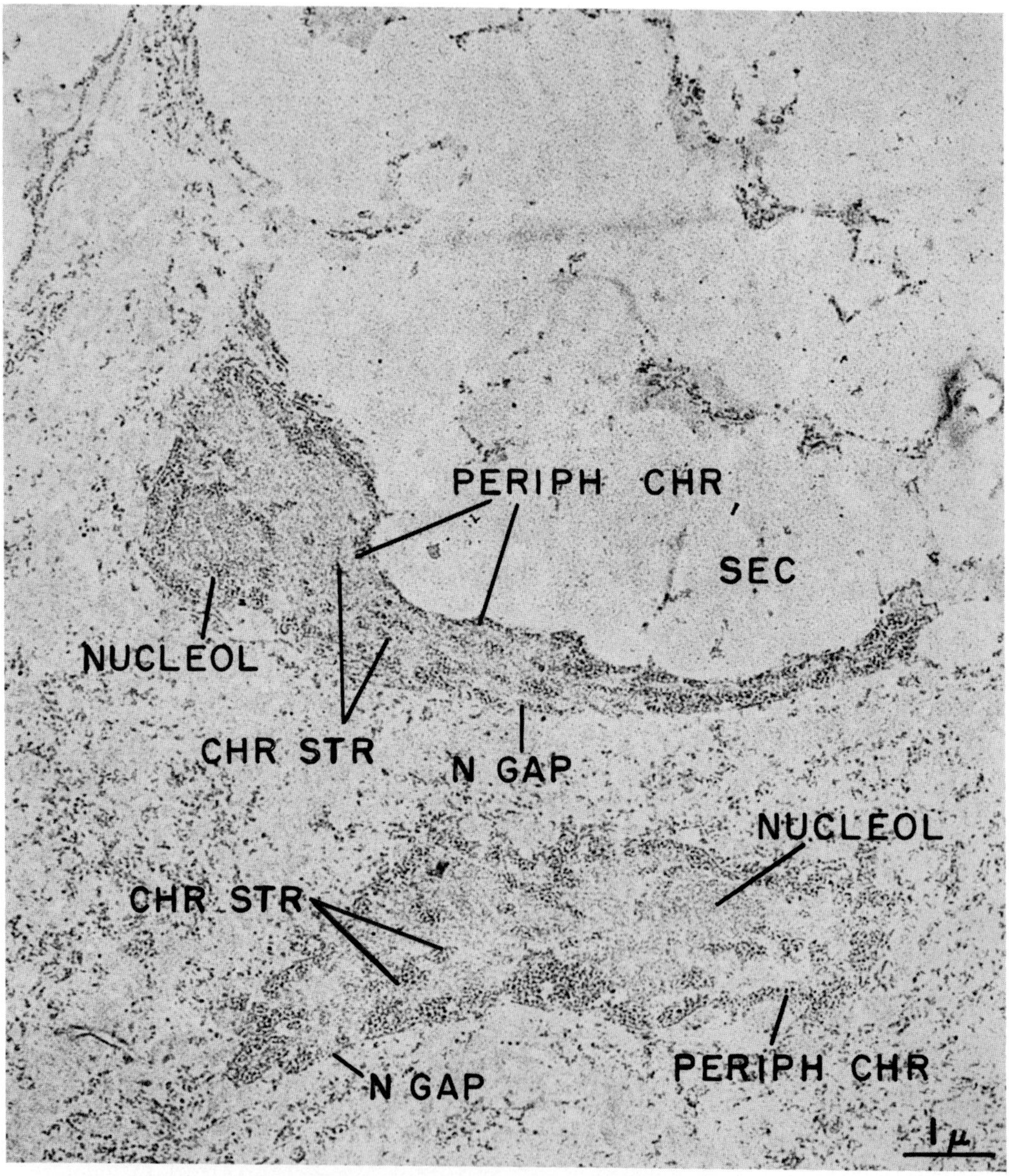

FIG. 11.23. A mucous cell deep in the mucous layer, and an adjacent cell of the stratified squamous epithelium. In the nucleus of the mucous cell, the peripheral chromatin (PERIPH CHR) and chromatin strands (CHR STR) are rather uniform. The nucleus of the nonmucous cell is similar, but the peripheral chromatin (PERIPH CHR) is irregular in thickness, and the nuclear gaps (N GAP) are more numerous. Both nuclei contain nucleoli (NUCLEOL).

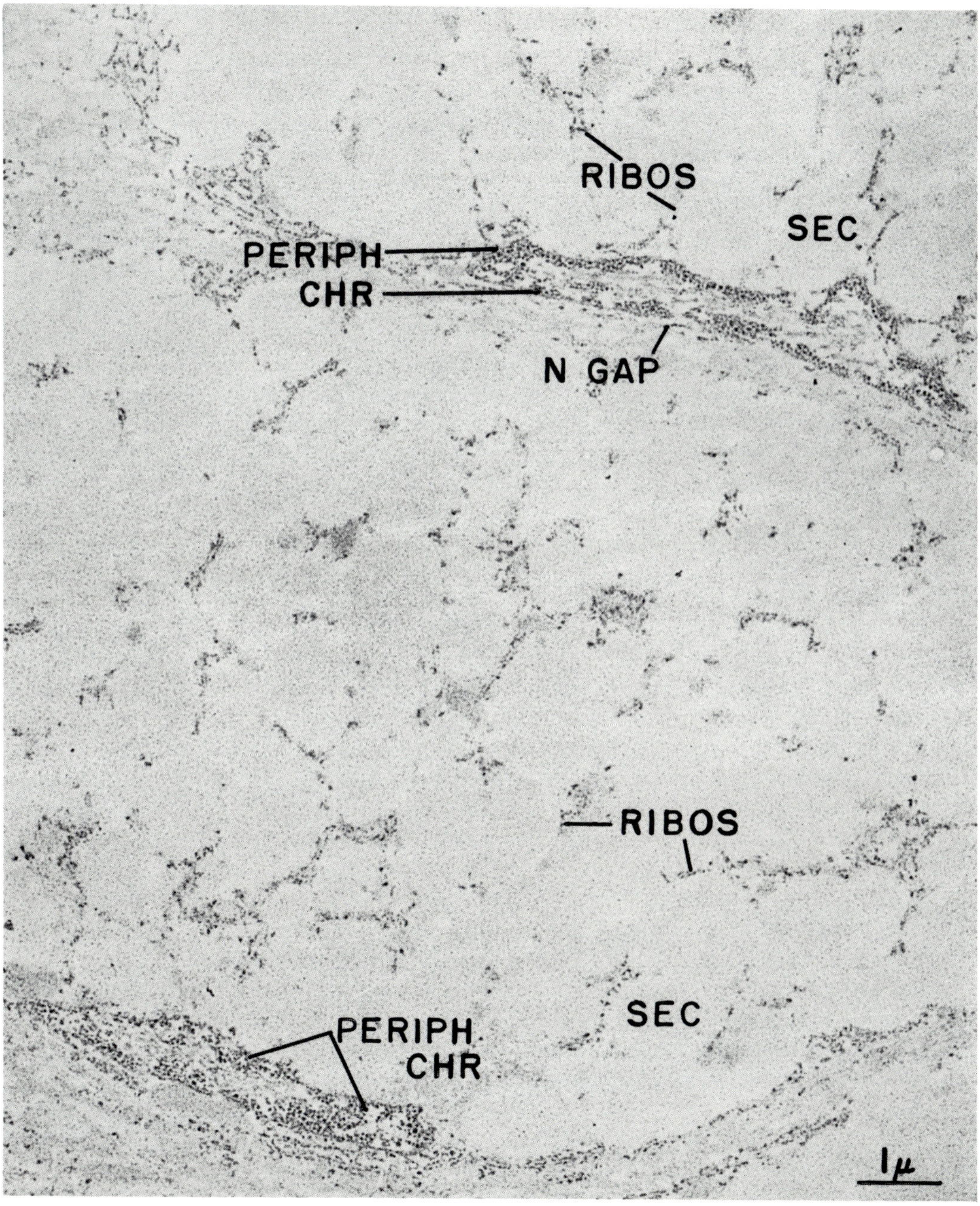

FIG. 11.24. Two superficial mucous cells, both containing numerous secretory granules (SEC) which are only partly outlined by cytoplasmic layers when they contain ribosomes (RIBOS). Both nuclei are irregular, with condensed peripheral chromatin (PERIPH CHR) which is more marked in the upper (more superficial) nucleus, where nuclear gaps (N GAP) are also larger.

The metaplastic transformation of epithelial cells producing keratin to cells producing mucigen involves a different process. Here a DNA program which is set for the synthesis of keratin is turned off, and a different program for the synthesis of a different protein is turned on. This takes place with an alteration in the DNA molecular pattern, especially in the cells which have produced and stored few mucigen granules. Later, when the cell is distended with numerous secretory granules, the DNA molecular pattern change is similar to that which takes place in maturing goblet cells. It would seem, then, that the second-order DNA helices of basal and intermediate cells responsible for cellular protein syntheses (including those for keratin) are exposed and potentially capable of activity. While the cistrons for keratin are effectively active, those for mucigen, if also exposed, are inactive or repressed by individual repressor molecules. Such cells, if not shed, may eventually develop into a highly stratified squamous epithelium in whose superficial cells nucleic acids may be lysed. If, however, in response to conditions not now known, the exposed second-order DNA loops are regrouped in heterochromatic regions and become inactive, the keratin program is repressed and turned off. If, at the same time, formerly "buried" second-order DNA coils are exposed and the pertinent cistrons are derepressed, the mucigen program is turned on, and the cellular synthetic activity could take a new pattern.

Summary

Epithelial cells of the mouse vagina which synthesize keratin may develop into nonnucleated keratin plates, or into mucus-producing cells. In the nuclei of the former, the second-order DNA coils become progressively more packed, and the nucleoplasm reduces in amount and becomes relatively free of RNA granules. Second-order DNA coils fragment. Then rapidly, in the time it takes one cell layer to form, DNA, RNA, and probably other protoplasmic components disappear, leaving nonnucleated keratin plates. In the mucus-producing cells, again rapidly, in the time it takes one cell layer to form, the pattern of organization of the second-order DNA coils changes, and develops like the nucleus of the mature goblet cell of the duodenum or colon. In this example, the organization of the second-order DNA helices changes abruptly from that characteristic of keratin-producing cells to that characteristic of mucin-producing cells. The findings are discussed in relation to the possible role of organization of second-order DNA coils in the expression of gene activity in somatic cells.

References

1. Baccarini, I. M. (1972). An autoradiographic study of the mouse vagina by serial biopsy. *J. Reprod. Med.* **8**, 122–127.
2. Bern, H. A., Alfert, M., and Blair, S. M. (1957). Cytochemical studies of keratin formation and of epithelial metaplasia in the rodent vagina and prostate. *J. Histochem. Cytochem.* **5**, 105–119.
3. Bernstein, I. A. (1964). Relation of the nucleic acids to protein synthesis in the mammalian

epidermis. *In* "The Epidermis" (W. Montagna and W. C. Lobitz, Jr., eds.), pp. 471–483. Academic Press, New York.

4. Brody, I., and Larsson, K. S. (1965). Morphology of mammalian skin; embryonic development of epidermal sublayers. *In* "Biology of the Skin and Hair Growth." (A. G. Lyne and B. F. Short, eds.), pp. 267–290. American Elsevier, New York.
5. Butcher, E. O., and Sognnaes, R. F. (1962). Fundamentals of keratinization. *Amer. Ass. Advan. Sci. Publ. No.* 70.
6. Cuadros, A., and Cooper, R. A. (1968). Ultrastructure of spontaneous vaginal keratinization in hanging-drop organ culture (Balb/c Crgl MICE). *Z. Zellforsch. Mikr. Anat.* **84,** 429–462.
7. Fell, H. B., and Mellanby, E. (1953). Metaplasia produced in cultures of chick ectoderm by high vitamin A. *J. Physiol.* **119,** 470–488.
8. Fell, H. B. (1957). The effect of excess vitamin A on cultures of embryonic chicken skin explanted at different stages of differentiation. *Proc. Roy. Soc. London.* **B146,** 242–256.
9. Hoffmeister, H., and Rupec, M. (1968). Elektronenmikroskopisch-zytochemischer Nachweis von Lysosomen im Vaginalepithel geschlechtsreifer Frauen, *Z. Geburtsh. Gynaekol.* **168,** 210–213.
10. Jarrett, A., and Spearman, R. I. C. (1964). "Histochemistry of the Skin-Psoriasis." English Universities Press, London.
11. Kahn, R. H. (1954). Effect of oestrogen and of vitamin A on vaginal cornification in tissue culture. *Nature (London)* **174,** 317.
12. Lawrence, D. J., and Bern, H. A. (1960). Mucous metaplasia and mucous gland formation in keratinized adult epithelium *in situ* treated with vitamin A. *Exp. Cell Res.* **21,** 443–446.
13. Lyne, A. C., and Short, B. F. (eds.). (1965). "Biology of the Skin and Hair Growth." Angus and Robertson, Sydney.
14. MacDonald, A. M., and Robson, J. M. (1939). The production of vaginal mucification with the synthetic oestrogen, triphenyl ethylene. *J. Pathol. Bacteriol.* **48,** 95–98.
15. Mercer, E. H. (1961). "Keratin and Keratinization." Pergamon, New York.
16. Meyer, R. K., and Allen, W. M. (1933). The production of mucified cells in the vaginal epithelium of certain rodents by oestrin and by corpus luteum extracts. *Anat. Rec.* **56,** 321–343.
17. Miyagawa, T., and Ogura, R. M. (1971). The biochemical properties and physiologic significance of deoxyribonuclease in the epidermis. I. Partial purification and properties of DNase isolated from cow snout epidermis, *J. Invest. Dermatol.* **57,** 330–336.
18. Montagna, W. (1961). Skin and integument and pigment cells. *In* "The Cell" (J. Brachet and A. E. Mirsky, eds.), Vol. V, pp. 267–322. Academic Press, New York.
19. Montagna, W., and Lobitz, W. C., Jr. (1964). "The Epidermis." Academic Press, New York.
20. Noble, R. L. (1939). Direct gynaecogenic and indirect oestrogenic action of testosterone propionate in female rats. *J. Endocrinol.* **1,** 184–200.
21. Novikoff, A. B. (1961). Lysosomes and related particles. *In* "The Cell" (J. Brachet and A. E. Mirsky, eds.), Vol. II, pp. 423–488. Academic Press, New York.
22. Olson, R. L., Nordquist, J., and Everett, M. A. (1968). Ultrastructural localization of aryl sulfatase in human epidermis. *Acta Dermato-Venereal.* **48,** 556–562.
23. Pelc, S. R. (1959). The participation of the cell nucleus and its DNA in the formation of keratin. *Exp. Cell Res. Suppl.* **6,** 97–104.
24. Rhodin, J. A. G., and Reith, E. J. (1962). Ultrastructure of keratin in oral mucosa, skin, oesophagus, claw, and hair. *In* "Fundamentals of Keratinization" (E. O. Butcher and R. F. Sognnaes, eds.), pp. 61–94. Amer. Ass. Advan. Sci., Washington, D.C.
25. Rowden, G. (1967). Ultrastructural studies of keratinized epithelia of the mouse. I. Combined electron microscope and cytochemical study of lysosomes in mouse epidermis and esophageal epithelium. *J. Invest. Dermatol.* **49,** 181–197.
26. Rowden, G. (1968). Ultrastructural studies of keratinized epithelia of the mouse. II. Aryl sulfatase in lysosomes of mouse epidermis and esophageal epithelium. *J. Invest. Dermatol.* **51,** 51–61.
27. Santoianni, P., and Rothman, S. (1961). Nucleic acid splitting enzymes in human epidermis and their possible role in keratinization. *J. Invest. Dermatol.* **37,** 489–495.

28. Steigleder, G. K., and Raab, W. P. (1962). The localization of ribonuclease and deoxyribonuclease activities in normal and psoriatic epidermis. *J. Invest. Dermatol.* **38**, 209–214.
29. Weiss, P., and James, R. (1955). Skin metaplasia *in vitro* induced by brief exposure to vitamin A. *Exp. Cell Res. Suppl.* **3**, 381–394.
30. Willis, R. A. (1958). "The Borderland of Embryology and Pathology," pp. 506–570. Butterworths, London.
31. Wolff, K., and Schreiner, E. (1970). Epidermal lysosomes. *Arch. Dermatol.* **101**, 276–286.
32. Zelickson, A. S. (1963). "Electron Microscopy of Skin and Mucous Membrane." Thomas, Springfield, Illinois.

12

Changes in Nuclear DNA Molecular Pattern of Spinal Ganglion Cells during Regeneration of Spinal Nerves of the Cat

Isidore Gersh and Peter J. Hand

The review by Hydén of the literature on the regeneration of peripheral nerves justifiably emphasized the recent contributions in this field by the Swedish school *(15)*. From their important researches, we have extracted some findings which point toward the critical and primary involvement of nuclear DNA in the process of regeneration. Neurosomal activity increases markedly during regeneration, as manifested by the increased synthesis of organic substances, including RNA, protein, and lipid *(4, 7–9, 14, 25)*. It should be added that base pair ratios of the RNA seem to remain constant during regeneration *(9)*. The nucleus is enlarged, and at some time may be partially surrounded by a shell of newly formed ribonucleoprotein. The nucleolus also becomes larger, and both it and the nucleus may be displaced. Besides these aspects of regeneration one must consider the marked changes which take place in the cytoplasmic RNA and protein content of nerve cells following mild or marked stimulation, and the effects on learning of inhibitors of RNA and protein synthesis. These findings are so consistent as to require involvement of nuclear activity, and in particular, gene action. Other reviews on chromatolysis do not state the problem as concisely or else ignore it entirely *(6, 19)*. Recent submicroscopic studies on chromatolysis have not shed light on the possible role of nuclear DNA on regeneration, at least as concerns nucleo–cytoplasmic relations, mainly because of inadequate fixation and staining. The chief points of agreement are summarized in the following two paragraphs, based on a number of references *(1–3, 5, 10–13, 16–18, 20–25)*.

The nucleus has been treated sparingly. Some reports merely acknowledge its existence, while others describe the nuclear membrane. Still others describe the

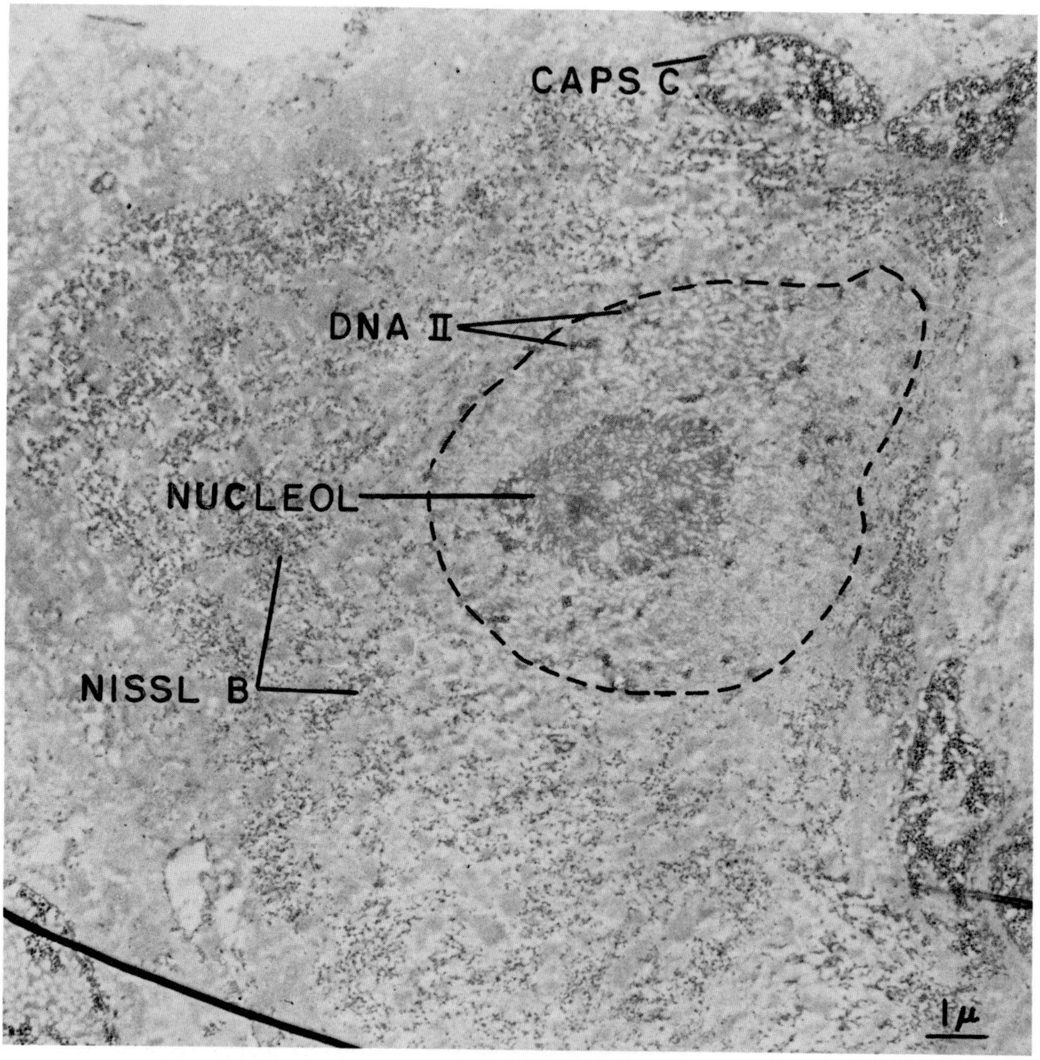

FIGS. 12.1–12.3. Six days after crushing T_{10}. Compare with Figs 12.4–12.7 of the unoperated control on the opposite side.

FIG. 12.1. Low power view. Nuclei of capsule cells (CAPS C) outline nerve cell in part. Nucleus is outlined with dotted lines. Nucleolus (NUCLEOL) shows typical spongelike structure. Attached to it are submicroscopic bits of DNA (DNA II), which are also scattered through the nucleoplasm. The cytoplasm contains numerous clumps of RNA aggregates, or Nissl bodies (NISSL B), which are mostly peripherally disposed. × 8000.

contents of the nuclear sac as granular, and, two have referred to them as presumably chromatin in nature. Nearly all reports describe the nucleolus as consisting of granules. During chromatolysis no consistent changes in the contents of the nuclear sac have been described. Some, however, have commented on indentations of the nuclear membrane which were observed more often in chromatolytic than in normal cells.

Nissl bodies are considered in all reports as aggregates of ribosomes (few of them occurring singly) organized in relation to endoplasmic reticulum. While some of the ribosomes are attached to these membranes as loops, rows, or spirals, most lie in the wide space between the membranes in small groupings called "rosettes." During chromatolysis these rosettes become smaller or disappear altogether as the number of ribosomes decreases. In later stages of chromatolysis, the compact rosettes are reconstituted in their usual relationship to endoplasmic reti-

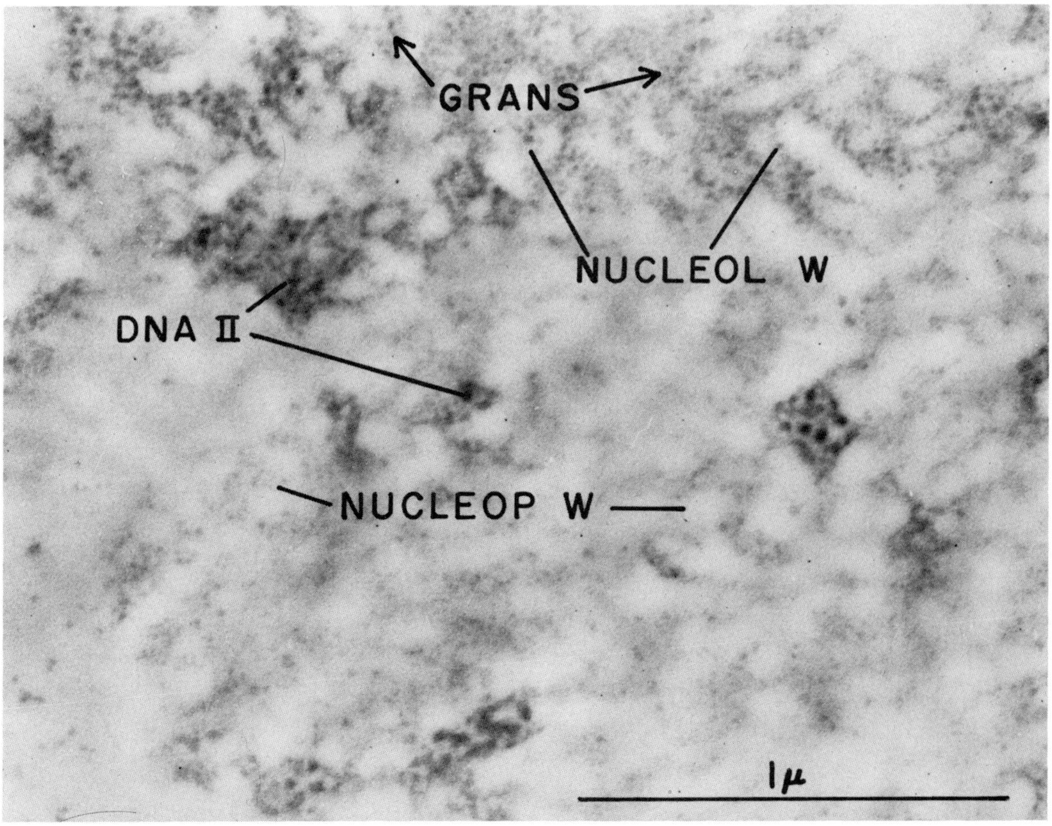

FIG. 12.2. Enlargement of a part of Fig. 12.1. Nucleolus is spongelike in structure, the walls (NUCLEOL W) containing numerous RNA granules (GRANS). At the periphery is a small aggregate of DNA coils (DNA II). Other second-order coils of DNA are in the nucleoplasm. The walls of the nucleoplasm (NUCLEOP W) contain numerous RNA granules. × 56,000.

culum, which has returned to normal from an interlude during which it was represented as a series of vacuoles. The Golgi apparatus is markedly hypertrophied during chromatolysis. Changes in mitochondrial morphology may take place, but these changes may reflect the quality of fixation rather than the effects of nerve regeneration.

The marked metabolic and morphological changes which take place in regenerating neurosoma resemble those of cells strongly stimulated to synthesize protein (this volume, Chapter 10) and, in a sense, are the reverse of changes which take

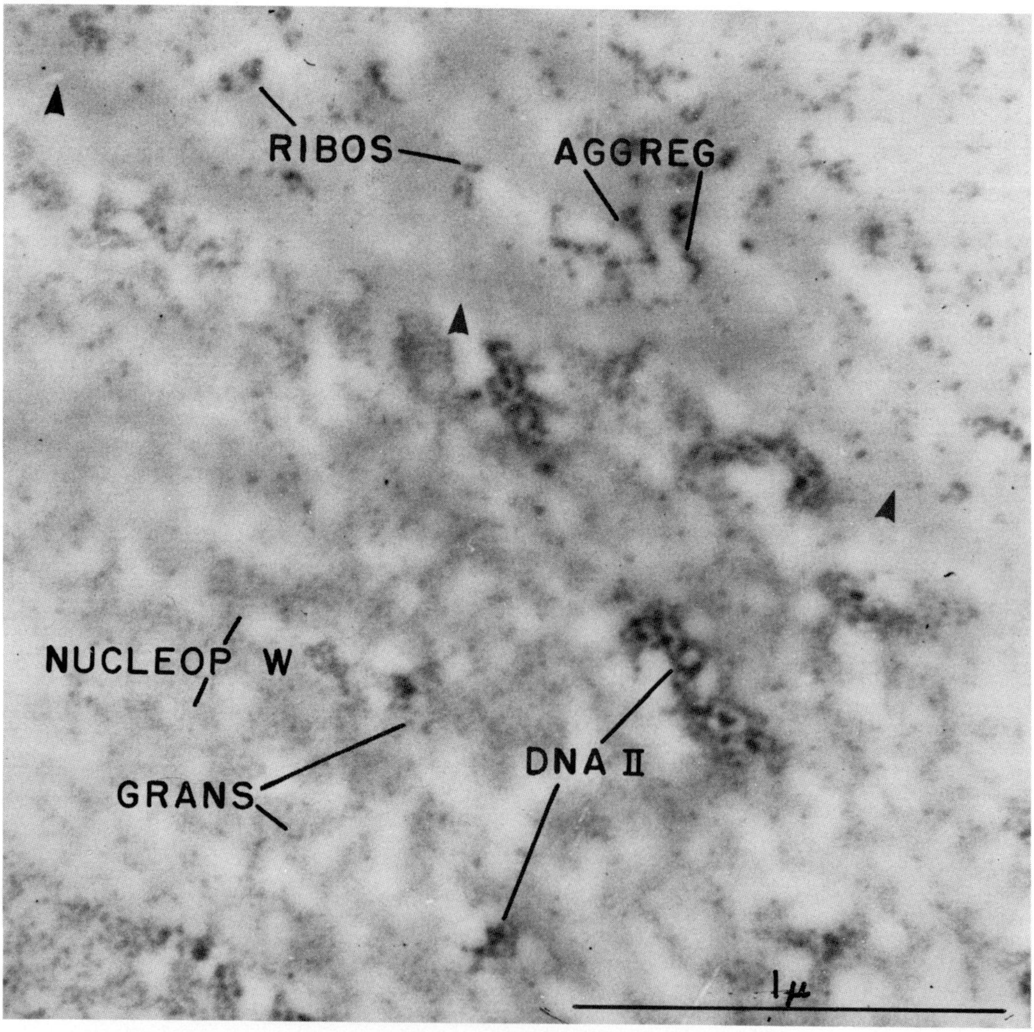

FIG. 12.3. Enlargement of a part of Fig. 12.1. Position of nuclear membrane is indicated by tips of arrowheads. Nucleoplasmic walls (NUCLEOP W) contain second-order DNA coils (DNA II) and numerous RNA granules (GRANS). Ribosomes (RIBOS) occur in perinuclear cytoplasm singly and in irregular sized and shaped aggregates (AGGREG). × 56,000.

place during maturation of certain cells (this volume, Chapter 9). Increased and decreased rates of protein synthesis in a variety of cells were found to be accompanied by altered patterns of organization of nuclear DNA. We anticipated that a notable change in the DNA molecular pattern would take place in nerve cells during regeneration, and sought such DNA alterations in the nucleus of spinal ganglion cells. These cells were selected for study because they can be suitably fixed. Cells in the cerebellum, cortex, or spinal cord cannot be frozen and dried satisfactorily, for unknown reasons. Cells of the sympathetic ganglia can also be well fixed, but the surgical problems are more complex.

Method

The material for this study was prepared by crushing certain thoracic nerves in two cats. The cats were anesthetized with Nembutal and the left thoracic nerves

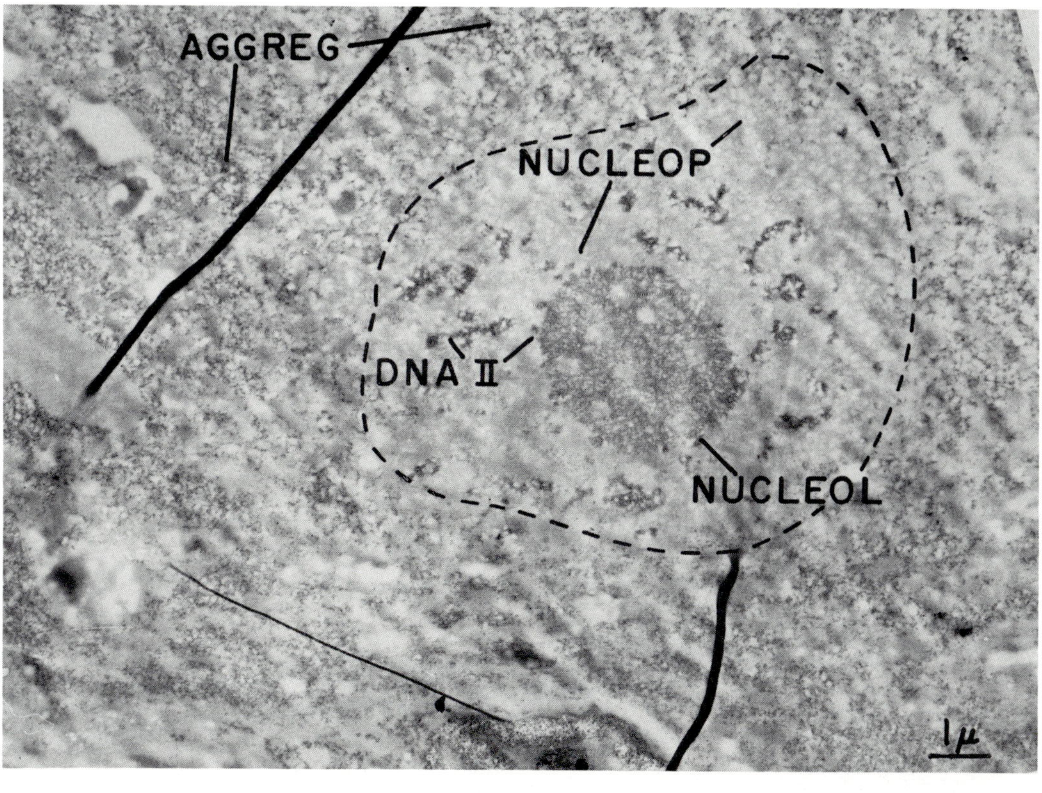

FIGS. 12.4–12.7. Unoperated control of T_{10}. Compare with Figs. 12.1–12.3 of operated T_{10}.

FIG. 12.4. Low power view. Nucleus is outlined with dotted lines. Nucleolus (NUCLEOL) shows typical spongelike structure. Attached to its periphery and in the nucleoplasm (NUCLEOP) are aggregates of second-order coils (DNA II) which are coarser than in nuclei of regenerating cells. Ribosomal aggregates (AGGREG) are distributed in the cytoplasm. × 8000.

were exposed by blunt dissection of the overlying epaxial musculature at the level of the intervertebral foramina with a hemostat. The nerves on the right side served as controls. This operation was performed successively on different nerves in one cat. On the day the experiment ended, the nerve fibers of T_{10} had been regenerating for 6 days, T_{11} for 10 days, T_{12} for 17 days, and T_{13} for 24 days. The control ganglia on the right side were from T_{10} and T_{13}. Ganglia were exposed

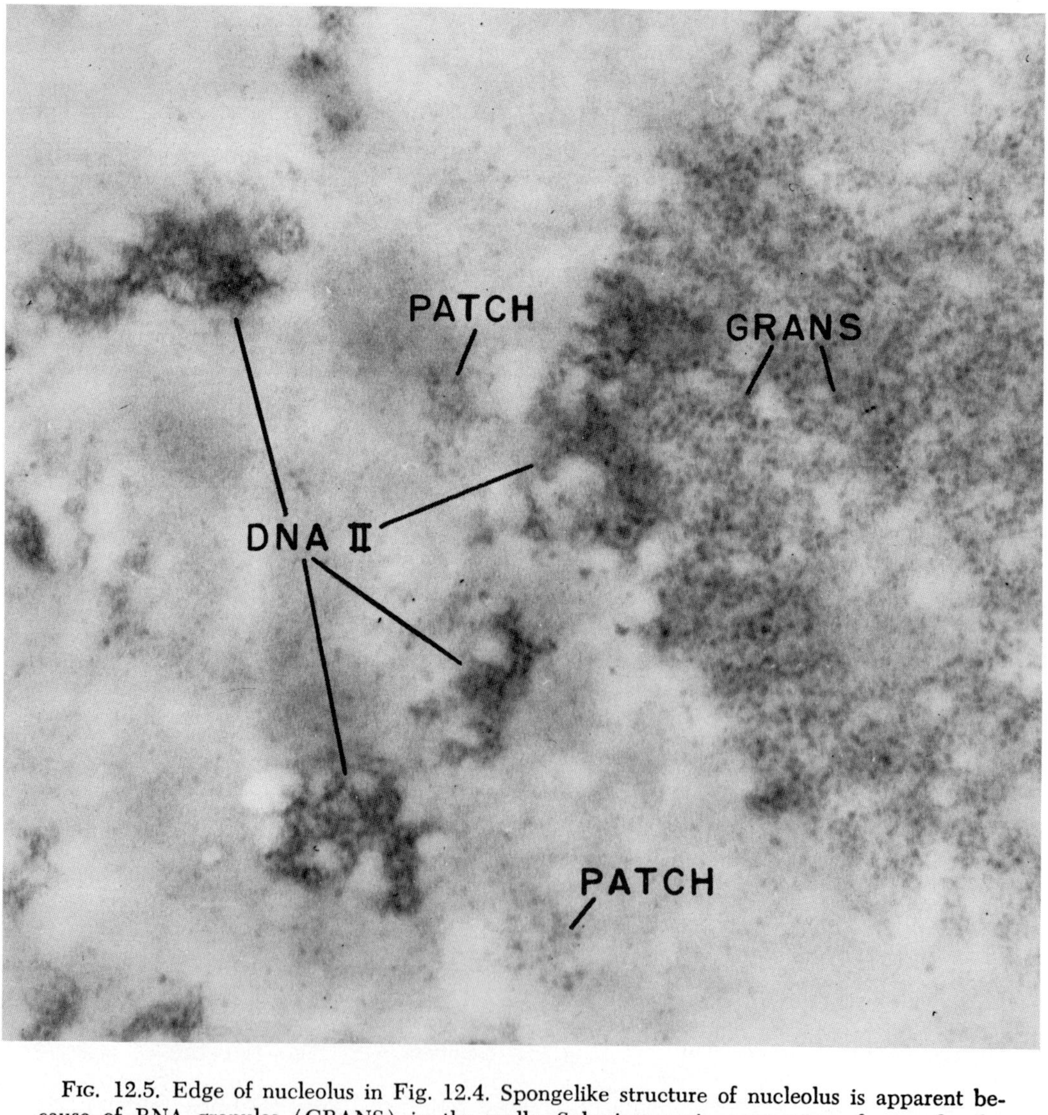

FIG. 12.5. Edge of nucleolus in Fig. 12.4. Spongelike structure of nucleolus is apparent because of RNA granules (GRANS) in the walls. Submicroscopic aggregates of second-order DNA coils (DNA II) are present at the edge of the nucleolus and in the nucleoplasm. Patches of RNA granules (PATCH) are shown in relation to some of these DNA coils. × 56,000.

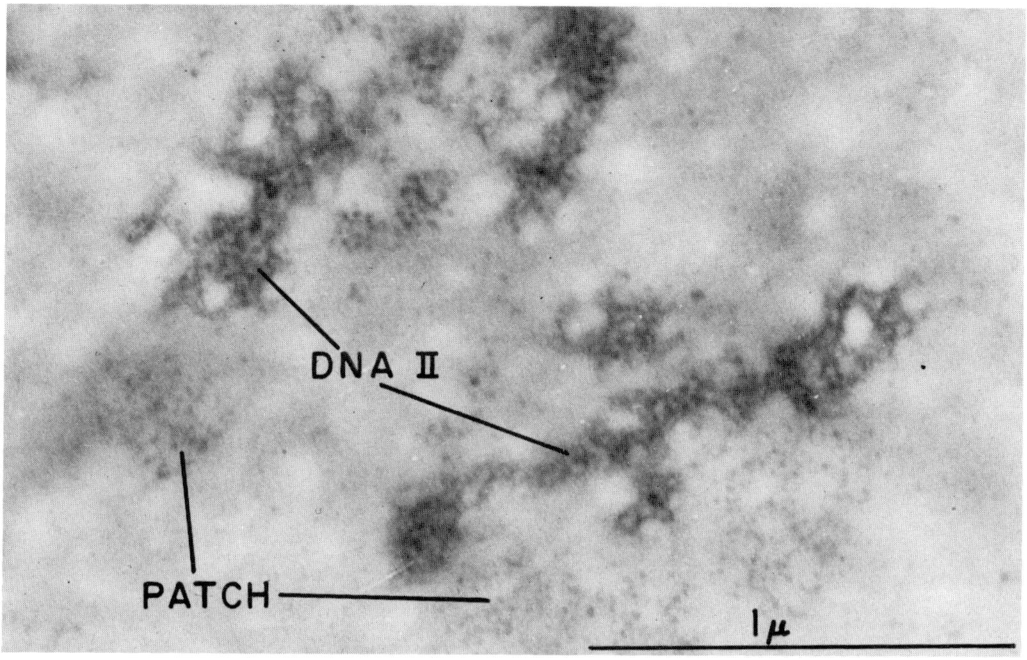

FIG. 12.6. Another portion of nucleoplasm of Fig. 12.4. Related to aggregates of second-order DNA coils (DNA II) are patches of RNA granules (PATCH). × 56,000.

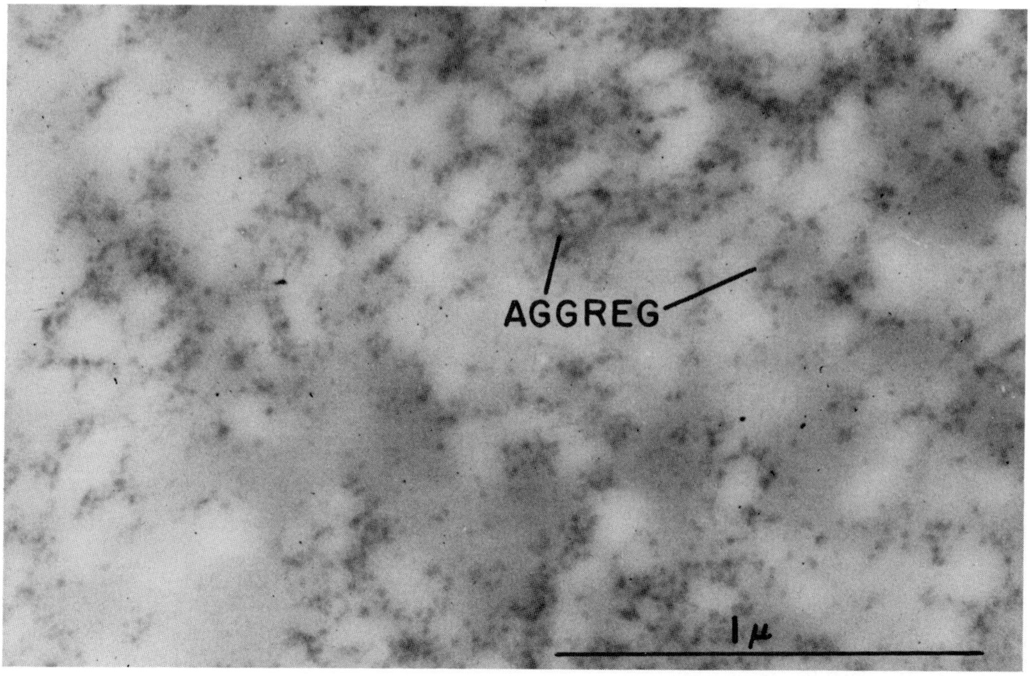

FIG. 12.7. A portion of cytoplasm of cell in Fig. 12.5. Ribosomes are aggregated (AGGREG), but the aggregates are so diffuse that Nissl bodies cannot be distinguished. × 56,000.

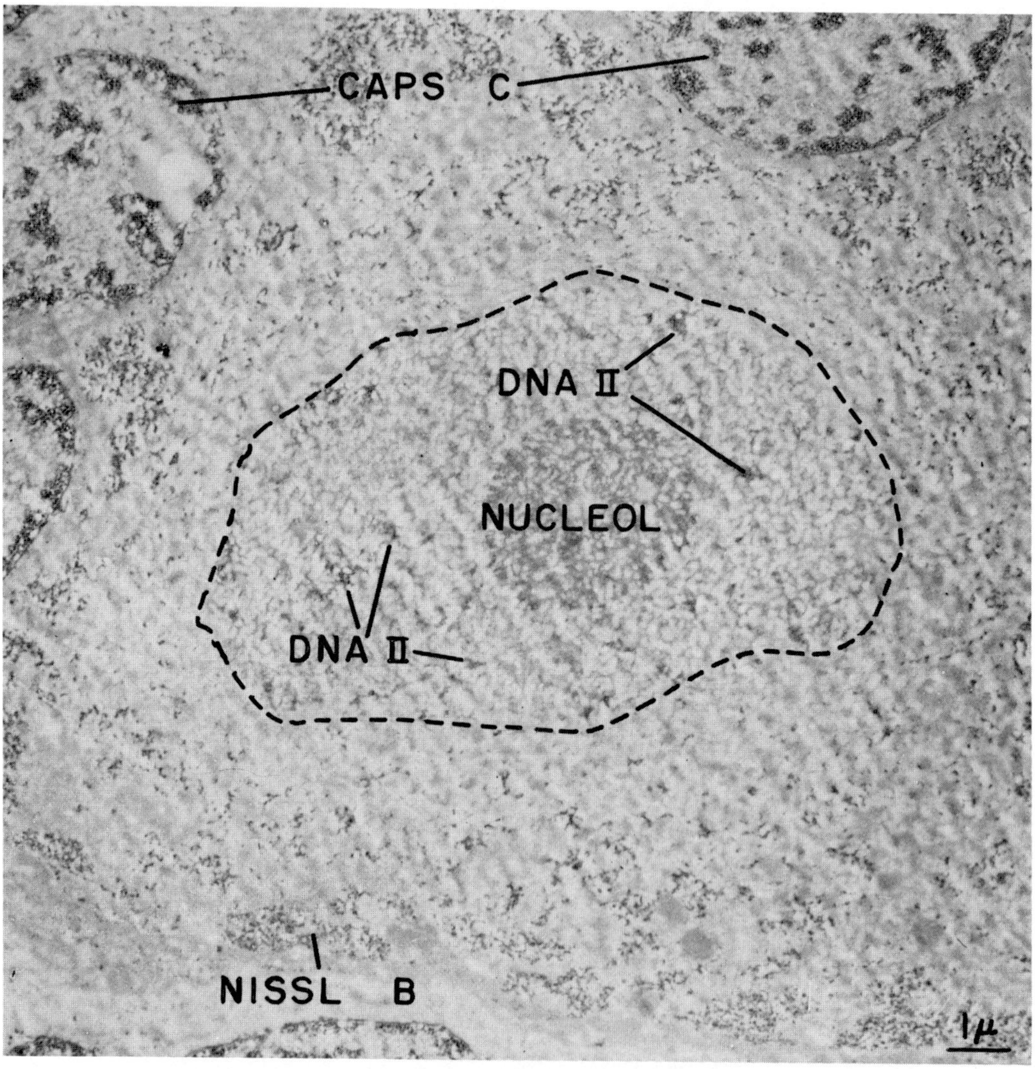

FIGS. 12.8–12.12. Ten days after crushing T_{11}. Compare with Figs. 12.1–12.3 (6 days after nerve crushing).

FIG. 12.8. Low power view. Nuclei of capsule cells (CAPS C) outline in part the nerve cell. Nucleus is outlined with dotted lines. Nucleolar walls emphasize spongelike structure of nucleolus (NUCLEOL). Small segments of second-order coils (DNA II) are distributed throughout the nucleoplasm. The cytoplasm contains a few peripherally disposed Nissl bodies (NISSL B). The remainder of the cytoplasm contains numerous ribosomal aggregates. × 8000.

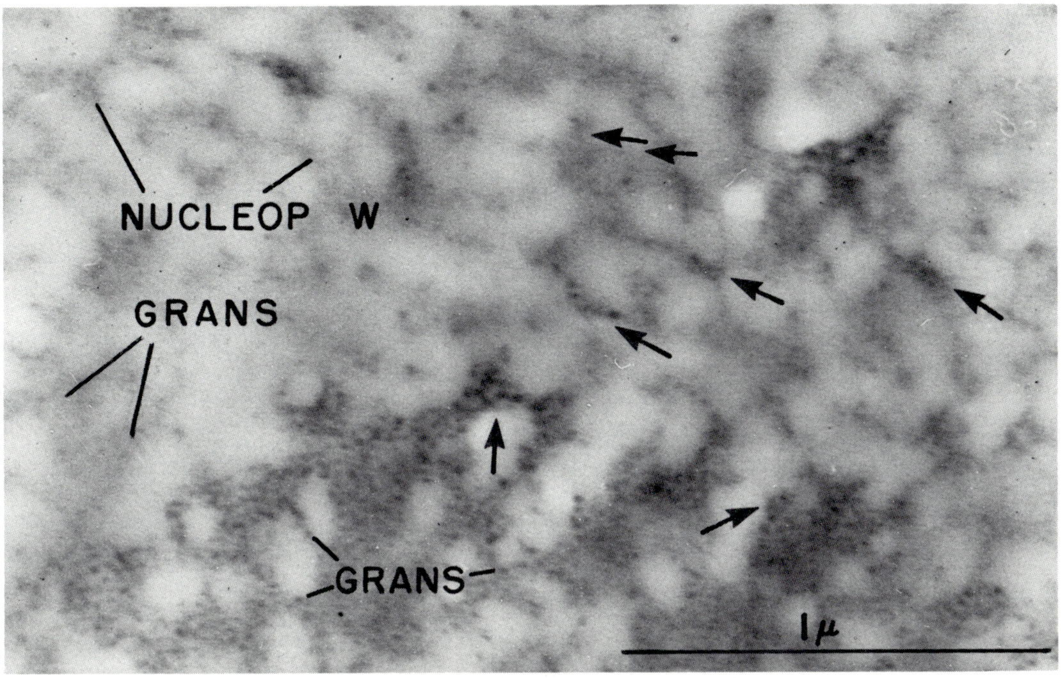

Fig. 12.9. Portion of the nucleolus of Fig. 12.8. The walls of the nucleolus (NUCLEOL W) are filled with RNA granules (GRANS). At the edge (arrows) are some DNA coils not apparent in Fig. 12.8. Other single DNA coils (arrows) are also present in the nucleoplasmic walls (NUCLEOP W), which also contain numerous RNA granules. × 56,000.

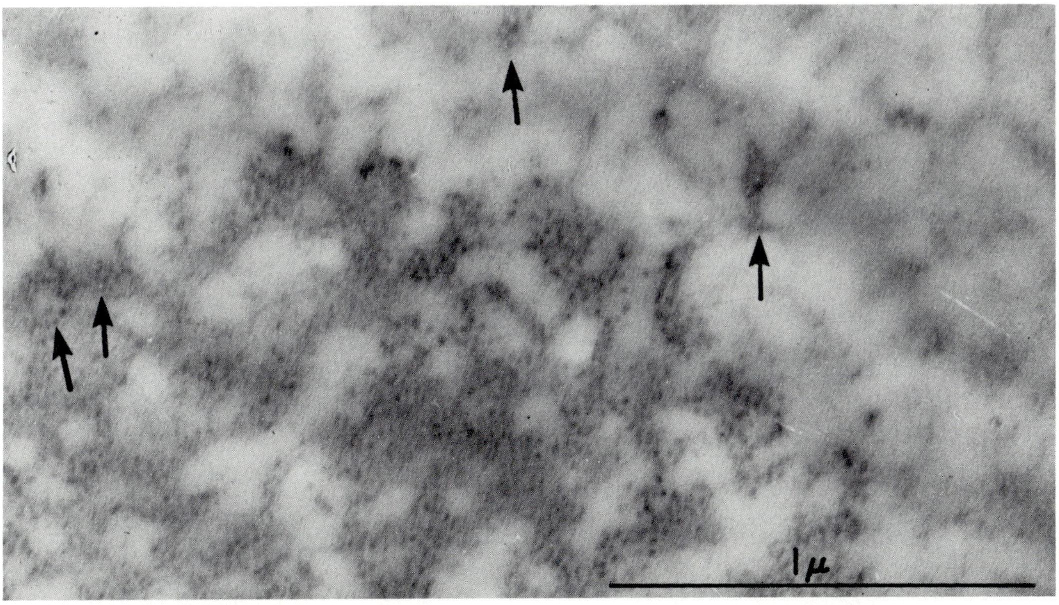

Fig. 12.10. Another portion of the nucleolus of Fig. 12.8 to show separate coils of DNA at the edge (arrows) and in the nucleoplasm. At one point, the DNA appears as straight or very gently curved filaments (double arrows).

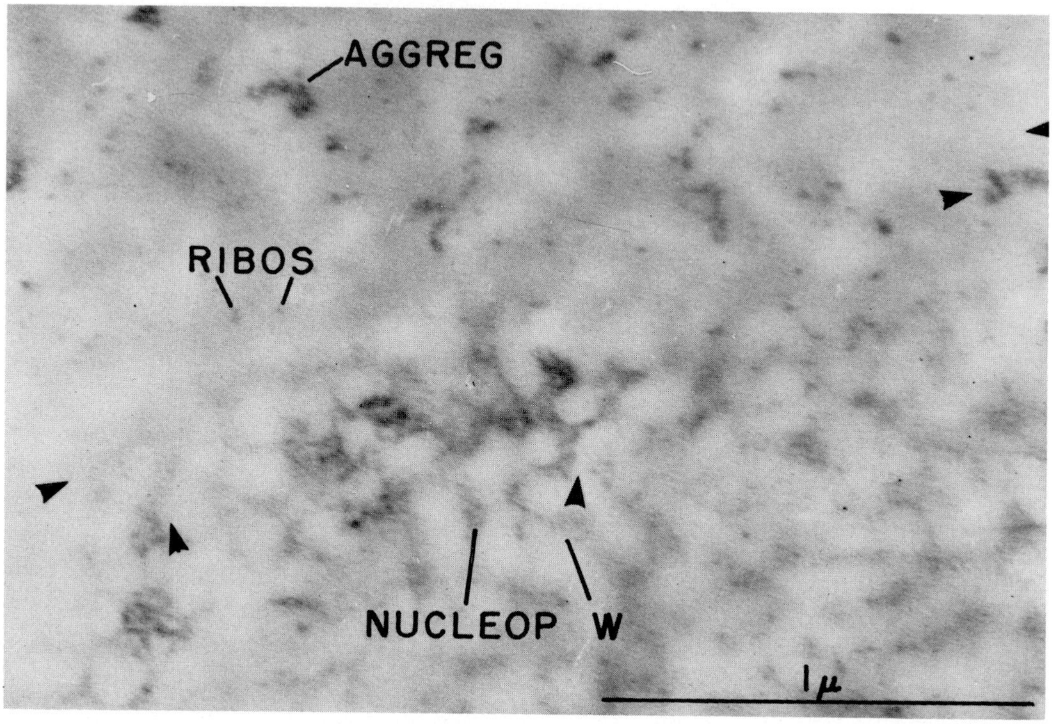

FIG. 12.11. A portion of the same nucleus shown in Fig. 12.8, near the nuclear membrane, indicated by arrow tips. The DNA (arrows) is present as single coils in the nucleoplasmic walls (NUCLEOP W). In the adjacent cytoplasm, ribosomes (RIBOS) occur singly, or in small or large aggregates (AGGREG). × 56,000.

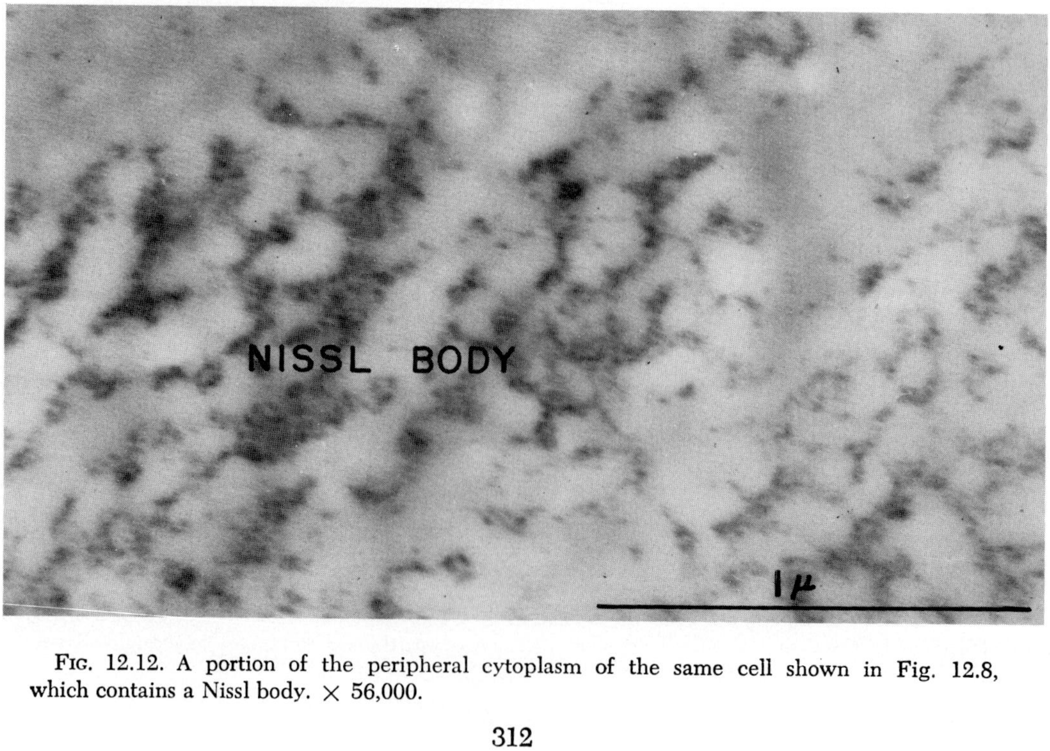

FIG. 12.12. A portion of the peripheral cytoplasm of the same cell shown in Fig. 12.8, which contains a Nissl body. × 56,000.

by an extensive dorsal laminectomy with the circulation and dura mater intact, were excised extradurally, and then transferred to a moist box singly and serially. Thin slices were frozen ultrarapidly and prepared for the study of nucleic acids with the electron microscope (see Chapter 3, this volume). The DNA molecular changes were later found to be so marked that it was necessary to show that the ganglion cells had been alive and recovering, and not moribund. This was tested on a second cat by crushing T_{10}, and comparing the DNA molecular pattern of the spinal ganglion cells 157 days later with the control nerve cells on the unoperated side. Thin slices of these ganglia were also frozen ultrarapidly and treated like the other specimens. All specimens were dried in a vacuum at a low temperature, postfixed *in vacuo* with vapors of mercury (II) hexafluoroacetylacetonate (Hg hfac), and stained for nucleic acids as described previously (Chapter 3, this volume). After infiltration with water-soluble Durcupan, the plastic was polymerized. Sections were cut, mounted on Formvar coated grids, and viewed with a Hitachi-11A electron microscope. Electron micrographs were made of ganglion cells sectioned through the nucleolus at magnifications of × 4000, × 5600, and × 28,000, and printed with a magnification of × 2.

In general, nucleic acids could be identified in all nuclei studied up to now by their morphology. DNA has been identified in electron micrographs as a thin thread about 45 A in diameter after staining with gallocyanin–chromalum. This first-order DNA helix appears in sections as part of a second-order helix, which has a diameter of about 400 A, except in chromosomes of somatic nuclei of the developing egg of *Drosophila,* when their dimensions are smaller and in sperm of the mouse. This has been identified further in some cell types by the use of DNase and RNase after the usual postfixation and before staining with the lake, for it persisted after the latter enzyme, and was no longer identifiable after the former enzyme. RNA has also been identified morphologically, since it occurs as granules or as an unresolvable background density (in prints at a magnification of × 56,000). These forms of RNA are also identifiable further by the use of DNase and RNase, since the background density is no longer identifiable after digestion with the latter enzyme, but persists in some altered form after the former enzyme. The background density, which probably represents, at least in part, transfer RNA (tRNA), sometimes obscures the RNA granules around which the tRNA is clustered. It may be removed, and at the same time its location may be identified, by treating the stained sections with RNase. All of these enzyme treatments were used in this study as aids in identifying DNA and RNA.

Observations

The following are descriptions of the distribution of nucleic acids in thoracic spinal ganglion cells.

T_{10}–Six Days Regeneration (Figs. 12.1–12.3). The aggregates of second-order DNA coils seem finer than in control cells which are described in the following paragraph. Some coarse granules are present between or attached to the second-order coils. The nucleoplasmic walls contain RNA granules throughout the

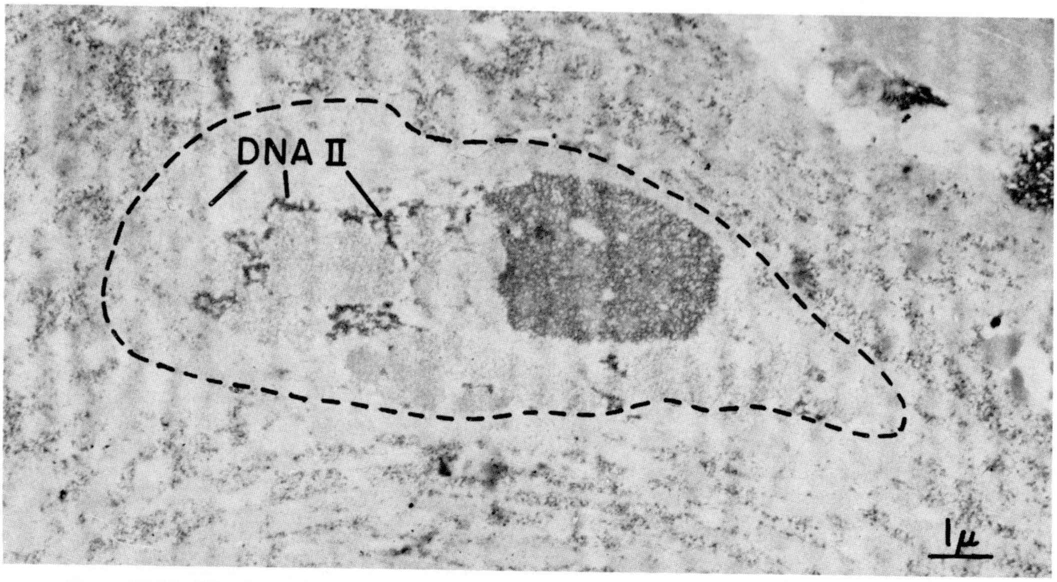

FIGS. 12.13–12.17. Seventeen days after crushing T_{12}.

FIG. 12.13. Low power view. Nucleus of ganglion cell outlined with dotted lines. Second-order DNA coils (DNA II) are scattered throughout the nucleus. Numerous small Nissl bodies are scattered throughout the cytoplasm. The separations between them are so small that it is doubtful that any Nissl bodies would appear as separate bodies in the light microscope. × 8000.

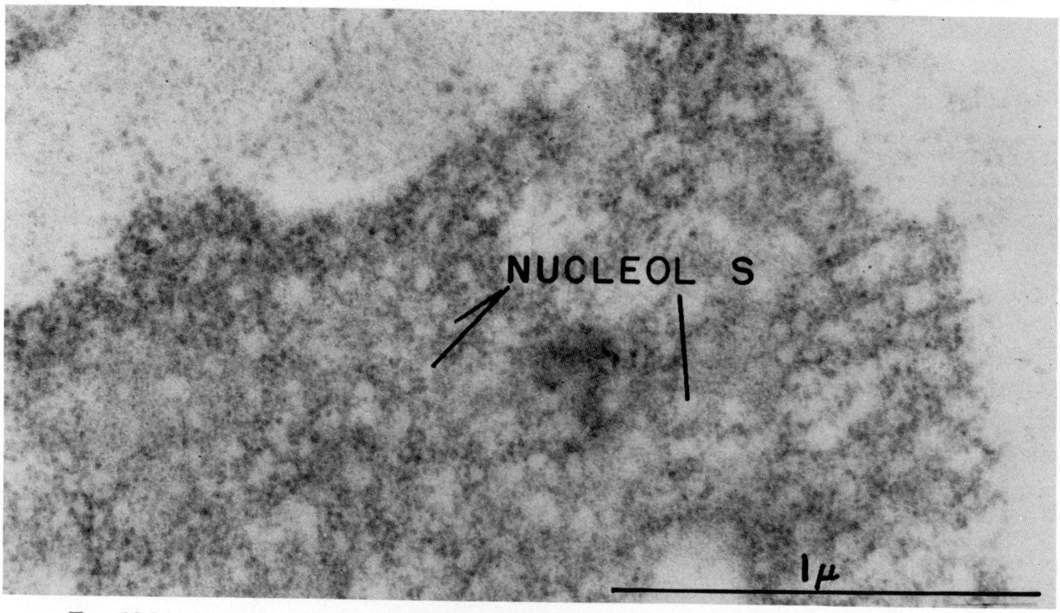

FIG. 12.14. A portion of the nucleolus in Fig. 12.13. The spongelike structure of the nucleolus is very closemeshed, and reflects the multiplicity of the RNA granules and the small size of the nucleolar spaces (NUCLEOL S). × 56,000.

nucleus. The ribosomes in the cytoplasm are scarce and scattered in the perinuclear region and are more concentrated at the periphery of the cell.

T_{10}—Six Days Unoperated Control (Figs. 12.4–12.7). The second-order DNA coils with a diameter of about 400 A form submicroscopic aggregates which are widely scattered in the nucleoplasm or attached to the nucleolus. The coils are in the walls of the nucleoplasm in the form of continuous helices, the apparent diameter of the aggregate depending on the plane of section of the walls. The walls and the cavities, which they enclose partially when viewed in three dimensions, are parts of a spongelike structure, the spaces being about 900 A and the walls about

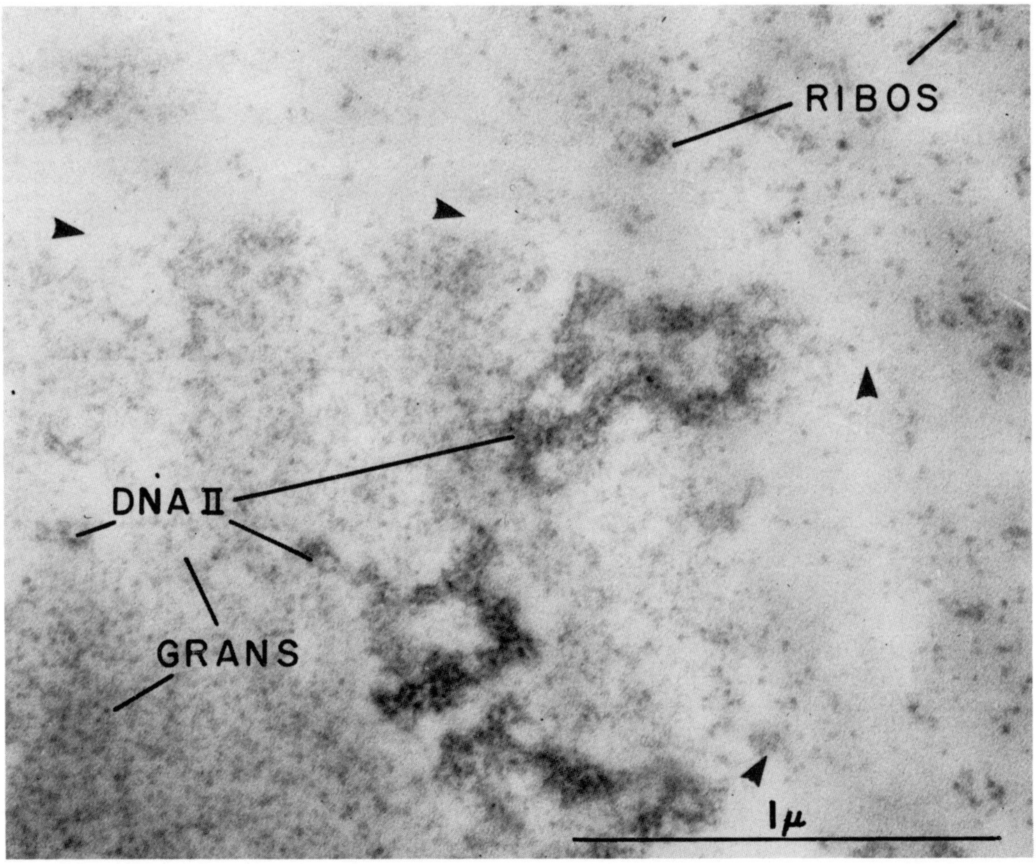

FIG. 12.15. A peripheral portion of the nucleus shown in Fig. 12.13. The second-order DNA coils (DNA II) are shown in relation to a cloud of RNA granules (GRANS) which form a transient nucleolus. Most of the nucleoplasm of Fig. 12.13 is occupied by such transient nucleoli. The position of the nuclear membrane is indicated by arrow tips. Perinuclear cytoplasm contains ribosomes (RIBOS) distributed as single granules or as small aggregates. × 56,000.

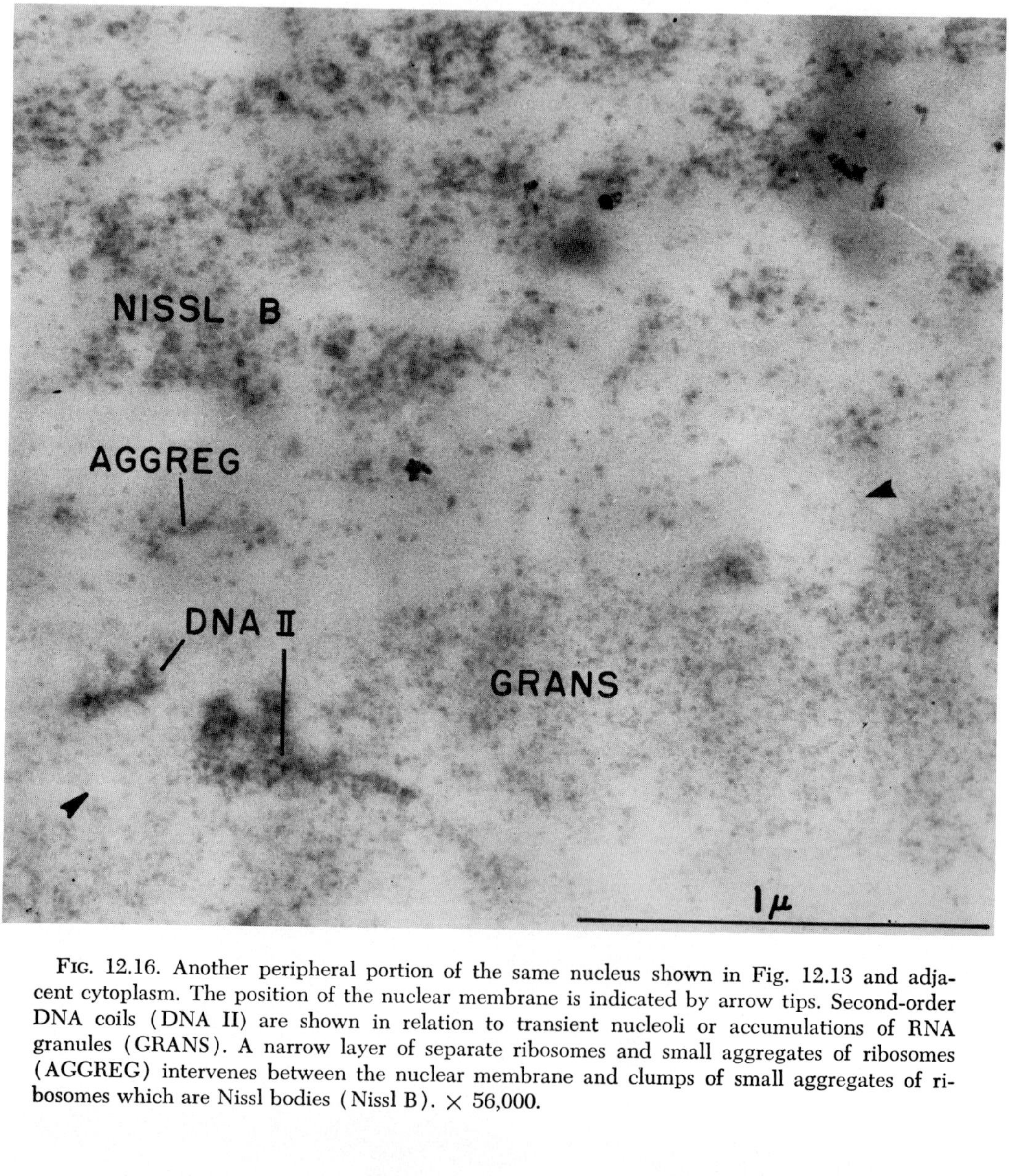

FIG. 12.16. Another peripheral portion of the same nucleus shown in Fig. 12.13 and adjacent cytoplasm. The position of the nuclear membrane is indicated by arrow tips. Second-order DNA coils (DNA II) are shown in relation to transient nucleoli or accumulations of RNA granules (GRANS). A narrow layer of separate ribosomes and small aggregates of ribosomes (AGGREG) intervenes between the nuclear membrane and clumps of small aggregates of ribosomes which are Nissl bodies (Nissl B). × 56,000.

400 A. The nucleoplasmic walls contain scattered RNA granules which are limited to the regions around the DNA aggregates. Some DNA aggregates are attached to the periphery of the nucleolus, whose spongelike structure may be deduced from the (RNA) granule studded walls and the spaces which they enclose partially. The walls of the nucleoplasm and nucleolus are continuous, as are the spaces. The cytoplasm is rich in ribosomes which are frequently arranged linearly.

T_{11}–Ten Days Regeneration (Figs. 12.8–12.12). The DNA strands attached to the nucleolus and in the nucleoplasmic walls appear to be more widely separated and to consist of single second-order helices. The nucleolar walls are very fine and are separated by wider spaces. Nucleoplasmic walls contain many RNA granules. In a wide sphere around the nucleus, ribosomes are scattered in small aggregates, but in the peripheral parts of the cell the ribosomes are arranged as clusters and aggregates in the cytoplasmic walls.

T_{12}–Seventeen Days Regeneration (Figs. 12.13–12.17). The DNA strands are like those of T_{10}. The walls of the nucleolus are even thinner than in the earlier stages of regeneration. In addition, the DNA helices are associated with clouds of RNA particles which could be regarded as temporary nucleoli which occupy most of the nucleoplasm (see p. 325 for discussion of temporary nucleoli).

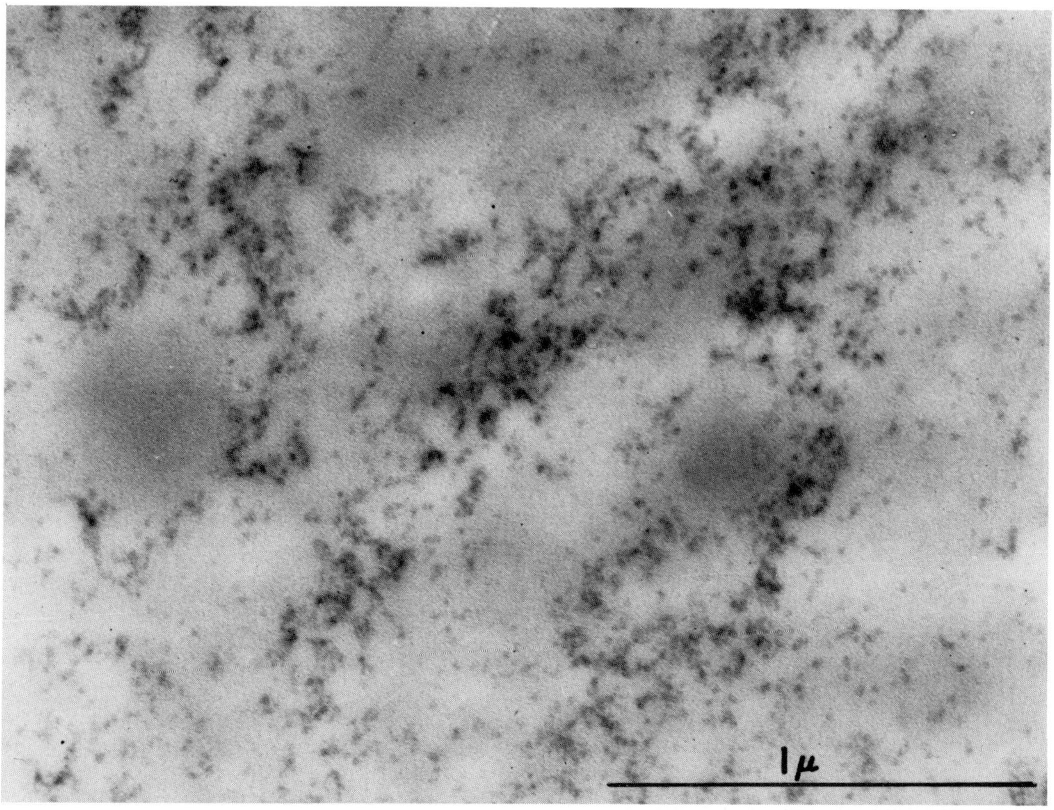

FIG. 12.17. A peripheral portion of the cytoplasm of the cell shown in Fig. 12.13. The aggregates are like those of Fig. 12.16. Widely spaced separate ribosomes and small aggregates of ribosomes intervene between the Nissl bodies. × 56,000.

Numerous coarse granules are attached to the DNA strands. Cytoplasmic ribosomes are concentrated in submicroscopic sheets between which isolated or small groupings of ribosomes are scattered.

T_{13}–Twenty-Four Day Regeneration (Figs. 12.18–12.25). The DNA is broadly scattered and appears mostly as single, straight filaments of molecular double helix, with a few occasional coils, both located within nucleoplasmic walls which also contain RNA granules. Cytoplasmic ribosomes are separate or are organized in aggregates which are localized in the walls of the spongelike structure which constitutes the cytoplasm. Between such ribosome-rich regions are others with irregularly branching networks, which are relatively free of ribosomes, and which may correspond with Golgi structures.

T_{13}–Twenty-Four Day Unoperated Control (Figs. 12.26 and 12.27). The DNA strands are thicker than on the operated side, and are composed of aggregates of second-order DNA coils not dissimilar to those described in the control for T_{10}.

T_{10}–157 Day Regeneration (Fig. 12.28) and Unoperated Control (Fig. 12.29). The DNA strands on both sides are indistinguishable from each other and from the control specimens described above. The ribosomal distribution is also similar.

DNase Treatment of Twenty-Four Day Regeneration Specimen (Figs. 12.30–12.31). The specimen was postfixed with vapors of Hg hfac, treated with DNase as described in Chapter 3 (this volume) and Chapter 5 (Vol. II), stained for nucleic acids with gallocyanin–chromalum, and embedded in plastic. No DNA strands (neither as filaments, coils, nor as aggregates) are visible in the nucleus. Nucleolar and nucleoplasmic RNA granules are visible, as are also cytoplasmic ribosomes. The density of the RNA-poor regions of cytoplasm between the ribosome-rich patches is increased as compared to specimens not treated with enzyme. (See p. 335 for possible explanation.)

RNase Treatment of Seventeen Day Regeneration Specimen (Figs. 12.32–21.35). The specimen was postfixed with vapors of Hg hfac, treated with RNase, stained for nucleic acids with gallocyanin–chromalum, and embedded in plastic. An extremely small fraction of RNA granules of the nucleolus can be recognized, and in the nucleoplasm even fewer can be identified. Ribosomes in the cytoplasm are also unidentifiable, though the regions of the cytoplasm which are nearly free of ribosomes in sections of untreated specimens have acquired an increased density. By contrast, the DNA strands appear as circular figures and filaments. Some stained granules which have resisted enzymic digestion are associated with the DNA filaments and coils.

RNase Treatment of Stained Sections *of T_{10} Unoperated Control Ganglion (Figs. 12.36–12.39).* The specimen was postfixed with vapors of Hg hfac, stained with gallocyanin–chromalum, and embedded in plastic; sections were treated with

RNase. RNA granules in the nucleus and nucleolus are larger than in untreated specimens, and they stand out more clearly against the low density background. The same is true also of cytoplasmic ribosomes. The coil structure of the second-order DNA helices is also more apparent than in most DNA strands in untreated specimens.

Discussion

While the observations on the morphology and distribution of the first-order DNA (double) helix during chromatolysis or nerve regeneration are interpretable to a certain degree, the significance of other findings is not obvious. These include

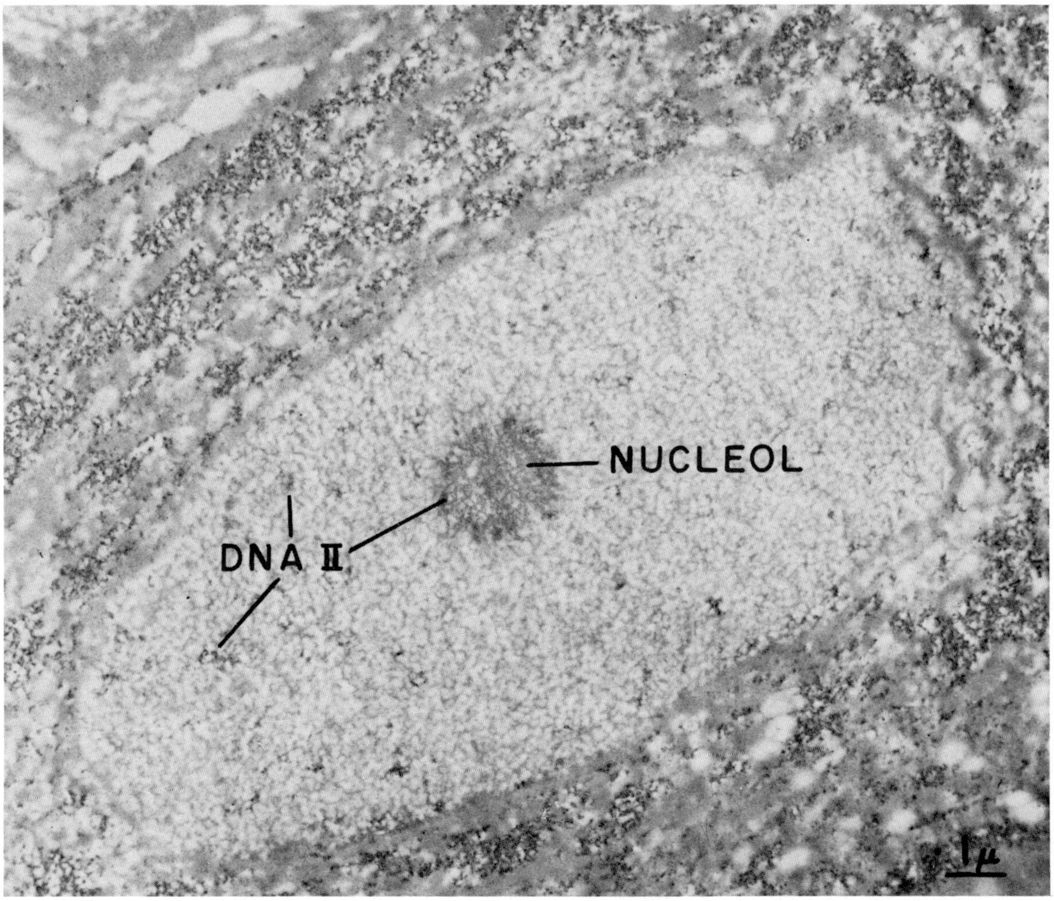

FIGS. 12.18–12.21. Twenty-four days after crushing T_{13}.

FIG. 12.18. Low power view of portion of a nerve cell. Small segments of strands of DNA (DNA II) are attached to the outside of the nucleolus (NUCLEOL) and are scattered through the nucleoplasm. Ribosome aggregates are distributed as small and large clumps. × 8000.

variations during chromatolysis in the structure of the Nissl bodies, in nucleolar structure, and in the density of RNA granules in the nucleoplasmic walls. Probably some of the difficulty is caused by the fact that there is no control over certain variations in "normal" cells, as in the Nissl pattern of spinal ganglion cells which may range at the microscopic level from diffuse and dustlike to granules, and which may be located peripherally, centrally, or evenly throughout the soma. Variations in nucleolar structure probably cannot be understood without quantitative data on size, density, RNA content of total nucleolus, and the number of RNA granules in control and experimental cells. The same applies also to a meaningful treatment of nucleoplasmic RNA particles. For such studies the fixation used in this report is the method of choice. For the study of nucleoplasmic RNA granules, the method used here is essential, as fixation acceptable in the study of Nissl bodies and nucleolus may result in the loss of a great part of the nucleoplasm.

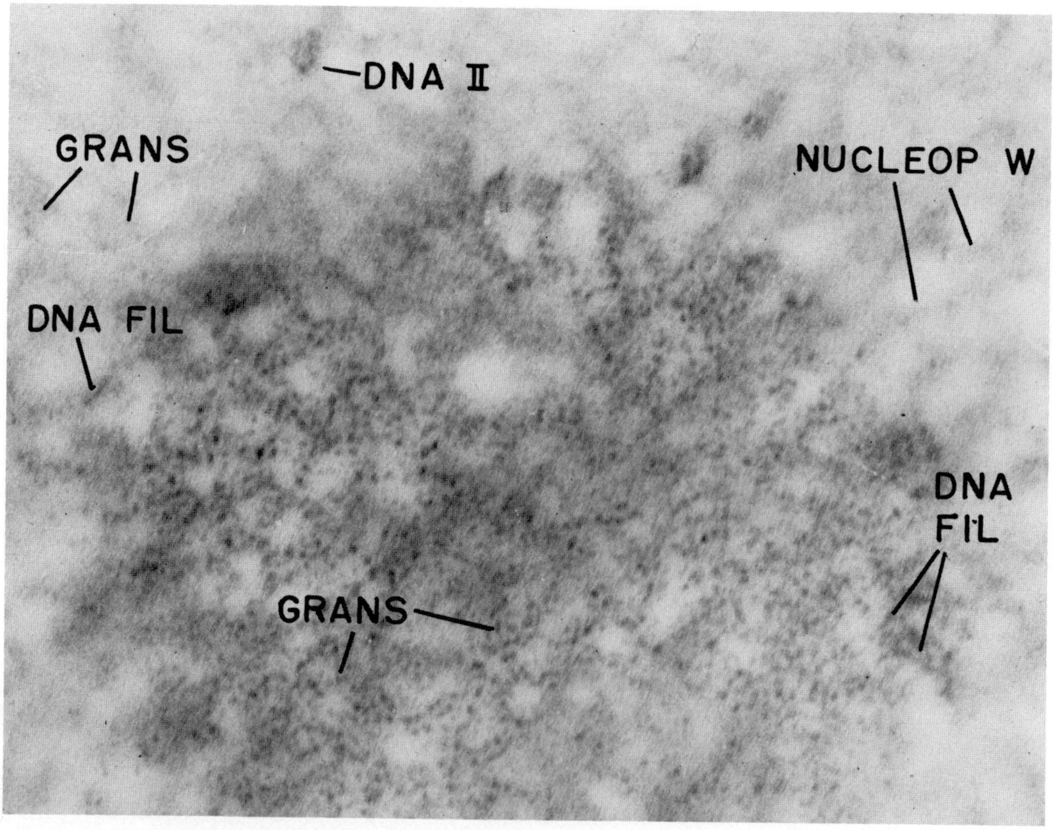

FIG. 12.19. Nucleolus and nucleoplasm of Fig. 12.18, enlarged to show spongelike structure and the attached DNA coils (DNA II) and filaments (DNA FIL). RNA granules (GRANS) are more numerous in the nucleolar walls and less numerous in the nucleoplasmic walls (NUCLEOP W). × 56,000.

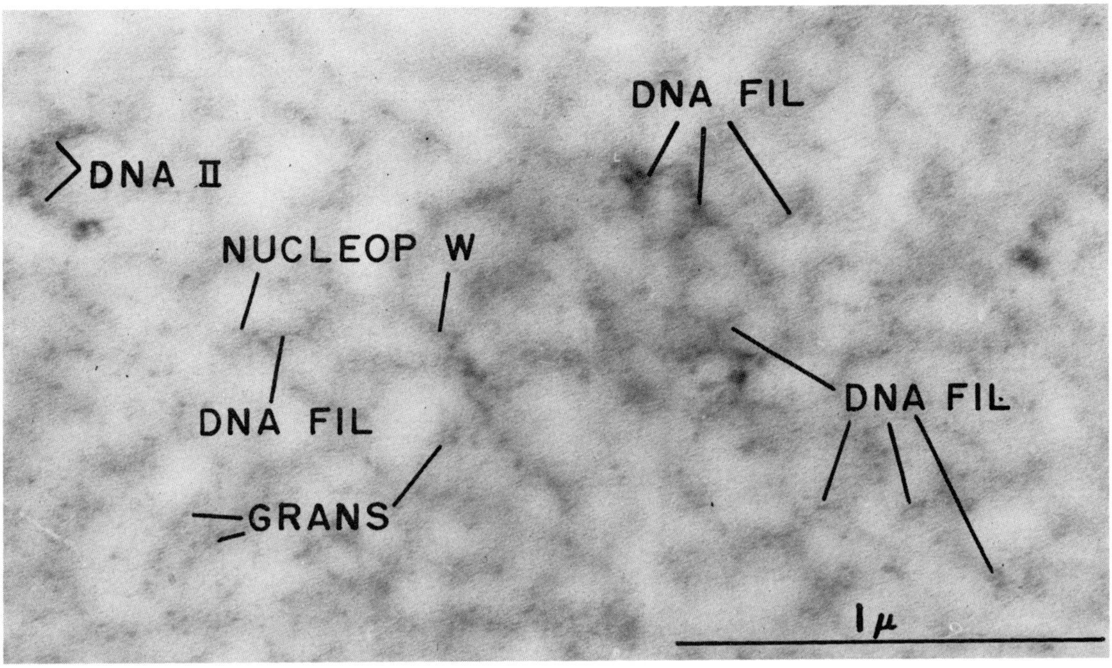

FIG. 12.20. A portion of nucleoplasm of the same nucleus as in Fig. 12.18. RNA granules (GRANS) are present in nucleoplasmic walls (NUCLEOP W), as are DNA filaments (DNA FIL). DNA II, second-order DNA coil. × 56,000.

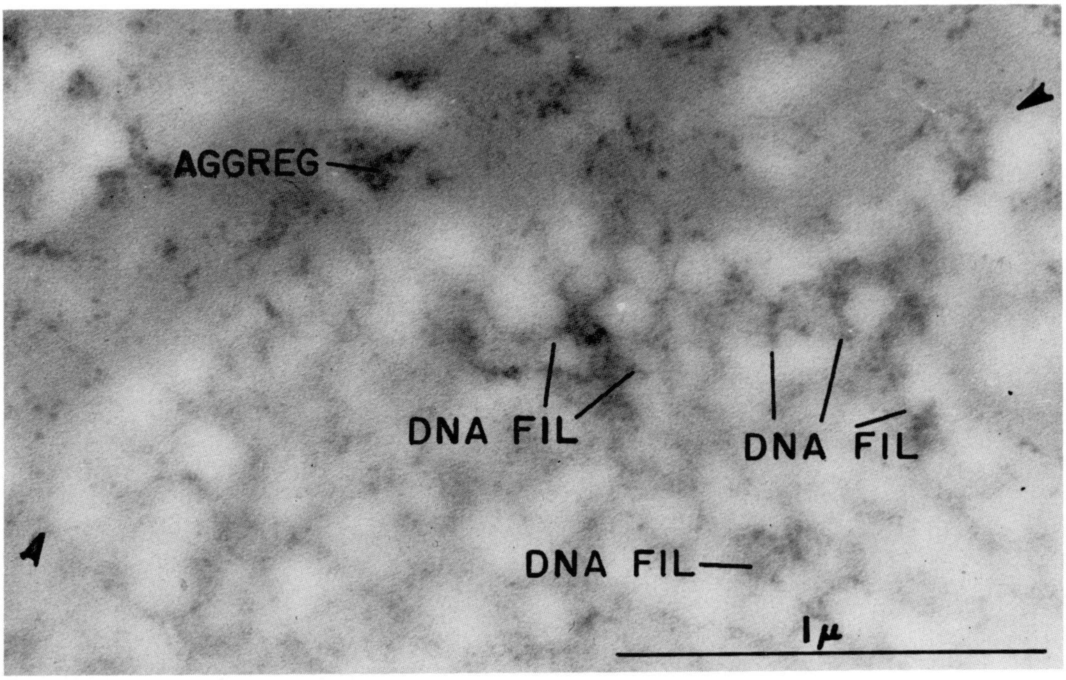

FIG. 12.21. A peripheral portion of the same nucleus as in Fig. 12.18, to show numerous DNA filaments (DNA FIL). Ribosome aggregates (AGGREG) are present near the nuclear membrane, which is indicated by arrowheads. × 56,000.

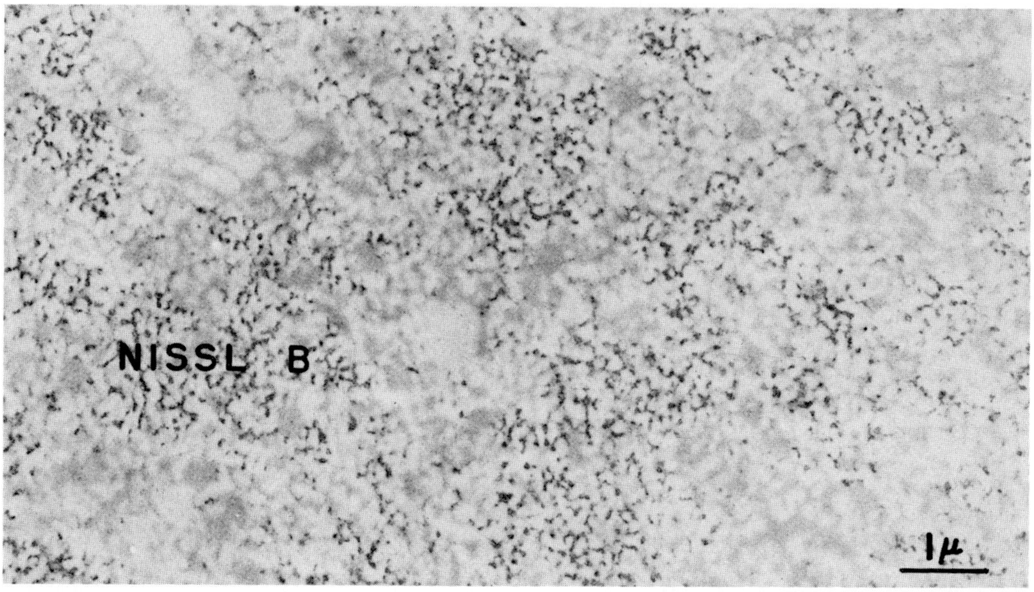

FIGS. 12.22–12.25. Portion of same cell in a nearby section to show arrangement of ribosomes in cytoplasm.

FIG. 12.22. Clumps of ribosomal aggregates or Nissl bodies (NISSL B) are separated by netlike structures containing strands of varying diameter, which have few ribosomes. These are probably parts of the Golgi apparatus. × 11,200.

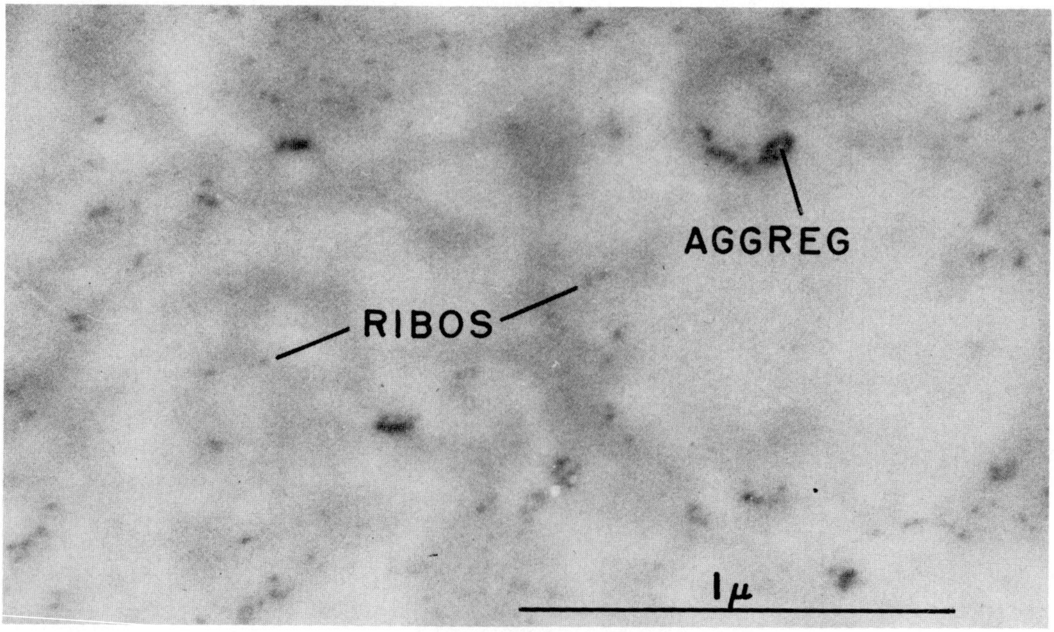

FIG. 12.23. A part of the pale portion of the cytoplasm to show separate ribosomes (RIBOS) and small aggregates of ribosomes (AGGREG) in the walls of the netlike structures. × 56,000.

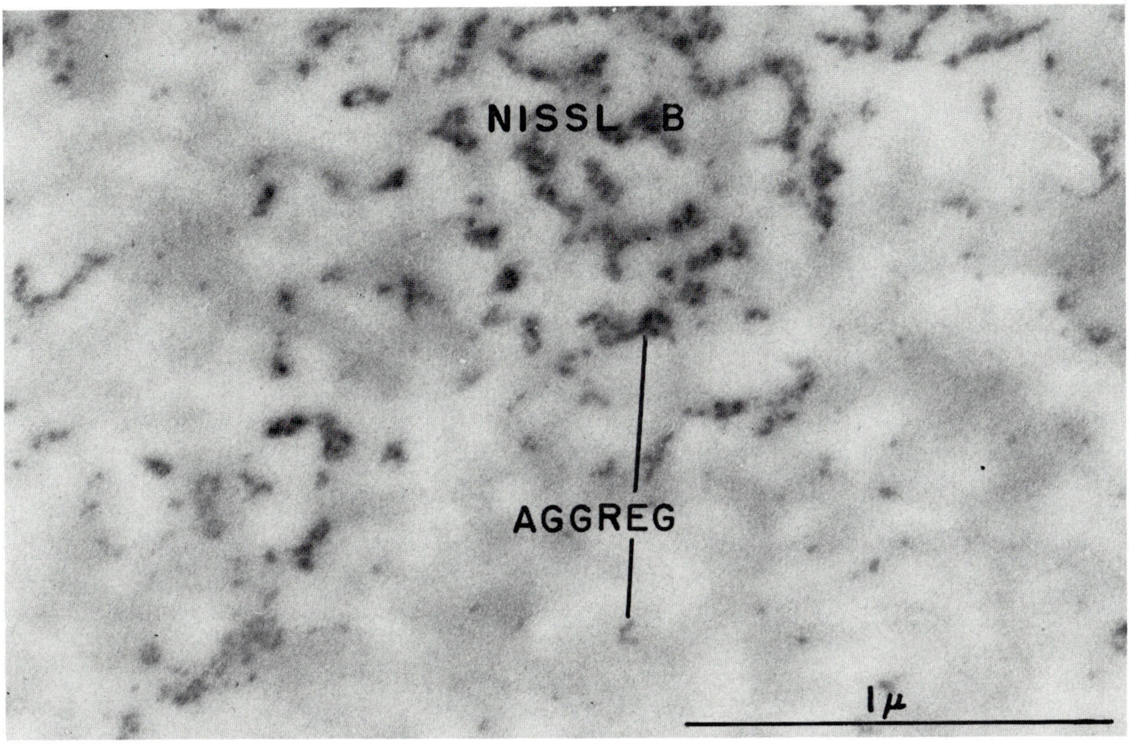

FIG. 12.24.

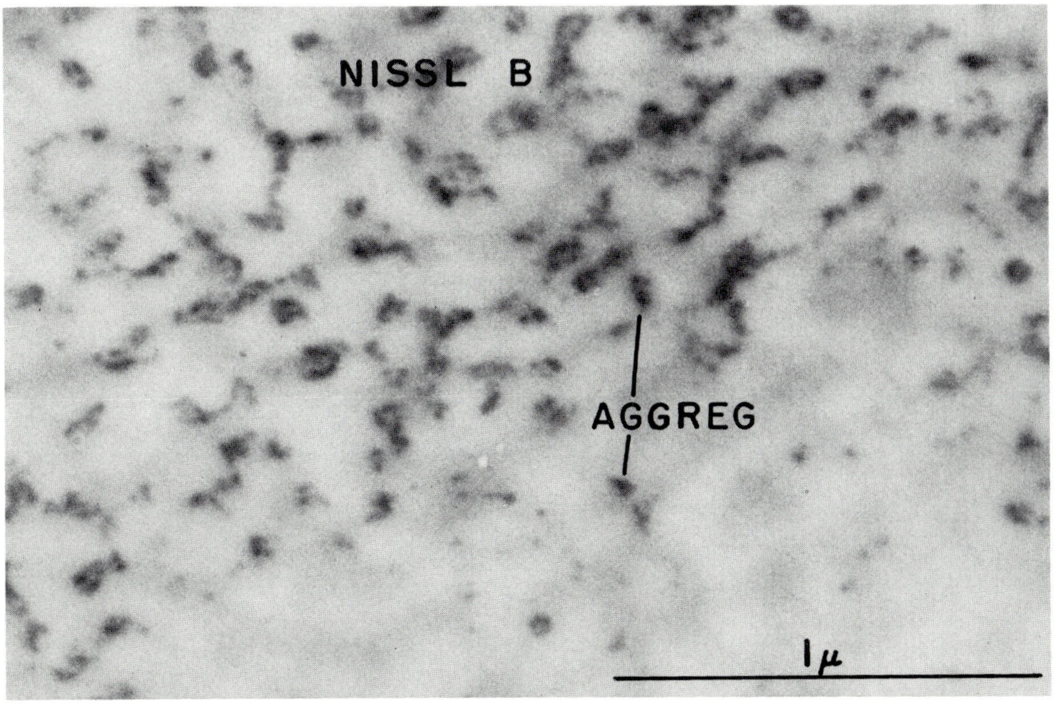

FIG. 12.25.

FIGS. 12.24 and 12.25. Other parts of the cytoplasm shown in Fig. 12.22, to show different netlike structures in the pale portions of the cytoplasm, and clumps of ribosomal aggregates (AGGREG) which are parts of Nissl bodies (NISSL B). × 56,000.

The findings on the morphology and distribution of the first-order DNA helix may be summarized briefly. In control or fully recovered neurons, nuclear DNA occurs as fine, mostly submicroscopic strands of DNA organized as aggregates of second-order helices with a diameter and period of about 400 A. Shortly after nerve crushing, these strands become thinner, presumably by separation of adja-

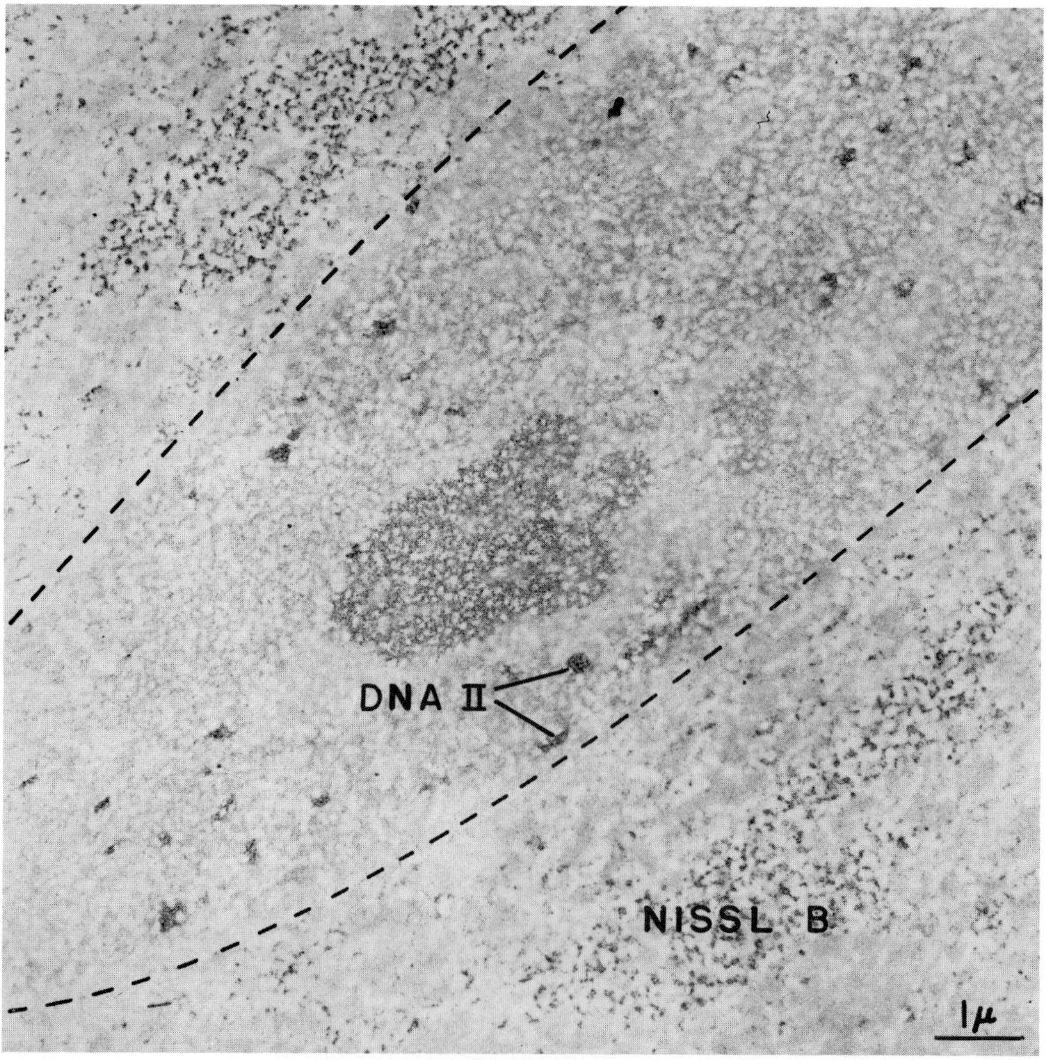

FIGS. 12.26 and 12.27. Unoperated control of T_{13}. Compare with Fig. 12.18 of operated T_{13}. Low power views of portions of nerve cells to show that the DNA strands (DNA II) are larger than those in the chromatolytic cell. Nissl bodies (NISSL B) also appear to be larger. × 11,200. Note that the magnification is greater than that in all preceding low power electron-micrographs.

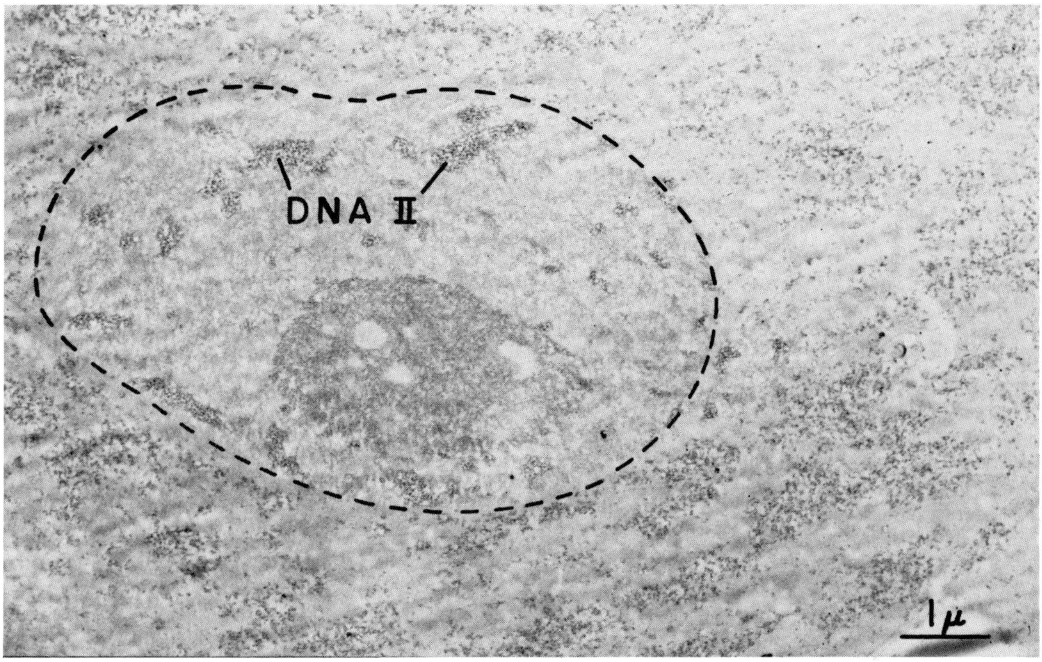

FIG. 12.27.

cent helices. At 17 days, these begin to extend, appearing now as fine filaments of first-order (or double) DNA helices. At 24 days, only a few second-order isolated coils are visible, most of the DNA appearing as sections of filaments. At about the same time, dense, more or less circumscribed accumulations of RNA granules, similar to but less crowded with granules than nucleoli, appear in relation to the DNA filaments. After 157 days, the normal state of aggregation of DNA in strands characteristic of control, unoperated neurons reappears.

These findings can be interpreted in the following way. The second-order DNA helices of normal ganglion cells are somewhat aggregated. They are less aggregated than in any other cell type (see this volume, Chapter 8) and the presumption is that more of the DNA is active in these nerve cells than in other cells. This is consistent with the recent study of this topic by Hahn and Laird *(11)*. Early during nerve regeneration, these helices separate from each other, thus achieving greater exposure to the nucleoplasm with which it can more efficiently exchange metabolites and metabolic products. Maximum exposure is achieved when the DNA double helix occurs primarily as extended filaments, with relatively few second-order coils. At this stage, RNA particles are produced and accumulate in such large numbers that transient nucleoli are formed. The RNA granules in the nucleoplasm probably contain rRNA and mRNA in condensed form. In the nucleolar and nucleoplasmic walls, the RNA granules are surrounded by soluble or sRNA, particularly that part, transfer or tRNA, which is combined with activated or acylamino acid. Both granular forms of RNA pass into the cytoplasm where, it

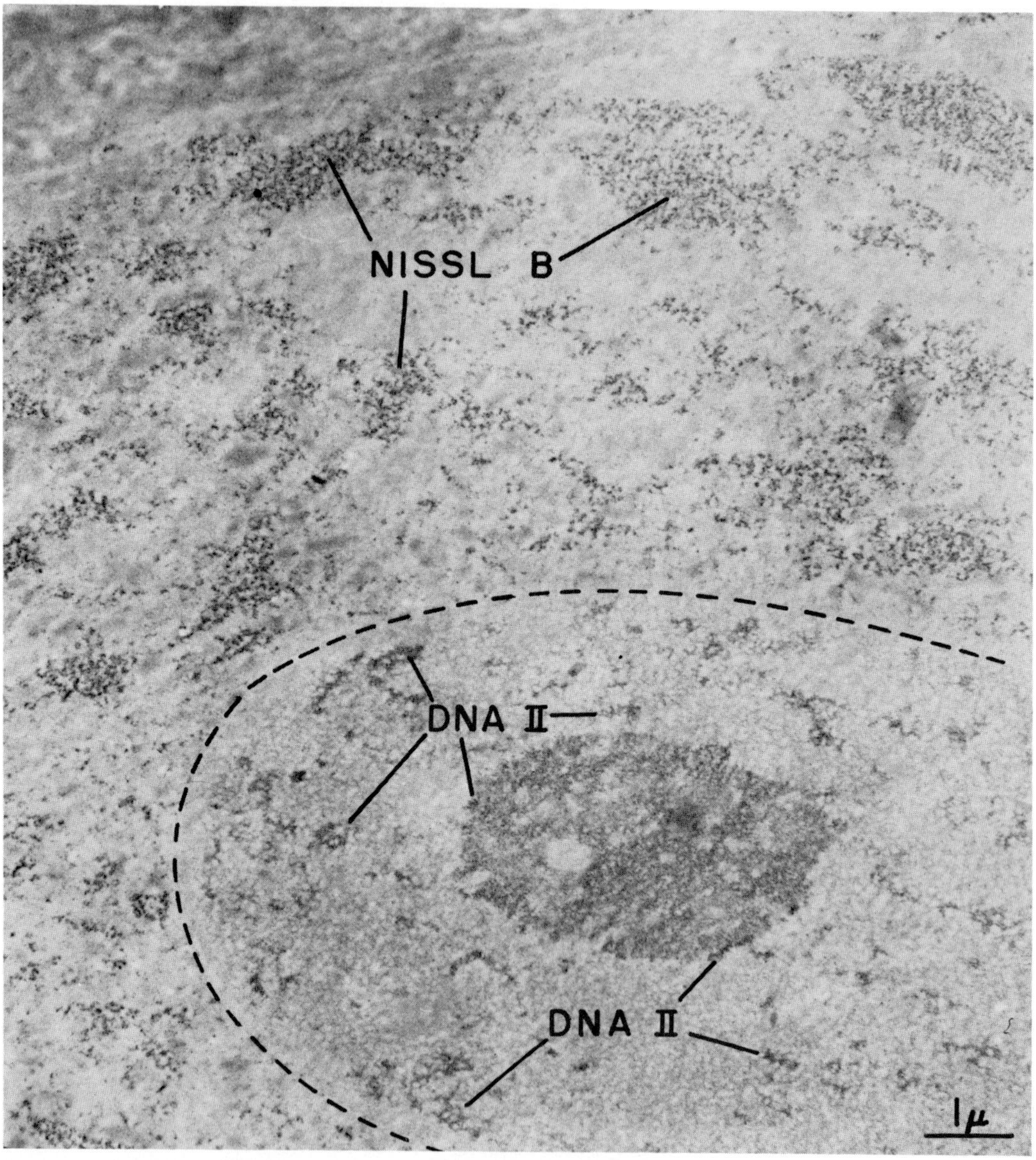

FIGS. 12.28 and 12.29. Low power views of portions of nerve cells of ganglion of nerve crushed 157 days earlier (Fig. 12.28), and of control unoperated ganglion of same thoracic seg-

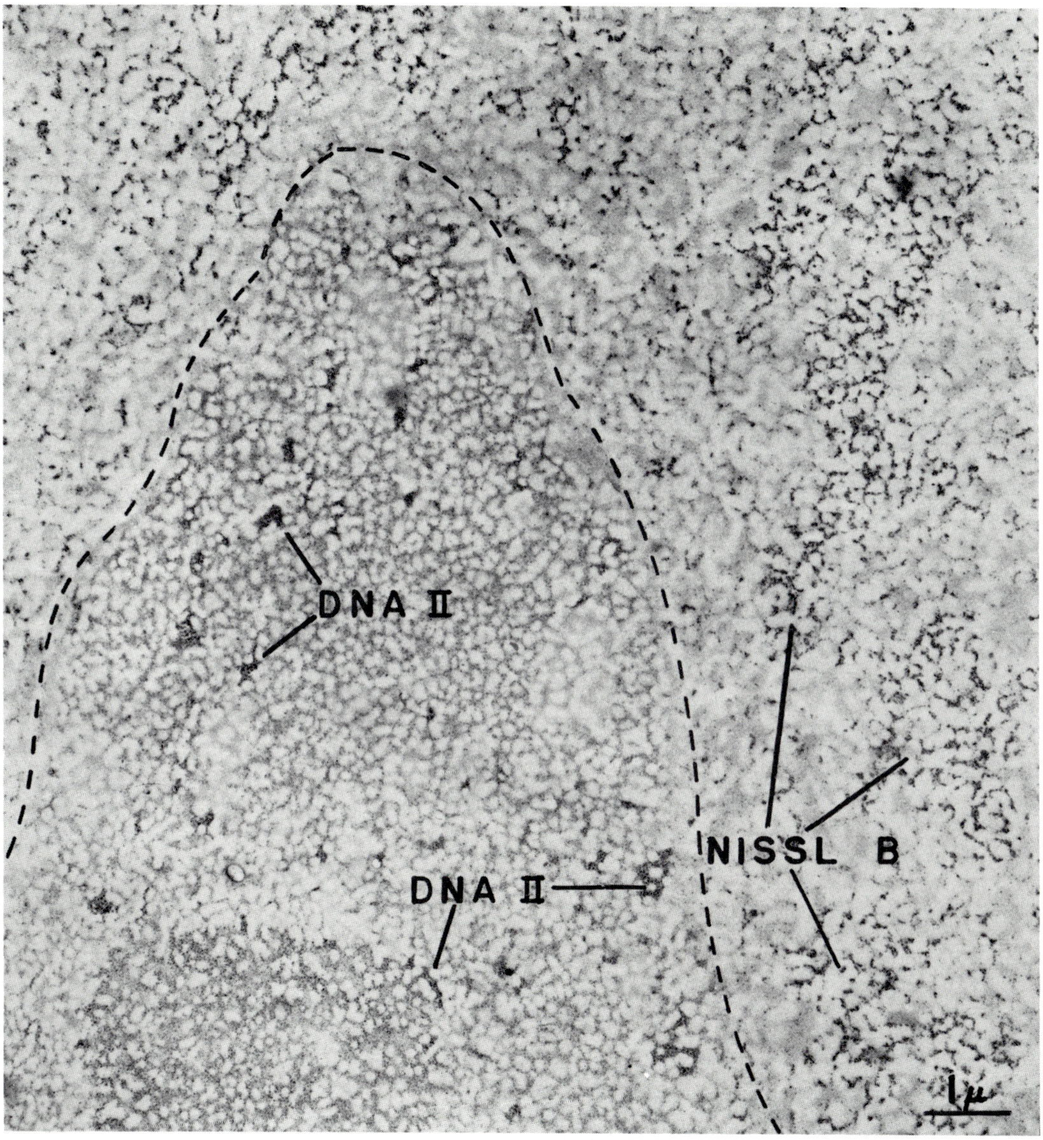

ment (Fig. 12.29). DNA strands (DNA II) in both cells are similar, and resemble those of Figs. 12.4, 12.26, and 12.27, which are also from normal ganglion cells. Nissl bodies (NISSL B) are clear in both cells. × 11,200.

is clear from the literature, they are organized partly in association with the endoplasmic reticulum and synthesize the large amount of protein generated in the recovery process.

The interpretation is consistent with studies on DNA in strongly stimulated cells of other tissues as described in an earlier chapter (Chapter 10), and is the reverse of changes which take place in certain cells during maturation (Chapter 9,

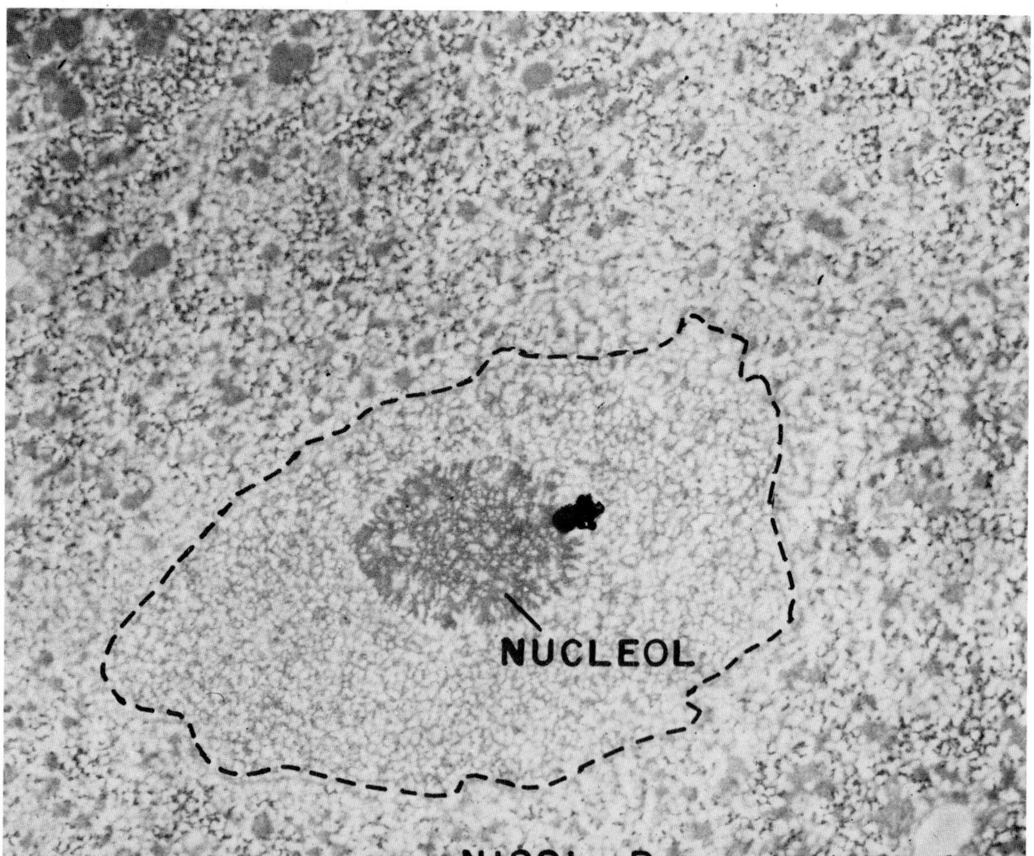

FIG. 12.30. Ganglion cell of specimen (24-day regeneration) treated with DNase after treatment *in vacuo* with vapors of the cross-linking reagent. The specimen was stained with gallocyanin-chromalum after treatment with DNase. The nucleus is outlined. Staining of nucleolus (NUCLEOL) and of Nissl bodies (NISSL B) persists, while there are no traces of DNA strands. Compare with cell not treated with DNase (Fig. 12.18). × 11,200.

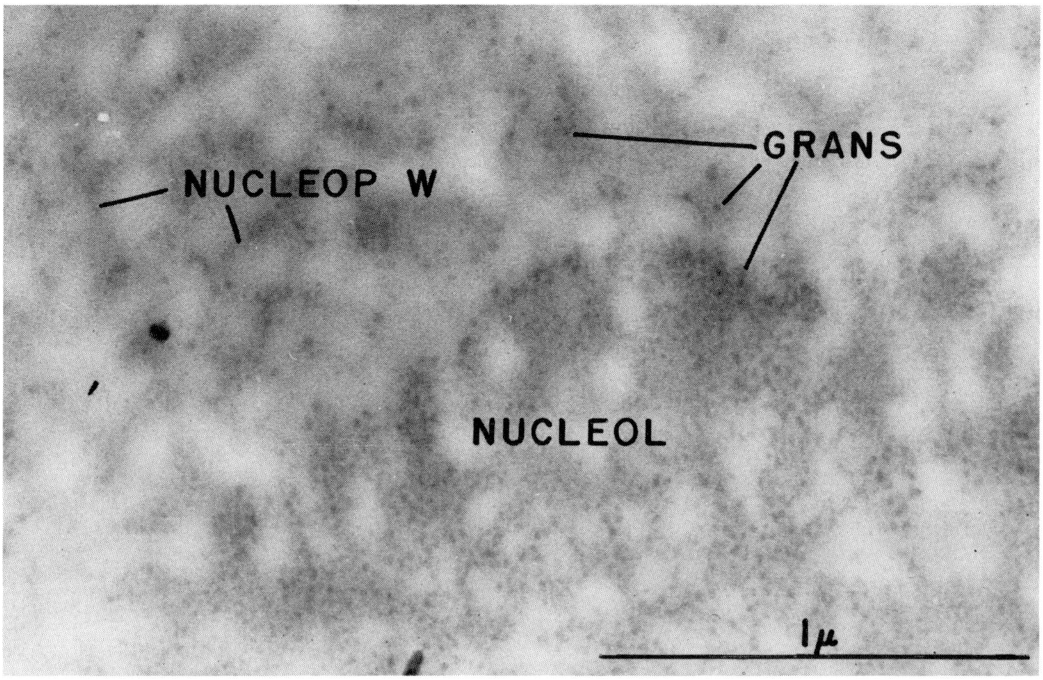

FIG. 12.31. Portion of same cell shown in Fig. 12.30. RNA granules (GRANS) in nucleolus (NUCLEOL) and nucleoplasmic walls (NUCLEOP W) are stained, with no evidence that DNA is present. × 56,000.

this volume). Throughout, in all of these cells in a great range of conditions, the DNA molecule remains intact, except during pyknosis (this volume, Chapter 11) or cell lysis (Vol. II, Chapter 5). The main difference between the DNA distribution in the nuclei of neurons and of other cells is that in the latter in the steady or resting state the second-order DNA coils are frequently more highly aggregated as heterochromatic strands or peripheral or perinucleolar chromatin, which are large enough to be visible in the light microscope, while in the spinal ganglion cells the aggregates of second-order DNA helices are so fine as to be nearly entirely submicroscopic in size. This implies that a larger proportion of the DNA of nerve cells is actually or potentially active as compared with most other cells of the adult organism. Nevertheless, excessive activity results in the same kinds of DNA changes in nonneural and neuronal nuclei—a progressive disaggregation and separation of second-order DNA helices. An extreme reaction has been observed only in neurons, where the second-order DNA helices are nearly abolished, being extended as filaments of first-order DNA (double) helices. When extended, the DNA helix is approximately 3.3 times longer. This achieves more extensive contact with the nucleoplasmic environment, and permits maximal exchange of DNA products (for example, mRNA) and nucleotidal metabolites.

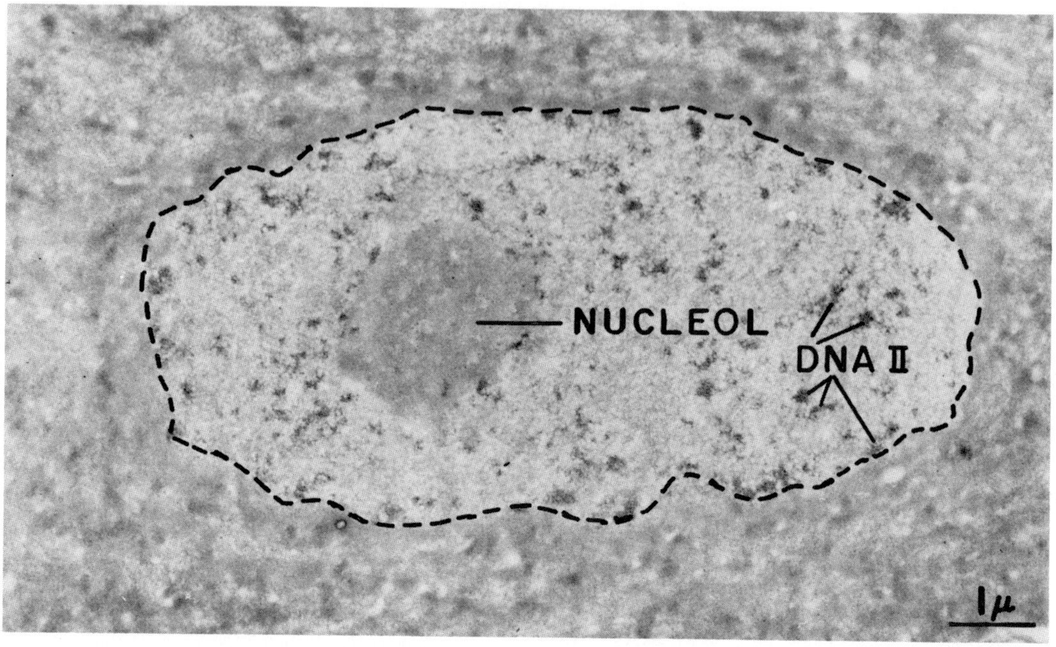

FIGS. 12.32–12.35. Ganglion cell of specimen of 11 days regeneration treated with RNase after treatment *in vacuo* with vapors of the cross-linking reagent and before staining with gallocyanin–chromalum.

FIG. 12.32. Most of the nearly specific staining of nucleolus (NUCLEOL) and of ribosome and ribosome aggregates is reduced or abolished. DNA strands (DNA II) persist. In addition, there are dense cytoplasmic granules. × 11,200.

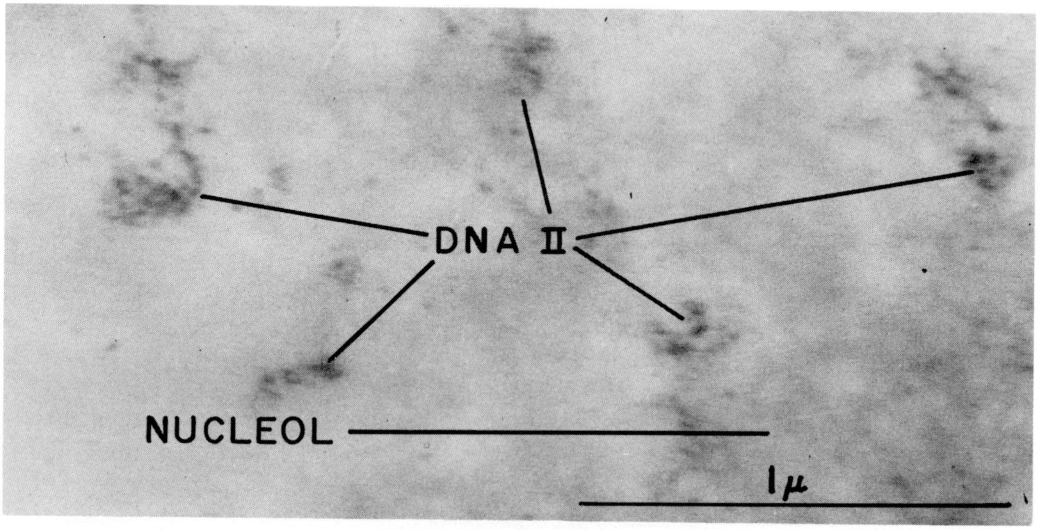

FIG. 12.33. Portion of nucleolus and nucleoplasm of same cell shown in Fig. 12.32. A few granules are present in the nucleolus and in the nucleoplasmic walls, both of very low density, while DNA strands (DNA II) attached to the nucleolus (NUCLEOL) and in the nucleoplasm are well stained, though somewhat distorted. × 56,000.

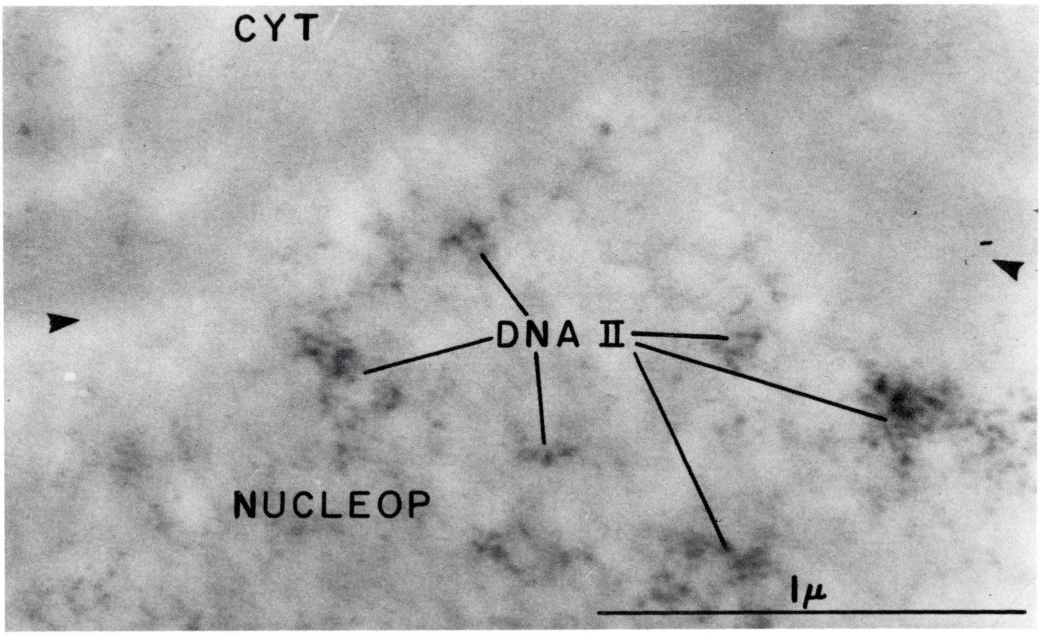

FIG. 12.34. Peripheral part of same nucleus as in Fig. 12.32. Position of nuclear membrane is shown by arrow heads. DNA strands (DNA II) in nucleoplasm (NUCLEOP). Few granules of low density are visible in the nucleoplasm (NUCLEOP) or in the cytoplasm (CYT). × 56,000.

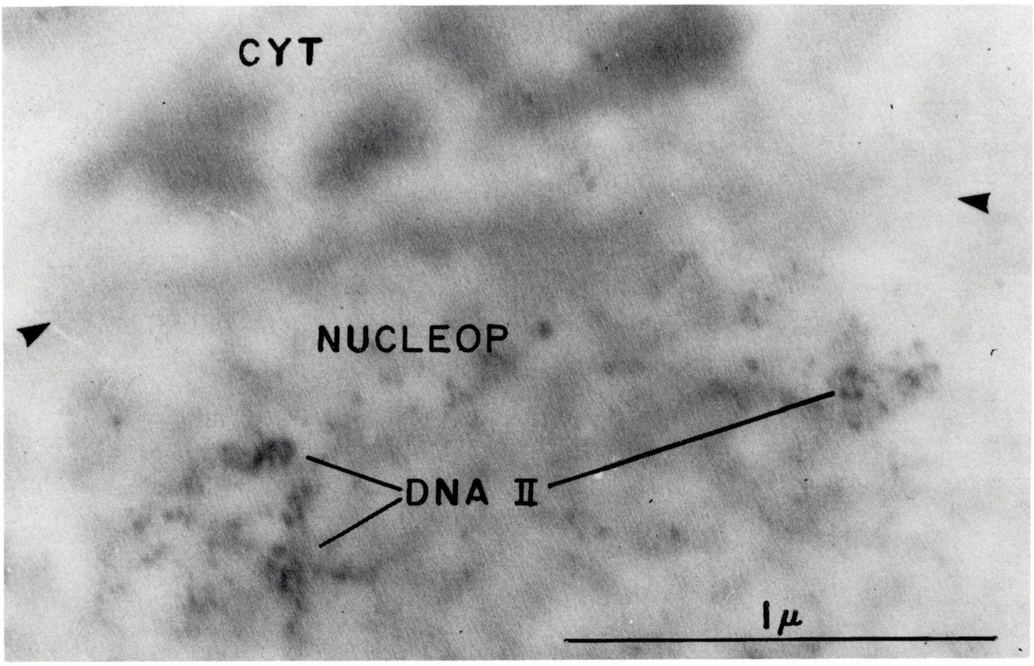

FIG. 12.35. Another peripheral part of same nucleus as in Fig. 12.32. Position of nuclear membrane is indicated by arrow heads. DNA strands (DNA II) in the nucleoplasm are stained. Less densely stained are a few granules in the nucleoplasm (NUCLEOP) and in the cytoplasm (CYT). The larger dense bodies in the cytoplasm are visible as such only after enzyme treatments. × 56,000.

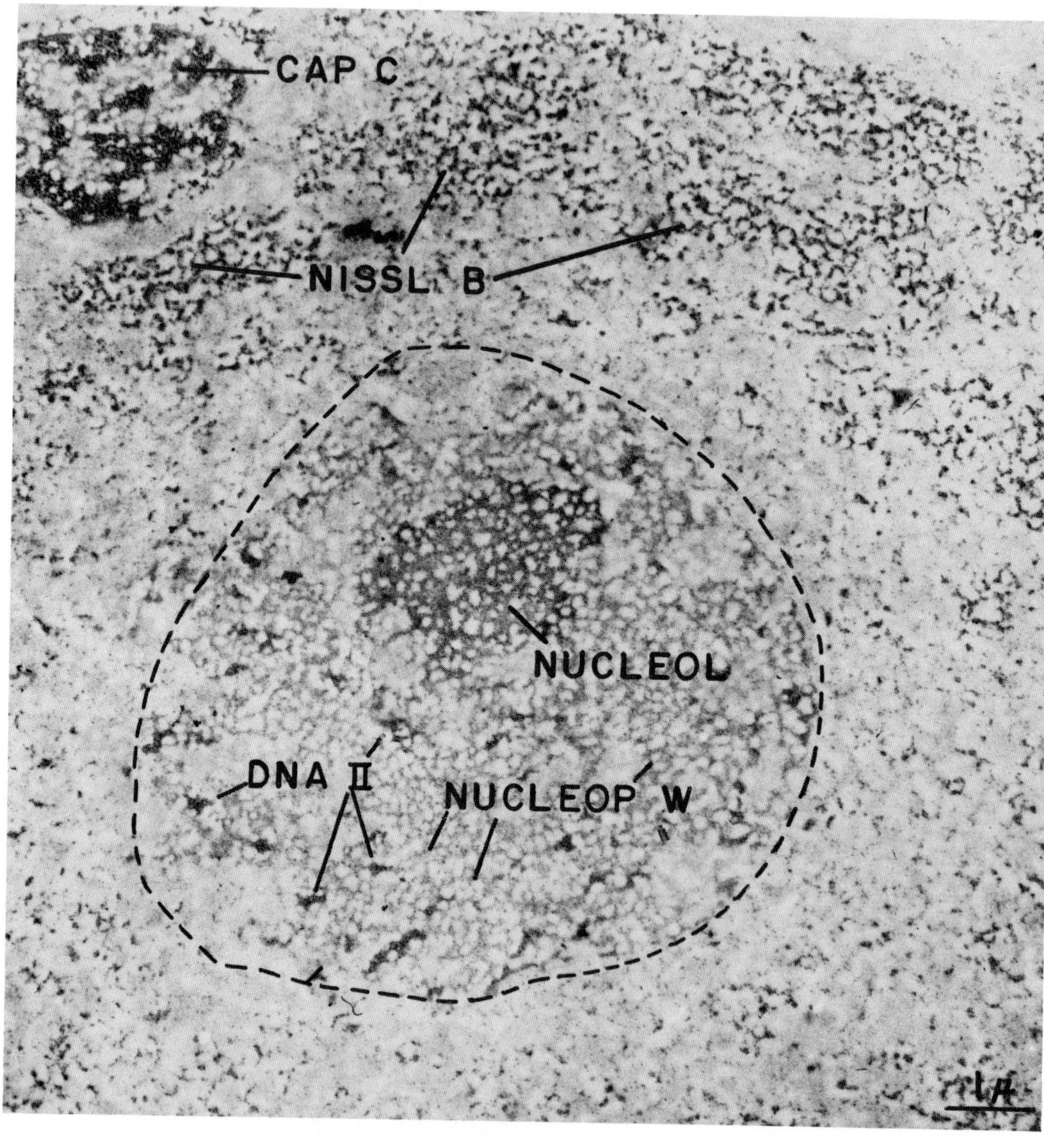

FIGS. 12.36–12.39. The specimen (control ganglion, unoperated, 157 days) was freeze-dried, postfixed *in vacuo* with vapors of cross-linking reagent, stained with gallocyanin–chromalum and embedded in water-soluble Durcupan. Then the ultrathin sections were treated with ribonuclease, washed, dried, and observed in the electron microscope.

FIG. 12.36. Portion of a ganglion cell with its nucleolus (NUCLEOL) and stained nucleoplasmic walls (NUCLEOP W) some of which contain DNA strands (DNA II). The nucleus is outlined. Nissl bodies (NISSL B) appear in the cytoplasm. Nucleus of a capsular cell (CAP C) is present. × 11,200.

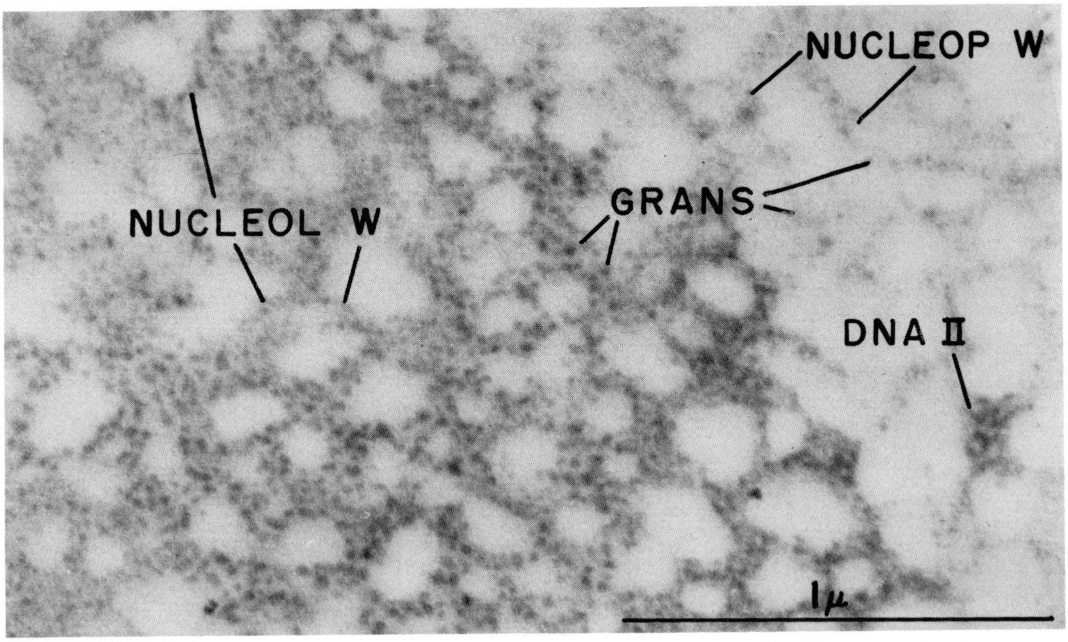

FIG. 12.37. Portion of same nucleolus shown in Fig. 12.36, with adjacent nucleoplasm. RNA granules (GRANS) of various sizes are present in both nucleolar (NUCLEOL W) and nucleoplasmic walls (NUCLEOP W). RNA granules are larger than in untreated control sections, and are more readily resolved. The DNA coils (DNA II) also stand out as coils more readily, probably because the background has been cleared of soluble forms of RNA. × 56,000.

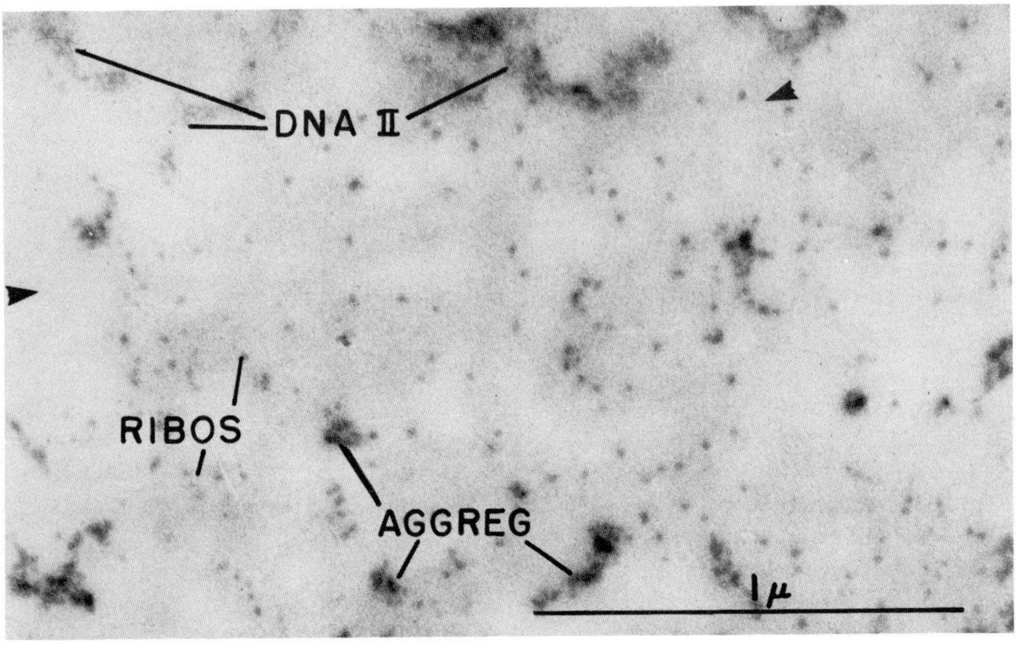

FIG. 12.38. Portion of same nucleus and adjacent cytoplasm. Position of nuclear edge indicated by arrowheads. The DNA coils (DNA II) near the nuclear membrane are again clear. Ribosomal granules (RIBOS), whether separate or as aggregates (AGGREG) are more discrete than in other kinds of preparations because the soluble RNA (probably tRNA) in the region of the granules has been removed by enzymic action. × 56,000.

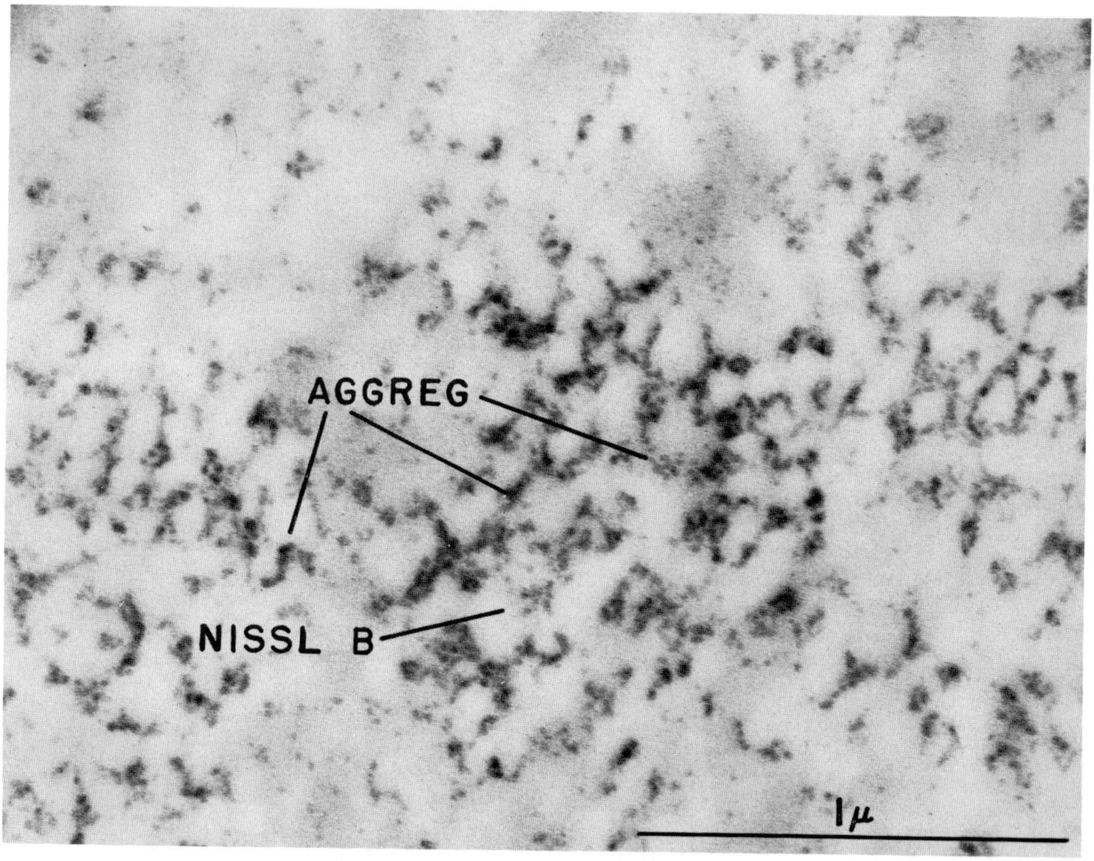

FIG. 12.39. Another portion of cytoplasm of same cell shown in Fig. 12.36. Nissl body (NISSL B) is composed mostly of aggregates of ribosomes (AGGREG), whose individual ribosomes are quite distinct, probably because the background density, representing chiefly tRNA, has been removed by the enzyme treatment. × 56,000.

If this interpretation is correct, then the ultimate effect of this change is to maximally expose the DNA cistrons from which are transcribed specific mRNA's required for specific proteins of nerve cells. At the same time, all the other cistrons which are necessary in other cell types for the synthesis of their specific proteins must remain inactivated, or inhibited or turned off by repressor substances of one sort or another.

In accordance with suggestions made earlier (see Chapter 8–10, this volume), the separation of second-order helices from each other (and from the nuclear membrane) requires that (hypothetical) coil-aggregating and membrane-attachment protein molecules which bind adjacent second-order helices to each other and to the nuclear membrane should be removed or physically or chemically altered so as to be ineffective as binding agents. In addition, the abolition of the second-order helix and the appearance of the DNA filament requires that a (hypothetical) coiling protein should be hydrolyzed or altered in some other way. During recovery, the attachment protein(s), coil-aggregating protein(s), and coiling protein(s) are presumably reconstituted or resynthesized, and the normal degree of aggregation of second-order helices is again achieved.

The analogy of nerve cells with gland cells was made by Brattgård, Edström, and Hydén, who wrote: "It seems not far-fetched to visualize the neuron as an enormous gland cell structure whose lively protein metabolism serves the specific nerve function." They also postulated a primary involvement of nuclear DNA in regenerating neurons when they wrote: "A nerve cell mechanism for the production of cytoplasmic nucleoproteins involving the activity of the nucleus and the cytoplasm simultaneously seems possible" and "The primary problem—lies at the level of the gene action on the cytoplasm and the action of external or cell milieu factors back on the genes of the nucleus."

Observations on the effects of enzyme treatments on nucleic acids in nerve cells fortify the identification of the ribo- and deoxyribonucleic acids based on purely morphological criteria. In general, RNase decreases or abolishes the staining of granular RNA in the nucleolus, nucleoplasm, or cytoplasm.

A second enzyme-associated phenomenon is that certain bodies appear in the cytoplasm after enzyme treatments as if their mass were increased as a result of the procedure employed. These bodies are visible in electron micrographs of untreated specimens as pale bodies of irregular form which are between the large aggregates of ribosomes. It is possible that the increased mass is caused by reaction of some component of the bodies with enzyme protein. It is possible that these represent the altered irregularly disposed three-dimensional net seen in untreated specimens between the Nissl bodies. It is difficult to identify the three-dimensional net, but it may represent some aspect of the Golgi apparatus as seen in other cell types in freeze-dried preparations.

Summary

The cytochemical distribution of nucleic acids was studied in normal spinal ganglion cells of the cat, and compared with that in ganglion cells, 6, 10, 17, 24, and 157 days after crushing the thoracic nerves. The cytochemical methods for nucleic acids are the same as those described in Chapter 3 (this volume) and Chapter 5 (Vol. II) and used to study the distribution of nucleic acids in a variety of normal mouse cells (this volume, Chapter 8) under conditions of maturation and decreased activity preliminary to senility and cell death (this volume, Chapter 9) as well as of increased activity following stimulation (this volume, Chapter 10). In normal cells or in cells which have recovered (157 days), the DNA double helix (or first-order helix) is organized as a second-order helix with a period and diameter of about 400 A. These course in the nucleoplasm for the most part singly or in very small aggregates which are submicroscopic in size, with very few aggregates large enough to be visible with the light microscope. RNA granules are crowded densely in the nucleolus, and occur also throughout the denser appearing walls of the nucleoplasm.

Shortly after the nerve is crushed, the DNA aggregates become smaller, and thereafter nearly all the DNA is in the form of separate second-order DNA coils (6–24 days). The RNA granules are most numerous in the nucleoplasm at 17 days, when they occur as aggregates called temporary nucleoli. At 10 days, and more markedly at 24 days, some of the second-order DNA coils are uncoiled, appearing as straight DNA filaments or first-order coils. This uncoiling has been observed in only two other cell types which were also presumably in very high states of activity [i.e., cartilage cells (Chapter 5, Vol. II) and syncytial somatic cells of the developing eggs of *Drosophila* (Chapter 13, this volume)].

The observations are interpreted in terms of increased synthesis of RNA during chromatolysis, associated with increased cytoplasmic synthesis. That is, the progressive exposure of second-order DNA coils as single helices and the uncoiling and extension of these as DNA filaments (first-order helix or DNA double helix) is probably associated with increased accessibility of DNA to transfer of metabolites to DNA, and of RNAs newly synthesized, to the nucleoplasm and then to the cytoplasm. These findings fall into the same pattern of organization of DNA in stimulated cells and, like them, are the obverse of the pattern of organization of mature cells destined for senility and cell death through pyknosis.

References

1. Barron, K. D., Doolin, P. F., and Oldershaw, J. B. (1967). Ultrastructural observations on retrograde atrophy of lateral geniculate body. *J. Neuropathol. Exp. Neurol.* **26**, 300–326.
2. Barton, A. A., and Causey, G. (1958). Electron microscopic study of the superior cervical ganglion. *J. Anat.* **92**, 399–407.
3. Bodian, D. (1964). An electron-microscopic study of the monkey spinal cord. *Bull. Johns Hopkins Hosp.* **114**, 13–119.
4. Brattgård, S.-O., Edström, J.-E., and Hydén, H. (1957). The chemical changes in regenerating neurons. *J. Neurochem.* **1**, 316–325.
5. Causey, G., and Hoffman, H. (1955). Cytoplasmic synthesis in nerve cells. *Brit. J. Cancer* **9**, 666–673.
6. Cole, M. (1968). Retrograde degeneration of axon and soma in the nervous system. *In* "The Structure and Function of Nervous Tissue" (G. H. Bourne, ed.), Vol. 1, pp. 269–300. Academic Press, New York.
7. Droz, B. (1969). Protein metabolism in nerve cells. *Int. Rev. Cytol.* **25**, 363–390.
8. Edström, J.-E. (1959). Ribonucleic acid changes in the motoneurons of the frog during axon regeneration. *J. Neurochem.* **5**, 43–49.
9. Edström, J.-E. (1964). Effect of spinal cord transection on the base composition and content of RNA in the Mauthner nerve fibre of the goldfish. *J. Neurochem.* **11**, 557–559.
10. Evans, D. H. L., and Gray, E. G. (1961). Changes in the fine structure of ganglion during chromatolysis. *Proc. Anat. Soc. Gr. Brit. Ireland,* pp. 71–74.
11. Hahn, W. E., and Laird, C. D. (1971). Transcription of nonrepeated DNA in mouse brain. *Science* **173**, 158–161.
12. Hartmann, J. F. (1954). Electron microscopy of motor nerve cells following section of axones. *Anat. Rec.* **118**, 19–33.
13. Hudson, G., Lazarow, A., and Hartmann, J. F. (1961). A quantitative electron microscopic study of mitochondria in motor neurones following axonal section. *Exp. Cell Res.* **24**, 440–456.

14. Hydén, H. (1943). Protein metabolism in the nerve cell during growth and function. *Acta Physiol. Scand.* **6**, *Suppl. 17*, 1–136.

15. Hydén, H. (1960). The neuron. *In* "The Cell" (J. Brachet and A. E. Mirsky, eds.), Vol. IV, pp. 215–323. Academic Press, New York.

16. LaVelle, A., and LaVelle, F. W. (1958). The nucleolar apparatus and neuronal reactivity to injury during development. *J. Exp. Zool.* **137**, 285–315.

17. Mackey, E. A., Spiro, D., and Wiener, J. (1964). A study of chromatolysis in dorsal root ganglia at the cellular level. *J. Neuropathol. Exp. Neurol.* **23**, 508–526.

18. Navarro, J. C. (1962). Elektronenmikroskopische Befunde an Spinalganglienzellen der Ratte nach Ischiadikotomie. *Proc. 4th Int. Congr. Neuropathol.*, Vol. 2, pp. 99–104. Thieme, Stuttgart.

19. Palay, S. L. (1964). The structural basis for neural action. *In* "RNA and Brain Function Memory and Learning" (M. A. B. Brazier, ed.), pp. 69–108. Univ. of Calif. Press, Berkeley, California.

20. Palay, S. L., and Palade, G. E. (1955). The fine structure of neurons. *J. Biophys. Biochem. Cytol.* **1**, 69–88.

21. Pannese, E. (1963). Investigations on the ultrastructural changes of the spinal ganglion neurons in the course of axon regeneration and cell hypertrophy. I. Changes during axon regeneration. *Z. Zellforsch. Mikrosk. Anat.* **60**, 711–740.

22. Porter, K. R., Bowers, M. B. (and 12 students). (1963). A study of chromatolysis in motor neurons of the frog *Rana pipiens*. *J. Cell Biol.* **19**, 56A and 57A.

23. Smith, K. R., Jr. (1961). The fine structure of neurons of dorsal root ganglia after stimulating or cutting the sciatic nerve. *J. Comp. Neurol.* **116**, 103–115.

24. Takano, I. (1964). Electron microscopic studies on retrograde chromatolysis in the hypoglossal nucleus and changes in the hypoglossal nerve, following its severance and ligation. *Okajimas Folia Anat. Jap.* **40**, 1–69.

25. Watson, W. E. (1965). An autoradiographic study of the incorporation of nucleic-acid precursors by neurones and glia during nerve regeneration. *J. Physiol.* **180**, 741–753.

13

Organization of DNA during the Mitotic Cycle in Eggs of Drosophila melanogaster

Eileen S. Gersh and Isidore Gersh

This report deals only with changes in the organization of DNA which take place during the mitotic cycle. The program for mitosis terminates in an equal division of the genetic material in such a way that the genetic linearity in the daughter nuclei is the same as in the original cell. The fixed (or "permanent") order of genes as derived from genetical studies is replicated and is the same in both daughter cells. These chemical and functional requirements may be restated morphologically to this effect: the amount of DNA is the same in both daughter nuclei as in the original nucleus, the continuity of the DNA double helix is retained at all times during the mitotic cycle, and during mitosis the DNA which has been replicated is arranged in chromosomes with such compactness as to facilitate their division between two cells equally. During interphase the requirements are different, for the cell must synthesize a duplicate set of DNA and an array of RNAs adequate for an appropriate increase of nuclear and cytoplasmic proteins and other chemical substances. These requirements are even more extreme than those of cells activated to secrete a single product or class of products (as treated in Chapters 8–10, this volume).

The eggs of *Drosophila* during cleavage divisions are especially advantageous for the morphological study of changes in the organization of DNA during the mitotic cycle *(10)*. It is the shortest known cycle, spanning about 10 minutes. About one-third of this cycle is spent in interphase (including G_1, S, and G_2), and the rest in mitosis. Because of the brevity of the interphase, one would expect that a maximal extension of the DNA double helix or of the second-order DNA helix would take place during interphase to provide sufficient opportunity for the exchange of metabolites and gene products with nucleoplasm and cytoplasm. The maximally extended state of DNA expected during interphase would thus be the obverse of the compactness achieved during mitosis.

There is only one published electron micrograph of a mitotic figure in *Drosophila* eggs *(2)*. The review article in which this appears was primarily concerned with technical aspects of freezing methods. These are considered in Chapter 1 of this volume.

The earliest divisions take place within the yolk; the nuclei then migrate to the surface where they form a syncytium. The divisions are remarkably synchronous, even after twelve duplications, when the number of nuclei in the peripheral syncytium is over 3000. This assures that all the nuclei are in very nearly the same state in the cycle. Thus the identification, in sections, of the mitotic stage with the light microscope carries over (within a few seconds) to all the several thousand nuclei in the egg. The total length of all four chromosomes is about 7.5 μ. The small size of the chromosomes is highly advantageous for studies with the electron microscope, for each ultrathin section of a chromosome contains a significant fraction of the total chromosome. The small number of chromosomes reduces the difficulty of interpreting their shape and relationships in electron micrographs.

Methods

The preparation of the eggs for the fixation and staining of nucleic acid is advantageous since all the nuclei of an egg are immobilized within fractions of a second. As described earlier (Chapter 3 this volume), the staining of the nucleic acids is achieved with minimal morphological disturbance and DNA can be distinguished morphologically from ribonucleic acids.

The eggs were collected in the manner described by Sonnenblick *(12)*, except that wooden spoons covered with corn meal-agar medium were used for the deposition and collection of eggs instead of banana skin. The age of the eggs at fixation, with few exceptions, was about 110 minutes ± 10 minutes with the incubation temperature at 24°C. The eggs were placed on small squares of thin aluminum foil, and frozen ultrarapidly in Freon 13 at about -175°C. The specimens were dried in a vacuum at a temperature below -40°C and treated *in vacuo* at room temperature with vapors of mercury (II) hexafluoroacetylacetonate in the presence of water vapor. After removal of excess cross-linking reagent, the specimens were infiltrated *in vacuo* with 95% alcohol. After hardening for a day, they were stained with aqueous gallocyanin–chromalum, dehydrated, and hardened again in 95% alcohol. They were infiltrated with water-soluble Durcupan, and the monomer mixture was induced to polymerize. Thin sections ($\frac{1}{4}$ μ) were made, mounted on slides, and stained with toluidine blue. When the stage was identified, all further sectioning was for study with the electron microscope. Photomicrographs and electron micrographs at various magnifications were made as described in Chapter 3 (this volume) and Chapter 5 (Vol. II) where details of specimen preparation are given in greater detail.

A total of 81 eggs was studied with the light microscope. Of these, 22 were prepared for study with the electron microscope. The description presented in this report is based on twelve of this number, which were studied most intensively.

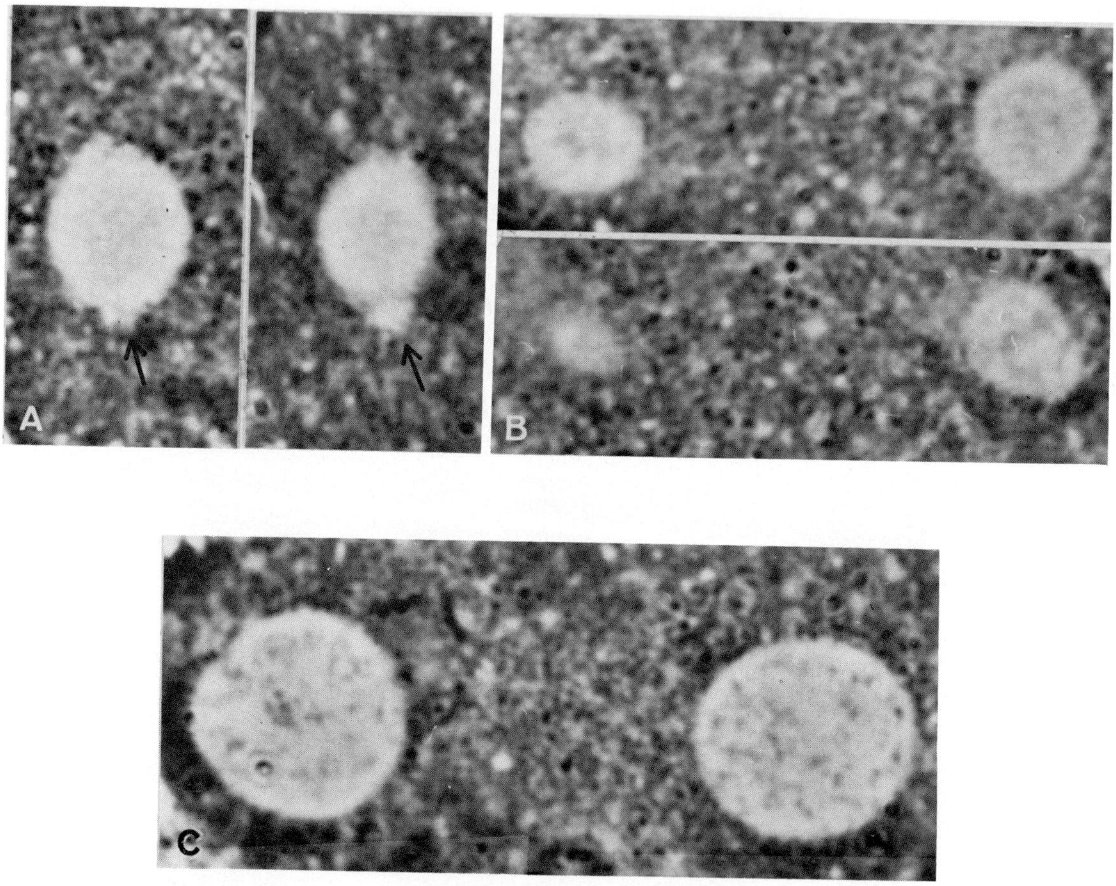

FIG. 13.1. Photomicrographs of eggs of *Drosophila melanogaster* in different stages of the mitotic cycle. Nucleic acids are selectively stained as described in Chapter 3 (this volume) × 3200. (A) Early interphase. Centriolar region appears at arrow. There are no resolvable chromatin granules or chromosomes. (B) Later interphase. Nuclear chromatin or chromosomes are not resolvable. Denser, vaguely indicated granules and masses are present. (C) Prophase. Chromosomes appear as rodlets or granules, all of them being sections of elongated chromosomes. (D) Metaphase. The mitotic apparatus is ovoid. Paired chromatids are aligned at the metaphase plate. They appear as granules which may be spherical, or elongated, depending on the plane of section through them. They may also be coiled. (E) Metaphase to early anaphase. While nearly all nuclei are in metaphase, in one nucleus at least two chromosomes are in early stages of separation. (F) Anaphase. The mitotic apparatus is elongated, and the chromosomes are outstretched at each pole. The chromosomes may be coiled, but if so the pitch and periodicity approach the limit of resolution. (G) Telophase to interphase. The appearance of the nucleus depends on the plane of section through the egg. The chromatin masses may be large and irregular, or minute, reaching to the limit of resolution.

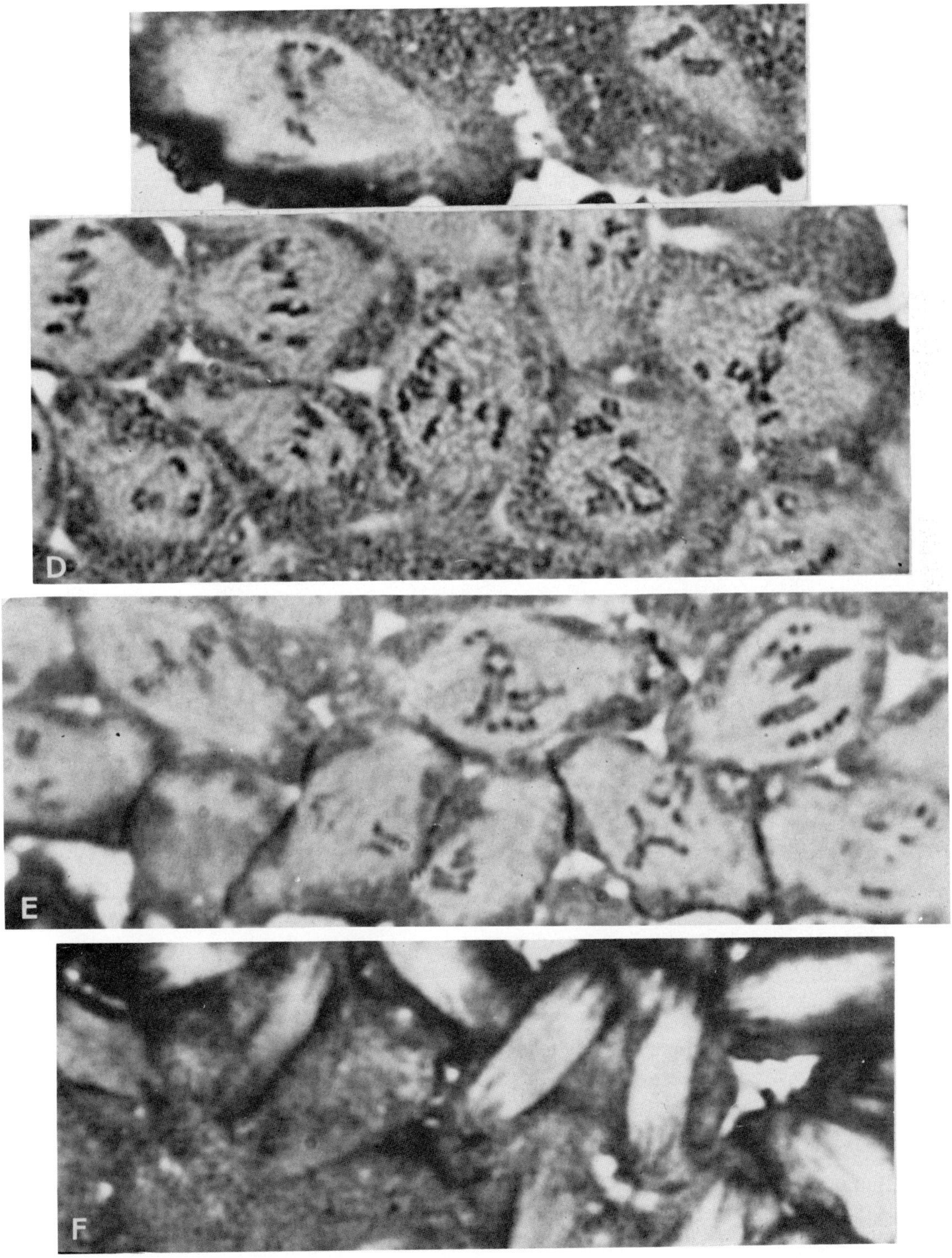
D
E
F

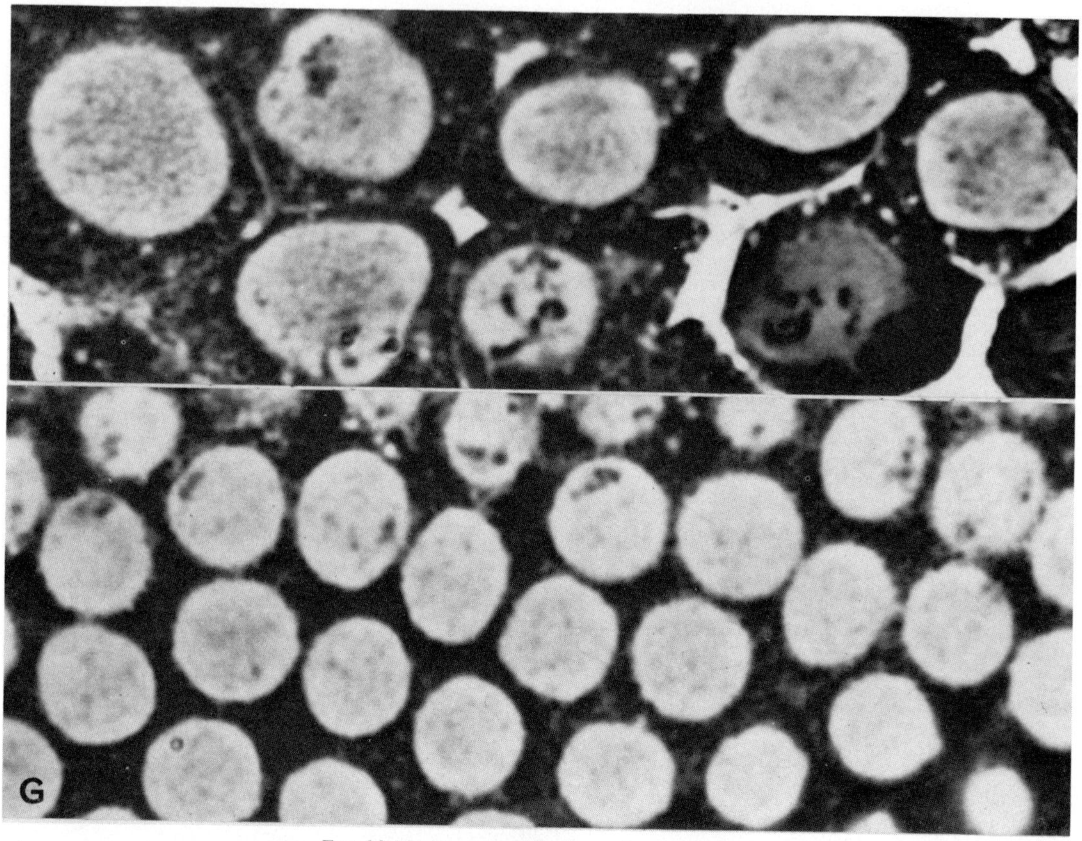

FIG. 13.1G. For complete legend see p. 340.

Observations

LIGHT MICROSCOPY

At interphase, the nuclei have no resolvable structure (Fig. 3.1A); in later stages some poorly resolvable masses, sometimes appearing cloudlike, are present (Fig. 13.1B). At prophase, granules are numerous and clearly resolvable (Fig. 13.1C). At metaphase, the chromosomes are paired and coiled, and arranged as a metaphase plate; the mitotic apparatus is ovoid (Fig. 13.1D). At anaphase, the chromosomes appear elongated, and eventually converge at the widely separated poles; the mitotic apparatus appears spindlelike (Fig. 13.1E and F). Telophase rapidly merges into interphase and a nucleus may have coarse chromatin masses, with very fine granules nearby, with no resolvable chromatin (or chromosomes) in the remainder of the nucleus, as at early interphase (Fig. 13.1G).

ELECTRON MICROSCOPY

The term first-order DNA coil is used for the Watson-Crick double helix. It appears as a line in specimens treated as described. It is arranged to form a second-order coil of about 200 A in diameter.

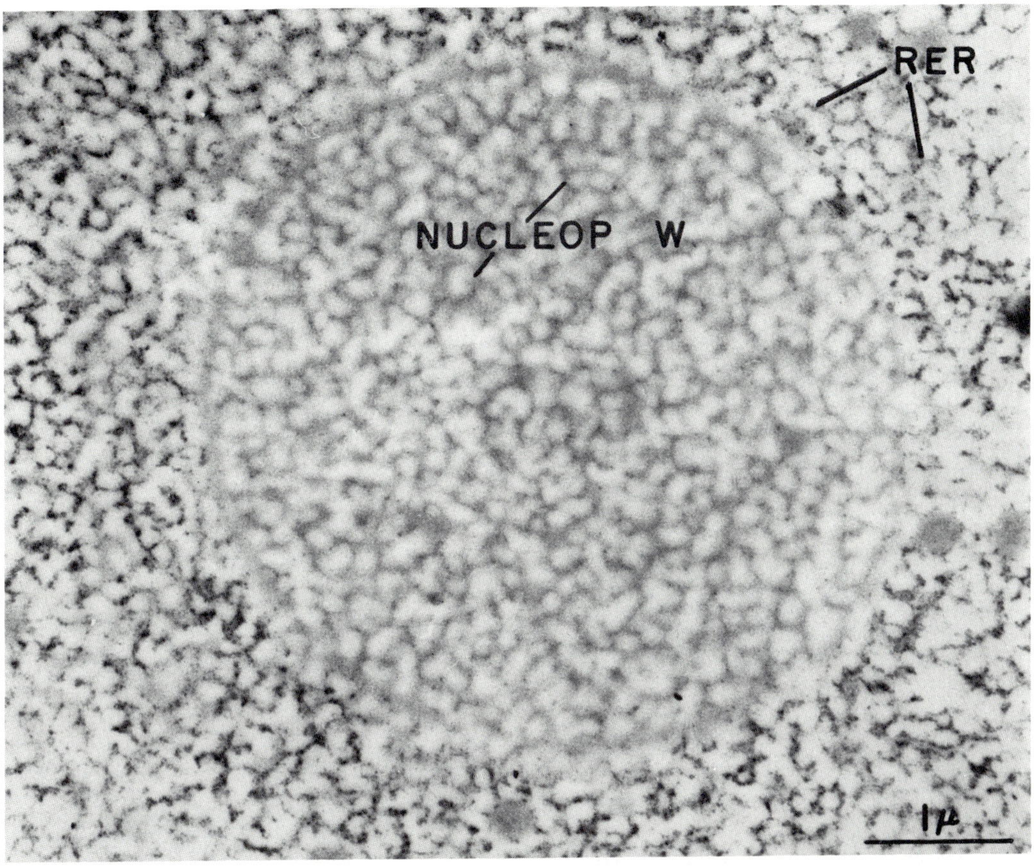

FIG. 13.2. Interphase nucleus and surrounding cytoplasm. Nucleoplasmic walls (NUCLEOP W) are relatively uniform in density and thickness. The DNA is too finely distributed to be visible clearly at this magnification. RER is densely stained in the cytoplasm. × 19,400.

At interphase, the DNA is quite widely distributed throughout the walls of the nucleoplasm, and can be recognized as occasional coils with a diameter of about 200 A. They differ only slightly in contrast from the protein component of the nucleoplasmic walls (Fig. 13.2). The latter vary in diameter and thickness, but not markedly. They could accommodate only a single second-order DNA helix. The scarce coils are more apparent in Fig. 13.3A, which is traced from a part of Fig. 13.3.

In late interphase, some nucleoplasmic walls are appreciably thicker and denser than others, forming broad zones of somewhat increased density (Fig. 13.4). Part of this increased density is attributed to the fact that more DNA coils are present in the nucleoplasmic walls, because the first-order helix is wound into a regularly repeating second-order coil with a diameter of 200 A and part to the fact that the coils are replicated and may be present in pairs. The period of 200 A is retained during the remaining stages of mitosis, and for this reason the stage is regarded as representing the earliest stage of chromosome formation. This could account for the increased thickness of the nucleoplasmic walls in certain regions. This is illustrated in Fig. 13.5, and in a tracing of a part of this in Fig. 13.5B (but not in 13.5A).

In prophase, the regions of increased density are more circumscribed, the patches reaching microscopic dimensions (Figs. 13.6 and 13.7). The dense patches contain DNA coils which are somewhat loosely aggregated in the nucleoplasmic walls as compared with later stages. The thickness of the walls can accommodate two second-order helices, which appear side by side in sections. Between these areas, DNA coils may be present in columns in the nucleoplasmic walls (Fig. 13.8). The dense patches are early and moderately late stages of chromosome formation which are due to an in-gathering and concentration of the second-order DNA coils from the earlier dispersed states. The chromosome bits may be so loosely packed and numerous in sections as to make it impossible to define

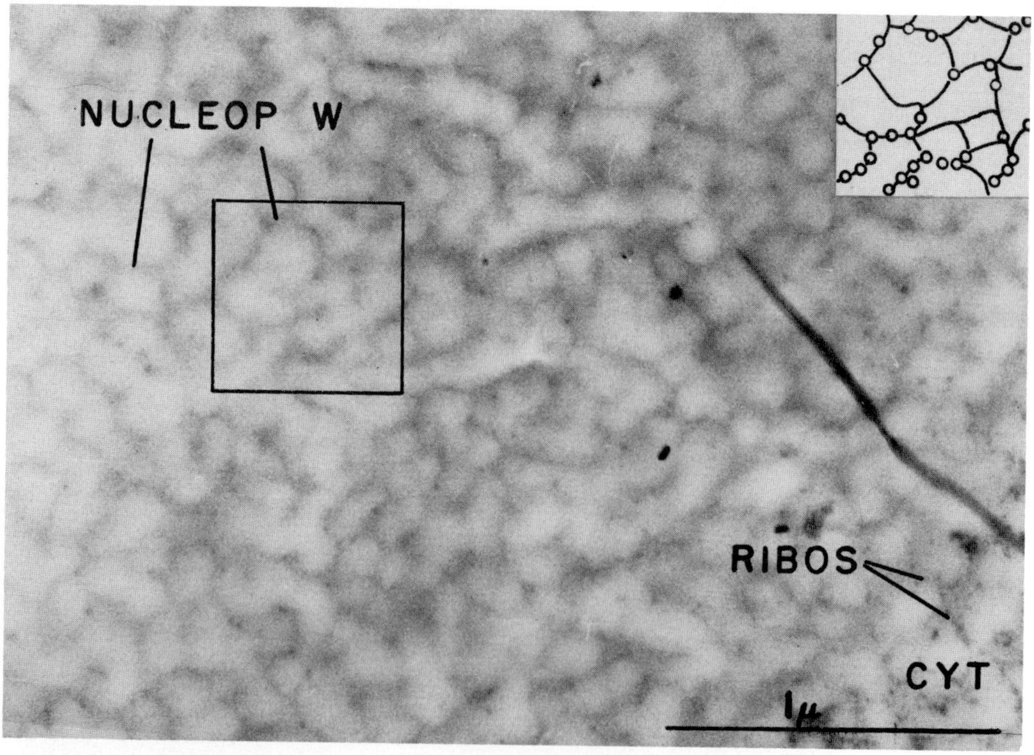

FIG. 13.3. Portion of another interphase nucleus at higher magnification. The second-order DNA coils are widely separated, and are difficult to distinguish in the figure. A tracing of a portion of the field from the print shows the location of discernible coils. These always occur in the nucleoplasmic walls (NUCLEOP W). × 44,000.

the presence of paired chromatids. These are, by contrast, readily observed during metaphase.

In metaphase, the chromatids consisting of packed second-order DNA coils are paired and may be lined up along the metaphase plate. The backbone of the chromatids seems to be a central core of tightly packed second-order DNA coils whose main orientation is longitudinal (Figs. 13.10, 13.11, and 13.13). The central core is surrounded by loops of second-order DNA coils which are in the nucleoplasmic walls immediately surrounding the core. Some of the latter are cut in a plane such that whole gyres of the second-order helix are retained in the section (Fig. 13.9). The length of the chromosome in sections also depends on the plane of section, and so varies from the transverse, to the oblique, to the longitudinal. Examples of the different planes are shown in a section through the chromosomal core in Fig. 13.9. Sometimes the section passes longitudinally through the chromosomal "core" (Figs. 13.10, 13.11, and 13.13), which may consist of as many as four strands of second-order helix. Sometimes the section passes through the peripheral loops (Figs. 13.9, 13.12, and 13.13), where generally two helices can be seen lying side by side (Figs. 13.12 and 13.14). These represent a third-order of coils which lie in the walls of pseudovacuoles. Sometimes an aggregation of second-order coils covers an area comparable to the surface of a pseudovacuole (Figs. 13.12 and 13.14). In some of the examples given above, chromatids did not appear coiled. When coiled, the chromatids may appear in section as a series of longitudinally or obliquely sectioned bodies more or less regularly interrupted (Figs. 3.10, 13.11, 13.12 and 13.14). The remainder of the nucleoplasmic walls contains numerous granules of RNA.

In anaphase, the second-order helices of the central core of the chromosome are commonly prominent (Figs. 13.15–13.17), and those which encircle the pseudovacuoles in the walls are less so. Besides, there are many aggregates of several second-order coils which project radially from the core (Figs. 13.15 and 13.17) and occupy about as much space as a pseudovacuole. The centromere regions of the chromosomes are attached to the nuclear membrane, which may be thickened (Fig. 13.17). Nucleoplasmic walls in the area may also be thickened (Fig. 13.18).

In telophase, the second-order DNA coils of the chromatin become less tightly packed than in anaphase. In all three specimens in telophase, large chromatin masses were present, consisting of densely aggregated second-order DNA coils (Fig. 13.19). In the same nuclei, were smaller, uniform aggregates which varied in size from the microscopic to the submicroscopic (Fig. 13.20). These were distributed around the larger chromatin masses. Also present in the same nuclei were large regions free of DNA aggregates of any sort, where second-order coils of DNA were present as single coils or helices (Fig. 13.21). An electron micrograph of an unusually thin section shows the second-order coils, with RNA granules interspersed between them (Figs. 13.22 and 13.23). Also present are numerous DNA filaments which course in the nucleoplasm as straight or curved filaments (Fig. 13.24). These are the first-order DNA helices or the double helix of Watson and Crick.

Discussion

During early interphase in *Drosophila* eggs, the second-order DNA helices are single and distributed through the nucleus, confined to the nucleoplasmic walls. Much of the DNA is even further extended as first-order DNA helices which appear as straight or gently curved filaments. Since the DNA is submicroscopic, it is no surprise that the nucleus appears homogeneous at this stage when viewed with the light microscope. The diameter of the second-order DNA coil is about 200 A in all egg nuclei studied. This is about one-half the diameter of the helix in cells of the imaginal disks of *Drosophila* (Chapter 6, this volume), the polytene chromosomes of the salivary glands of *Drosophila* larvae (Chapter 4, this volume), and

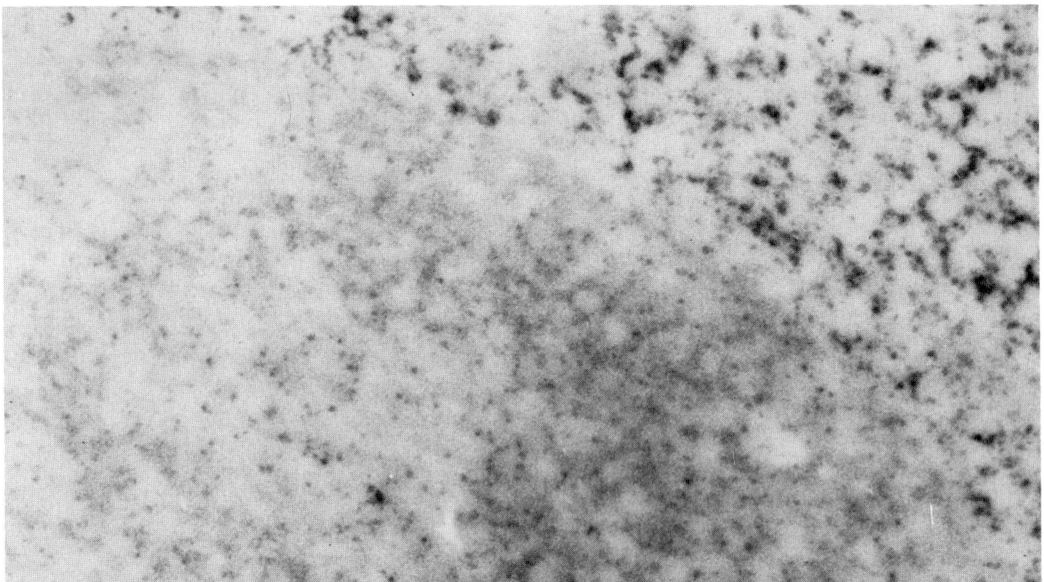

FIG. 13.4. Late interphase nucleus. In patches of nucleoplasm, the walls are thicker than in other regions, where the density and thickness are as in early interphase. × 19,400.

of somatic cells of the mouse (Chapters 3, 7, and 8, this volume, and Chapter 5, Vol. II). The extensive contact with the nucleoplasm of the separated and disaggregated second-order DNA coils and first-order DNA filaments in the interphase nucleus is assumed to permit maximal activity of the nucleus during a limited period of intensive growth; for although it appears that some enzymes are not synthesized until the time of hatching *(14)*, extensive gene activity is surely required earlier for the conversion of yolk materials to new protoplasmic substance.

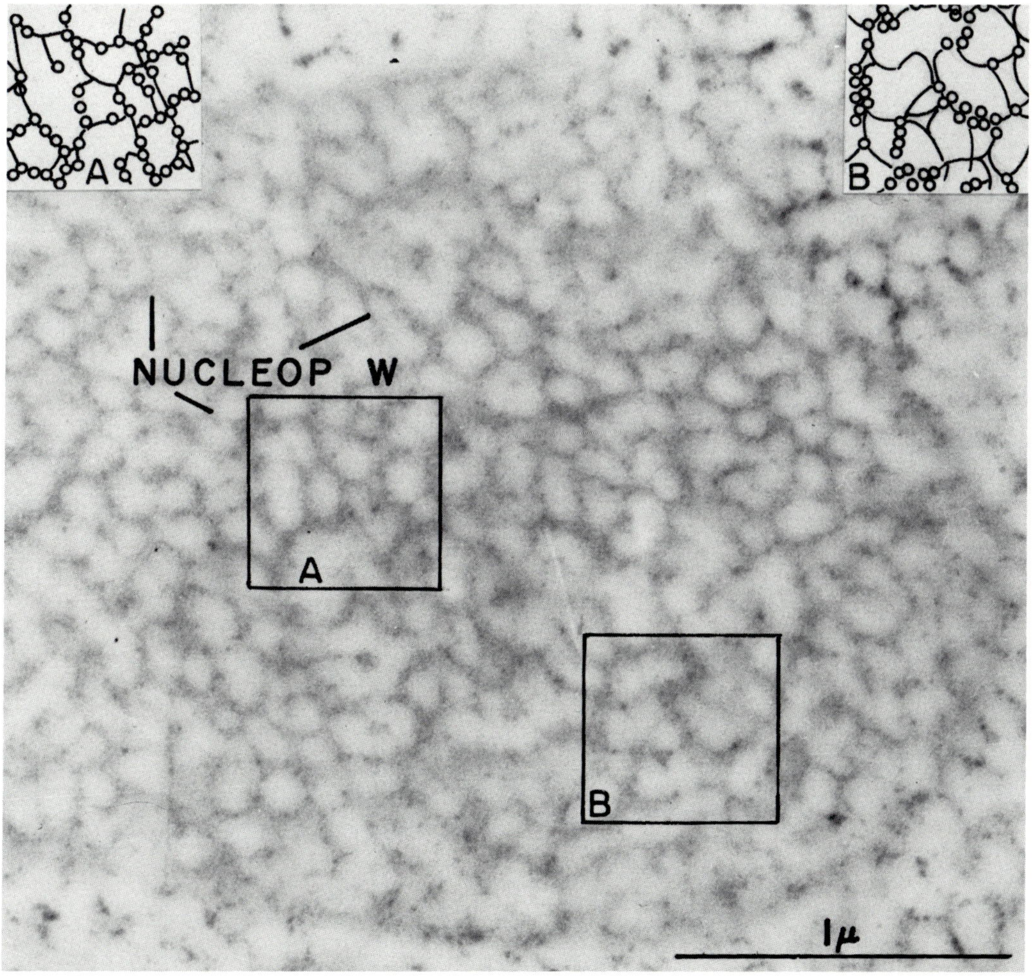

Fig. 13.5. A portion of another nucleus from the same egg. The dense nucleoplasmic walls (NUCLEOP W) often contain two sets of second-order DNA coils. These are shown in one part of a tracing of a portion of the nucleus (B). There is still one set of coils in the other part (A). × 44,000.

The proximal heterochromatic regions of the chromosomes are consistently heteropycnotic at telophase, interphase, and prophase in many, though not all *(5)*, tissues of *Drosophila*. It seems remarkable that in spite of the exceptional brevity of the cleavage division interphases, estimated at 2 or 3 minutes *(12)*, these regions unwind completely from the compact supercoiled condition achieved during mitosis. It would seem that this euchromatization would indicate that genes in these regions have important functions during the cleavage divisions. We suggest that a significant proportion of the genes controlling the events of mitosis are located in these proximal regions *(13)* and that the time of action of these genes is normally in a fleeting period before the initiation of mitosis.

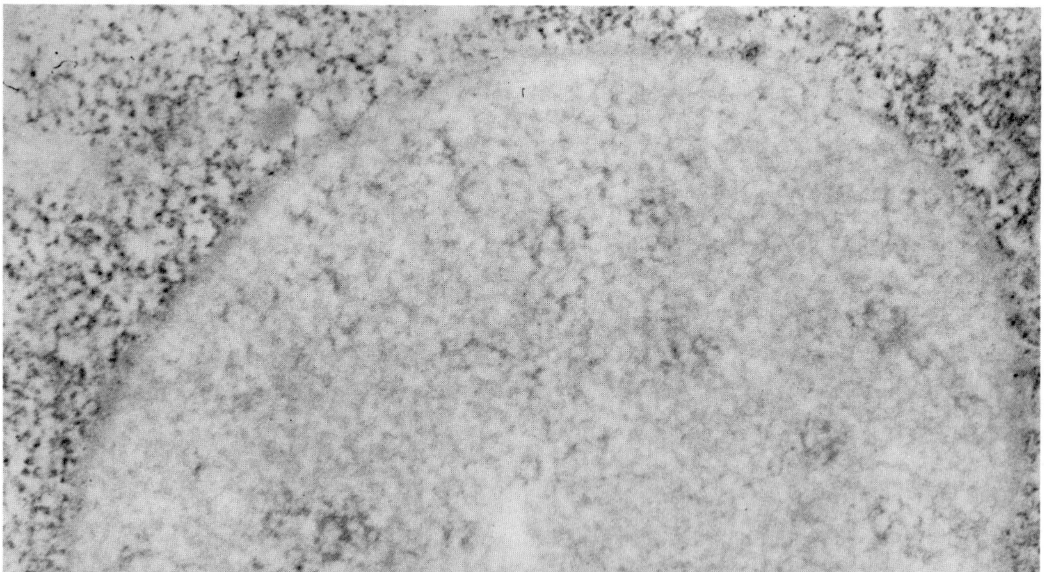

FIGS. 13.6–13.8. Prophase. In some parts of the nucleus the second-order DNA coils have coalesced to form aggregates which range from the microscopic to the submicroscopic, as seen in Fig. 13.6. In other parts of the nucleus, the coalescence is more advanced, as in Fig. 13.7. It is of interest that the most condensed DNA aggregates are near the surface of the nucleus. At a higher magnification in Fig. 13.8, are separate DNA helices (COIL DNA—1, —2, —3, —4). Figures 13.6 and 13.7, × 19,400; Fig. 13.8, × 44,000.

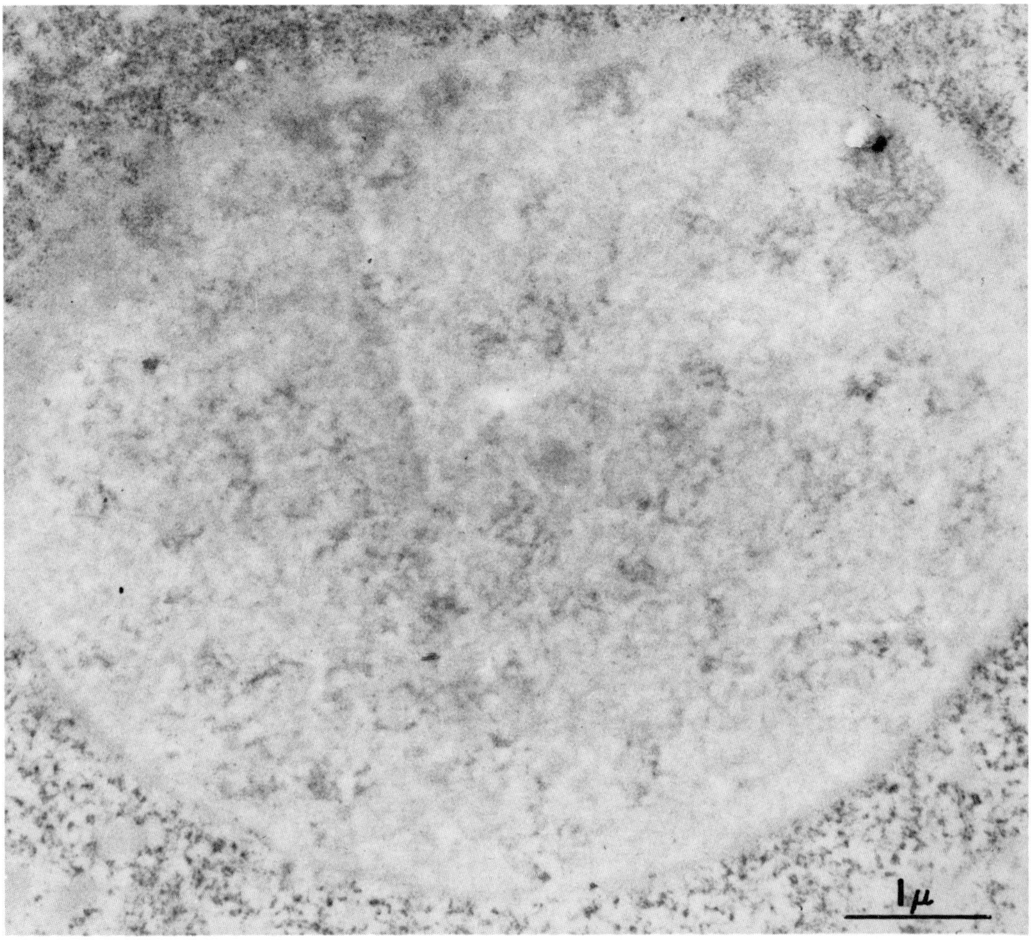

FIG. 13.7. ▲

▼ FIG. 13.8.

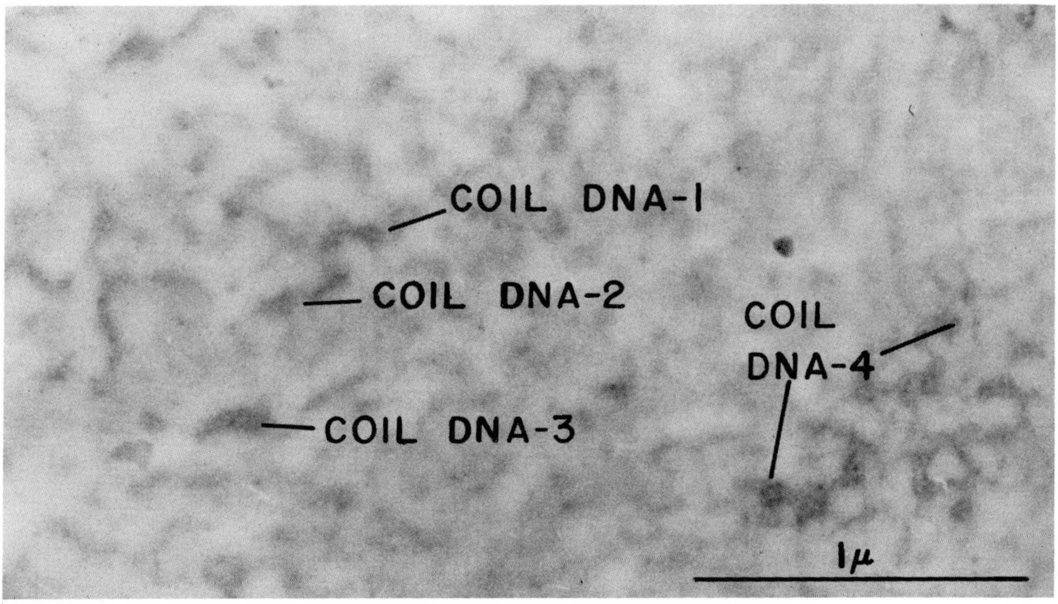

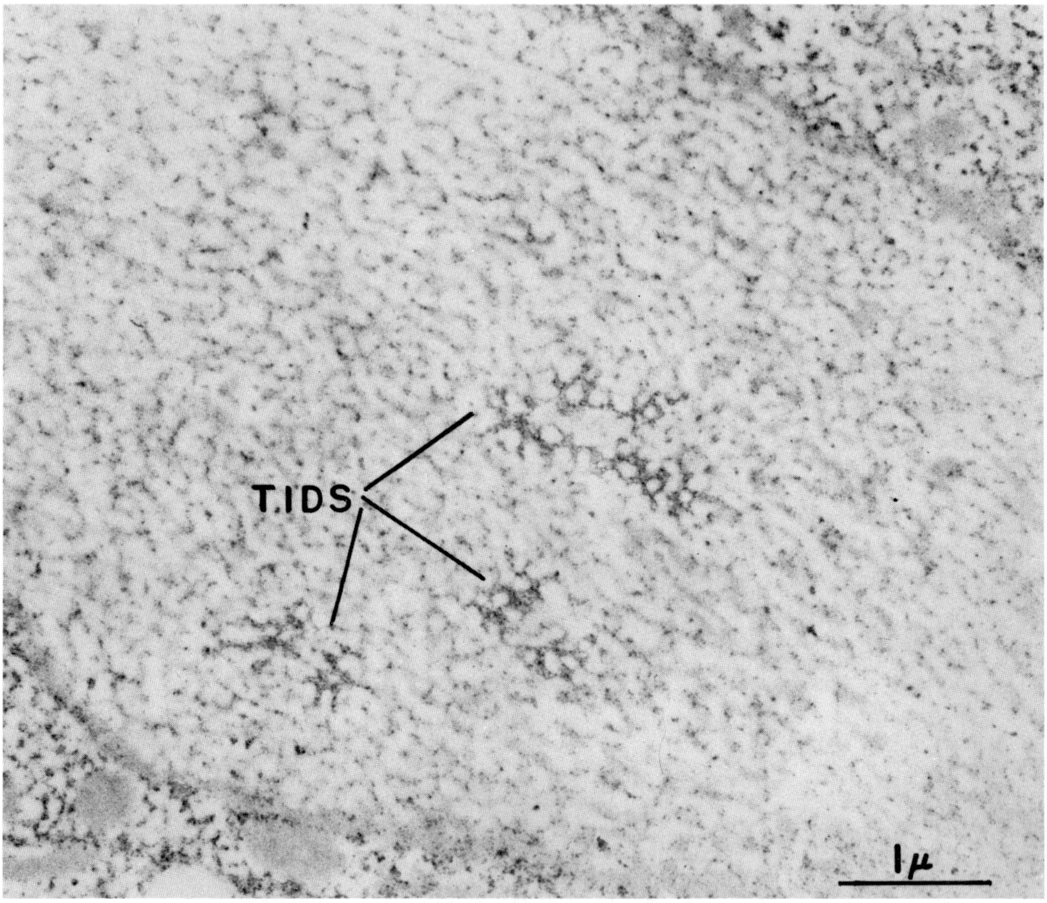

FIG. 13.9.

FIGS. 13.9–13.14. Metaphase chromosomes. In Fig. 13.9, sections through three regions of paired chromatids (TIDS) are present—two sectioned more or less transversely, one unequally obliquely. The two parts of each chromosome are separated by one or two nucleoplasmic walls with intervening spaces. In Fig. 13.10, the chromatid pairs are sectioned more nearly longitudinally. The section includes the chromosome "core" (CHROM C) and the interruption in the upper half probably reflects the period of the fourth-order helix. Figure 13.11 is the same chromatid pair at higher magnification to show the circular figures of the second-order DNA coils. Nucleoplasmic walls (NUCLEOP W) contain numerous RNA granules (GRANS). Figure 13.12 is a similar section through a chromatid pair but through the part of the chromosome (CHROM P) where the second-order DNA coils course through the nucleoplasmic walls. Interruptions at A and B probably reflect the period of the fourth-order helix. Figure 13.13 shows the "core" of the longest section of a metaphase chromatid observed. The "coils" seem to be more or less longitudinally oriented. RNA granules (GRANS) in the nucleoplasmic walls (NUCLEOP W) are rather numerous. Figure 13.14 shows a section through a coiled chromatid with a relatively short fourth-order coil period. The coils are numbered. RNA granules (GRANS) in the nucleoplasmic walls are rather numerous. Figure 13.9 × 19,400; Fig. 13.18, × 22,600; Figs. 13.11–13.14, × 44,000.

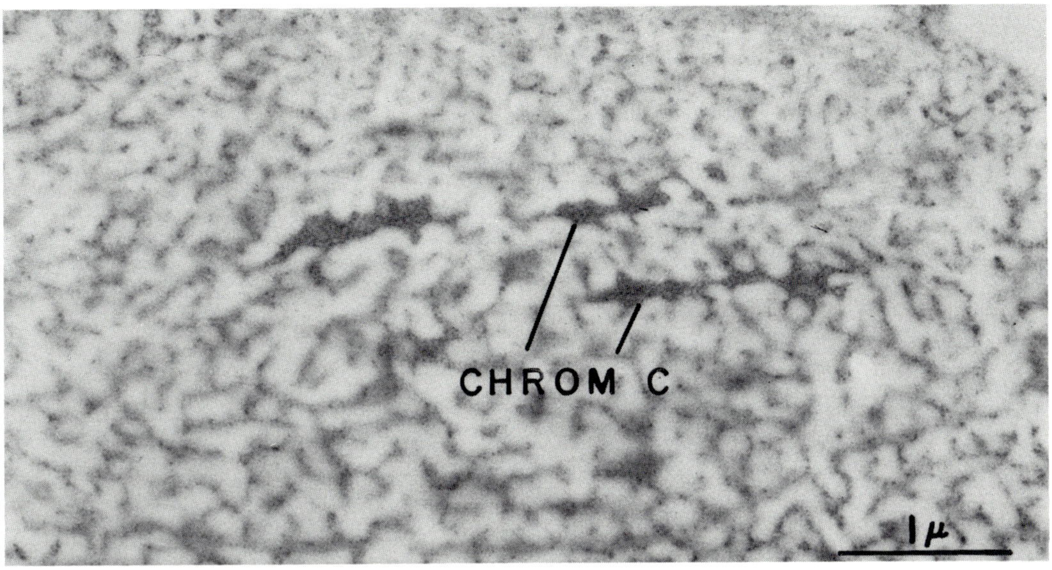

FIG. 13.10. ▲

▼ FIG. 13.11.

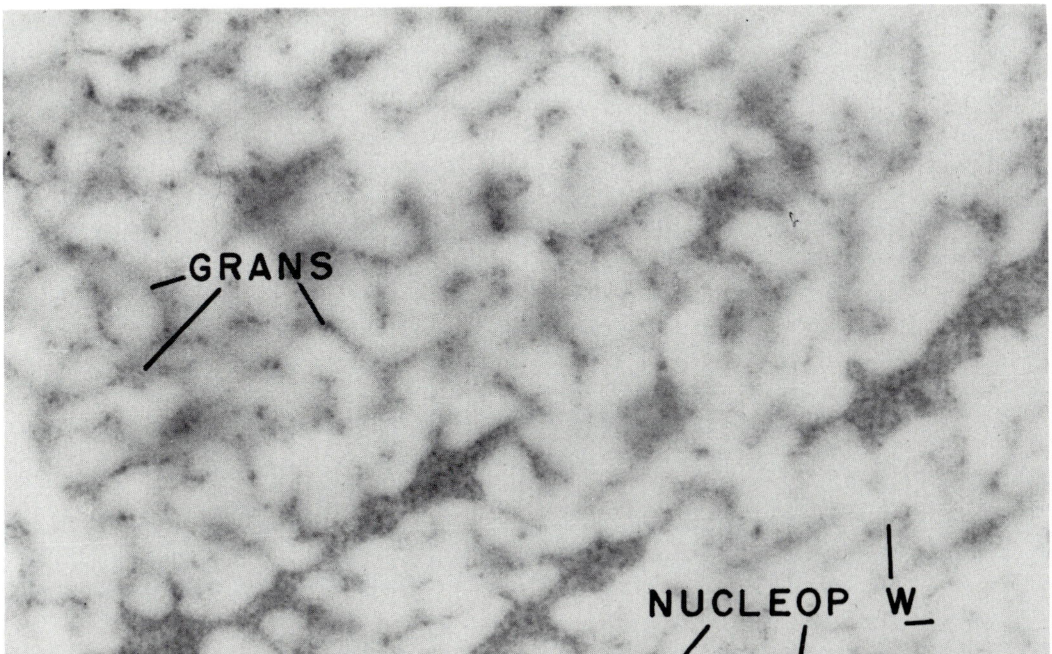

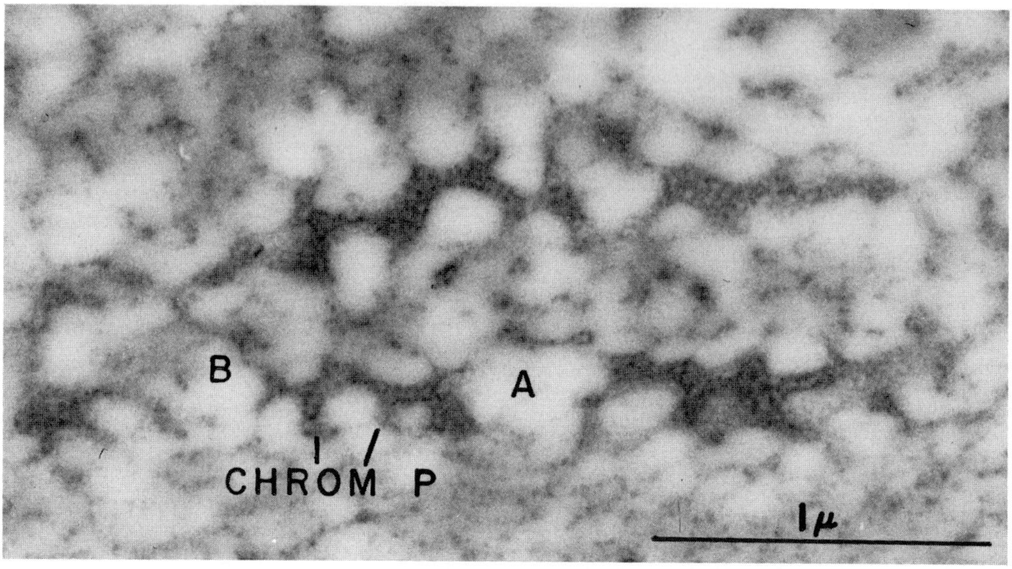

FIG. 13.12. See p. 350 for complete legend.

There is circumstantial evidence for this hypothesis. Nuclei, which are in rapid division or are stimulated to rapid division, commonly become completely euchromatic as mentioned in the preceding paragraphs. The proximal heterochromatic regions, or parts of them, are underrepresented in polytene nuclei of *Drosophila;* they are not replicated to the same degree as other regions *(11)*. The genes of the unreplicated regions presumably play little or no part in the development of the cells which, while they cease to divide at an early embryonic stage, continue to increase in volume throughout larval life. The failure to replicate of genes in the proximal chromosome regions may be intimately related to the loss of capacity for division. Then there is evidence for the presence in the proximal regions of "collochores" which cause the association of chromosome pairs at meiosis *(3)* and their subsequent orientation and regular disjunction *(9)*. There is also evidence of elements in these regions which control chromosome movement at meiotic anaphase *(8)*.

If this hypothesis is correct, all nuclei about to divide should show a premitotic euchromatization of chromatin regions that contain genes controlling mitosis. Perhaps the occasional large neuroblast nucleus of *Drosophila* [a cell type that is capable of division but shows no sign of heteropycnosis *(5)*] is at this stage. Perhaps the diffuse stage in the meiotic prophase of many organisms *(1)* is another instance of this stage, but displaced in time in relation to the initiation of prophase.

These suggestions are based on results of the studies on changes in the DNA molecular pattern of certain activated cells (see Chapters 10 and 12, this volume and Chapter 5, Vol. II). In such cells, more second-order DNA coils were dispersed into the nucleoplasm, depleting the heterochromatic aggregations of heter-

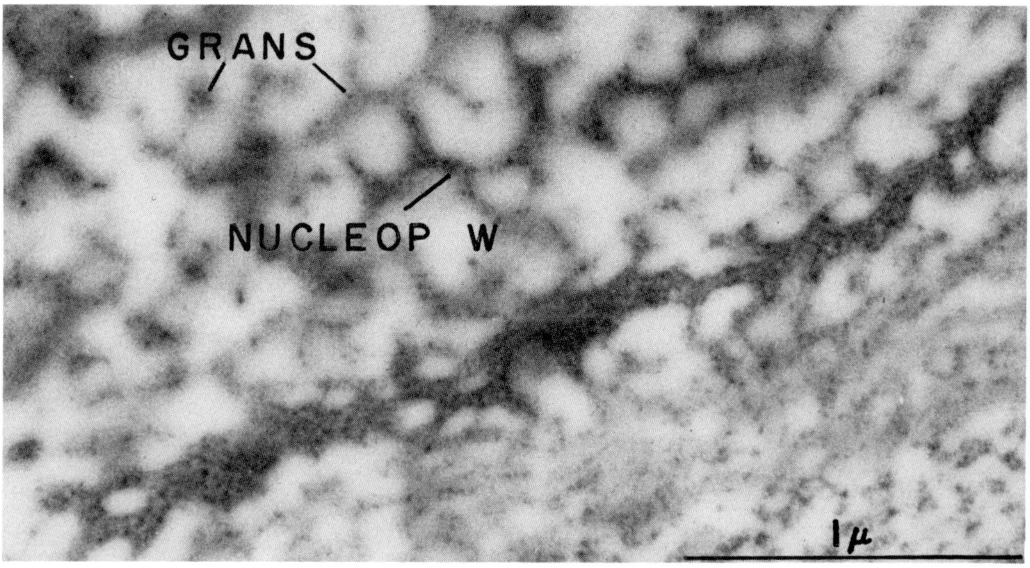

FIG. 13.13. ▲

▼ FIG. 13.14.

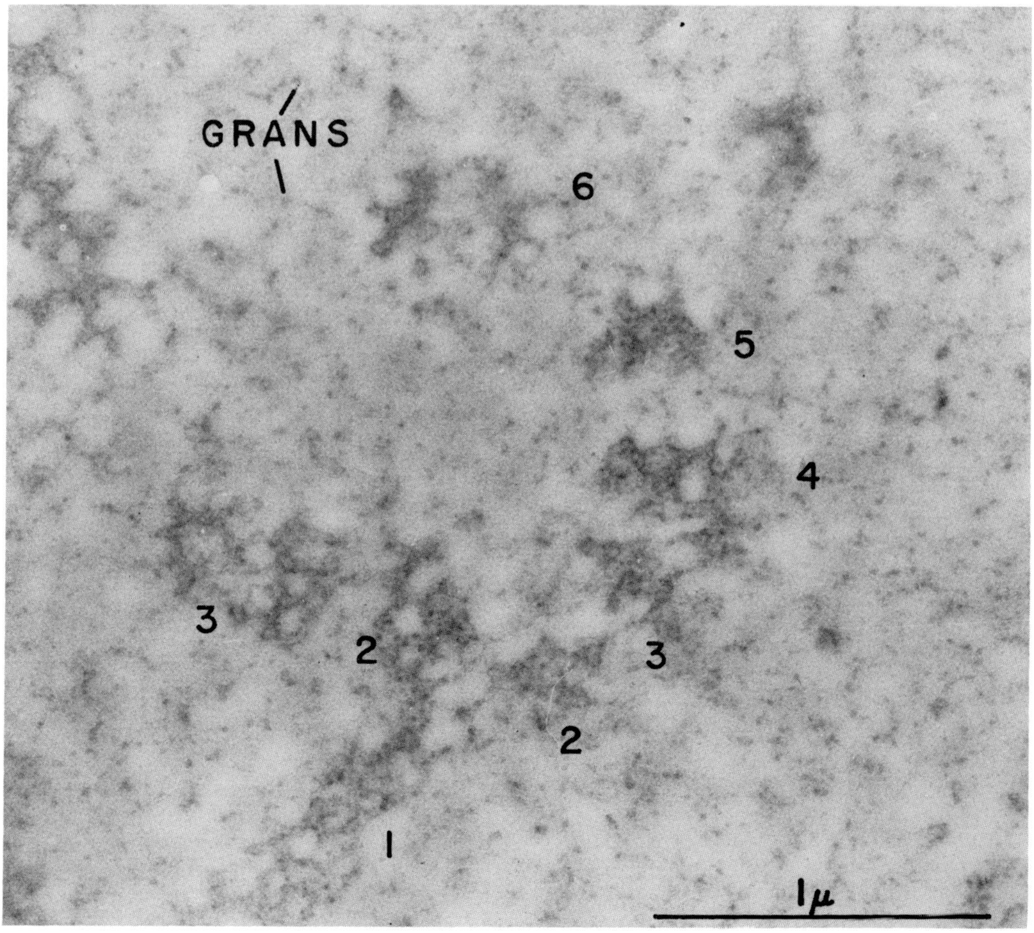

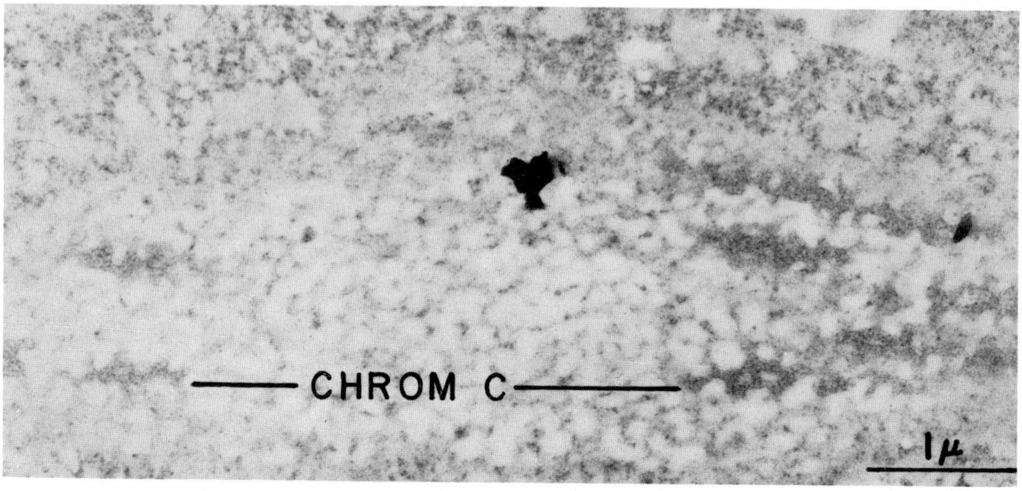

FIG. 13.15.

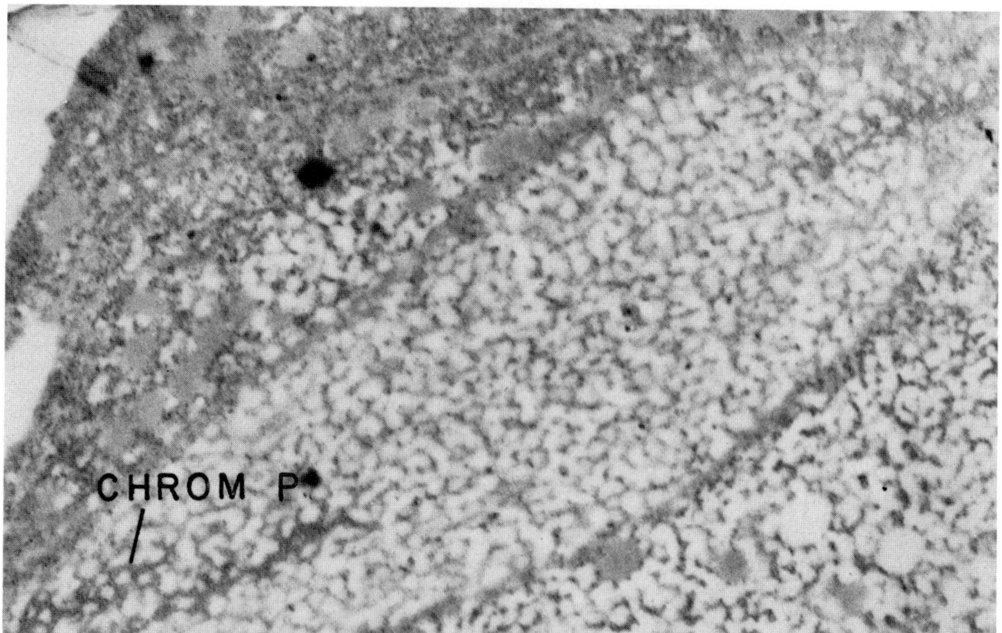

FIG. 13.16.

FIGS. 13.15–13.18. Anaphase chromosomes. Figures 13.15 and 13.16 show chromosomes at the poles in two different nuclei. In Fig. 13.15, the plane of section encloses chromosomal "cores" (CHROM C), while in Fig. 13.16, the plane of section includes more third-order coils surrounding pseudovacuoles (CHROM P). Cytoplasmic ribosomes are dense. Figure 13.17 shows the site of attachment of the centromere regions of chromosomes (CENT CHROM) to the nucleo–cytoplasmic boundary. This site of attachment also appears in Fig. 13.18 which is

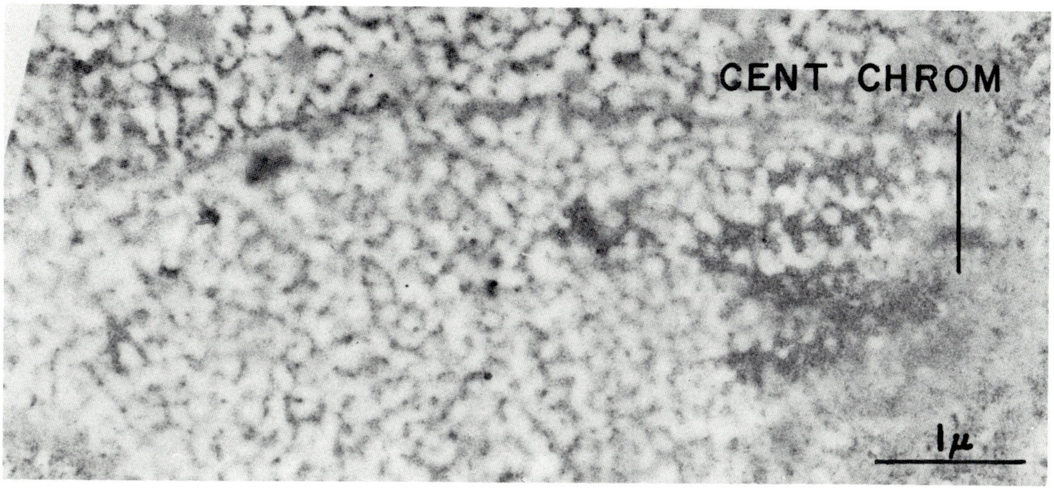

Fig. 13.17.

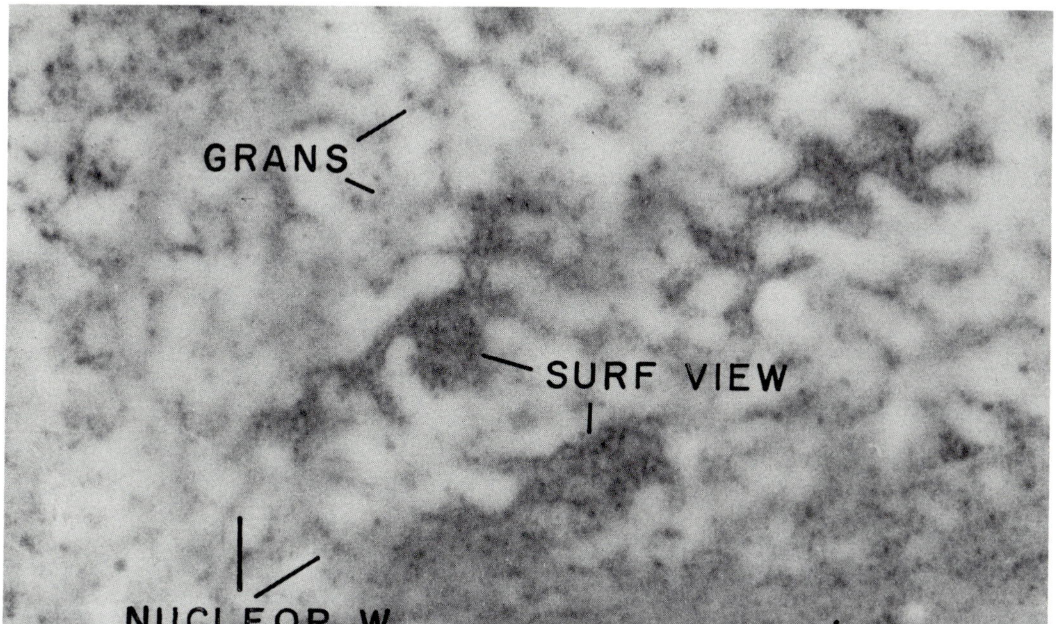

Fig. 13.18.

designed to show also the helical arrangement of the second-order DNA coils. In both Figs. 13.17 and 13.18 there are examples of surface views of pseudovacuoles whose walls are filled with second-order DNA coils (SURF VIEW). Ribosomes are resolvable in the cytoplasm. Numerous RNA granules (GRANS) are present in the nucleoplasmic walls (NUCLEOP W). Figures 13.15–13.17, × 19,400; Fig. 13.18, × 44,000.

ochromatic regions, and exposing more free paths for rapid exchange between nucleo- and cytoplasm. These findings are supplemented by the observation of fully extended or first-order DNA helices which appear in electron micrographs as filaments. Such figures have been observed in highly activated cartilage cells (Vol. II, Chapter 5) and in nerve cells at the height of the chromatolysis reaction (this volume, Chapter 12). Further support for the suggestion is proffered from preliminary observations of nuclei of rapidly dividing cells of tumors derived from hamster

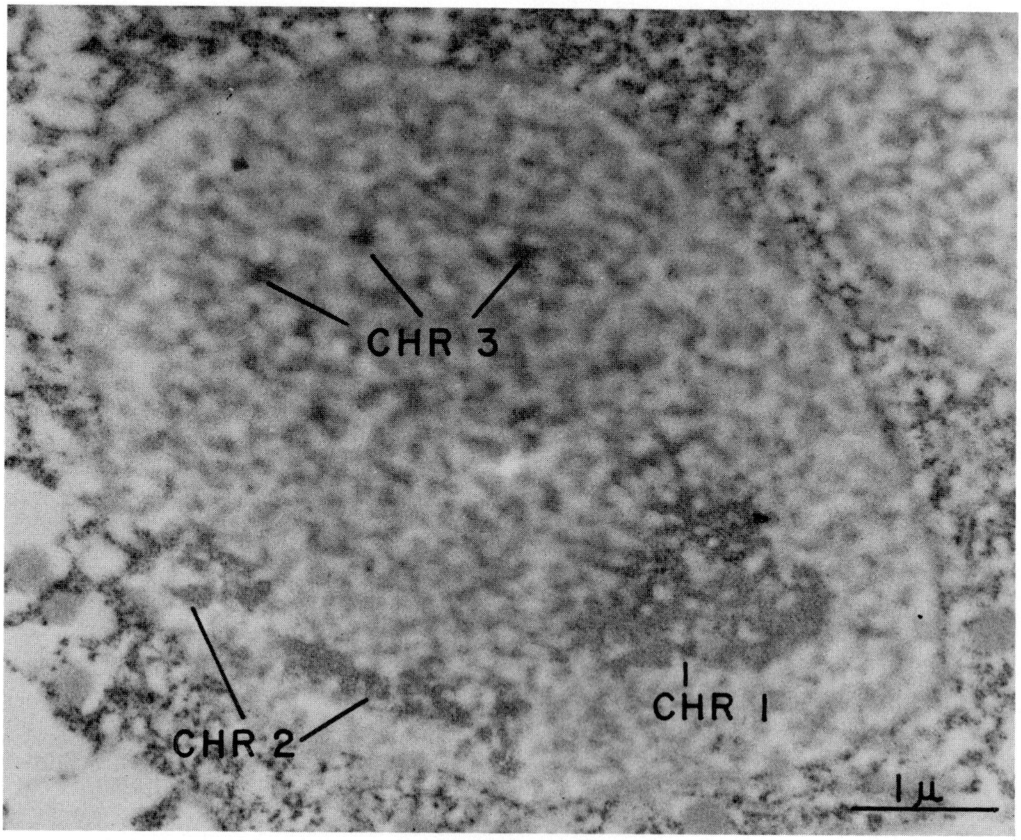

FIG. 13.19.

FIGS. 13.19–13.24. Telophase nucleus. In the nucleoplasmic walls of Fig. 13.19 are microscopic and submicroscopic chromatin masses, some of which, as at CHR 1 are composed of a tightly bound aggregate of second-order DNA coils, while at other parts, such as CHR 2, the aggregates are more loosely bound, and at still other parts, such as CHR 3, the aggregates are submicroscopic. Several of the latter are shown at higher magnification, in Fig. 13.20, where the second-order DNA coils (COIL DNA) are clearly distinguishable. In Fig. 13.21, the second-order DNA helices (COIL DNA) are still further extended with separated groups of 200 A coils marked with arrows. In Figs. 13.22 and 13.23, the partly loosened aggregate is shown at higher magnification to show the second-order DNA coils as circles, with RNA granules associated with them. In Fig. 13.24, the DNA is present as filaments (DNA FIL) which represent the extended first-order DNA helix or the double helix of Watson and Crick. Figure 13.19, × 19,400; Fig. 13.20, × 54,000; Fig. 13.21, × 44,000; Figs. 13.22–13.24, × 90,600.

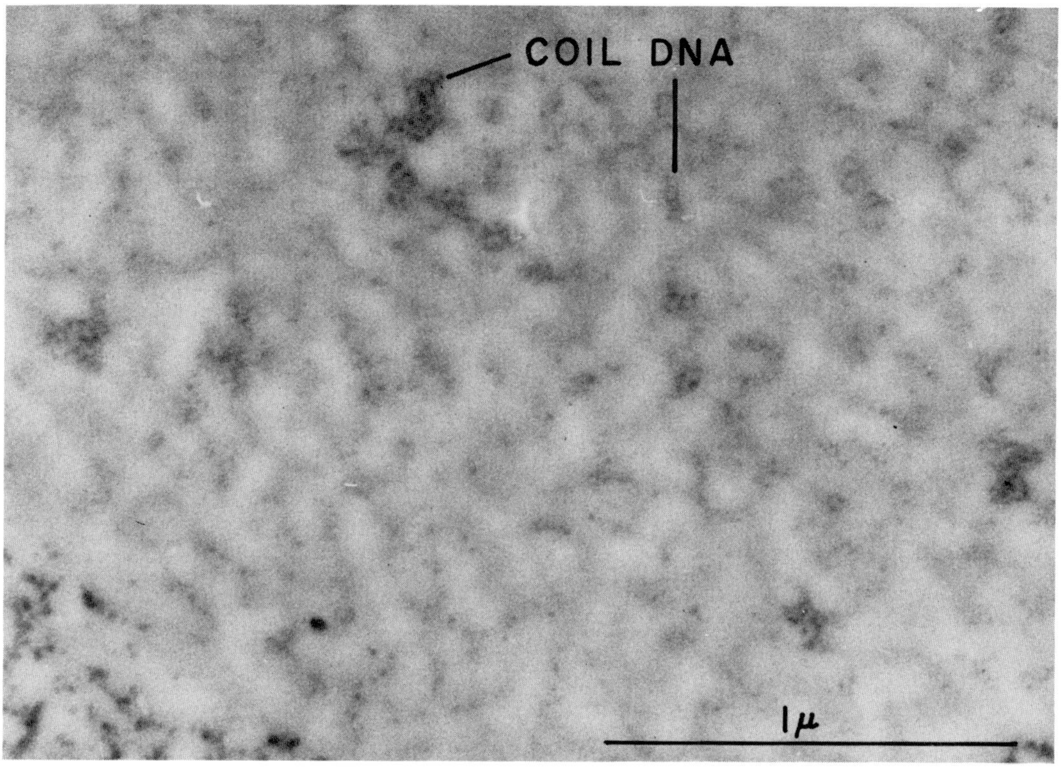

FIG. 13.20.

cheek pouch epithelium treated with dimethylbenzanthracene, as well as from preliminary studies of rapidly proliferating embryonic mesenchymal cells. In both instances, growth is rapid, and the second-order DNA helices are distributed throughout the nucleoplasm during interphase.

Later, in *Drosophila* cleavage interphase, the DNA coils are in-gathered or concentrated and about the same time they are replicated. This probably represents an overlap between the late G_1, and S or early G_2 periods of interphase and is characterized morphologically by the short distance between the adjacent second-order DNA coils and the occurrence of two layers of such helices in some thickened nuclear walls. These changes take place over large areas (G_2 phase), while other areas (in sections) are still pale, as in early interphase (G_1).

The process of in-gathering of second-order DNA coils continues during prophase. The coils aggregate to form submicroscopic and later microscopic clumps. The latter form the chromosomes. These are long and loosely arranged, and in thin sections observed with the light microscope, they appear mostly as rodlets. In ultrathin sections, the appearance seems even more confusing, and we have been unable to distinquish the two parts of each chromosome. Because the two chromatids are clearly visible during metaphase with the light and electron microscope, we deduce that the DNA helix in each of the two chromatids have become

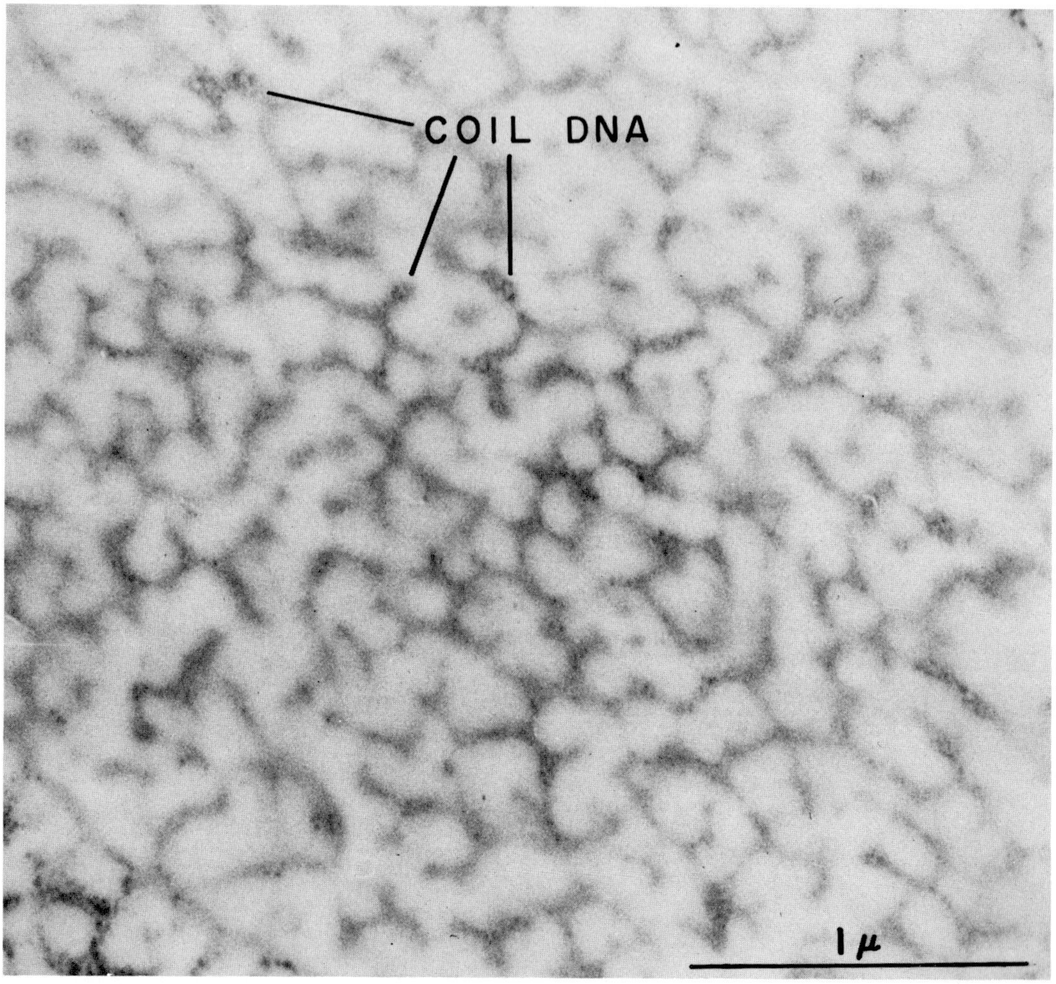

Fig. 13.21.

condensed simultaneously and separately during prophase. The major requirement for such a condensation is that each chromatid should be separately anchored at one end to a fixed surface, such as a nuclear membrane, and that a protein should react relatively firmly, if temporarily, with certain fixed periodic points on the second-order DNA coil, in the manner of cross-linking agents. The occurrence of random movement in the coil, and the restriction of the coils to the nucleoplasmic walls (which constitute about 60% of the nuclear volume) would, in time, result in a coalescence of widely dispersed second-order DNA coils to form visible chromosomes.

In the course of this process, the second-order DNA helices are disposed around the pseudovacuoles, in their walls, as gyres of a third-order helix whose inner diameter is approximately 1800 A. The diameter and course of this helix are determined by the facts that the DNA is confined to the nucleoplasmic walls, and that these are arranged as in tight packing, without any tendency toward alignment of the nucleoplasmic spaces and walls.

The metaphase chromosome is coiled microscopically, as shown in Figs. 13.1

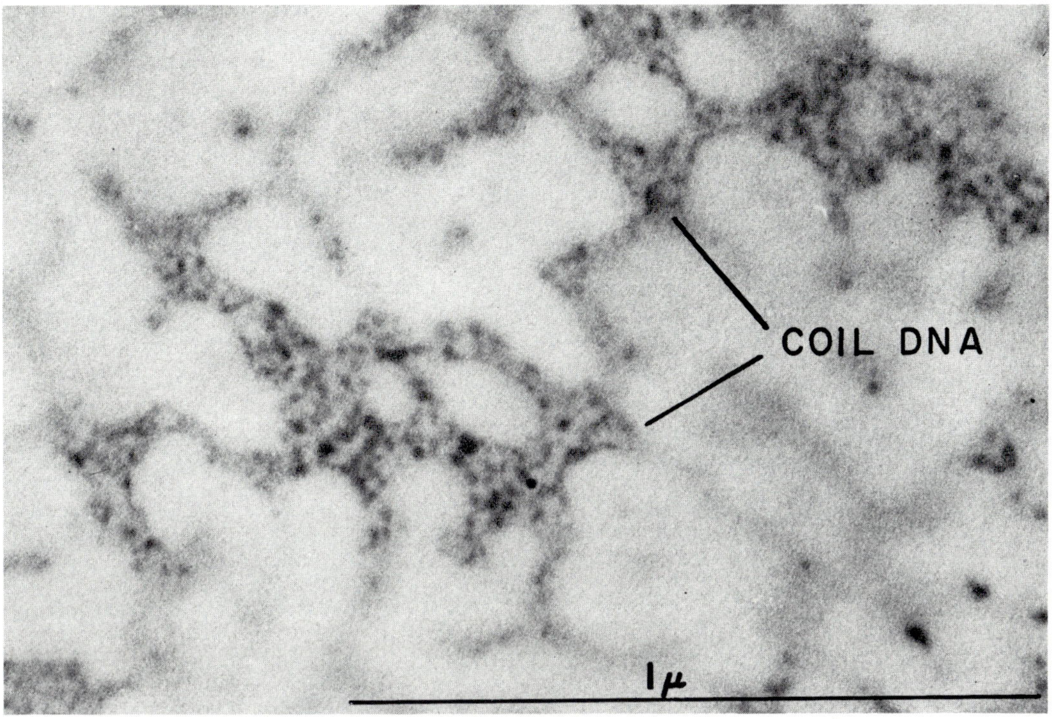

FIG. 13.22. ▲ See p. 356 for complete legend. ▼ FIG. 13.23.

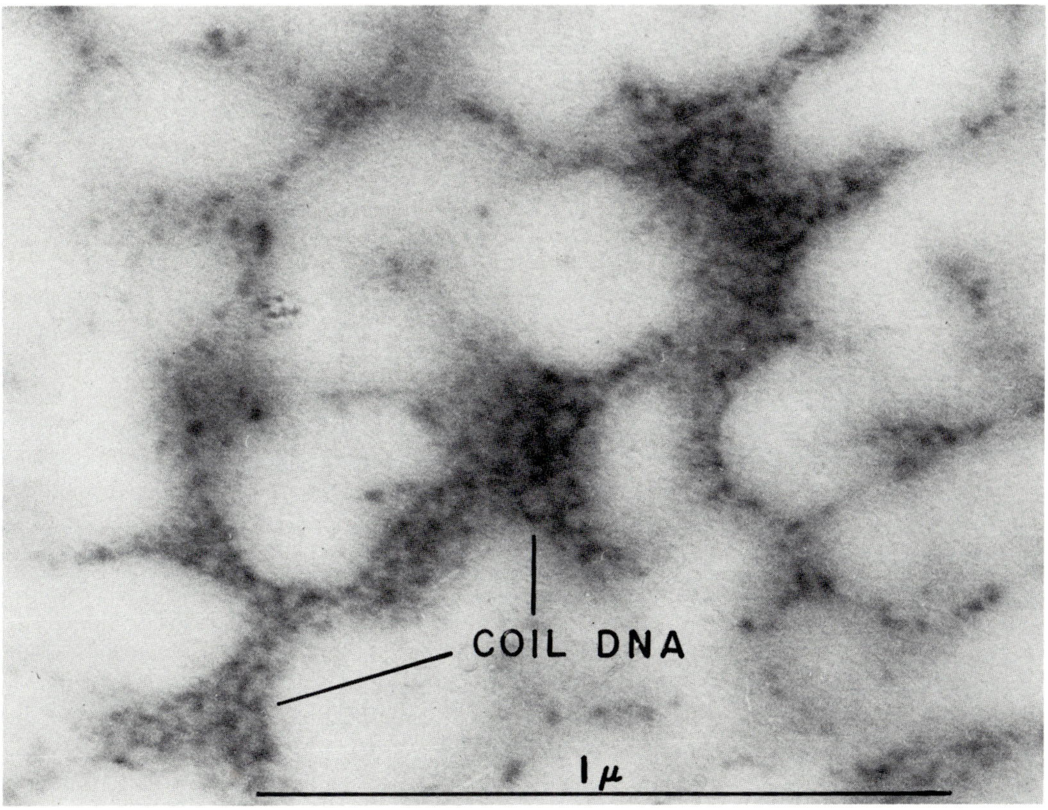

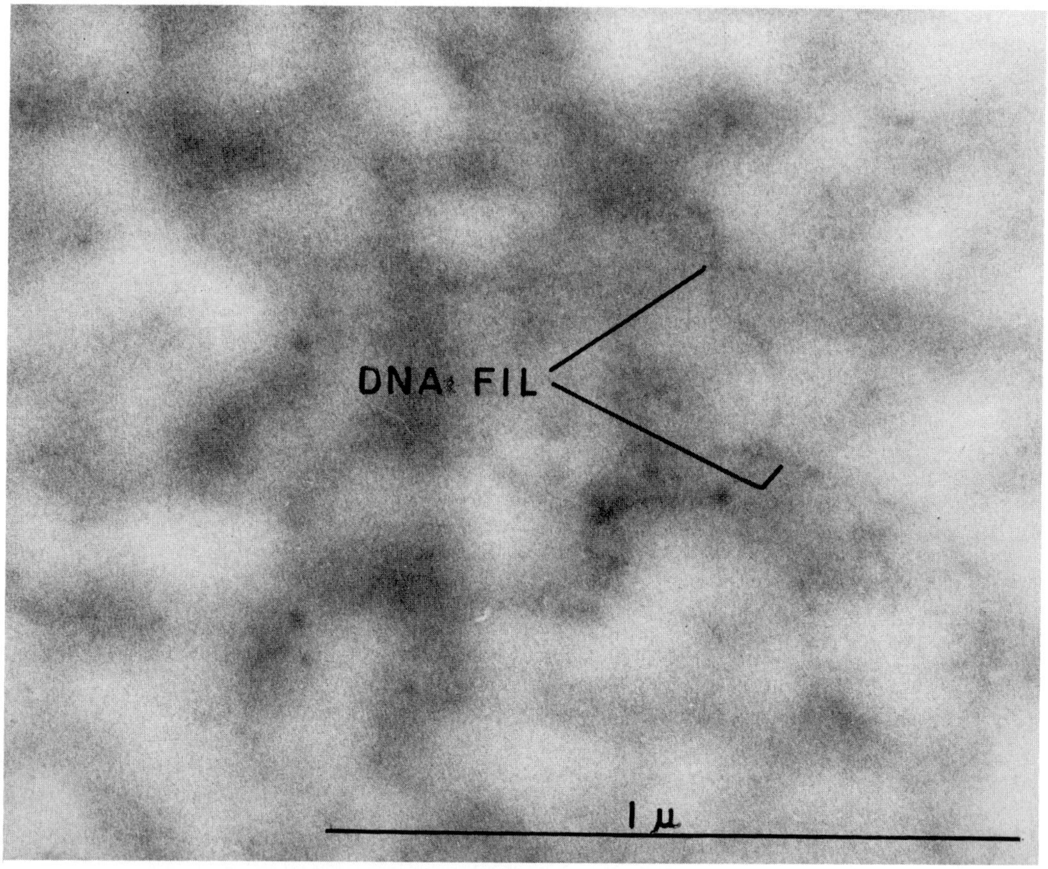

FIG. 13.24. See p. 356 for complete legend.

D and E, in gyres of the order of ⅔ μ in diameter and with approximately the same periodicity. These gyres are derived by coiling of the microscopically visible threads which constitute the prophase chromosomes and which have a diameter of 0.2 to 0.3 μ. It would seem that the third-order coils already described characterize the prophase chromosome, while a fourth-order helix is involved in the formation of the contracted and thickened metaphase chromosome. While the periodicity of this helix in the photomicrograph is about ⅔ μ, that of the large coil in electron micrographs varies from about ½ to 1 μ or more. One may assume that these too are fourth-order coils and that they do in fact vary in pitch and diameter, spatially and temporally, during the rapid transition from prophase through metaphase to anaphase. The presence of two second-order helices side by side in the walls of the pseudovacuole from prophase to metaphase can be interpreted in various ways. One possibility is that the chromatid already consists of two first-order helices; that is, it is two-stranded. Another possibility is that a single second-order helix winds around in the walls of a pseudovacuole in a series of close-packed gyres of the third order and then winds back relationally over the same course to its point of origin (see Fig. 13.25 and Chapter 4, this volume, p. 195). The patches of aggregated coils in which more than two second-order helices participate might then be interpreted as surface views of pseudovacuoles covered by closely packed third-order coils. The appearance of a "core" of varying thickness might likewise be produced according to the plane of the sections, which, it must be remembered, are only about 500 to 600 A thick. A tangential section through a succession of pseudovacuoles, whose walls were packed with DNA, could give the picture of a core without peripheral coils (Figs. 13.10 and 13.13). A more cen-

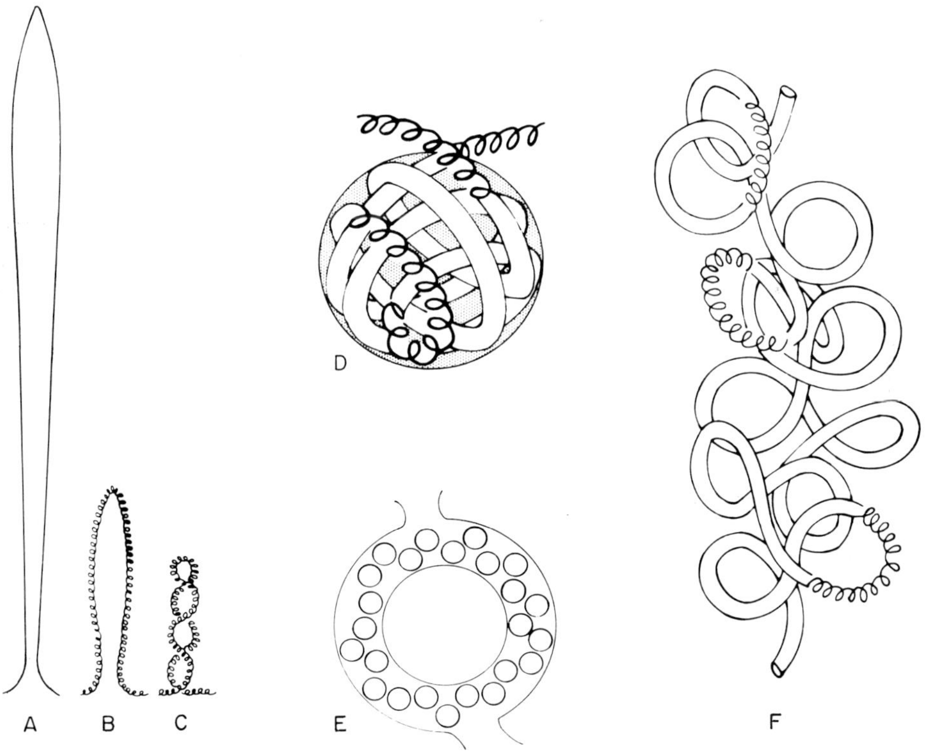

FIG. 13.25. Diagrams to show arrangement of DNA in the metaphase chromosome. (A) A loop of first-order DNA double helix of Watson and Crick. (B) The DNA double helix is organized as a second-order coil, with a corresponding degree of condensation of DNA. (C) A further condensation is effected by a relational coiling. This is a third-order of coiling. In this arrangment, shown in (D), the second-order coils are looped in two layers in the walls of the pseudovacuoles. This diagram includes about $\frac{1}{4}$-$\frac{1}{6}$ of the number of relational coils estimated to be present; the remainder were excluded in the interest of simplifying the concept. The number estimated to be present is more clearly shown in (E), which is a transverse section through a pseudovacuolar wall. The degree of compaction achieved by these two modes of coiling is better appreciated when it is realized that the scale of magnification of (D) and (E) is about four times greater than in (A), (B), and (C). In the diagram of a portion of the metaphase chromosome shown in (F) each third-order coil represents the whole complex of relational coils, shown in (D) and (E), which course in the walls of the pseudovacuoles.

tral section could take in complete or partial cross sections of pseudovacuoles, as well as tangential sections of the proximal portions of some, giving the impression of a central core with peripheral loops.

In anaphase, the process of coalescence of second-order DNA coils continues. The increase of aggregated second-order coils, with a decrease in pseudovacuoles surrounded by coils, suggests that the pseudovacuoles collapse, bringing together the surrounding coils. At the end of this period the chromosomes appear firmly attached to the nuclear membrane at their centromere regions.

In telophase, parts of the chromosome becomes less dense as the second-order DNA coils are dispersed over a larger number of nucleoplasmic walls, but great lumps of chromatin are still visible with both light and electron microscopes. The process of disaggregation of second-order DNA coils is faster at the surface farther removed from the nuclear membrane, so that microscopic and submicroscopic aggregates and finally single second-order and first-order DNA helices appear. The latter state of the DNA approaches that of interphase nuclei. The major requirement for this process of orderly disaggregation of the second-order DNA coils is that the coiling protein (whose existence was proposed as a requirement for chromosome formation) should be hydrolyzed or so altered as to be no longer effective as a coil-binding agent, thus releasing the second-order DNA coil from restraint. This process must be systematic if the second-order DNA helices are to be fully and separately extended throughout the dense phase of the nucleoplasm without being tangled.

Thus the relations of the second-order DNA helix are seen to be constantly changing during the mitotic cycle, but throughout these processes the linearity of the first-order DNA (double) helix persists. It is difficult to see how the first-order DNA helix can give rise to a second-order coil without assuming the presence of a "coiling protein," which at certain regular spacings attaches to and binds the second-order coil, the energy for the coil formation arising in part from random intramolecular movements of the first-order coil and the coiling protein. Similarly, the formation of the third-order coil first requires a hitching together of two widely separated parts of the second-order coil, with a subsequent continued relational, not necessarily regular, spacing of the intervening stretch of second-order coils in the pseudovacuolar walls. Finally, the condensation of the third-order coils in the walls of the pseudovacuoles to form fourth-order coils could be maintained by a coil-aggregating protein, which may be similar to or the same as the coil-aggregating protein postulated in the next chapter for the formation of other kinds of heterochromatin. The maintenance of the relations of chromosomes to each other and to the chromatin masses of interphase nuclei require still another protein, the membrane attachment protein. These are discussed in greater detail in the following chapter.

The continuity of the double helix is visualized during metaphase as in the models presented in Fig. 13.25. The degree of packing of the double helix achieved by these arrangements can be estimated as about 760 times. This figure is arrived at in the following way. The second-order helix, having a diameter and pitch of about 200 A, provides a packing factor of 3.3 × for the DNA double helix. The second-order helix is, in turn, wound doubly in closely packed gyres as a third-order helix around a pseudovacuole whose inner diameter is about 1800 A, giving an additional packing factor of 2 × 35. The total degree of packing is thus 3.3 × (2 × 35) × 3.3 or 762 ×. This figure is about 15 × too low, according to the data available, which demand a packing factor of 10,000 × in order to accommodate a 91 mm length of DNA *(6)* in approximately 9.1 μ of total chromosome length *(4)* at mitotic metaphase. (Figures apply to the haploid chromosome set, assuming the anaphase to S period chromosome to be single-stranded.) It is possible that during metaphase the fourth-order gyres become packed in contact with one another, which would increase the degree of packing by a factor of 5 ×, bringing it within less than an order of magnitude of the required value. Because of our small sample of eggs at the various stages, we could have missed such a stage.

It is difficult or impossible with the material at hand to determine unequivocally the number of chromonemata per chromosome. The structures observed in all stages of the mitotic cycle, as well as the model of metaphase and anaphase chromosomes required for their interpretation only one chromonema—a single thread of first-order DNA (double) helix—per chromatid *(7)*. It is obvious, however, that the chromosome must be at least double from the S period to metaphase, even though unequivocal evidence of doubleness has been observed only in late interphase and at metaphase.

Summary

The mitotic cycle of the syncytium of the developing egg of *Drosophila* is the shortest one known, and is synchronized for all nuclei. These factors, as well as the small size of the chromosomes, make them very advantageous for the study of the manner in which DNA is made reversibly compact with each cycle. In the interphase, DNA is distributed, in part, as a first-order helix (the double helix of Watson and Crick) and, in part, as a second-order coil, which consists of the first-order helix wound as a helix with a diameter of about 200 A. The first-order helices coalesce during prophase to form more compact aggregates which eventually end as tightly wound second-order helices with a period of about 200 A. These helices are coiled, in turn, to form a somewhat irregular relational coil in the walls of nucleoplasmic vacuoles. Cumulatively, these constitute the chromosome visible with the light microscope at prophase. By metaphase, there is a further drawing together of these coils so as to form what seems like a core of second-order helices, with a more loosely related periphery whose detailed morphology is again related to the walls of the pseudovacuoles. On the whole structure is superimposed a fourth-order coil about ⅔ μ in diameter. In anaphase, the whole chromosome appears as an elongated core which may be formed partly by the collapse of the pseudovacuoles and aggregation of the surrounding second-order helices. The chromosomes become attached at the centromere regions to the nuclear wall. During telophase, the DNA coils unwind to form smaller aggregates, which separate into single second-order DNA coils. Many of these unwind completely to form the DNA filaments or first-order DNA helices. During this process the linearity of the first-order DNA helix is maintained. The observations are discussed in the following chapter in relation to several kinds of proteins which may be involved in maintaining a relationship of the DNA coils to each other and to the nuclear membrane.

References

1. Barry, E. G. (1969). The diffuse diplotene stage of meiotic prophase in Neurospora. *Chromosoma* **26**, 119–129.
2. Bullivant, S. (1970). Present status of freezing techniques. *In* "Some Biological Techniques in Electron Microscopy" (D. F. Parsons, ed.), pp. 101–146. Academic Press, New York.
3. Cooper, K. W. (1944). Analysis of meiotic pairing in *Olfersia* and consideration of the reciprocal chiasmata hypothesis of sex chromosome conjunction in male *Drosophila*. *Genetics* **29**, 537–568.
4. Cooper, K. W. (1950). Normal spermatogenesis in *Drosophila*. *In* "Biology of *Drosophila*." (M. Demerec, ed.), pp. 1–61. Wiley, New York.
5. Cooper, K. W. (1959). Cytogenetic analysis of major heterochromatic elements (especially Xh and Y) in *Drosophilla melanogaster*, and the theory of "heterochromatin." *Chromosoma* **10**, 535–588.
6. DuPraw, E. J. (1968). "Cell and Molecular Biology," p. 521. Academic Press, New York.
7. Laird, C. D. (1971). Chromatid structure: relationship between DNA content and nucleotide sequence diversity. *Chromosoma* **32**, 378–406.
8. Lindsley, D. L., and Novitski, E. (1958). Localization of the genetic factors responsible for the kinetic activity of X chromosomes of *Drosophila melanogaster*. *Genetics* **43**, 790–798.
9. Lindsley, D. L., and Sandler, L. (1958). The meiotic behavior of grossly deleted X chromosomes in *Drosophila melanogaster*. *Genetics* **43**, 547–563.
10. Rabinowitz, M. (1941). Studies on the cytology and early embryology of the egg of *Drosophila melanogaster*. *J. Morphol.* **69**, 1–49.
11. Rudkin, G. I. (1964). The structure and function of heterochromatin. *Genet. Today Proc. 11th Int. Congr. Genet.*, Vol. 2, pp. 359–374. Macmillan, New York.
12. Sonnenblick, B. P. (1950). The early embryology of *Drosophila melanogaster*. *In* "Biology of *Drosophila*" (M. Demerec, ed.), pp. 62–167. Wiley, New York.
13. Suzuki, D. T. (1970). Temperature-sensitive mutations in *Drosophila melanogaster*. *Science* **170**, 695–706.
14. Wright, T. R. F. (1970). The genetics of embryogenesis in *Drosophila*. *Advan. Genet.* **15**, 261–395.

14

Persistence of Individuality of Chromosomes during Interphase, and the Role of the Nuclear Membrane

Peter S. Amenta, Isidore Gersh, and Eileen S. Gersh

Much of the DNA of most interphase nuclei is attached to the nuclear membrane. With his customary insight, Wilson *(42)* summarized the significance of the extensive literature on this relationship in this way: " . . . we can hardly avoid the conclusion that even vegetative nuclei of the ordinary type must consist of definite areas or regions, each the product of a single chromosome and each the fundamental basis of a future chromosome." This summary has been updated by Vanderlyn *(39, 40)*, Comings *(13)*, and Mosolov *(25)*, especially in relation to DNA and the cell membrane in bacteria *(32)*. Doubt has been cast on the validity of some of the data presented by Comings in later work by others *(16)*, and an alternative explanation has been offered *(1)*. They reestablish the fundamental relationship of hereditary material and the nuclear membrane, and show that this relationship is the very basis of the first appearance and orderly arrangement, during mitosis, of chromosomes as well as the continuity and activity of the genome during interphase. It is extraordinary to what degree all these writers, with their widely divergent backgrounds, deal in their own ways with essentially the same problems: the origin and persistence of a DNA–membrane relationship, the relatively constant chromatin pattern in unstimulated cells during interphase, the nature of heterochromatic and euchromatic regions, and the activity of the genome.

We propose in this chapter to discuss some aspects of the problems mentioned above beginning with films of cells in mitosis, and to supply a morphological basis for these phenomena at a molecular level, much of which is speculative. The bias in these speculative suggestions derives from observations reported in earlier chapters of this work on DNA in rapidly dividing nuclei of *Drosophila* eggs (Chapter 13), in interphase nuclei of the mouse in unstimulated (Chapter 8) and activated cells (Chapter 10), and in cells undergoing differentiation and maturation (Chapters 6 and 9).

Transformation of Interphase Chromatin to Prophase Chromosomes, and of Telophase Chromosomes to Interphase Chromatin, in Living Cells

Material and Methods

The cells, which were photographed with dark-medium phase contrast optics (Tri-X reversal 16 mm film, × 430), were in monolayer areas of atrial explants of *Triturus viridescens (8)*. They were cultured in a medium of one part chicken plasma and nine parts amphibian Tyrode solution.

Observations and Discussion

In interphase, the nucleus appears nearly structureless, except for the nucleoli (Fig. 14.1). The earliest indication of mitosis is the appearance of inhomogeneities (Fig. 14.2) which become more prominent (Fig. 14.3). When the nucleoli have disappeared from view, the definitive prophase chromosomes appear, like twisted dense stripes, on the inner surface of the nuclear membrane (Fig. 14.4). These shorten and with "dissolution" of the nuclear membrane, enter metaphase. The two sets of chromosomes separate, and each set develops into anaphase (Figs. 14.5 and 14.6). Telophase to interphase changes are more clearly illustrated by frames from a second film. After telophase, the chromosomes spread out and become irregularly dense on the interface between the nucleus and cytoplasm (Fig. 14.7). The nuclear membrane is fully restored at this point *(26, 27, 29, 31, 36)*. After rapid enlargement of individual chromosomes, some thin connections appear between them (Fig. 14.7). These changes take place on the nuclear membrane. Soon the chromosomes are no longer recognizable as such, being replaced at the nuclear membrane by numerous granules of smaller size and variable density (Fig. 14.8). Even these vaguely defined chromatin bodies become less apparent in time (Fig. 14.9).

Thus the morphological changes from interphase chromatin to prophase chromosome appear to be very similar to the transformation of telophase chromosome to interphase chromatin, but in the reverse order. Most of the changes appear to take place at or near the nuclear membrane or nucleo–cytoplasmic interface. Although frames from only two films are illustrated, the same phenomena were observed to a greater or lesser degree in all films examined.

This description differs in no essential way from those collected, illustrated, and supplemented by Hughes *(22)*, who remarks that "the chromosome threads are closely applied to the inner surface of the nuclear membrane." The stripes of chromatin or chomosomes on the inner surface of the nuclear membrane in these two critical stages of mitosis had been described earlier by cytologists from fixed and stained preparations of plant and animal cells *(3, 10, 17, 20, 28, 37, 42)*. The early cytologists believed the evidence suggested that chromosomes retained their individuality during interphase. Boveri *(3)* noted that chromosomes were "autonomous individuals that retained this autonomy even in the resting nucleus." His convincing morphological observations on the nuclei of *Ascaris* fully justify this statement. Similar conclusions follow from the very recent work with tritiated thymidine *(7, 23, 24)*. Changes in nuclear membrane conforming to the general pattern were described in numerous fixed and freeze-etched species of plant cell cultures *(30)*.

Distribution of DNA in Chromatin in Unstimulated Somatic Cells

In a sense, the association of the Barr body with the nuclear membrane in most cells and sometimes with the nucleolus is a kind of model of how a chromosome is associated regularly with a restricted area of the nuclear membrane. Unfortunately, this relationship gives no clues as to which chromosomes regularly abut on the Barr body in interphase nuclei. A very recent article deals with this very topic *(18)* (see also Chapter 7, this volume). In interphase of most somatic cells of the mouse, most of the chromatin is distributed, probably in specific regions, on the inner surface of the nuclear membrane. When examined at the molecular level, the DNA occurs in fine or coarse aggregates comprising a series of loops of second-order DNA helices, each about 400 A in diameter, with a periodicity of about the same order. This major component is called the peripheral chromatin. A minor fraction of nuclear DNA in most somatic cells extends, as central strands of chromatin, from one portion of the peripheral chromatin to another, or extends as nucleolar-associated chromatin into the nucleoplasm as one or more mushroom-like bodies (each of which partially encloses one or more nucleoli). When the DNA is tightly aggregated, it is regarded as heterochromatic and inactive, as the tight packing reduces the access of DNA to the nucleoplasm and thus limits the synthesis and transfer of particulate RNA molecules. When loosely packed, or extended into the nucleoplasm as single or double strands of second-order DNA helices, DNA is regarded as euchromatic and possibly active, as the DNA molecules are more freely exposed to the nucleoplasm. When the number of strands is greater (four or more) then some second-order DNA helices become less accessible or inaccessible for free exchange to take place with nucleoplasm. In certain nuclei, which are thought to be in states of extreme activity [cartilage cells (see Vol. II, Chapter 5), chromatolytic nerve cells (this volume, Chapter 12), and rapidy dividing interphase of the *Drosophila* egg (this volume, Chapter 13)], the DNA is fully extended as the first-order coil (the double helix). The pattern of distribution of these forms of DNA in nuclei of unstimulated cells was shown in Chapter 8 (this volume) to be characteristic of each cell type.

It has been suggested in an earlier chapter that the form of the second-order DNA coil is maintained by a coiling protein. When this function is discontinued, it is possible that the DNA is uncoiled as the first-order coil (the DNA double helix), which appears in electron micrographs as a line. We have also suggested that the second-order DNA helix is organized in chromatin as loops attached to the nuclear membrane by another protein, the attachment protein. The bond between the nuclear membrane and the DNA loops would seem to be rather strong, for it

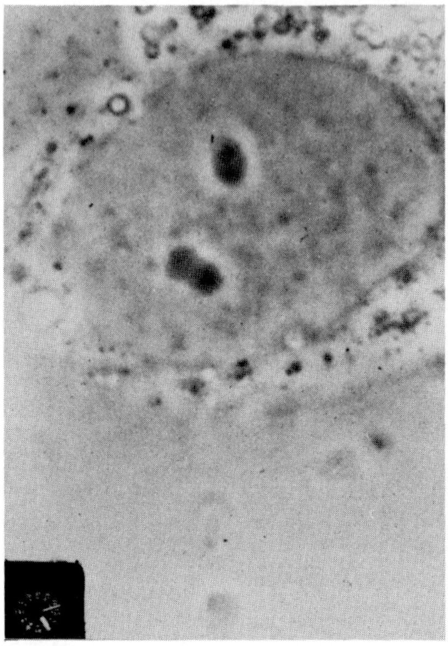

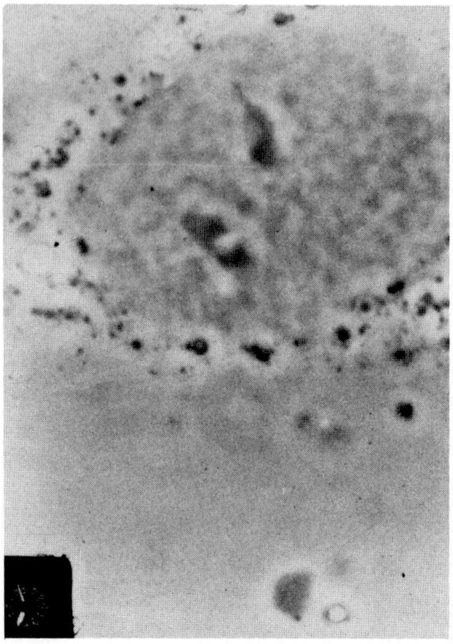

FIGS. 14.1–14.9. Outgrowths from atrial transplant of *Triturus viridescens*.
FIG. 14.1. Interphase, 1:58 PM. × 916. FIG. 14.2. Early prophase, 3:43 PM. × 916.

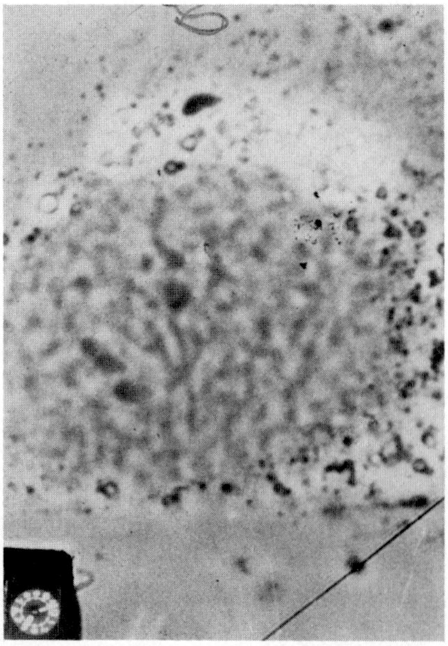

FIG. 14.3 Prophase, 3:56 PM. × 916.

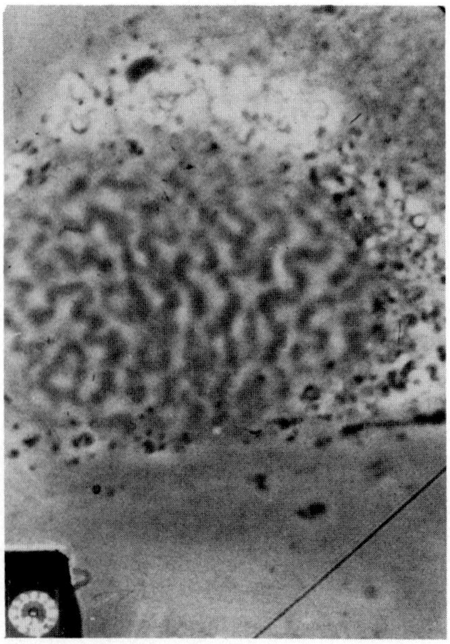

FIG. 14.4. Prophase, 4:13 PM. × 916.

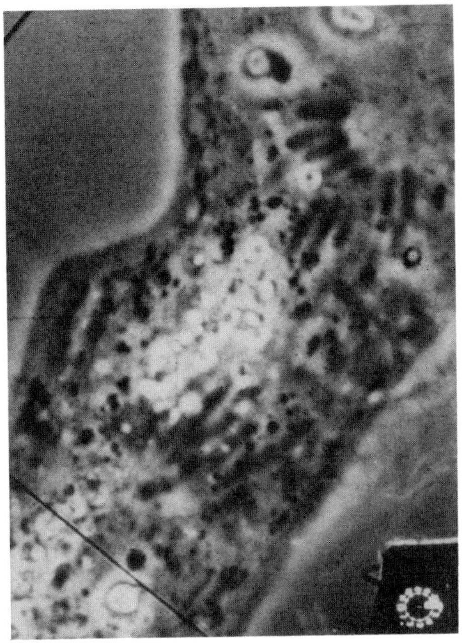

FIG. 14.5. Late anaphase, 6:30 PM. × 916.

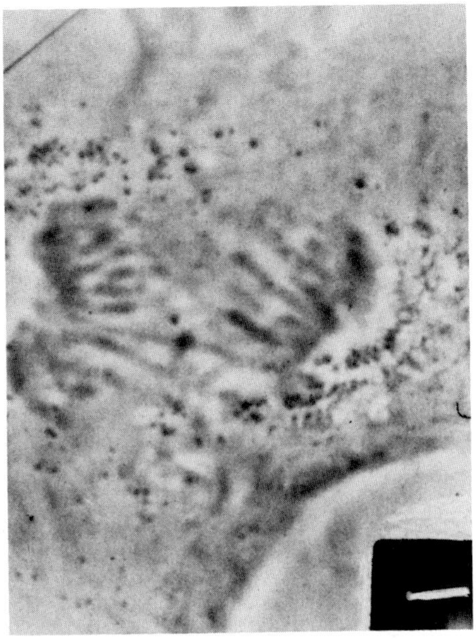

FIG. 14.6. Late anaphase, 12:00 noon. × 1000.

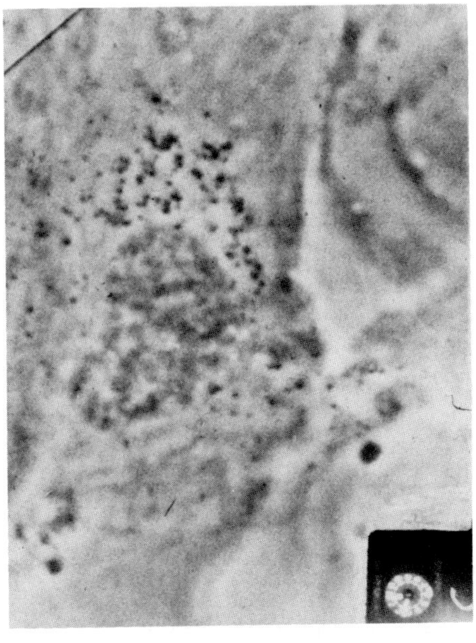

FIG. 14.7. Post-telophase reconstruction, 12:13 PM. × 1000.

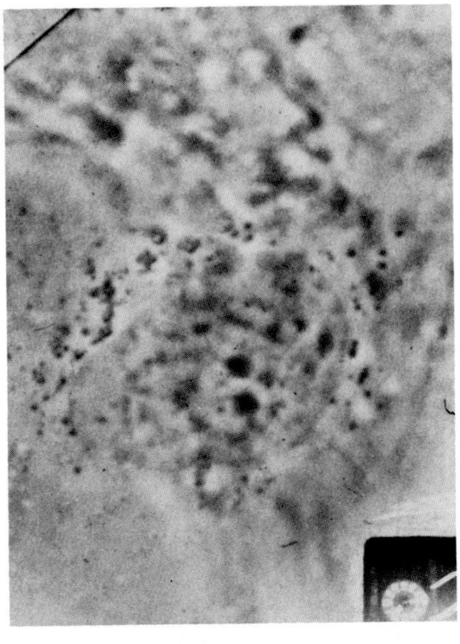

FIG. 14.8. Early interphase, 1:28 PM. × 1000.

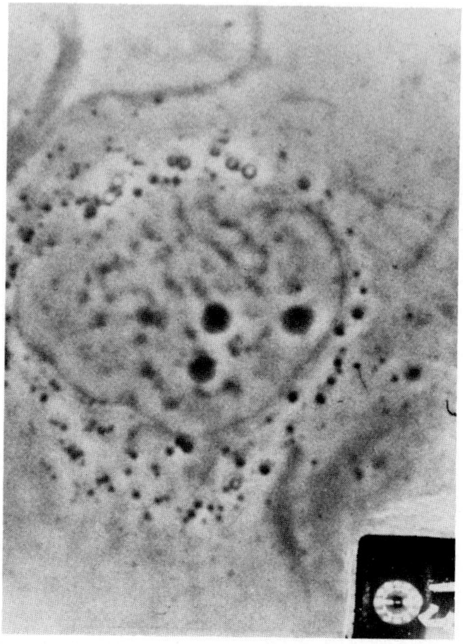

FIG. 14.9. Interphase, 6:00 PM. × 1000.

can withstand the giant forces generated during ultracentrifugation *(5, 6, 11)*. We have also suggested that loops are also attached to each other by coil-aggregating proteins.

We should emphasize that the very naming of the proteins (coil, aggregating, attachment) points to their speculative nature. Whether the attachment and aggregating proteins are different is also unknown. Moreover, with so little information on this subject available, mostly for the sake of simplicity of presentation, we have not involved other possibilities for the formation and release of coils and attachments, such as various kinds of RNA (this volume, Chapter 4), bound water *(12)*, the role of charges and of various ions *(34)*, and other forms of cross-linking. Whether the coiling, aggregating, and attachment proteins, if they are indeed separate proteins, are histones or acidic proteins cannot be resolved by the methods we have used. However, it seems likely that histones are implicated, since these are the chief proteins of nuclei *(9)*. It has also been suggested that histones be considered as coiling proteins while other more specific and localized roles are assumed by acid proteins *(19)*. Several reviews on the role of nuclear proteins have been published recently *(2, 4, 35)*.

Distribution of DNA during the Change of Interphase Chromatin to Prophase Chromosome and of Telophase Chromosome to Interphase Chromatin

In the S period of interphase, DNA is presumably replicated in the fully extended state or as a second-order helix. In either case, all the DNA eventually or rapidly becomes organized into second-order helices. The euchromatic loops of second-order DNA are reattached to the heterochromatic loops, possibly by the aggregating and attachment proteins. The heterochromatic portions are further condensed and reorganized, with attachment points to the nuclear membrane continuously reestablished. The creeping of the chromosomal loops along the nuclear membrane seems to require that there be a continual breaking and remaking of their attachments to the membrane. At some point, these loops become sufficiently dense to become visible in the light microscope, as seen in Figs. 14.3 and 14.4. As the process of condensation continues, with further packing of the DNA second-order helices, the DNA of each chromatin area on the nuclear membrane now shrinks to an individual chromosome. The process of condensation seems to require that each chromatid of each chromosome remain attached to the nuclear membrane, but that the attachment proteins become temporarily ineffective, reestablishing new binding sites progressively closer to one another as the prophase chromosomes assume form. The prophase chromosomes are released by "dissolution" of the nuclear membrane and enter metaphase.

During the transformation of telophase chromosome to interphase chromatin, DNA changes in the reverse order. At telophase, the individual chromosomes are tightly packed in a region such that the centromeres and centriole region are close, and the former is attached to the nucleocytoplasmic interface or to the nuclear membrane. Each chromosome begins to lengthen, and each one preempts a portion of the newly formed nuclear membrane while retaining the same positions relative to one another, as is observed in Fig. 14.7. Comings and others *(13–15, 21, 41)*, give further evidence in support of the suggestion that chromosomes are arranged nonrandomly, probably in specific relationships to each other, in the interphase nucleus. While chromosomes in mature sperm of the grasshopper were described as being arranged in tandem in random order *(38)* there may be other interpretations of the findings which do not exclude an orderly arrangement of chromosomal material in the nuclei of somatic cells. Each chromosome (or chromatin) region continues to remain attached to the nuclear membrane, and subsequent changes entail a continual renewal of the attachment points. During this process, the denser DNA condensations are eliminated, and the second-order DNA helices are arranged in loops, as perhaps in Figs. 14.1 and 14.9. While most of the latter remain attached as peripheral chromatin, some extend as strands into the central part of the nucleus, returning either to the same region or to another region of the same peripheral chromatin mass; others project as nucleolar-associated DNA. Those parts of the second-order DNA helices, which extend as single or double helices from the heterochromatic regions into the nucleoplasm, become potentially euchromatic. Thus the normal complement and distribution of heterochromatic and euchromatic parts of chromatin of somatic nuclei is achieved.

Changes in Distribution of DNA in Chromatin of Activated Cells

Regardless of the mechanism by which cells are stimulated, the outstanding change is an increase in the euchromatic DNA, that is, of single or double loops of second-order DNA helices. Concomitantly, the heterochromatic DNA is depleted. We suggest that some of the loops of second-order DNA helices are released from the nuclear membrane through inactivation of the hypothetical attachment and coil-aggregating proteins. At extremes of activity [as in the interphase of rapidly dividing somatic nuclei of the *Drosophila* egg (this volume, Chapter 13), in regenerating neurons (this volume, Chapter 12), and in hypertrophic cartilage cells (Vol. II, Chapter 5)], the second-order DNA helices disappear and the double helix DNA molecule is fully extended. We suggest that when this takes place, the coiling protein is ineffective. In most instances, the reduction of heterochromatic DNA aggregates automatically extends the nuclear gaps, and there is more opportunity for transfer of substances between nucleus and cytoplasm.

Changes in Distribution of DNA in Somatic Cell Nuclei during Differentiation and Maturation

During differentiation, as in wing disks of *Drosophila* larvae, the pattern of second-order DNA helices changes (this volume, Chapter 6). The end result in the mouse (i.e., the specific patterns of DNA coils in the varied differentiated cells) demonstrates this clearly (this volume, Chapter 8). We suggest that in such instances some euchromatic loops of second-order DNA helices rejoin the aggregated heterochromatic regions, while others are released, both processes possibly requiring antecedent changes in the appropriate attachment and loop proteins. Similarly, during cell maturation, a reorganization of the DNA loops takes place, the prime change being a retraction of euchromatic DNA into the peripheral heterochromatic region (see Chapter 9, this volume). This change also requires changes in the attachment and coil-aggregating protein.

The processes suggested above could account for the retention of the same spatial distribution of chromosomes (as chromatin) in successive generations of differentiated cells. To provide for the continuation in successive cell generations of particular patterns of active and inactive genes characteristic of each cell type, it is neccesary that specific patterns of euchromatic and heterochromatic chromatin be formed in somatic cells after cell division. That this takes place is suggested by the recent report showing a persisting pattern of tritiated thymidine in euchromatic and heterochromatic regions of successive generations *(24)*. This high degree of specificity could be achieved if one assumes that two mechanisms of genetic control are activated shortly after telophase while the chromosomes are extending along the nuclear membrane: (1) the activation of genes which control the formation of attachment, coil-aggregating, and coiling proteins (which together determine the structure and pattern of heterochromatic and euchromatic regions of chromatin in this specific cell type) and (2) the derepression of specific gene sites in the euchromatic loops. The first system acts as a coarse genetic control device for inactivating masses of genes, and freeing some for possible activity, and the second acts as a fine genetic control device for activating a relatively small number of genes in the DNA loops not entrapped in heterochromatic regions.

Persistence of Specific Orientation of Chromosomes during Metaphase

From the foregoing, it seems reasonable that chromosomal DNA is distributed in an orderly, uniform manner characteristic of each cell type in interphase. This relationship is retained while the nucleus goes into prophase, and is reestablished when going from telophase. In order to complete the cycle, it is necessary to establish that chromosomes are arranged so as to occupy the same relative position in metaphase. This was claimed to be so for many cells by Schrader *(33)*, and this claim was supported by numerous cytological and genetical findings summarized by Comings *(13)*.

Summary

Data are presented which bear on the transformation of mitotic chromosomes to chromatin regions and back again. Cinematographical evidence is presented which shows that interphase chromatin is mobilized on the nuclear membrane during prophase as separate chromosomes, and that telophase chromosomes spread out on the nuclear membrane as interphase chromatin. A review of the literature provides further support for the hypothesis proposed during the last century that chromosomes retain their individuality even in interphase, despite the fact that only chromatin masses are observed, without any clear identification of chromosomal domains. The transformations of chromosomes and chromatin are interpreted in terms of the packing of first- and second-order DNA helices. Essentially the same interpretation was extended to a variety of phenomena: the different distributions of DNA aggregates in differentiated cells, the relations of DNA helices in heterochromatic and euchromatic regions, and the changes in the distribution of DNA molecules during stimulation of cellular activity, and during development and maturation. The mechanism underlying the changes in distribution of DNA molecules in all the conditions mentioned above as well as during mitosis involve the presence of three postulated kinds of proteins whose reactions with DNA result in (1) coiling of the first-order DNA (double) helix to form second-order DNA coils, and uncoiling of the latter as first-order DNA helix, (2) aggregation and disaggregation of second-order helices, and (3) attachment and reattachments of the latter to the nuclear membrane.

References

1. Alfert, M., and Das, N. K. (1971). Evidence for control of the rate of nuclear DNA synthesis by the nuclear membrane in eukaryotic cells. *Proc. Nat. Acad. Sci. U.S.* **63**, 123–128.
2. Allfrey, V. G. (1971). Functional aspects of DNA-associated proteins. *In* "Histones and Nucleohistones." (D. M. Phillips, ed.), pp. 241–294. Plenum Press, New York.
3. Baltzer, F. (1967). "Theodor Boveri." Univ. California Press, Berkeley, California.
4. Baserga, R., and Stein, G. (1971). Nuclear acidic proteins and cell proliferation. *Fed. Proc.* **30**, 1752–1759.

5. Beams, H. W., and Kessel, R. G. (1968). Properties of the Ehrlich ascites tumor cell as determined by electron microscopy, ultracentrifugation, and hydrostatic pressure. *Cancer Res.* **28**, 1944–1951.
6. Beams, H. W., and Mueller, S. (1970). Effects of ultracentrifugation on the interphase nucleus of somatic cells with special reference to the nuclear envelope-chromatin relationship. *Z. Zellforsch. Mikrosk. Anat.* **108**, 297–308.
7. Blondel, B. (1968). Relation between nuclear fine structure and ^{3}H-thymidine incorporation in a synchronous cell culture. *Exp. Cell Res.* **53**, 348–356.
8. Bloom, W., Zirkle, R. E., and Uretz, R. B. (1955). Irradiation of parts of individual cells. III. Effects of chromosomal and extrachromosomal irradiation on chromosome movements. *Ann. N.Y. Acad. Sci.* **59**, 503–513.
9. Bonner, J., and Ts'o, P., (eds.). (1964). "The Nucleohistones." Holden-Day, San Francisco, California.
10. Boveri, T. (1888). Zellen-Studien. *Jena. Z. Naturwissensch.* **22**, 685–882.
11. Brenner, S. (1953). The chromatic nuclear membrane. *Exp. Cell Res.* **5**, 257–260.
12. Cole, A. (1962). A molecular model for biological contractility: implications in chromosome structure and function. *Nature (London)* **196**, 211–214.
13. Comings, D. E. (1968). The rationale for an ordered arrangement of chromatin in the interphase nucleus. *Amer. J. Human Genet.* **20**, 440–460.
14. Comings, D. E., and Riggs, A. D. (1971). Molecular mechanisms of chromosome pairing, folding and function. *Nature (London)* **233**, 48–50.
15. Ellison, J. R., and Barr, H. J. (1972). Quinocrine fluorescence of specific chromosome regions. Late replication and high A:T content in *Samoaia leonensis*. *Chromosoma* **36**, 375–390.
16. Fakan, S., Turner, G. N., Pagano, J. S., and Hancock, R. (1972). Sites of replication of chromosomal DNA in a eukaryotic cell. *Proc. Nat. Acad. Sci. U.S.* **69**, 2300–2305.
17. Flemming, W. (1882). "Zellsubstanz, Kern und Zelltheilung." F. C. W. Vogel, Leipzig.
18. Gillot, S. (1968). Hétérogenéités fonctionelles dans l'ADN de noyaux géants. Étude autoradiographique sur la glande séricigène de *Bombyx mori* L. *Exp. Cell Res.* **50**, 388–402.
19. Hart, R. G. (1968). A model of the chromosome. *Advan. Biol. Med. Phys.* **12**, 138–161.
20. Heuser, E. (1884). Beobachtungen über Zellkerntheilung. *Bot. Zentralbl.* **17**, 154–157.
21. Hsu, T. C., Cooper, J. E. K., Mace, M. L. Jr., and Brinkley, B. R. (1971). Arrangement of centromeres in mouse cells. *Chromosoma* **34**, 73–87.
22. Hughes, A. (1952). "The Mitotic Cycle." Academic Press, New York.
23. Kuroiwa, T., and Tanaka, N. (1970). Fine structure of interphase nuclei. II. The behavior of the chromatin during G_1 period revealed by electron microscopic autoradiography. *J. Electron Microsc.* **19**, 63–69.
24. Kuroiwa, T., and Tanaka, N. (1971). Asynchronous dispersion of chromosomes from telophase to G_1 period as revealed by electron microscopic autoradiography. *Exp. Cell Res.* **65**, 177–185.
25. Mosolov, A. N. (1968). Model of the functioning and packing of DNA in the chromosome taking into account the association between the genome with the nuclear membrane. *Genetika* **4**, 135–145 (Russian with English summary).
26. Murray, R. G., Murray, A. S., and Pizzo, A. (1965). The fine structure of mitosis in rat thymic lymphocytes. *J. Cell Biol.* **26**, 601–619.
27. Porter, K. R., and Machado, R. D. (1960). Studies on the endoplasmic reticulum. IV. Its form and distribution during mitosis in cells of the onion root tip. *J. Biophys. Biochem. Cytol.* **7**, 167–180.
28. Rabl. C. (1885). Ueber Zelltheilung. *Morphol. Jahrb.* **10**, 214–330.
29. Robbins, E., and Gonatas, N. K. (1964). The ultrastructure of a mammalian cell during the mitotic cycle. *J. Cell Biol.* **21**, 429–463.
30. Roberts, K., and Northcote, D. H. (1971). Ultrastructure of the nuclear envelope; structural aspects of the interphase nucleus of sycamore suspension culture cells. *Microsc. Acta* **71**, 102–120.

31. Roth, L. E., and Daniels, E. W. (1962). Electron microscopic studies of mitosis in amebae. II. The giant ameba *Pelomyxa carolinensis*. *J. Cell Biol.* **12**, 57–78.
32. Ryter, A. (1968). Association of the nucleus and the membrane of bacteria: a morphological study. *Bacteriol. Rev.* **32**, 39–54.
33. Schrader, F. (1944). "Mitosis," pp. 80–81. Columbia University Press, New York.
34. Steffensen, D. M. (1961). Chromosome structure with special reference to the role of metal ions. *Int. Rev. Cytol.;* **12**, 163–197.
35. Stein, G., and Baserga, R. (1972). Nuclear proteins and the cell cycle. *Advan. Cancer Res.* **15**, 287–330.
36. Stevens, B. J. (1965). The fine structure of the nucleolus during mitosis in the grasshopper neuroblast cell. *J. Cell Biol.* **24**, 349–368.
37. Strasburger, E. (1882). Ueber den Theilungsvorgang der Zellkerne und das Verhältniss der Kerntheilung zur Zelltheilung. *Arch. Mikrosk. Anat.* **20**, 476–590.
38. Taylor, J. H. (1964). The arrangement of chromosomes in the mature sperm of the grasshopper. *J. Cell Biol.* **21**, 286–289.
39. Vanderlyn, L. (1948). Somatic mitosis in the root-tip *Allium cepa*—a review and a reorientation. *Bot. Rev.* **14**, 270–318.
40. Vanderlyn, L. (1949). The heterochromatin problem in genetics as related to other branches of investigation. *Bot. Rev.* **15**, 507–582.
41. Wagenaar, E. B., and Sadasivaiah, R. S. (1969). End to end, chain-like associations of paired pachytene chromosomes of *Crepis capillaris*. *Can. J. Genet. Cytol.* **11**, 403–408.
42. Wilson, E. B. (1947). "The Cell in Development and Heredity," 3rd ed., p. 895. Macmillan, New York.

Author Index

Numbers following the names of authors indicate the pages where complete references are listed. Numbers in italics refer to reference listings for senior or sole authors. Roman numbers refer to listings for co-authors.

N

O

P

R

Subject Index

P

R

S